Mechanical Engineering Series

Series Editor

Vish Prasad, Department of Mechanical and Energy Engineering, University of North Texas, Denton, TX, USA

The Mechanical Engineering Series presents advanced level treatment of topics on the cutting edge of mechanical engineering. Designed for use by students, researchers and practicing engineers, the series presents modern developments in mechanical engineering and its innovative applications in applied mechanics, bioengineering, dynamic systems and control, energy, energy conversion and energy systems, fluid mechanics and fluid machinery, heat and mass transfer, manufacturing science and technology, mechanical design, mechanics of materials, micro- and nano-science technology, thermal physics, tribology, and vibration and acoustics. The series features graduate-level texts, professional books, and research monographs in key engineering science concentrations.

Stephen P. Radzevich
Editor

Gear Accuracy

A Treatise on Gear Noise Excitation, Vibration Generation, and Dynamics of Operation

Springer

Editor
Stephen P. Radzevich
Machine Design Department
National Technical University of Ukraine
"Igor Sikorsky Kyiv Polytechnic Institute"
Sterling Heights, MI, USA

ISSN 0941-5122 ISSN 2192-063X (electronic)
Mechanical Engineering Series
ISBN 978-3-031-74797-7 ISBN 978-3-031-74798-4 (eBook)
https://doi.org/10.1007/978-3-031-74798-4

This Springer imprint is published by the registered company Springer Nature Switzerland AG
The registered company address is: Gewerbestrasse 11, 6330 Cham, Switzerland

Gear industry
Gear science
NO,
thank you !!
We are
too busy !!

Introduction

Knowledge is power (Scientia potestas est). Francis Bacon (1561–1626)

Historical Background

Gears are the means by which power (a rotation and torque) is transmitted from an input shaft to an output shaft. Gears and gear drives have a long-lasting history of evolution. Made out of wood, ancient cogwheels evolved to a high precision involute gear capable of transmitting a huge amount of power at high rotations.

Present-day gear designs are based on the latest accomplishments in the theory of gearing. Ultimately, these yield a very high predictability of the designed gear-sets.

Despite of enormous amount of the research in the field has been carried out in the past, the gear science, and gear practice, have been further evolved in the recent decades.

Gears and gearing become more and more sophisticated, and the problems that arise, are getting more challenging. Noise emission in gearing, and gear vibration generation, is among the most challenging problems of nowadays designs of gear transmission.

Uniqueness of This Publication

The noise excitation and vibration in gear mesh is an important issue that deserves to be investigated to the best possible extent. This challenging problem is poorly covered in the most of advanced sources. In much, the gear noise and vibration generation is still remained a dark area in the contemporary engineering.

This book deals with gears and with the gear noise and vibration in gearing in particular. *Parallel-axes gearing* (P_a*–gearing*), *intersected-axes gearing* (I_a*–gearing*), and *crossed-axes gearing* (C_a*–gearing*) are covered in the text that follows.

The noise emission and vibration generation in both geometrically accurate and approximate gear pairs are considered.

Intended Audience

The book is written for mechanical engineers, those involved in design, manufacture, inspection, and application of gears of all kinds, namely, of (a) *parallel-axes gearing*, (b) *intersected-axes gearing*, and (c) *crossed-axes gearing*. High power density gears (or, in other words, *high power-to-mass ratio gears*) and *high RPM–RPM-gears* are the particular area of gearing that is most vulnerable to the excessive noise emission and vibration generation. Most of the gear engineers and gear researchers from the industry as well as graduate students will be benefited by the book.

Organization of This Book

This book is composed of seven chapters. The list of the chapters, along with a brief description of each chapter, can be found out immediately below. Also, several appendices are appended at the end of the book.

Chapter 1, "Accuracy of Gears: Noise Emission and Vibration Generation in Gearing," is contributed by Prof. *Stephen P. Radzevich* (USA).

This section of the book deals with the accuracy of gears and of the impact of the gear accuracy onto the noise emission and vibration generation when the gear transmission is operated. Here, in this text, the accuracy of the gears themselves along with the accuracy of the gear configuration in relation to one another is covered by the term "gear accuracy." A variety of the root causes for the displacements of the gear axes of rotation are considered. To attract the readers' attention to a complex problem of gear noise and vibration is the main goal of the text below.

Chapter 2, "Tooth Contact Analysis (TCA) of Cylindrical Gears with Tooth Surface Deviations," is contributed by Dr. *Michał Batsch* (Poland).

A novel method of tooth contact analysis (TCA) of cylindrical helical gears to determine the transmission error excitation is presented in this chapter. The transmission error was calculated based on the geometry of mating teeth surfaces with topography errors. These errors can be introduced for the purpose of analysis of measurements of actual gears. The obtained geometrical transmission error is then used to calculate its frequency spectrum (*Fourier transformation*). Next, the transmission error spectrum was compared with the spectrum of the actual gear case acceleration signal in the power circulating test rig to evaluate its components and the influence on the gearbox's operation.

Chapter 3, "Influence of Tooth Profile Deviation Due to Wear on Fluctuation of Gear Ratio," is contributed by Dr. *Pavlo M. Tkach* (Ukraine).

This chapter deals with heavy loaded spur gears operating at severe conditions that have three main sources:

(i) Instability of center distance and misalignment of axes of pinion and gear due to the variability of load and cantilever mounting of the pinion
(ii) Dynamical forces in teeth meshing caused by the manufacturing errors and the surges of external load
(iii) Presence of the abrasive particles and water in the lubricant

For these gears, even in the early stage of service, the operating fluctuation caused by kinematical error of tooth flanks due to their wear is much more intensive than the one caused by manufacturing errors.

The dependence of tooth thickness on the operating life was defined based on the results of periodical tooth thickness measurements. The appropriate predicting procedure with the calculation results is given. Using the profilography method for the tooth flanks, the nature of the profile form deviation was defined. The analysis of the profile forms obtained on the different stages of the wear progress shows that the intensiveness of the deviation of tooth flank from the involute on the early stage is not equal for different segments of tooth flank. Later, the form of profile becomes constant and only tooth thickness changes. Based on the real tooth profiles the fluctuation of gear ratio during the meshing is defined for different degrees of wear, taking into account the positive deviation of center distance.

The obtained results will allow to define the dynamic component of load which, when taken into account during the laboratory tests on wear resistance, can provide more reliable data that will allow to choose better materials, heat treatment conditions, and finishing processes for heavy loaded spur gears.

Chapter 4, "Elements of Dynamics of Geometrically-Accurate Crossed-Axes Gear Pair with Line Contact Between the Tooth Flanks: *R-gearing*," is contributed by Prof. *Stephen P. Radzevich* (USA).

This chapter of the book deals with geometrically accurate crossed-axes gearing that features line contact between the tooth flanks of a gear and of a mating pinion. The gearing of this design is commonly referred to "*R−gearing*." *R−gearing* is the only possible kind of crossed-axes gearing, in which the tooth flanks of a gear and of a mating pinion are in contact with one another. Contact motion characteristics and the key elements of dynamics of *R−gearing* are concisely outlined in this chapter. The tooth profile sliding and the sliding in the lengthwise direction of the gear tooth are covered at the beginning of the chapter in the section titled *Contact Motion Characteristics*. The key elements of dynamics of *R−gearing* are covered in the rest of the sub-sections of the book. Here, analytical solution to the problems under consideration is presented. The obtained results of the research are also valid for gear pairs that operate on intersected axes of rotation of a gear and of a mating pinion. This becomes evident if one assumes the center-distance in the gear pair is equal to zero. The discussed results of the research form a foundation for the further analysis of dynamics of real gear pairs that operate on crossing axes of rotation of a gear and of a mating pinion.

Chapter 5, "On the Lack of Understanding of the *Scientific Theory of Gearing* by the Majority of Gear Experts," is contributed by Prof. *Stephen P. Radzevich* (USA).

An insufficient training and the lack of understanding of the foundation of the scientific theory of gearing by the majority of gear experts all over the world are briefly outlined in this chapter of the book.

Chapter 6, "Improving the Accuracy of Generation of Face Screw Surfaces of Teeth of Hobs," is contributed by Prof. *Oleksandr Okhrimenko* (Ukraine), Prof. *Michael Storchak* (Germany), Prof. *Yurii Danylchenko* (Ukraine), and Associate Professor *Olha Kholiavik* (Ukraine).

Currently, many gear transmissions in use mainly consist of cylindrical gears. Among the existing methods for cutting cylindrical gears, the process of cutting with worm milling cutters has gained the widest popularity.

Research and practical experience in the use of worm milling cutters have shown that the accuracy of the profile of worm milling cutters depends on the geometry of their cutting part. In turn, the performance of worm milling cutters depends on the geometry of the cutting part. Thus, the tasks of increasing the accuracy and performance of worm milling cutters are interconnected. To address these tasks, it is necessary, first and foremost, to analyze possible errors that occur during the gear cutting process and the reasons for their occurrence, as well as the role of design methods for worm milling cutters in their accuracy and performance. When substituting the generating surfce of the milling cutter to the real instrument choose the surfaces of tooth of milling cutter, one of that is a re-sharping, and the second is a not re-sharping, setting initial position of these surfaces the parameters of geometry of cutting part in the instrumental system of coordinates.

However, due to the peculiarities of forming these surfaces during the manufacture of the cutting part of a worm milling cutter, the resulting cutting edge takes the form of a spatial line and does not lie on the generating surface of the milling cutter that was used as the basis for the design.

This leads to additional errors in the profile of the gear cut with such a cutter. The question of the impact of the shape of the cutting edge of the milling cutter, obtained in this way, on the parameters of the gear's surface has not been considered up to this point due to the absence of work in this direction.

The influence of the shape of the cutting edge on the surface of the gear is considered through the parameters of the accuracy of the manufacture of these surfaces, which form the cutting edge and are standardized by the corresponding deviations. The shape of the rake surface of the tooth, which is regulated by the relevant standard, plays a vital role, so it is necessary to first consider the features of the formation of re-sharpening surfaces of worm milling cutters—the rake surface.

The chapter discusses methods for forming the front helical surface of worm milling cutters, and based on the results of theoretical and practical research, recommendations are provided for increasing the accuracy of such milling cutters by studying the formation of these surfaces of this tool.

Chapter 7, "Analysis of Gear Pairs Composed of Eccentrically Mounted Circular Gears (EM Gears)," is contributed by Dr. *Boris M. Klebanov* (Israel, USA) and Dr. *Yuriy V. Shekhtman* (USA). This chapter is devoted to gear pairs formed by two identical circular involute gears mounted eccentrically on their shafts. The issues of kinematics and dynamics of such gear drives are considered. The method of designing EM gears is proposed, including the choice of eccentricity, dimensioning of gears, and determination of module and number of teeth depending on the load, rotational speed, and strength of the material the gear rim is made of. The text is accompanied by design formulas and diagrams that make the design process relatively easy and quick. An example of design and calculations contributes to better understanding of proposed method.

Chapter 8, "Envelopes in Gearing: A Novel Accomplishment in the Classical Differential Geometry," is contributed by Prof. *Stephen P. Radzevich* (USA). This chapter of the book deals with envelope surfaces used in gear design and in gear machining processes, more specifically, with those utilizing *gear generating* principle when rotations either of two gears in mesh, or of a gear and the gear cutting tool are timed with one another. The key features of conventional approach of generation of envelope surfaces [developed in both (i) the differential geometry of surfaces and (ii) the kinematic geometry of surfaces] are briefly outlined. It is stressed here that conventional approach of generation of envelope surfaces is valid in the cases when there is no correlation between the parameters of the relative rolling motion of the components, and between curvature of the interacting surfaces. It is shown that in case of certain correlation between the rolling motion, and between the curvature of two interacting surfaces, conventional approach of generation of envelope surfaces is invalid. Thus, a corresponding approach for the determination of envelope surfaces in the case under consideration is required to be developed.

The core of the proposed method of determination of envelope surfaces in gear design and in gear machining processes is disclosed in the chapter. The problem of determining envelope surfaces in gear design and in gear machining processes is correctly formulated, and then solved by the author of this chapter at around 2008. All three kinds of gearing [i.e., (a) *parallel-axes gearing* (P_a–*gearing*); (b) *intersected-axes gearing* (I_a–*gearing*); and (c) *crossed-axes gearing* (C_a–*gearing*)] are covered by the derived solution to the problem. [(A) *L. Euler* (~1760) is credited for the determination of envelope interacting tooth flanks in P_a–*gearing*; (B) determination of the envelope interacting tooth flanks in I_a–*gearing* was started by *G. Grant* (1887) and it was completed by Prof. *S.P. Radzevich* (~2088); (C) Prof. *S.P. Radzevich* (~2088) is credited for the determination of envelope interacting tooth flanks in C_a–*gearing*.]

The appendices are contributed by Prof. *Stephen P. Radzevich*.

It is likely that this book is not free from omissions or mistakes, or that it is as clear and ambiguous as it should be. If you have any constructive suggestions, please communicate them to the editors via e-mails: radzevich@usa.com.

Sterling Heights, MI, USA
April 2, 2024

Stephen P. Radzevich

Contents

About the Editor and Contributors

About the Editor

Stephen P. Radzevich (USA) is Professor of Mechanical Engineering and Professor of Manufacturing Engineering. He received M.Sc. (1976), Ph.D. (1982), and Dr.(Eng)Sc. (1991)—all in mechanical engineering. Dr. Radzevich has extensive industrial experience in gear design and manufacture. He has developed numerous software packages dealing with Computer-Aided Design (*CAD*) and Computer-Aided Engineering (*CAM*) of precise gear finishing for a variety of industrial sponsors. His main research interest is *Kinematic Geometry of Surface Generation*, particularly with the focus on (a) precision gear design, (b) high power density gear trains, (c) torque share in multi-flow gear trains, (d) design of special purpose gear cutting/finishing tools, (e) design and machining (finishing) of precision gears for low-noise/noiseless transmissions of cars, light trucks, etc. Dr. Radzevich has spent about 45 years developing software, hardware, and other processes for gear design and optimization. Besides his work for industry, he trains engineering students at universities and gear engineers in companies. He has authored and co-authored over 40 monographs in the field, handbooks, and textbooks. The monographs entitled *Generation of Surfaces* (2001), *Kinematic Geometry of Surface Machining* (CRC Press, 2008), *CAD/CAM of Sculptured Surfaces on Multi-Axis NC Machine: The DG/K-Based Approach* (M&C Publishers, 2008), *Gear Cutting Tools: Fundamentals of Design and*

Computation (CRC Press, 2010, 2nd edition 2017), *Precision Gear Shaving* (Nova Science Publishers, 2010), *Dudley's Handbook of Practical Gear Design and Manufacture* (2nd edition, CRC Press, 2012, 3rd edition 2016, 4th edition 2021), *Theory of Gearing: Kinematics, Geometry, and Synthesis* (CRC Press, 2012; 2nd edition 2018), *Geometry of Surfaces: A Practical Guide for Mechanical Engineers* (Wiley, 2013; 2nd edition, Springer, 2019), *Advances in Gear Design and Manufacture* (CRC Press, 2019), *High-Conformal Gearing: Kinematics and Geometry* (CRC Press, 2015; 2nd edition, Elsevier, 2020), *Advances in Gear Design and Manufacture* (CRC Press, 2019), *Recent Advances in Gearing – Scientific Theory and Applications* (Springer, 2021) are among the recently published volumes. He has also authored and co-authored about 400 scientific papers and holds over 260 patents on invention in the field, both US patents and patents of other countries.

About the Contributors

Michał Batsch (Poland), B.Eng., M.Sc., Ph.D., has gained the title of Engineer at the Faculty of Mechanical Engineering and Aeronautics at the Rzeszów University of Technology in 2012 (field of automatic control and robotics). In 2013, he graduated in the field of mechanics and mechanical engineering (specialty mechanical drives), gaining the title of Master of Science. In 2015 he received the title of Doctor of Engineering for the thesis entitled "Analysis of the Contact Region Geometry of the Cylindrical Convexo-Concave Novikov Gear" within the discipline of design and exploitation of machines. Since then, he has been employed as an assistant professor in the Department of Mechanical Engineering at the Rzeszów University of Technology and conducts classes in Gear Geometry and Kinematics, Machine Elements, and Computer-Aided Design. His scientific interests cover the design and manufacturing technology of gears with unconventional teeth profiles. He is also employed in the Gear Research Laboratory where he is involved in research of spiral bevel gear grinding technology. He is an author of over 25 scientific publications in the field of gear drives. In addition, he has participated as a manager or member of the research team in many research projects, in which he was responsible for developing the original constructional solutions implemented in the industry.

Yurii M. Danylchenko (Ukraine) is Professor of Mechanical Engineering, and Head of the Department of Machine Design, Igor Sikorsky Kyiv Polytechnic Institute (KPI). He received the degree of specialist in 1982, Ph.D. in 1987, and Dr.Sc.tech. in 2004. Dr. Danylchenko has extensive experience studying processes and technological equipment for complex geometry parts machining, including gears. His main scientific interest lies in the field of kinematics of forming and metal-cutting machines kinematic structure, in particular with an emphasis on the contribution of elastic systems components deformations of machines on machining accuracy.

Dr. Danylchenko has authored a monograph *Precision Spindle Assemblies on Rolling Bearings (Theory and Practice)* (NTUU KPI, 2003). He is the author and coauthor of 30+ scientific papers and patents on inventions in the field of gear cutting.

Olha V. Kholiavik (Ukraine) holds a Ph.D. degree in engineering. She is an associate professor at the Aircraft Manufacturing Technologies Department of the National Technical University of Ukraine "Igor Sikorsky Kyiv Polytechnic Institute." She is Deputy Director Education and Research Institute of Mechanical Engineering National Technical University of Ukraine "Igor Sikorsky Kyiv Polytechnic Institute." She received her M.Sc. in 2001 and Ph.D. in 2012, all in Metal Forming Processes and Machines. She is a complex surface processing specialist. Her main research interests are as follows: Machining of Complex Surfaces, Metal Forming Processes and Machines, Intensification of Metal Forming Processes.

Olha Kholiavik is the author/coauthor of 1 monograph and more than 60 scientific publications and holds 12 patents on inventions in the field of Mechanical Engineering.

Boris M. Klebanov (Israel/USA) is a professional designer of machines and mechanisms with 60 years of experience in heavy and aviation industries. He received his M.Sc. (1959) and Ph.D. (1969) degrees in mechanical engineering. He is co-author of four monographs: "Gears in Mechanical Engineering," "Mashinostroenie," Leningrad, 1978; "Calculation and Design of Reduction Gears," "Politecnica," Sanct-Peterburg, 1993; "Machine Elements: Life and Design," CRC Press, Boca Raton–London–New York, 2008; "Power Mechanisms of Rotational and Cyclic Motions," CRC Press, Boca Raton–London–New York, 2016.

Dr. Boris M. Klebanov has authored dozens of scientific articles on design and calculation of machine elements for strength and rigidity and participated in collective collections of articles "Advances in Gear Design and Manufacture," CRC Press, Boca Raton–London–New York, 2019, and "Dudley's Handbook of Practical Gear Design and Manufacture," Fourth Edition, CRC Press, 2022. He also has many inventions in the field of machinery.

Oleksandr A. Okhrimenko (Ukraine) is Doctor of Technical Sciences, Professor of the Mechanical Engineering Technology at the National Technical University of Ukraine "Igor Sikorsky Kyiv Polytechnic Institute" since 2019. He received his master's degree in 2001, Ph.D. in 2004, and Dr.Sc.tech in 2012. Dr. *Okhrimenko* is chief of the integrated manufacturing engineering department of National Technical University of Ukraine "Igor Sikorsky Kyiv Polytechnic Institute," and from 2021 to present Head of the Department of Mechanical Engineering Technology at the IME, National Technical University of Ukraine "Igor Sikorsky Kyiv Polytechnic Institute." His professional activity area is design and automation of the manufacture of tools, technologies for the manufacture of parts for general mechanical engineering and aviation technology, design of mechanical engineering structures gear, gearbox, and digitalization of machine-building production. He has authored and coauthored one monograph and a textbook.

Stephen P. Radzevich (USA) is Professor of Mechanical Engineering, and Professor of Manufacturing Engineering. He received M.Sc. (1976), Ph.D. (1982), and Dr.(Eng)Sc. (1991)—all in mechanical engineering. Dr. Radzevich has extensive industrial experience in gear design and manufacture. He has developed numerous software packages dealing with *CAD* and *CAM* of precise gear finishing for a variety of industrial sponsors. His main research interest is *Kinematic Geometry of Surface Generation*, particularly with the focus on (a) precision gear design, (b) high power density gear trains, (c) torque share in multi-flow gear trains, (d) design of special purpose gear cutting/finishing tools, (e) design and machining (finishing) of precision gears for low-noise/noiseless transmissions of cars, light trucks, etc. Dr. Radzevich has spent about 45 years developing software, hardware, and other processes for gear design an optimization. Besides his work for industry, he trains engineering students at universities and gear engineers in companies. He has authored and co-authored over 40 monographs in the field, handbooks, and textbooks. The monographs entitled *Generation of Surfaces* (2001), *Kinematic Geometry of Surface Machining* (CRC Press, 2008), *CAD/CAM of Sculptured Surfaces on Multi-Axis NC Machine: The DG/K-Based Approach* (M&C Publishers, 2008), *Gear Cutting Tools: Fundamentals of Design and Computation* (CRC Press, 2010; 2nd edition 2017), *Precision Gear Shaving* (Nova Science Publishers, 2010), *Dudley's Handbook of Practical Gear Design and Manufacture* (2nd edition, CRC Press, 2012, 3rd edition 2016, 4th edition 2021), *Theory of Gearing: Kinematics, Geometry, and Synthesis* (CRC Press, 2012; 2nd edition 2018), *Geometry of Surfaces: A Practical Guide for Mechanical Engineers* (Wiley, 2013; 2nd edition, Springer, 2019), *Advances in Gear Design and Manufacture* (CRC Press, 2019), *High-Conformal Gearing: Kinematics and Geometry* (CRC Press, 2015; 2nd edition, Elsevier, 2020), *Advances in Gear Design and Manufacture* (CRC Press, 2019), *Recent Advances in Gearing – Scientific Theory and Applications* (Springer, 2021) are among the recently published volumes. He has also authored and co-authored about 400 scientific papers and holds over 260 patents on invention in the field, both US patents and patents of other countries.

Yuriy V. Shekhtman (USA) is an expert in mathematical modeling and stress analysis. He has received M.Sc. (1963) and a Ph.D. (1967) degrees in Physics and Mathematics. Working as head of the Static Strength Department at the Central Institute of Aviation Motors (Moscow), he made significant contributions to the study of reinforced shell stability, structural analysis of aviation engine parts, methods for their calculation, and shape optimization.

Drawing upon over 40 years of experience, he has created a number of computer programs based on Finite Element Approach (FEA) and other numerical methods. He is a software developer for AKGears. Dr. Shekhtman is also the author of many scientific articles on the calculation and optimization of geared systems and coauthored the monograph "Aviation Gears and Speed Reducers. Reference book" (Mashinostroenie, Moscow, 1981). Dr. Yuriy V. Shekhtman passed away before the publication of this book. His untimely passing is deeply felt by all who knew him, and his contributions to the field will be profoundly missed.

Michael Storchak (Germany) received his Dipl.-Eng. in 1974, Dr.-Eng. (Ph.D.) in 1984 in mechanical engineering, and Dr.Sci.tech. in 1994 in manufacturing engineering. His research areas and scientific interests cover three main sections: gear processing, study of cutting processes, and research in the state of metal cutting machines. In the field of gear processing, the main activities of Dr. Storchak are: study, optimization, and synthesis of technological processes for gears finishing; design and development of cutting tools equipped with hard alloys and synthetic super-hard materials; development of dressing tools with complex periodic profiles; software development for calculating the profile; and geometric parameters of tools for gear machining.

The results of research and development of Dr. Storchak are presented in one monograph and more than 150 scientific papers, and he holds over 50 patents on inventions.

Pavlo M. Tkach (Ukraine) is a Ukrainian scientist whose professional research is focused on the reduction spur and helical gears for variety of application. The topic of his Ph.D. thesis (defended in 2004) is the lifetime extension of reduction spur gears through tooth crowning, as well as tooth profile modification. He has authored over 100 scientific publications, including four monographs: *Highly Loaded Cylindrical Gears with Biconvex and Biconcave Teeth* (V. Dahl EUNU Publishing, 2005), *Intro to Machines Theory* (V. Dahl EUNU Publishing, 2009), *Short Course on Machine Elements* (V. Dahl EUNU Publishing, 2010), and *Machine Elements: Competitive Tasks* (V. Dahl EUNU Publishing, 2014). Also, he is granted with 15 patents on inventions. Currently, his research is focused mainly on the wear resistance increase, and the tooth strength improvement of spur gears utilizing for this purpose non-standard tooth profiles.

From 2002 to 2014 Dr. Tkach occupied a position of assistant professor, and, later on, a position of associate professor at Volodymyr Dahl East Ukrainian National University (city of Luhansk, Ukraine). In the meantime, he is the Head of the Science Management Department at E.O. Paton Electric Welding Institute of the National Academy of Sciences of Ukraine (city of Kyiv, Ukraine).

Contributors

Michał Batsch Faculty of Mechanical Engineering and Aeronautics, Department of Mechanical Engineering, Rzeszow University of Technology, Rzeszów, Poland

Yurii M. Danylchenko National Technical University of Ukraine "Igor Sikorsky Kyiv Polytechnic Institute", Institute of Mechanical Engineering, Kyiv, Ukraine

Olha V. Kholiavik National Technical University of Ukraine "Igor Sikorsky Kyiv Polytechnic Institute", Institute of Mechanical Engineering, Kyiv, Ukraine

Boris M. Klebanov Aerospace Industry, Los Angeles, CA, USA

Oleksandr A. Okhrimenko National Technical University of Ukraine "Igor Sikorsky Kyiv Polytechnic Institute", Institute of Mechanical Engineering, Kyiv, Ukraine

Stephen P. Radzevich Machine Design Department, National Technical University of Ukraine "Igor Sikorsky Kyiv Polytechnic Institute", Sterling Heights, MI, USA

Yuriy V. Shekhtman Aerospace Industry, Los Angeles, CA, USA

Michael Storchak Institute for Machine Tools, Universityof Stuttgart, Stuttgart, Germany

Pavlo M. Tkach E.O. Paton Electric Welding Institute of the National Academy of Sciences of Ukraine, Kyiv, Ukraine

Chapter 1
Accuracy of Gears: Noise Emission and Vibration Generation in Gearing

Stephen P. Radzevich

1.1 Introduction

Noise emission and vibration generation in gearing can be viewed as an integral response to the impact of numerous root-causes, each of which contributes to performance of a gear transmission. Accuracy of tooth flanks of a gear and of a mating pinion is the first candidate to be considered in this regard.

Accuracy of gear teeth is an important consideration when designing, and when manufacturing gears. Even perfectly designed gears feature manufacturing errors of certain values, and, therefore, the actual tooth flanks, $\mathcal{G}_r$ and $\mathcal{P}_r$, of a gear, and of a mating pinion deviate from their desirable geometries, $\mathcal{G}$ and $\mathcal{P}$. For inspection purposes, tooth flanks, $\mathcal{G}$ and $\mathcal{P}$, of geometrically accurate gears are used as the datum surfaces. Otherwise, no reliable results of the inspection can be obtained. In the present-day practice of inspection of gears for parallel-axes gearing, either involute tooth profile or the screw involute surface is used as the datum surface – and this is correct. However, for inspection of gears for intersected-axes gearing, as well as for crossed-axes gearing, no geometrically accurate tooth flanks, $\mathcal{G}$ and $\mathcal{P}$, are used as the datum surfaces. Instead, approximate surfaces are used for the inspection purposes, which is incorrect, especially in cases when gears with low tooth count are inspected.

Modern CMMs and GMMs (i.e., computer measuring machines and gear measuring machines, correspondingly) are capable of taking measurements when inspecting the accuracy of a surface of arbitrary geometry. When inspecting the accuracy of gears, features of the tooth flank geometry can be taken into account. The latter is helpful to simplify the method of the inspection, to improve the accuracy of the inspection, and to make the results of the inspection reliable.

S. P. Radzevich (✉)
Machine Design Department, National Technical University of Ukraine "Igor Sikorsky Kyiv Polytechnic Institute", Sterling Heights, MI, USA

S. P. Radzevich (ed.), *Gear Accuracy*, Mechanical Engineering Series,
https://doi.org/10.1007/978-3-031-74798-4_1

1.2 Gear Accuracy: Key Features

In this section of the book, the readers' attention is focused mainly on the features of the tooth flanks, $\mathcal{G}$ and $\mathcal{P}$, geometry that can be taken into account when developing novel methods of and means for inspection of the accuracy of real gears.

1.2.1 Inspection of Accuracy of Gear Tooth Flanks: State of the Art

Gear accuracy is an important consideration in designing, in production, and in application of gears and gear pairs, in design of various mechanisms and machines. The performance of gearboxes, the gear noise emission, and vibration generation, the dynamics of gearing, and so forth, are strongly correlated with the gear accuracy, including, but not limited to the accuracy of functional portions of tooth flank of the gear.

Numerous design parameters are inspected in production of precision gears. The transverse gear tooth profile angle, ϕ_t, the tooth pitch helix angle, ψ_g, and the gear circular pitch, p, are among these design parameters. Each of these design parameters (namely: ϕ_t, ψ_g, and p) is inspected to a high level of accuracy in order to ensure the actual values of these gear design parameters are within corresponding tight tolerance bands.

Consider a parallel-axes gear pair (or just P_a–*gearing*, for simplicity) as an illustrative example. It is proven that two involute gears that feature equal base pitches (base pitch, $p_{b.\ g}$, of the gear, and that, $p_{b.\ p}$, of the mating pinion, must be equal to each other) can be engaged in proper mesh with one another. Therefore, instead of inspecting three independent design parameters (ϕ_t,ψ_g, and p) in a gear, it is sufficient to inspect the equality of the base pitches of mating gears ($p_{b.\ g} = p_{b.\ p}$) [4]. Practically, it is preferred to inspect a single design parameter ($p_{b.\ g}$), rather than inspecting several independent design parameters (ϕ_t,ψ_g, and p).

The equality, $p_{b.\ g} = p_{b.\ p}$, is valid only with respect to geometrically accurate involute gears. This equality is not true when the gears are displaced in relation to one another, and due to the displacements, the inequality $p_{b.\ g} \neq p_{b.\ p}$ is observed.

In this latter case (namely, when the gears are displaced in relation to one another), base pitch of the gear, $p_{b.\ g}$, and that, $p_{b.\ p}$, of the mating pinion, each of them, is required to be equal to an operating base pitch, $p_{b.\ op}$, of the gear pair. This results in that the requirement ($p_{b.\ g} = p_{b.\ p}$) is enhanced to a set of two equalities, namely: $p_{b.\ g} = p_{b.\ op}$ and $p_{b.\ p} = p_{b.\ op}$.

In cases of geometrically accurate intersected-axes gearing (or just I_a–*gearing*, for simplicity), as well as of geometrically accurate crossed-axes gearing (or just C_a–*gearing*, for simplicity) the equality, $p_{b.\ g} = p_{b.\ p}$, is replaced with an analogous equality $\varphi_{b.\ g} = \varphi_{b.\ p}$ of the angular base pitches, $\varphi_{b.\ g}$ and $\varphi_{b.\ p}$, of the gear and of the mating pinion, correspondingly.

Again, the equality, $\varphi_{b.\ g} = \varphi_{b.\ p}$, is valid only with respect to geometrically accurate intersected-axes gearing (I_a*−gearing*), as well as with respect to geometrically accurate crossed-axes gearing (C_a*−gearing*). This equality is not true when the gears are displaced in relation to one another, and due to the displacements, the inequality $\varphi_{b.\ g} \neq \varphi_{b.\ p}$ is observed.

In this latter case (namely, when the gears are displaced in relation to one another), angular base pitch of the gear, $\varphi_{b.\ g}$, and that, $\varphi_{b.\ p}$, of the mating pinion, each of them, is required to be equal to an operating angular base pitch, $\varphi_{b.\ op}$, of the gear pair. This results in that the requirement ($\varphi_{b.\ g} = \varphi_{b.\ p}$) is enhanced to a set of two equalities, namely: $\varphi_{b.\ g} = \varphi_{b.\ op}$ and $\varphi_{b.\ p} = \varphi_{b.\ op}$. This latter set of two equations is valid in case of intersected-axes gear pairs with the displaced axes of rotation, O_g and O_p, of the gear and of the mating pinion (I_a*−gearing*). It is also valid in case of crossed-axes gear pairs with the displaced axes of rotation, O_g and O_p, of the gear and of the mating pinion (C_a*−gearing*).

1.2.2 Root Causes for Base Pitch Deviation in a Gear Pair

Uniform transmission of a uniform rotation from the input shaft to the output shaft is possible only by means of geometrically accurate gears. In geometrically accurate gears of all designs (i.e., in P_a−, in I_a−, as well as in C_a*−gearing*), base pitch of a gear, and that of a mating pinion, are exactly equal to operating base pitch of the gear pair. Because of this, operating base pitch is commonly omitted, and just equality of base pitches of a gear and of a mating pinion is taken into account.

It should be realized that no set of gears can ever be made to absolute precision, and each of the individual gear teeth will have geometrical variations. In reality, in gears of all designs (i.e., in P_a*−gearing*, in I_g*−gearing*, as well as in C_a*−gearing*), base pitch of a gear, and that of a mating pinion, deviates from operating base pitch of the gear pair. Because of this uniform transmission of the rotation from the input shaft to the output shaft gets impossible: equality of base pitches of a gear, and of a mating pinion, to operating base pitch in the gear pair is of critical importance to guarantee high performance of gears.

Two main root causes for base pitch variation in a gear pair are taken into account in the analysis below.

Manufacturing errors (when machining/finishing the gear and the mating pinion tooth flanks) are *the first root cause* for base pitch variation. When the actual gear tooth flank, $\mathcal{G}^{act}$, deviates from the desirable tooth flank geometry, $\mathcal{G}$, then violation of the condition of equal base pitches becomes inevitable.

Linear and angular displacements of the gear and the mating pinion in relation to one another under the influence of various factors are *the second root cause* for base pitch variation.

In reality, accuracy of gear tooth flanks is inspected within *functional portions* of the tooth flanks, $\mathcal{G}$ and $\mathcal{P}$, of a gear, and of a mating pinion.

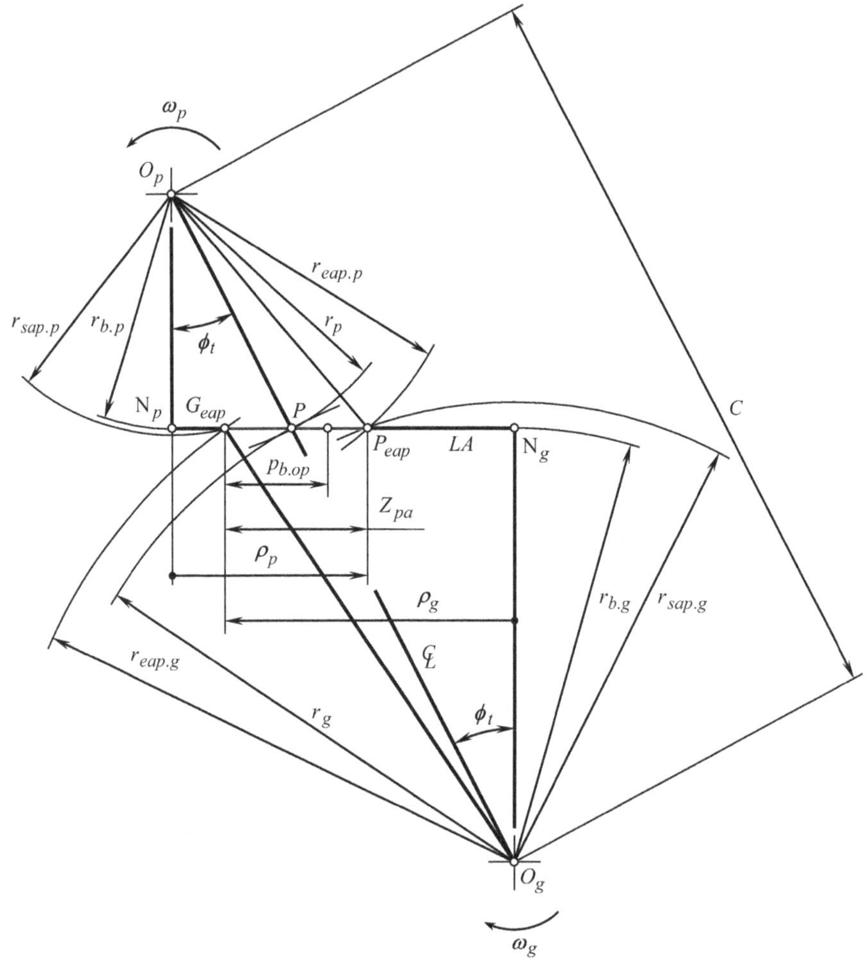

Fig. 1.1 Section of a geometrically accurate parallel-axes gear pair by a transverse plane

As illustrated in Fig. 1.1, the functional portion of tooth flanks, $\mathcal{G}$, in a gear, is bounded by (a) the *start-of-active profile* circle (or just *SAP−circle*, for simplicity), (b) the *end-of-active profile* circle (or just *EAP−circle*, for simplicity), and (c) the functional face width, F_{pa}, in the gear pair (see Fig. 1.2).

The construction of the design parameter Z_{pa} is illustrated in Fig. 1.1.

The following designations are adopted in Figs. 1.1 and 1.2:

O_g and O_p – are the axes of rotation of the gear and of the mating pinion, correspondingly

℄ – is the center-line

C – is the center-distance

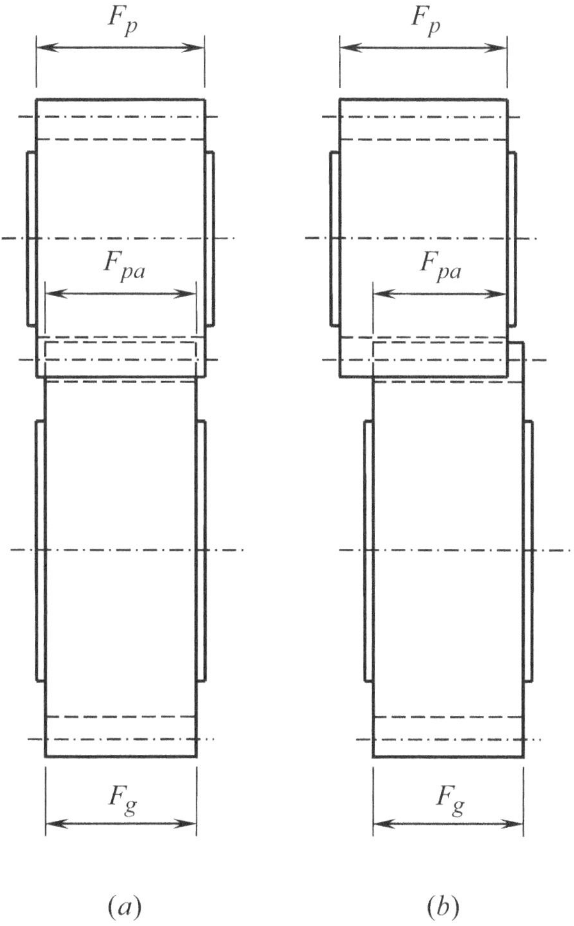

Fig. 1.2 Functional face-width, F_{pa}, in geometrically accurate parallel-axes gear pair

P – is the pitch point

ϕ_t – is the transverse pressure angle

$r_{b.\ g}$ and $r_{b.\ p}$ – are the base radii of the gear and of the mating pinion, correspondingly

LA – is the line of action in the gear pair

r_g and r_p – are the pitch radii of the gear and of the mating pinion, correspondingly

$r_{sap.\ g}$ – is the *start-of-active-profile* radius of the gear

$r_{sap.\ p}$ – is the *start-of-active-profile* radius of the gear

$r_{eap.\ g}$ – is the *end-of-active-profile* radius of the gear

$r_{eap.\ p}$ – is the *end-of-active-profile* radius of the gear

$p_{b.\ op}$ – is the operating base pitch in the gear pair

ρ_g – is the radius of curvature of the gear tooth profile at the $r_{eap.\ g}$ circle of the gear (at the point G_{eap} within the circle of the radius $r_{eap.\ g}$)

ρ_p – is the radius of curvature of the pinion tooth profile at the $r_{eap.\ p}$ circle of the gear (at the point P_{eap} within the circle of the radius $r_{eap.\ p}$)

ω_g and ω_p – are the rotations of the gear and of the mating pinion, correspondingly

F_g and F_p – are the face widths of the gear and of the mating pinion, correspondingly

Z_{pa} – is the length of the plane of action, *PA*

F_{pa} – is the face width of the plane of action, *PA*

In geometrically accurate parallel-axes gearing, two mating gears are rotated about their axes O_g and O_p with angular velocities, ω_g and ω_p, correspondingly.

1.3 Inspection of Accuracy of Gears for Parallel-Axes Gear Pairs

Tooth flanks in geometrically accurate gears for parallel-axes gear pairs are shaped in form of screw involute surfaces. (In case of spur gears, a screw involute surface, $\mathcal{G}$, is reduced to surface of an involute cylinder). Here, in Fig. 1.3, important features of the geometry of screw involute surface, $\mathcal{G}$, are discussed.

1.3.1 Concept of a Novel Method of Inspection of Involute Gear Tooth Flank

A conventional method of inspection of accuracy of involute gear tooth profile is based on the kinematic properties of involute of a circle. The method is illustrated in

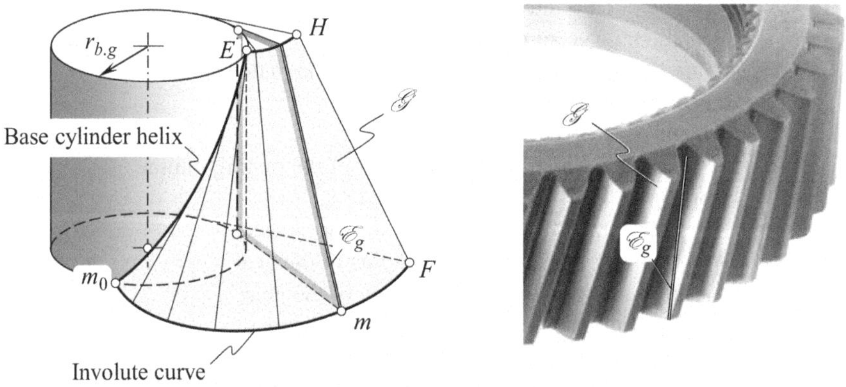

Fig. 1.3 Features of the geometry of tooth flank, $\mathcal{G}$, of a helical involute gear

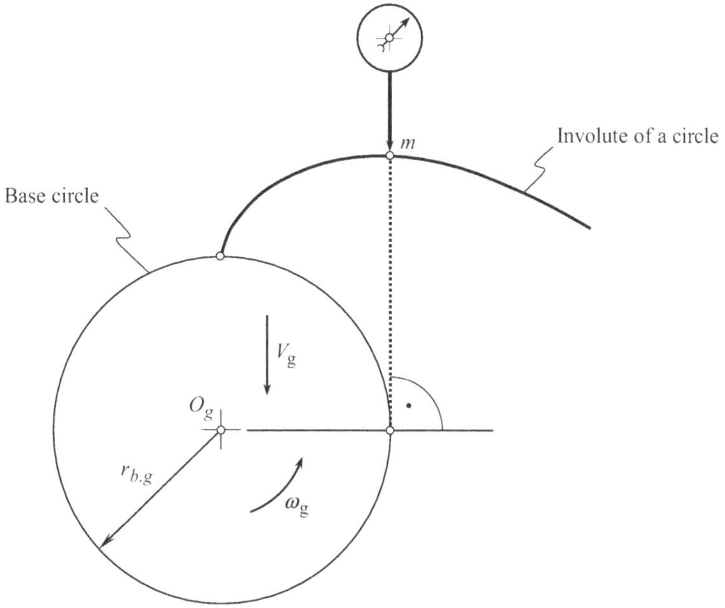

Fig. 1.4 Schematic of conventional method of inspection of accuracy of involute gear tooth profile (in section by a transverse plane)

Fig. 1.4. For the inspection purpose, axis of the stylus of an indicator of dial type is set tangential to the base cylinder of a base radius, $r_{b.\ g}$, of the involute gear to be inspected. Then the gear and the dial-type indicator are given relative motion so as to make the axis of the indicator of dial-type rolling with no slippage over the base circle of a radius, $r_{b.\ g}$. The required relative motion can be obtained as superposition of a rotation, ω_g, of the gear about its axis of rotation, O_g, and a translation either of the gear or of the stylus, along the stylus axis, as it is shown in Fig. 1.4. The gear travels straight with a linear velocity, V_g. The rotation velocity, ω_g, and the translation velocity, V_g, are synchronized with one another so as to ensure rolling with no slippage of the stylus axis over the base circle. If the stylus travels straight then the direction of the translation, V_s, is reciprocal to the translation of the gear, V_g.

In the relative motion, the stylus point traces a true involute profile in a reference system associated with the gear. When inspecting an involute gear, the stylus follows the gear tooth profile. If the arrow of the dial-type indicator does not change its angular position, that is, the readings are *zero*, this means that the gear tooth profile is of a high accuracy (with zero deviation from the true involute profile). If the arrow of the dial-type indicator changes its position, that is, readings are *nonzero*, this means that the gear tooth profile deviates from true involute profile. In this second case, a decision has to be made whether the deviations are reasonably small and all of them are within the tolerance band for the specified gear, or the deviations of the involute tooth profile are unacceptably large, and the gear is machined not in compliance to the blueprint.

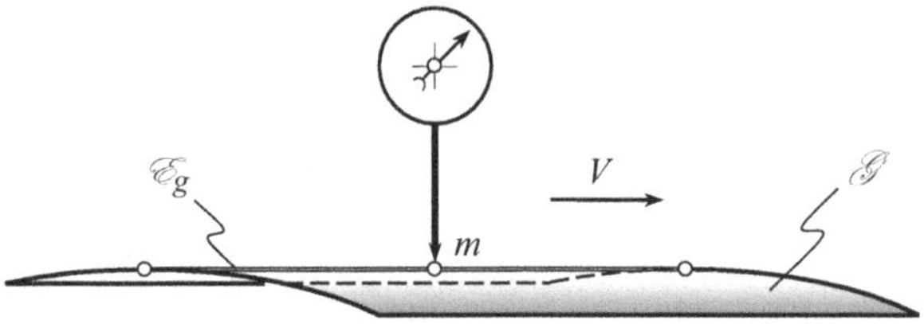

Fig. 1.5 Schematic of a novel method of inspection of accuracy of involute gear tooth flank

After the accuracy of the involute tooth profile is inspected, then the accuracy of the helix angle is inspected. Common methods of and means for the inspection of the accuracy of the helix angle are used in this case.

The features of the involute gear tooth flank geometry are used in novel method for the inspection of the accuracy of involute gear tooth flanks. The concept of this novel method is briefly discussed immediately below.

In a section of a screw involute surface by a normal plane, the line of intersection is a straight line. This means that a screw involute surface can be generated by a straight line (generator), $\mathcal{E}_g$, that performs a corresponding screw motion about the axis of the screw involute surface. In such a motion, the straight line, $\mathcal{E}_g$, is permanently tangent to the base helix of the screw involute surface (see Fig. 1.3).

Concept of the method of inspection of involute gear tooth flank is illustrated in Fig. 1.5. The straight line, $\mathcal{E}_g$, through point of interest, m, is entirely situated on the screw involute surface, $\mathcal{G}$. The stylus of the dial-type indicator is traced along the straight generator, $\mathcal{E}_g$, of the gear tooth flank, $\mathcal{G}$. [This straight line, $\mathcal{E}_g$, through point of interest, m, is also shown in Fig. 1.3]. In the case of geometrically accurate screw involute surface, $\mathcal{G}$, the readings of the dial-type indicator are zero. If the actual gear tooth flank deviates from the desirable geometrically accurate screw involute surface, $\mathcal{G}$, readings of the dial-type indicator are of nonzero value. The smaller the readings of the dial-type indicator, the smaller the deviations of the actual gear tooth flank from the geometrically accurate screw involute surface, $\mathcal{G}$. After the straightness and a proper configuration of one straight line, $\mathcal{E}_g$, on the gear tooth flank is complete, the gear is indexed, and the straightness and a proper configuration of the other straight line on the gear tooth flank are performed. The meaning of the term *indexed* in this particular case becomes clear from Fig. 1.3.

1.3.2 Span Measurement

One more method of inspection of involute gears is based on the following consideration.

The normal sectional view of the auxiliary generating rack, $\mathcal{R}_g$ (of the basic rack, in other words), is depicted in Fig. 1.6 with relation to the helical involute gear. An elementary trigonometrical analysis immediately returns an equation [1]:

$$W_k = d_{b.g}\frac{\tan\phi_t}{\sin\psi_{b.g}}1 \tag{1.1}$$

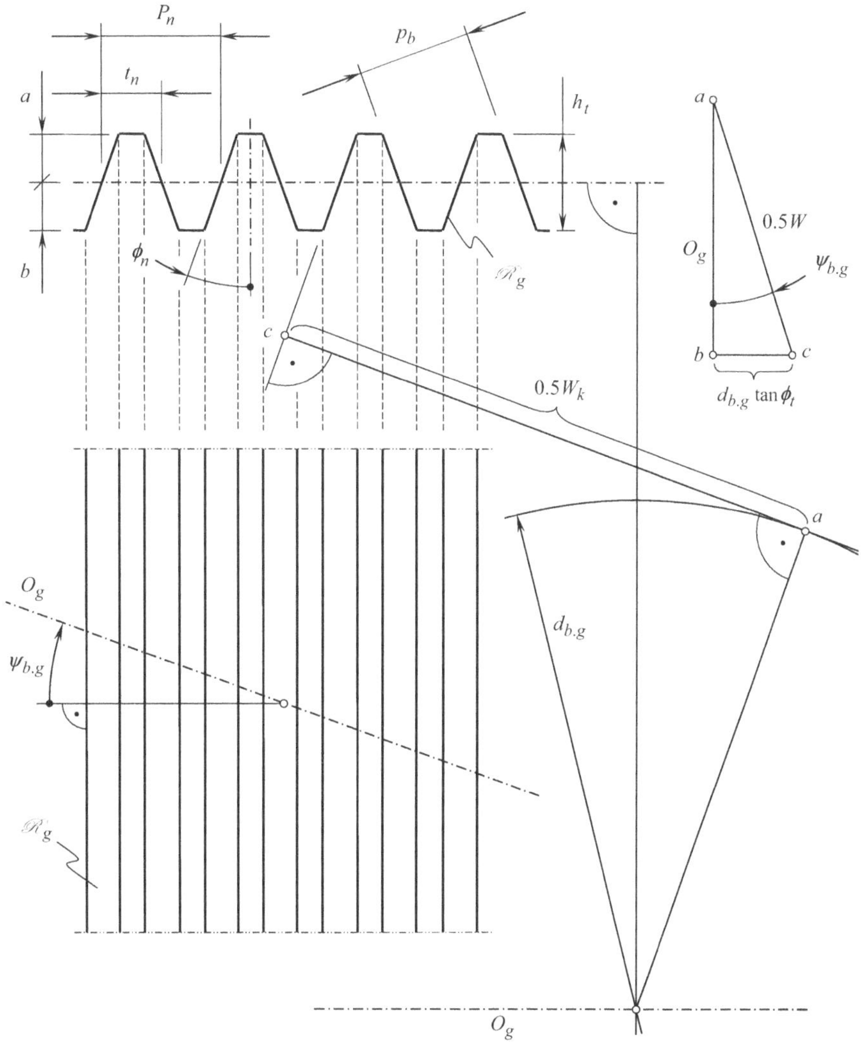

Fig. 1.6 Calculation of span measurement, W_k, in a helical involute gear

for the span measurement, W_k.

In Eq. (1.1) is designated:

$d_{b.\,g}$ – is the base diameter of the helical gear
ϕ_t – is the transverse profile angle at the pitch diameter of the gear
$\psi_{b.\,g}$ – is the base helix angle of the gear

The base tangent length is the distance between two parallel planes tangential to two opposite tooth flanks, namely, to a left-hand flank and to a right-hand flank. It is

Fig. 1.7 Equidistant base thickness of an involute gear tooth

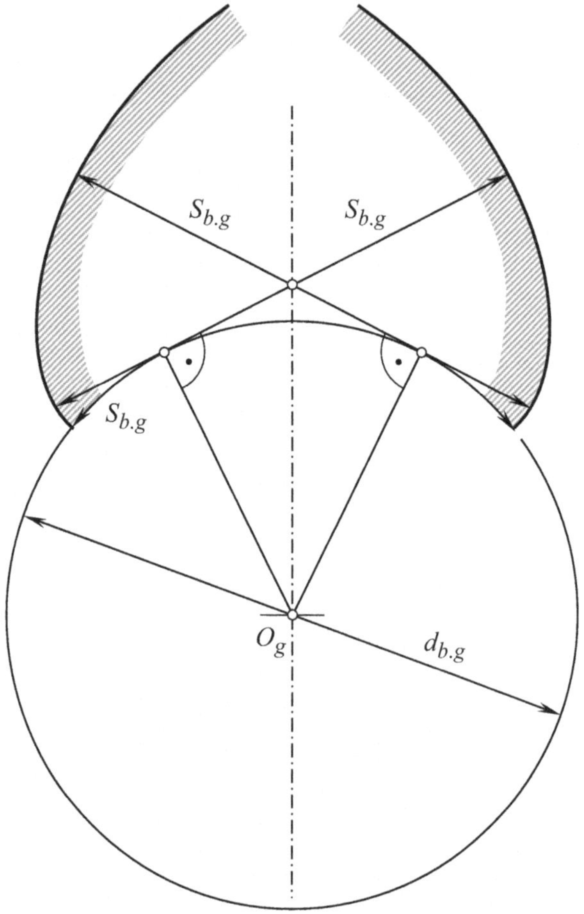

an indirect measure of tooth thickness [1]. This makes use of the property of the involute, that the points of intersection of a tangent to the base circle with a right- and left-hand involute flank are equidistant irrespective of the position of the tangent. In the case of the opposed involutes forming a tooth, this constant distance is the transverse base tooth thickness, s_b, and is equal to the length of the arc between the origins of the involutes on the base circle, as illustrated in Fig. 1.7.

The involute helicoid flanks on helical tooth have the same properties. The parallel planes tangential to the tooth flanks are at an angle, ψ_b, to the axis of rotation of the gear, and the distance between them is s_{bn}. In practice, the measurement has to be carried out over tooth flanks spanning a number, *k*, of teeth instead of a single tooth as shown in Fig. 1.8. The number *k* depends on the tooth geometry, namely, on the pressure angle, the number of teeth, and the addendum modification coefficient.

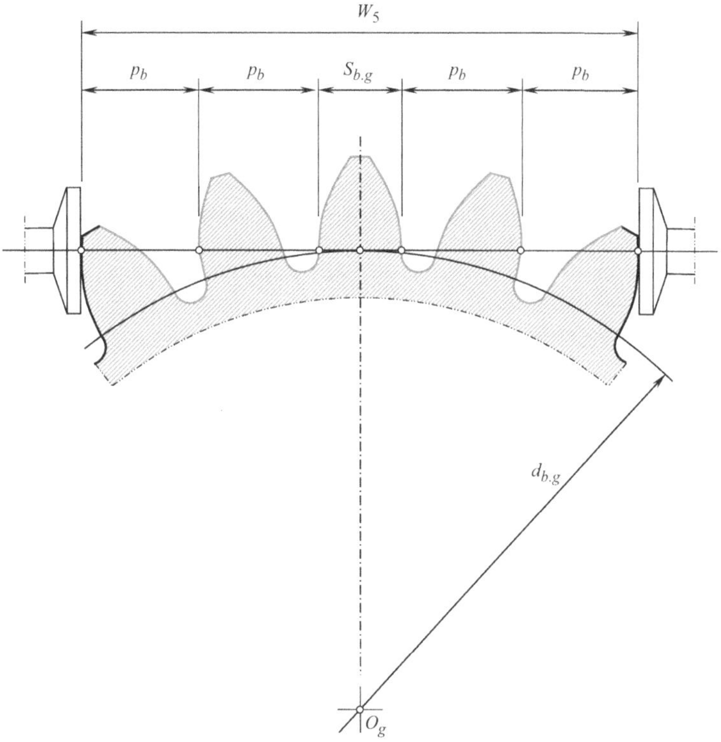

Fig. 1.8 Base tangent length, W_k, on a spur gear

The base tangent length, W_k, (the subscript k after W specifies the number of teeth between the flanks measured), on spur, or helical gears is composed of the normal base tooth thickness, s_b, and a number of normal base pitches, p_b. The number of teeth included in the measurement should be chosen so that there is some latitude in the position in which the measuring instrument can be applied to the flanks.

This characteristic is also true for helical gears.

Calculation of the base tangent length, W_k [1]:

1. The tooth thickness measured along the arc of the reference cylinder on the spur and helical gears is:

$$S_n = m\left(\frac{\pi}{2} + 2x\tan\phi\right) \tag{1.2}$$

2. The base tooth thickness measured along the arc of the base cylinder is on:

   spur gears:

$$s_b = Nm\cos\phi\left(\frac{S_n}{Nm} + \text{inv}\,\phi\right) \tag{1.3}$$

   helical gears:

$$s_b = Nm\cos\phi\left(\frac{S_n}{Nm} + \text{inv}\,\phi_t\right) \tag{1.4}$$

The base tangent length (without taking into account any tolerances) for gears without backlash then becomes on [1]:

spur gears:

$$W_k = s_b + (k-1)p_b \tag{1.5}$$

helical gears:

$$W_k = s_{bn} + (k-1)p_b \tag{1.6}$$

The combined formula for the base tangent length, W_k, is therefore as follows:

$$W_k = m\left[(k-0.5)\pi\cos\phi + N\,\text{inv}\phi_t\cos\phi + 2x\sin\phi\right] \tag{1.7}$$

The number of teeth, k, can be calculated from the following formulas:

$$k = \frac{S_x - W_1}{\pi m\cos\phi} + 1 \tag{1.8}$$

[the calculated value from Eq. (1.8) is rounded off to the nearest integer number]:

$$S_x = \frac{d_b\tan\phi_x}{\cos\psi_b} \tag{1.9}$$

$$\cos\phi_x = \frac{d_b}{d_o - 2m} \tag{1.10}$$

$$W_1 = m\left(\frac{\pi}{2}\cos\phi + N\,\text{inv}\,\phi_t\cos\phi + 2x\sin\phi\right) \tag{1.11}$$

The above formulae apply to external spur and helical gears and also to the tooth space profile of internal spur and helical gears, although the base tangent length on such gears only has a largely theoretical significance.

On internal gears, the tooth thickness is measured by taking measurements between balls.

On external gears, the actual base tangent length, AW_k, is less than the theoretical dimension, W_k, for zero backlash by the amount of the normal backlash allowance, Δj_n. On internal gears, the base tangent length is increased by the amount of the backlash allowance [1]. Therefore, the base tangent length is equal:

on external gears:

$$AW_k = W_k - \Delta j_n \tag{1.12}$$

on internal gears:

$$AW_k = W_k + \Delta j_n \tag{1.13}$$

Tooth thickness of a gear can also be inspected based on the discussed features (see Fig. 1.6) of the geometry of an involute helical gear. Two approaches are used for determining the tooth thickness of an involute gear. One of them is based on span measurement over two or more gear teeth. The other one is based on measurement over balls/pins.

When the span measurement, W_k, over two or more gear teeth are given, for the calculation of circular tooth thickness, t_n, of the helical gear, the following approach can be used.

The dimension, W_t, in the transverse section that corresponds to the span measurement, W_k, of the helical gear (see Fig. 1.9) and is calculated from the formula [1]:

$$W_t = \frac{W_k}{\cos \psi_b} = W_k \frac{\sin \phi_t}{\sin \phi_n} \tag{1.14}$$

The calculated value of the dimension, W_t, allows for the calculation of transverse circular tooth thickness, t_t, at pitch diameter of the gear:

$$t_t = d\left(\frac{M_t}{d_b} - \frac{\pi N_s}{N_g} - \operatorname{inv} \phi_t\right) \tag{1.15}$$

Here, N_s is the number of tooth spaces.

Ultimately, for the calculation of normal circular tooth thickness, t_n, the formula:

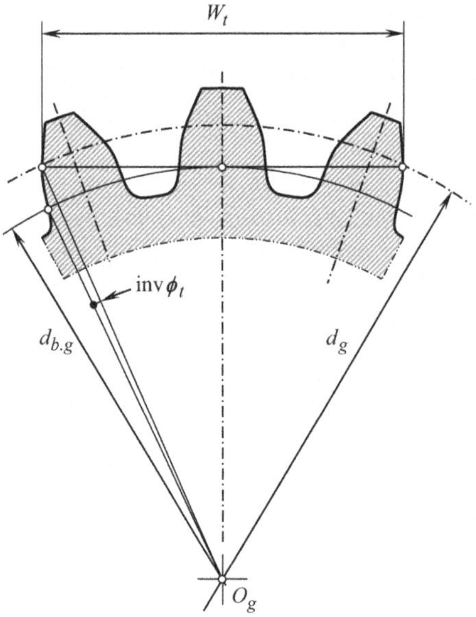

Fig. 1.9 Span measurement, W_t, over two or more teeth in transverse section of a helical gear

$$t_n = t_t \cos \psi \tag{1.16}$$

is used.

The number of teeth in a span is calculated from the approximate formula:

$$k = \frac{N_g}{\cos^3 \psi} \cdot \frac{\phi_t}{180^\circ} + 0.5 \tag{1.17}$$

Here ψ is the pitch helix angle of the gear.

Following another approach, either balls or pins are used for the measurement of the tooth thickness of a spur gear. For the measurement of a helical gear tooth thickness, balls are commonly used.

When the dimension over two balls or pins is given, the following approach is used for the calculation of the circular tooth thickness, t_n, of the helical gear.

For the calculations, the normal width of space between teeth is used as an input parameter. The normal space width, w_n, is equal to $w_n = p_n - t_n$.

The measurements are performed using balls of certain standard diameter. Once a gear is specified, the approximate diameter of the ball is equal:

$$d_{ball} \cong 1.728 / P_n \tag{1.18}$$

Then, the calculated value of the diameter, d_{ball}, is rounded to the nearest standard value.

The transverse profile angle, ϕ_m, to the center of the ball/pin is calculated from the equation:

$$\mathrm{inv}\,\phi_m = \mathrm{inv}\,\phi_t + \frac{d_{ball} - w_n \cos\phi_n}{N_g \cos\phi_n} P_n \tag{1.19}$$

With the input parameters, w_n, d_{ball}, and ϕ_m, calculated, the dimension over two balls, D_{Me}, for a gear with an even tooth number is calculated from the formula [1]:

$$D_{Me} = \frac{d_{b.g}}{\cos\phi_m} + d_{ball} \tag{1.20}$$

For gears with an odd number of teeth, the dimension over two balls is equal to [1]:

$$D_{Me} = \frac{d_{b.g}\cos\left(90^\circ/N_g\right)}{\cos\phi_m} + d_{ball} \tag{1.21}$$

The same formulas are used for the measurement of a spur gear. The only difference is that the transverse profile angle, ϕ_t, and the normal profile angle, ϕ_n, in Eq. (1.19), are equal to each other [1].

1.3.3 Constant Chord Tooth Thickness

Two opposite tooth profiles are simultaneously generated by a straight-sided generating rack, $\mathcal{R}$. In this scenario, a design parameter that relates to two involute tooth profiles in a transverse section of the gear is distinguished. This is the so-called *constant chord tooth thickness* [1].

Refer to Fig. 1.10 in order to determine the *constant chord tooth thickness*.

Here, in Fig. 1.10, a transverse section of a helical involute gear is shown. This schematic is also valid for spur gear. In Fig. 1.10, point C is the point of intersection of the line of symmetry of the involute tooth by the pitch circle of a radius r_g. Two straight lines, AC and BC, through point C are tangent to the base circle of a radius $r_{b.\,g}$ (at points A and B, correspondingly). These straight lines intersect the gear tooth profiles at points D and E, correspondingly. The straight-line segment, DE, is the constant chord to be determined.

A performed analysis of the schematic in Fig. 1.10 reveals that the following equalities are valid:

$$AE = BD = r_{b.g} \cdot (\phi_t + \mathrm{inv}\phi_t) \tag{1.22}$$

$$AC = BC = r_g \cdot \sin\phi_t \tag{1.23}$$

$$CD = CE = r_{b.g} \cdot (\phi_t + \mathrm{inv}\phi_t) - r_g \cdot \sin\phi_t \tag{1.24}$$

$$\angle CDE = 180^\circ - 2 \cdot \phi_t,\ \cos(\angle CDE) = -\cos 2\phi_t \tag{1.25}$$

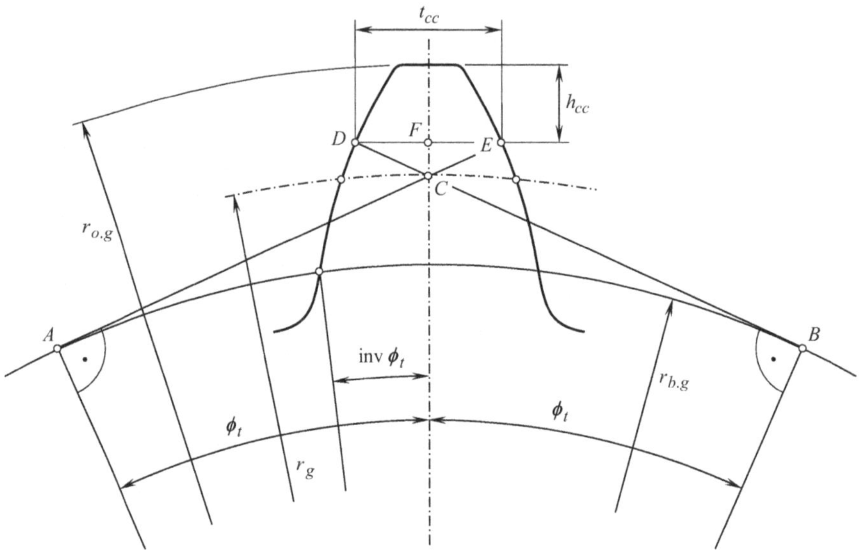

Fig. 1.10 Constant chord, t_{cc}, and height, h_{cc}, above constant chord in transverse section of external helical gear

Law of sines returns (from the analysis of the $\triangle CDE$):

$$DE = \sqrt{CD^2 + CE^2 - 2 \cdot CD \cdot CE \cdot \cos(\angle CDE)} \tag{1.26}$$

$$DE = \left[r_{b.g} \cdot (\phi_t + \text{inv}\phi_t) - r_g \cdot \sin\phi_t\right] \cdot \sqrt{2 \cdot (1 + \cos 2\phi_t)} \tag{1.27}$$

$$\sqrt{2 \cdot (1 + \cos^2\phi_t - \sin^2\phi_t)} = \sqrt{2 \cdot (\cos^2\phi_t + \cos^2\phi_t)} = 2 \cdot \cos\phi_t \tag{1.28}$$

$$DE = 2 \cdot \left[r_{b.g} \cdot (\phi_t + \text{inv}\phi_t) - r_g \cdot \sin\phi_t\right] \cdot \cos\phi_t \tag{1.29}$$

$$t_{cc} = 2 \cdot \left[r_{b.g} \cdot (\phi_t + \text{inv}\phi_t) - r_g \cdot \sin\phi_t\right] \cdot \cos\phi_t \tag{1.30}$$

The *constant chord tooth thickness* does not depend on the tooth count of an involute gear and is equal in all gears with the same tooth profile angle, and the profile shift.

The height above constant chord in transverse section of external helical gear:

$$CF = CD \cdot \sin\phi_t; \quad CF = \left[r_{b.g} \cdot (\phi_t + \text{inv}\phi_t) - r_g \cdot \sin\phi_t\right] \cdot \sin\phi_t \tag{1.31}$$

$$h_{cc} = r_{o.g} - r_g - \left[r_{b.g} \cdot (\phi_t + \text{inv}\phi_t) - r_g \cdot \sin\phi_t\right] \cdot \sin\phi_t \tag{1.32}$$

The design parameters t_{cc} and h_{cc} can be directly measured in a gear.

1.3.4 Tooth Thickness Measurement in Internal Gear

For the calculation of tooth thickness in an internal gear, measurement between two pins or balls is used. Balls are used for measuring both spur and helical gears, while pins are used for measuring spur gears only.

The required dimension between two balls can be calculated in the following way.

Normal space width, w_n, ball diameter, d_{ball}, and transverse profile angle, ϕ_t, to the center of the ball are used as the input parameters for the calculations. The normal space width is calculated from the equation [1]:

$$w_n = p_n - t_n \tag{1.33}$$

The approximate ball diameter is equal to $d_{ball} \cong 1.44/P$. The calculated value of the diameter, d_{ball}, is rounded then to the nearest standard value.

Transverse profile angle, ϕ_t, to the center of the ball can be calculated from the equation:

$$\operatorname{inv}\phi_t = \operatorname{inv}\phi_t - \frac{d_{ball} - w_n \cos\phi_n}{N_g \cos\phi_n} P_n \tag{1.34}$$

For gears with even number of teeth, the dimension between two balls, $D_{M\,i}$, can be expressed by:

$$D_{M\,i} = \frac{d_{b.g}}{\cos\phi_t} - d_{ball} \tag{1.35}$$

For gears with odd tooth count, the dimension between two balls ($D_{M\,i}$) can be expressed by:

$$D_{M\,i} = \frac{d_{b.g} \cos\left(90^\circ/N_g\right)}{\cos\phi_t} - d_{ball} \tag{1.36}$$

Those same formulas are used for the measurements of a spur gear. The only difference is that the transverse profile angle, ϕ_t, and the normal profile angle, ϕ_n, in Eq. (1.34), are equal to each other.

1.4 Inspection of Accuracy of Gears for Intersected-Axes Gear Pairs

The present-day practice of inspection gears for intersected-axes gear pairs is poor and inconsistent. Primarily, this because the approximate datum surface is used for inspection purposes. This is the main reason for the inconsistencies in the inspection of gear accuracy. The actual datum surface significantly deviates from the desirable datum surface, as it is generated as an envelope to a family of consecutive positions of the crown straight-sided rack in its motion in relation to the gear being inspected. Especially this becomes clear when gears with a low tooth count are inspected – in this case, the inconsistency of the known method for the inspection gears becomes more evident. As the actual datum surface itself deviates from the desirable geometrically accurate datum surface, the output of the gear inspection cannot be accurate by nature.

For the inspection of gears for intersected-axes gear pairs the methods of and means for the inspection that are based on the features of the tooth flanks geometry can be developed.

1.4.1 Novel Concept of Inspection of Accuracy of Bevel Gear Tooth Flank Geometry (In Section by Transverse Sphere)

For the inspection of bevel gears, a novel method is proposed. The determination of how much the actual tooth flank of a bevel gear deviates from the nominal (desirable) tooth flank geometry is the main goal of the proposed method of gear inspection. The smaller the deviation, the more accurate bevel gear is and vice versa.

Selection of a proper reference surface for the inspection of bevel gears for intersected-axes gearing is of critical importance, especially in cases of inspection bevel gears with a low tooth count (with 12 teeth and a fewer).

In the present-day practice, tooth flank of a crown gear is used as a reference surface. The tooth flank of crown gear is simulated by the cutting edges of the gear cutting tool in the gear machining operation. When pitch plane of the crown gear rolls with no sliding over pitch cone of the bevel gear being inspected, the tooth flank of the bevel gear is generated as an envelope to a family of consecutive positions of the tooth flank of the crown gear. In the present-day practice, this envelope surface serves as the reference surface for the inspection of bevel gears. This is a wrong practice to use the envelope to a family of crown surfaces as the datum surfaces for inspection purposes.

Similarly, the tooth flank of crown gear is simulated by the corresponding motions of the stylus when inspecting the gear. When inspecting a bevel gear, pitch plane of a crown gear rolls with no sliding over the pitch cone of the bevel gear being inspected. The anticipated tooth flank of the bevel gear is generated as an envelope to a family of consecutive positions of the tooth flank of the crown gear.

In known methods of inspection of bevel gears for intersected-axes gearing, the stylus travels over the tooth flank to be inspected so as to be permanently at the kinematically generated anticipated tooth flank of the bevel gear. Tooth flank of an actual gear being inspecting deviates from the tooth flank of the desirable geometry. These deviations cause the corresponding displacements of the stylus. Such a practice of inspection of a tooth flank of a bevel gear is poor.

An improper datum surface is one of the main disadvantages of known methods of inspection of bevel gears for intersected-axes gearing. Because of this, the results of the measurements do not correctly reflect the actual accuracy of the bevel gear. Ultimately, accurate gears can be recognized as of poor accuracy, and inversely, poor quality gears can be mistakenly recognized as accurate gears.

The use of the proper datum surface is one of the possible ways to improve the accuracy of the inspection of gears for intersected-axes gear pairs.

To proceed with the development of the method of inspection of bevel gears for intersected-axes gearing, the base cone of the gear has to be inspected, and the plane of action has to be constructed.

When inspecting a bevel gear (see Fig. 1.11), a certain relative motion of the gear to be inspected and the plane of action is performed. The gear is rotated about its axis of rotation, O_g. The plane of action, *PA*, is rotated about its axis of rotation, O_{pa}. The rotations, ω_g and ω_{pa}, are synchronized with one another.

When a bevel gear tooth flank is inspected (see Fig. 1.12), the rotations, ω_g and ω_{pa}, are synchronized so as to keep the stylus at current point of interest, *m*, on the desirable bevel gear tooth flank. The point, *m*, is permanently located within the plane of action, *PA*. The line *a*, along which the deviation is measured, is within the

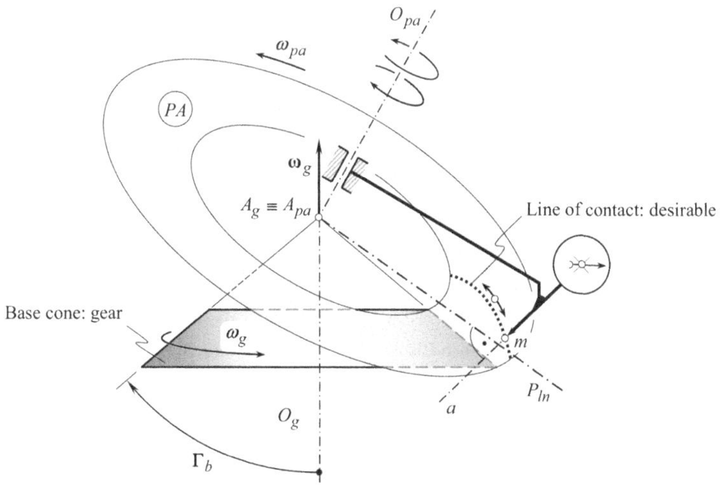

Fig. 1.11 Kinematics of a method of inspection of gear for intersected-axes gear pair

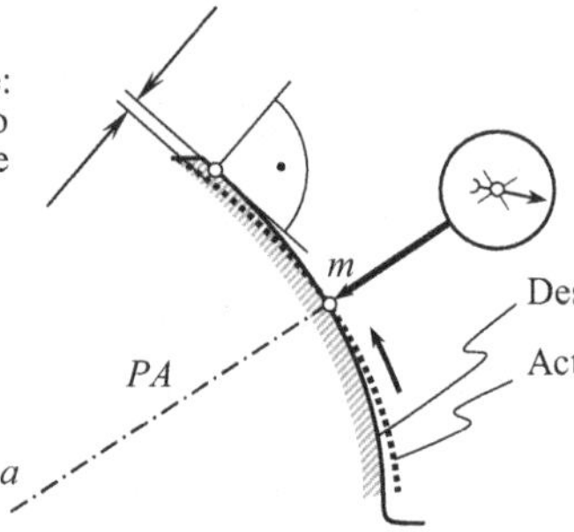

Fig. 1.12 Inspection of accuracy of tooth flank tooth flank geometry in gear for intersected-axes gear pair

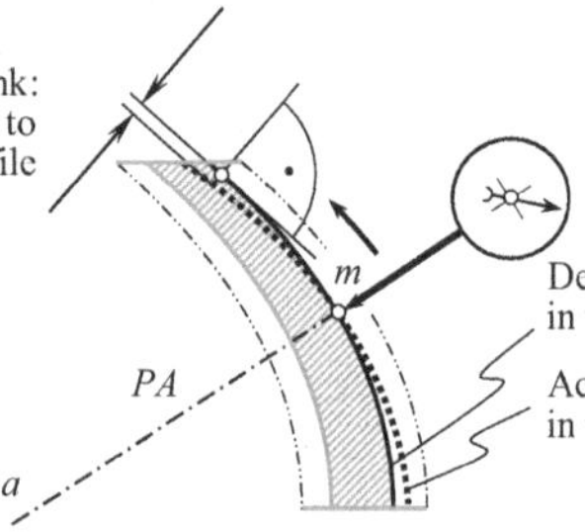

Fig. 1.13 Inspection of accuracy in the lengthwise direction of a gear for intersected-axes gear pair

plane of action, *PA*, and it is perpendicular to the bevel gear tooth flank at point, *m*. The stylus tip is motionless in relation to the plane of action, *PA*, when geometry error of the bevel gear tooth flank is inspected.

1.4.2 Novel Concept of Inspection of Accuracy of Bevel Gear Tooth Flank Geometry (In Lengthwise Direction of Gear Teeth)

When inspecting the tooth flank geometry of a bevel gear in the lengthwise direction (see Fig. 1.13), the stylus tip travels along the line of intersection of the desirable tooth flank by the plane of action, *PA*. At current point of interest, *m*, the straight line $a - a$, along which the deviation is measured, is within the plane of action, *PA*, and it is perpendicular to the bevel gear tooth flank at the point, *m*.

A detailed discussion on inspection of accuracy of tooth flank geometry of bevel gear for intersected-axes gearing can be found in the immediately following section of the chapter, where inspection of accuracy of tooth flank geometry of bevel gear for crossed-axes gearing is considered. This discussion on C_a-*gearing* is also applicable for I_a-*gearing* under the evident assumption that the center-distance in the C_a-gear pair shrunk to zero ($C = 0$).

1.5 Inspection of Accuracy of Gears for Crossed-Axes Gear Pairs

Below in this chapter, geometry of tooth flanks of geometrically accurate bevel gears for crossed-axes gearing is considered to the best possible extent. The discussion begins from the features of geometry of tooth flanks of the reference of bevel gear.

1.5.1 *Reference Surface of Bevel Gear*

Commonly, the reference surface of a gear to be inspected is generated by means of a virtual straight-sided crown rack,$\mathcal{R}_c$. For this purpose, the gear vector diagram of the mesh "gear-to-crown rack" is employed. When rolling in relation to the reference system, $X_gY_gZ_g$, associated with the gear to be inspected, the auxiliary crown-rack, $\mathcal{R}_c$, occupies a plurality of consecutive positions. Envelope surface to a family of consecutive positions of the auxiliary crown-rack, $\mathcal{R}_c$, is assumed to be the reference surface of the gear.

As it is discussed in numerous publications (see [1, 4], and others) that this is incorrect. The auxiliary crown-rack, $\mathcal{R}_c$, and the reference surface of the gear, are the two different surfaces of completely different nature. They are not congruent to one another, and the deviations of the envelope surface from the desired reference surface are not known. Neither the actual value of the deviations nor the directions (inward or outward the bodily side) in which the deviations are measured are not known. Such a surface cannot be used as reference surface when inspecting precision gears.

Without going into details of this discussion, the following statement can be drawn up: *No better alternatives are known for reference surfaces for the inspection precision gears that operate on crossing axes of rotation*. Instead of a strictly defined reference surface, another poorly defined surface is extensively used in computer software for modern CMMs and GMMs. This is because the straight-sided crown rack, $\mathcal{R}_c$, is not conjugate to the reference surface of the gear.

Having no reference surface, it is impossible to make correct measurements, and all readings when inspecting precision gears for gear pairs that operate on crossing axes of rotation cannot be correct.

In order to proceed with the further analysis, the key elements of the kinematics of precision crossed-axes gearing are required to be concisely outlined.

1.5.2 Key Elements of the Kinematics of Geometrically Accurate Crossed-Axes Gearing

Crossed-axes gear pairs are used to transmit a rotary motion from an input shaft to an output shaft, axes of which intersect. Commonly, the pinion is a driver, and the gear is a driven component of the gear pair.

As it is illustrated in Fig. 1.14a, the gear is rotated, ω_g, about its axis of rotation, O_g, while the pinion is rotated, ω_p, about its axis of rotation, O_p. The axes of rotation, O_g and O_p, are at a center-distance, C, apart from one another. In other words, the distance between the gear *base-cone-apex*, A_g, and between the pinion *base-cone-apex*, A_p, equals to C. The axes, O_g and O_p, form the shaft angle, Σ. More precisely, the shaft angle, Σ, is commonly specified as $\Sigma = \angle(\boldsymbol{\omega}_g, \boldsymbol{\omega}_p)$.

Here, in Fig. 1.14a, an image of crossed-axes gear pair is overlapped by a corresponding *gear vector diagram*. The *gear vector diagram*, constructed for this particular gear pair, is shown in Fig. 1.14b.

The axis of instant rotation, P_{ln}, is pointed along the vector of instant rotation, $\boldsymbol{\omega}_{pl}$, where $\boldsymbol{\omega}_{pl} = \boldsymbol{\omega}_p - \boldsymbol{\omega}_g$.

Three rotation vectors: $\boldsymbol{\omega}_g$, $\boldsymbol{\omega}_p$, and $\boldsymbol{\omega}_{pl}$, form the core of the *gear vector diagram*. These rotation vectors are employed to specify four principal planes associated with the gear pair. They are (see Fig. 1.15):

- The *pitch-line plane* (or just *P_{ln}–plane*, for simplicity). The *P_{ln}–plane*, is a plane through the vector of instant rotation, $\boldsymbol{\omega}_{pl}$, and through the center-line, ℄;
- The *center-line plane* (or just *C_{ln}–plane*, for simplicity). The *C_{ln}–plane*, is a plane through the center-line, ℄, perpendicular to the vector of instant rotation, $\boldsymbol{\omega}_{pl}$;
- The *normal-line plane* (or just *N_{ln}–plane*, for simplicity). The *N_{ln}–plane*, is a plane through the plane-of-action-apex, A_{pa}, perpendicular to the center-line, ℄;

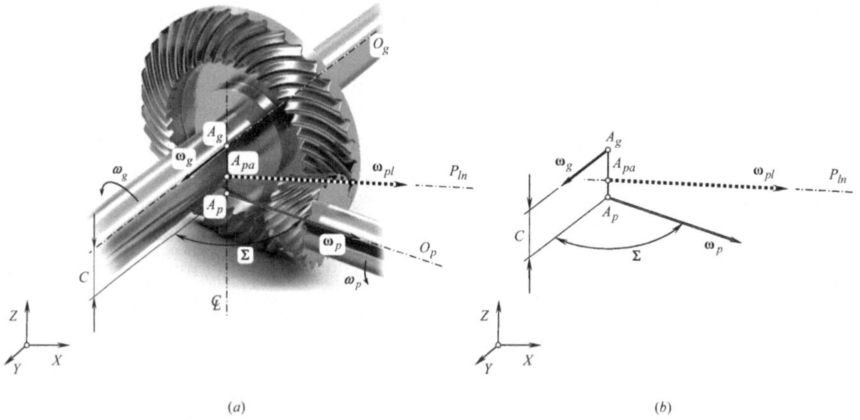

Fig. 1.14 Schematic of crossed-axes gear pair (**a**); and corresponding gear vector diagram (**b**)

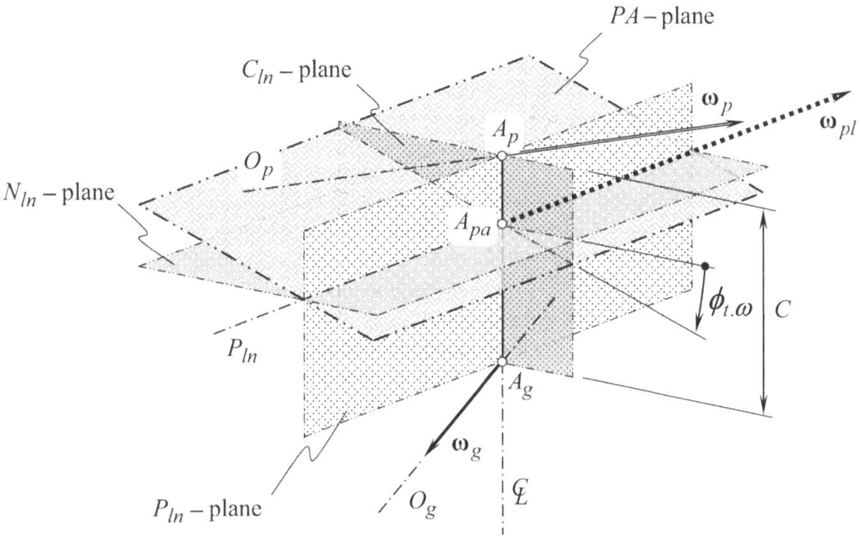

Fig. 1.15 A set of principal planes of crossed-axes gear pair associated with *gear vector diagram*

- The *plane-of-action plane* (or just *PA–plane*, for simplicity). The *PA–plane*, is a plane through the vector of instant rotation, $\boldsymbol{\omega}_{pl}$, that forms a transverse pressure angle, $\phi_{t.\ \omega}$, with the N_{ln}*–plane*.

The further discussion can be significantly simplified by means of usage of the gear vector diagrams (see Fig. 1.14) together with the set of principal planes associated with the gear pair (see Fig. 1.15).

1.5.3 Tooth Flank Geometry in Geometrically Accurate Gears for Gear Pairs That Operate on Crossing Axes of Rotation

There are many similarities between the proposed method of inspection of precision gears for gear pairs that operate on crossing axes of rotation of the gears and between the method of generation of the gear tooth flanks in *R–gearing*. *R–gearing* is the only kind of geometrically accurate crossed-axes gearing with line contact between the tooth flanks $\mathcal{G}$ and $\mathcal{P}$. Therefore, it would be helpful to outline the key points of derivation of the gear tooth flank geometry in *R–gearing*.

Designing gears for *R–gearing*, begins with the specification of a desirable line of contact, LC_{des}. As an example, let us assume that the desirable line of contact, LC_{des}, is shaped in the form of a circular arc, LC_{circ}, of certain radius R_{lc}, that is, $LC_{\text{des}} \equiv LC_{\text{circ}}$, as shown in Fig. 1.16. By definition, the line of contact, LC_{circ}, is entirely situated within the plane of action, *PA*. Despite the desirable line of contact,

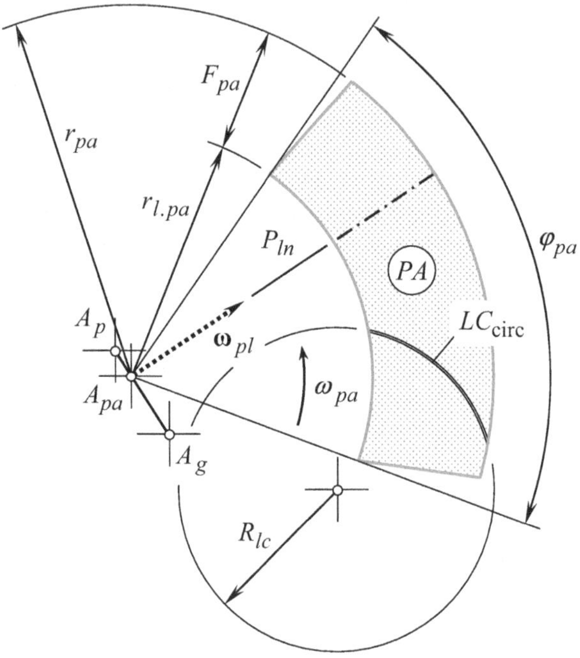

Fig. 1.16 Configuration of the desirable line of contact, LC_{circ}, within the plane of action, PA, in geometrically accurate crossed-axes gear pair

LC_{des}, is already specified, no tooth flanks, $\mathcal{G}$ and $\mathcal{P}$, of the gear, and of the mating pinion teeth a generated yet.

In crossed-axes gearing, the plane of action, PA, is tangent from the opposite sides to base cones of the gear, and that of the mating pinion. It is helpful to consider the sketch in Fig. 1.16 together with the schematic of the set of principal planes, and the gear vector diagram depicted in Fig. 1.15. It is also recommended to involve into this analysis the gear vector diagram along with the schematic of the gear pair shown in Fig. 1.14.

Consider a *Cartesian coordinate system*, $X_gY_gZ_g$, associated with the gear. Another *Cartesian coordinate system*, $X_pY_pZ_p$, is associated with the mating pinion. The base-cone-apexes, A_g and A_p, are at distances C_g and C_p from the plane-of-action-apex, A_{pa}, correspondingly, as illustrated in Fig. 1.17.

When the plane of action, PA, is rotated, ω_{pa}, about the *plane-of-action-apex*, A_{pa}, the *Cartesian coordinate systems*, $X_gY_gZ_g$ and $X_pY_pZ_p$, are rotated, and the rotations, $\boldsymbol{\omega}_g$ and $\boldsymbol{\omega}_p$, are properly timed with one another. In such a relative motion two family of the lines of contact, LC_{circ}, are generated. The first family of the lines LC_{circ} is generated in the reference system $X_gY_gZ_g$. These lines are entirely situated within the gear tooth flank, $\mathcal{G}$, and, therefore can be employed as reference for inspection the gear tooth flank, $\mathcal{G}$. Another family of the lines LC_{circ} is generated in the reference system $X_pY_pZ_p$. These lines are entirely situated within the pinion tooth flank, $\mathcal{P}$, and, therefore can be employed as reference for inspection the pinion tooth flank, $\mathcal{P}$.

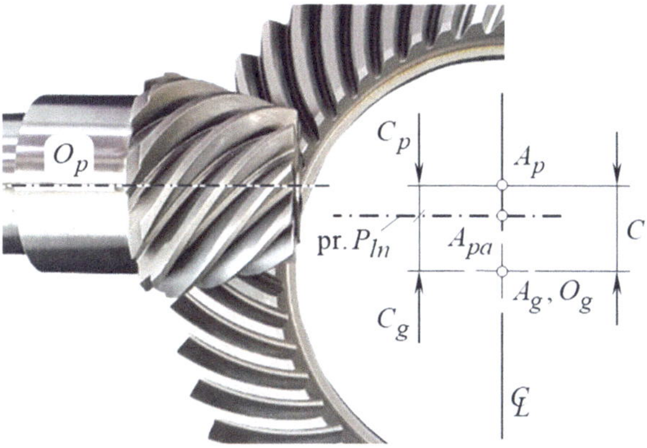

Fig. 1.17 The *gear-center-distance*, C_g, and the *pinion-center-distance*, C_p, in crossed-axes gear pair

1.5.4 Kinematics of Inspection of Geometrically Accurate Gears for Crossed-Axes Gearing

When geometrically accurate gears for crossed-axes gearing are inspected, the stylus of dial type indicator has to travel properly in relation to the gear to be inspected. For this purpose, the gear is rotated about its axis of rotation, O_g. The plane of action, PA, is rotated about its axis of rotation, O_{pa} (the axis of rotation, O_{pa}, is a straight line through the plane-of-action-apex, constructed perpendicular to the plane of action, PA). The rotations, ω_g and ω_{pa}, are synchronized with one another. In addition to the rotations, ω_g and ω_{pa}, of the gear, and of the plane of action, the stylus travels either along the gear tooth (in the lengthwise direction of the gear tooth), or across the tooth.

The Concept of Inspection of Gear in Lengthwise Direction of Gear Tooth In the course of inspection of a gear for real crossed-axes gear pair, intersection of the tooth flank, $\mathcal{G}$, by the plane of action, PA, is considered. In Fig. 1.18a, the line of intersection through arbitrary point within the tooth flank is labeled as LG_i. When the gear is rotated simultaneously with the plane of action, PA, the geometry of the line of intersection, LG_i, of the gear to be inspected alters. A plurality of the lines of intersection, LG_i, on the actual gear tooth flank, those generated during the meshing cycle of a pair, is situated within a bend of an angular width, δ, in the plane of action, PA. The angular deviation, δ_i, at a current configuration of the gear in relation to the plane of action, PA, can be physically measured in a gear/pinion. Current value of the angular deviation, δ_i, can be also calculated.

When tooth flank geometry of a gear in the lengthwise direction of the gear tooth is inspected (see Fig. 1.18b), the stylus tip travels along the line of intersection of the desirable tooth flank by the plane of action, PA. At current point of interest, m, the line, a, along which the deviation is measured, is situated within the plane of action, PA, and it is perpendicular to the bevel gear tooth flank at this point, m.

The considered concept of inspection of gear tooth flank accuracy can be employed for the inspection of the variation of the angular base pitch in a gear.

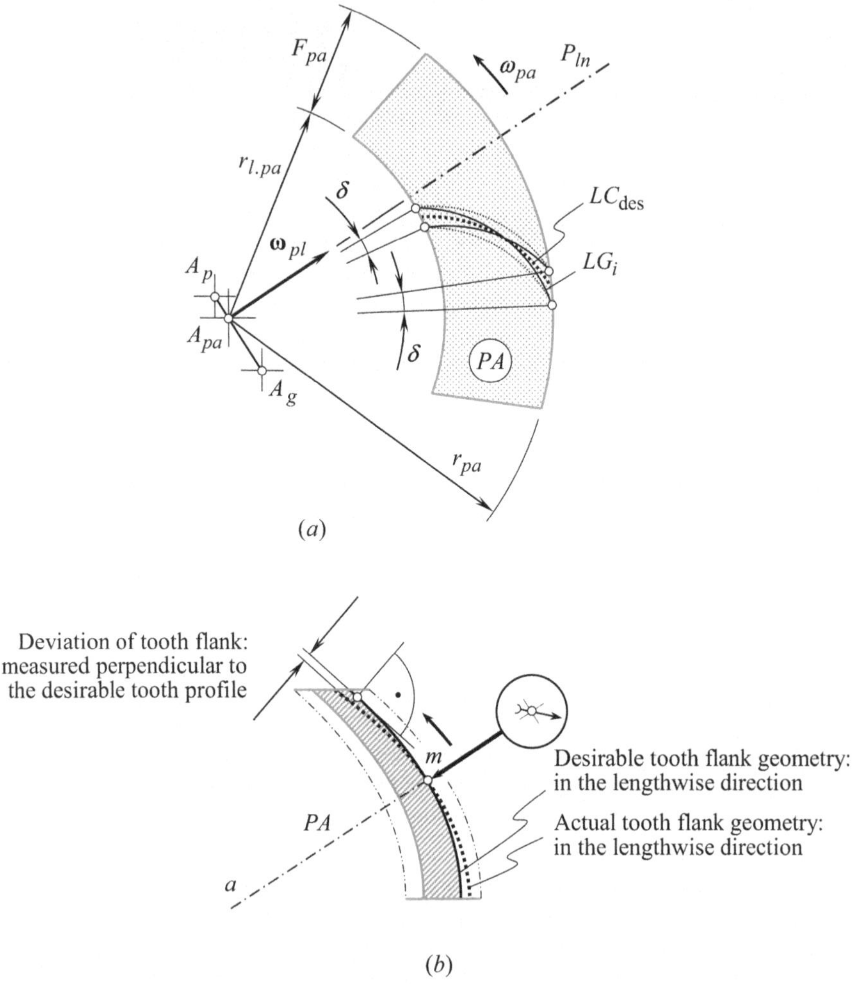

Fig. 1.18 Deviation of tooth flank of a real gear from tooth flank, $\mathcal{G}$, of a corresponding geometrically accurate gear (**a**); and inspection of accuracy in the lengthwise direction of a gear for crossed-axes gear pair (**b**)

The Concept of Inspection of Gear Tooth Profile When a gear for intersected-axes gear pair is inspected (see Fig. 1.19a), the gear is rotated about its axis of rotation, O_g. The plane of action, *PA*, is rotated about its axis of rotation, O_{pa}. When crossed-gears are inspected (see Fig. 1.19b), the axes of rotation, O_g and O_{pa}, are at a certain distance, C_g, from one another.

The rotations, ω_g and ω_{pa}, are synchronized with one another so, as to keep the stylus at current point, *m*, on the desirable gear tooth flank, $\mathcal{G}$. Point of interest, *m*, is situated within the plane of action, *PA*. The line *a*, along which the deviation is measured, is entirely located within the plane of action, *PA*, and it is perpendicular to

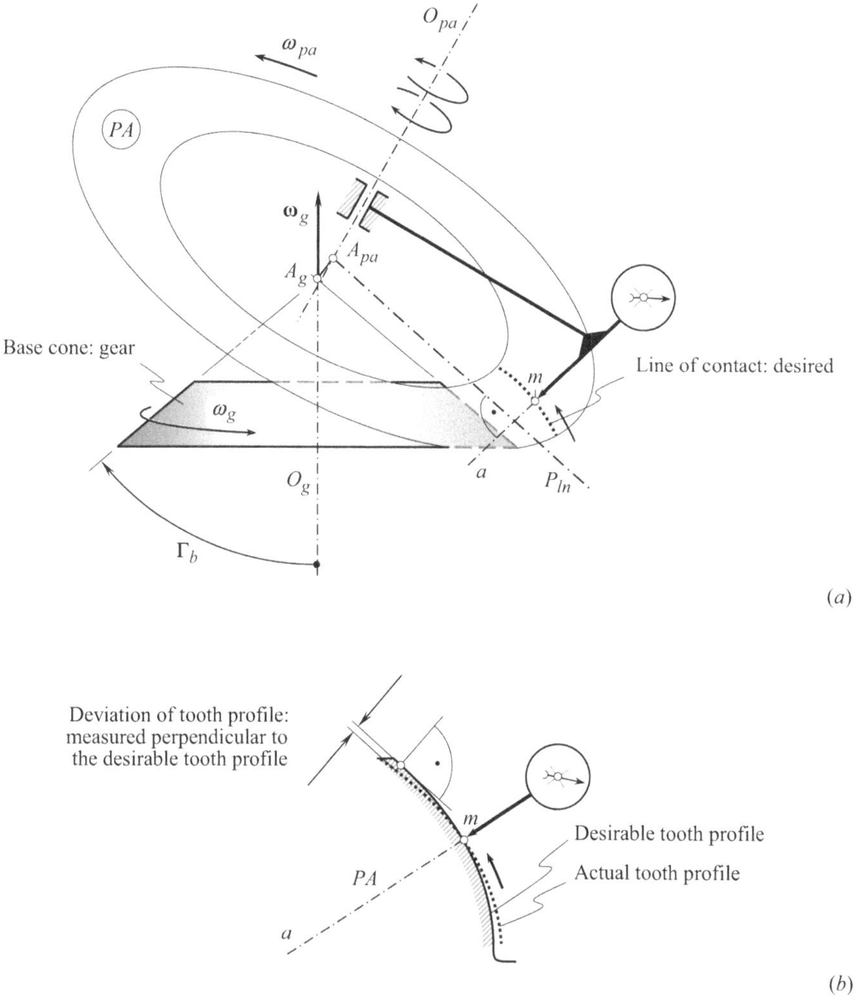

Fig. 1.19 Inspection of the accuracy of tooth profile in a gear for crossed-axes gear pair (**a**); and kinematics of a method of inspection gears for crossed-axes gear pair (**b**)

the gear tooth flank at the point, *m*. The stylus tip is motionless in relation to the plane of action, *PA*, when inspecting the tooth profile error of the gear tooth flank.

When the gear tooth flank, $\mathcal{G}$, is inspected, the reference surface in both cases is a kind of conjugate surface, namely, it can be conjugate both to the tooth flank of the crown rack, $\mathcal{R}$ (not straight-sided rack in this particular case), and to the tooth flank of a mating pinion, $\mathcal{P}$.

The tooth flank of a geometrically accurate gear for gear pairs that operate on crossing axes of rotation of a gear, and of a mating pinion is schematically shown in Fig. 1.20.

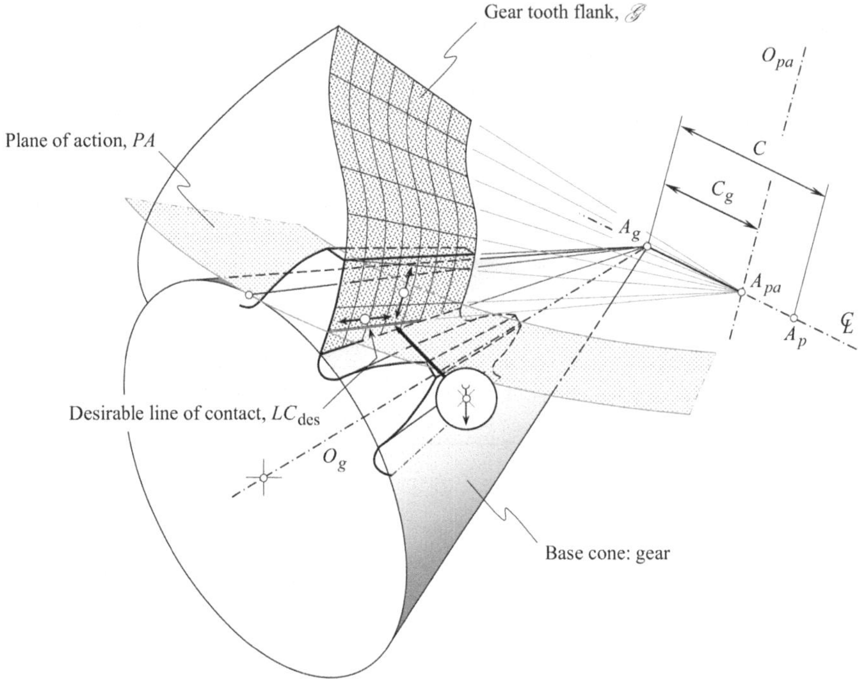

Fig. 1.20 Tooth flank, $\mathcal{G}$, in a precision gear for crossed-axes gear pair: The stylus tip traces in the lengthwise direction of the gear tooth, and in its transverse section, form a net

The net formed of stylus tip traces in the lengthwise direction of the gear tooth, and in its transverse section, is entirely situated on the reference surface.

The Concept of Inspection of Operating Angular Base Pitch in Crossed-Axes Gearing Operating base pitch, $\varphi_{b.\ op}$, is defined as the angular distance between points of intersection of two adjacent line of contact, LC_i and LC_{i+1}, by a common circular arc that is centered at the plane-of-action apex, A_{pa}. This is illustrated in Fig. 1.21.

The actual value of the operating angular base pitch, $\varphi_{b.\ op}$, does not depend on the value of the circular arc radius: $r_{pa.1}$, $r_{pa.2}$, and so forth. For a specified gear pair, the operating angular base pitch, $\varphi_{b.\ op}$, is of a constant value ($\varphi_{b.\ op} = const$).

The angular base pitch in a gear, $\varphi_{b.\ g}$, must be equal to the operating base pitch, $\varphi_{b.\ op}$, in a gear pair ($\varphi_{b.\ g} = \varphi_{b.\ op}$). Similarly, the angular base pitch in a mating pinion, $\varphi_{b.\ p}$, must be equal to the operating base pitch, $\varphi_{b.\ op}$, in a gear pair ($\varphi_{b.\ p} = \varphi_{b.\ op}$).

Three angular base pitches, namely: $\varphi_{b.\ g}$, $\varphi_{b.\ p}$, and $\varphi_{b.\ op}$, are of fundamental importance for geometrically accurate gearing.

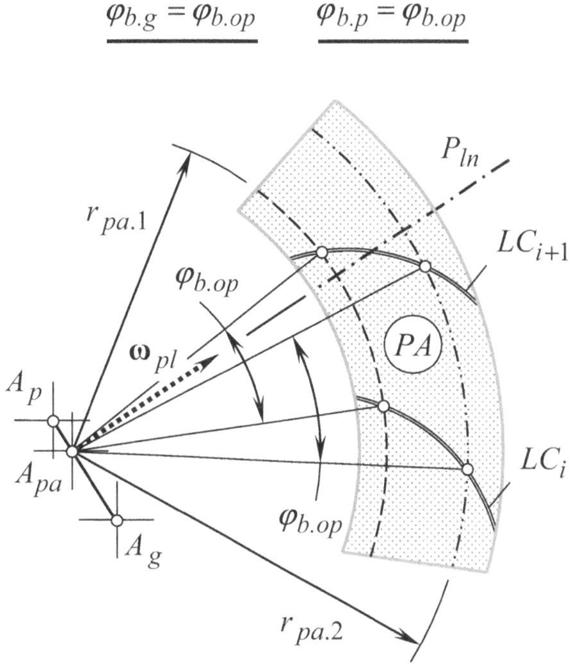

Fig. 1.21 Operating angular base pitch, $\varphi_{b.\,pa}$, in gearing that operates on crossing axes of rotation

1.6 Mounting Distance: Accuracy of Axial Location of Gears in the Gear Housing

The performance of intersected-axes and of crossed-axes gear pairs strongly depends on the accuracy of the axial location of the gears – on the accuracy of the mounting distance. Precision gearing is especially sensitive to the accuracy of mounting of gears in the gear housing.

Mounting distance in gears for intersected-axes gear pairs, as well as for crossed-axes gear pairs, is a design parameter that cannot be inspected directly, as the base cone apex of the gear does not exist physically. This is the main reason for the significant inconveniences when inspecting the accuracy of mounting distance. Therefore, indirect methods for the inspection of the accuracy of the mounting distance are used in practice.

1.6.1 On the Correlation Between Tooth Flank Geometry and the Mounting Distance

The mounting distance is a vital design parameter for intersected-axes gearings, as well as for crossed-axes gearings. The specification, inspection, and tolerancing of the mounting distance are challenging to understand by many practical gear engineers.

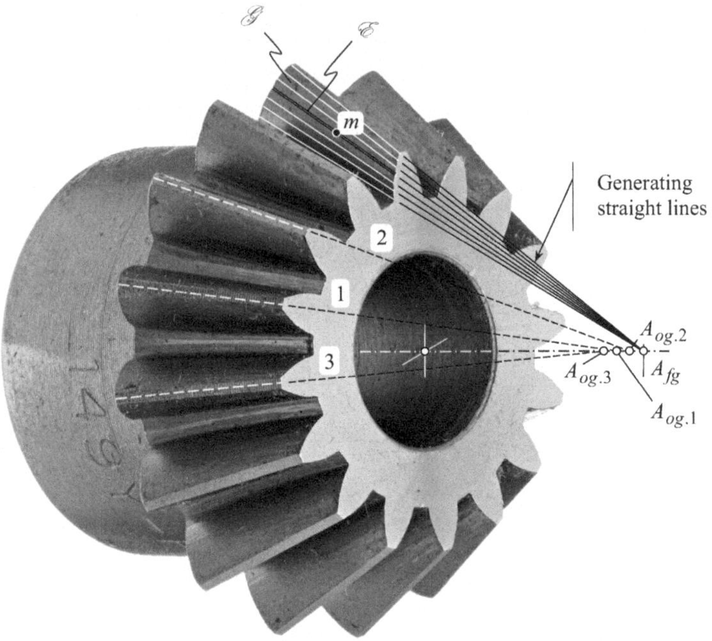

Fig. 1.22 A simplified model that is used in the derivation of the definition of the term *tooth flank apex*, A_{fg}, in intersected-axes gearing, and in crossed-axes gearing

To make things clear, it is necessary to outline to the best possible extent what the term mounting distance stands for. As an example, consider tooth flank of a straight bevel gear[1] as illustrated in the simplified model, shown in Fig. 1.22. The gear tooth flank, $\mathcal{G}$, can be construed as a family of straight lines through each point of the gear tooth profile and the gear tooth flank apex, A_{fg} (the gear tooth flank apex, A_{fg}, is an equivalent to the gear base cone apex, A_{bg}). Only the gear tooth flank apex, A_{fg}, should be used for the specification and inspection of the mounting distance, *MD*. Commonly, the outer cone apex of the gear, $A_{og.1}$, is inward in relation to the gear tooth flank apex, A_{fg}, as shown for the straight generating line 1 in Fig. 1.22. However, this is not a must as the gear outer cone apex, $A_{og.2}$, can be located either outward, or the gear outer cone apex, $A_{og.3}$, can be located inward in relation to the apex $A_{og.1}$, as it is illustrated in Fig. 1.22 for the straight generating lines 2 and 3 of the outer cone of the gear. The actual location of the gear outer cone apex (either $A_{og.1}$, or $A_{og.2}$, or $A_{og.3}$) does not affect the conditions of meshing of the gear, $\mathcal{G}$, and the pinion, $\mathcal{P}$, tooth flanks. The conditions of meshing of the gear, $\mathcal{G}$, and the pinion, $\mathcal{P}$, tooth flanks depend only on the actual location of the gear/pinion tooth flank apex, A_{fg}. Therefore, the tolerance for the mounting distance (the mounting distance is measured in relation to the gear tooth flank apex, A_{fg}) is tight, while the configuration of the apex of the outer cone of the gear, A_{og}, is a free dimension with no specified tolerance.

[1]This concept can be easily extended to bevel gears of any and all other designs.

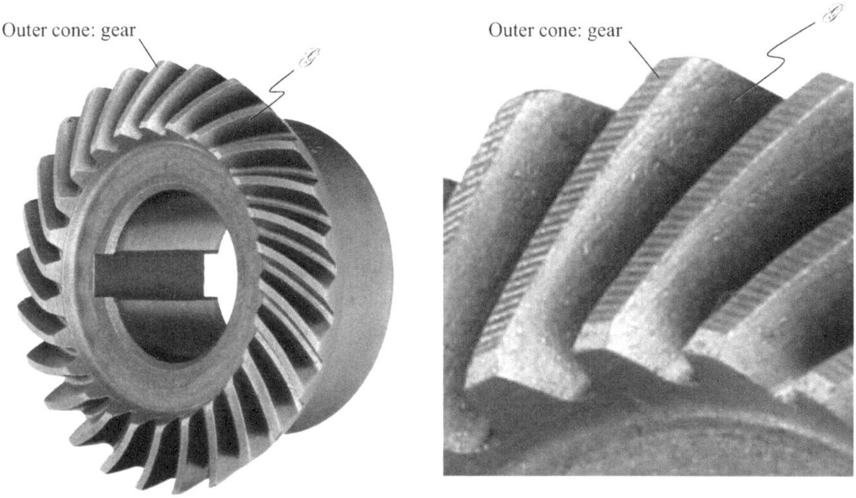

Fig. 1.23 Outer cone surface of a gear for intersected-axes gear pair can be used as a datum surface for inspection of the mounting distance when the cone surface is finished-cut in one setup with the gear tooth flank, $\mathcal{G}$ (a similar is valid with respect to crossed-axes gear pair)

Depending on the accuracy of the setup parameters when cutting straight bevel gears, the axial position of the tooth flank, $\mathcal{G}$, can be changed. After being assembled, the gear tooth flank apex, A_{fg}, must be coincident with the centerlines in the gearbox housing. Therefore, the mounting distance of a gear is the distance from the gear tooth flank apex, A_{fg}, to the common reference surface of the gear. A similar definition is valid with respect to the mounting distance of a pinion.

Then, it should be taken into account that the crown is not geometrically associated with the gear tooth flank, $\mathcal{G}$. The performed analysis of Fig. 1.23 reveals that depending on the accuracy of the setup parameters when machining straight bevel gears, the axial position of the tooth flank, $\mathcal{G}$, and the gear tooth flank apex, A_{fg}, can be variable.

A gear tooth flank in intersected-axes, as well as in crossed-axes gearing, must be correctly specified in relation to the common reference surface. For this purpose, arbitrary point, m, either within the tooth flank or within an extension of this surface can be used. The tooth flank apex, A_{fg} (see Fig. 1.22), is convenient to be used for this purpose. It is a wrong practice to use the wedge, E, and the dimension, F (see Fig. 1.23), to specify the mounting distance in I_a- *gearing*, and in C_a-*gearing*.

Direct measurements of the mounting distance[2] are necessary. This can be done, for example, by means of laser scanning of the machined tooth flank, and creating a

[2] ***Important*:** In addition to the mounting distance, in C_a-gearing, one more design parameter must be inspected. This is an angle that a gear/pinion axis of rotation forms with the centerline, ℄. A tolerance for this angle needs to be specified, and the actual deviation of the angle from required value of it needs to be inspected. An appropriate tolerance for the center distance is also required.

CAD model of the actual tooth flanks on the premise of that. Comparison of two CAD models, that is, of the desirable and of the actual tooth flanks, returns an exact number of the mounting distance in a machined gear in I_a- *gearing*, and in C_a- *gearing*. Simpler methods of and means for the direct inspection of the actual mounting distance in intersected-axes and crossed-axes gearings can be developed.

Reliable methods for the calculation of the tolerances for the mounting distance in intersected-axes, as well as crossed-axes gearings, need to be developed. Tooth flank of a gear can be used for the inspection of the accuracy of the mounting distance. The problem in question can be resolved if an alternative equivalent surface can be used for the inspection. The outer cone surface of a gear to be inspected can be used for the inspection of accuracy of the mounting distance if this surface is finish-cut in one setup with finish machining of the tooth flanks of the gear. This can be done either if the full profile cutters are used in the design of the cutter-heads, or additional one, or if the two cutters are mounted on the cutter-head of the conventional design.

Inspection of the mounting distance and of the contact pattern are the two main approaches for the inspection of the correctness of the axial location of the gears in intersected-axes and crossed-axes gear pairs.

1.6.2 Mounting Distance in Intersected-Axes Gearing

To operate properly, gears in intersected-axes gear pairs must be properly configured in relation to one another [7].

At the beginning, let's clarify what the term ***mounting distance*** stands for.

- ***First***, and the most critical, by means of mounting distances the location of tooth flanks in the axial direction of a gear and a mating pinion are specified in relation to the centerlines in the gear housing. Apex point of a tooth flank of the gear, A_{fg}, and the pinion, A_{fp}, must be coincident with the point of intersection of the centerlines in the gear housing in I_a-gearing. The pitch plane apex is commonly designated as A_{pp}, and the plane of action apex is commonly designated as A_{pa}. These two conditions can be expressed as follows: $A_{fg} \equiv A_{pp} \equiv A_{pa}$ and $A_{fp} \equiv A_{pp} \equiv A_{pa}$
- ***Second***, the pitch cone apex of a gear, A_g, (and of a mating pinion, A_p) must be coincident with the apex point of a tooth flank of the gear, A_{fg}, and the pinion, A_{fp}. These two conditions can be expressed as follows: $A_g \equiv A_{fg}$ and $A_p \equiv A_{fp}$
- ***Third***, the base cone apex of a gear, A_{bg}, (and of a mating pinion, A_{bp}) m ust be coincident with the apex point of a tooth flank of the gear, A_{fg}, and the pinion, A_{fp}. These two conditions can be expressed as follows: $A_{bg} \equiv A_{fg}$ and $A_{bp} \equiv A_{fp}$
- ***Fourth***, there are two outer cones. One of them is of the gear, and another one is of the mating pinion. There are also two root cones. Again, one of them is of the gear and another one is of the mating pinion. Neither the outer cones nor the root cones directly affect the actual value of the mounting distance. However, these cones need to be mentioned here, as in the present-day practice outer cones are commonly used when determining the mounting distance.

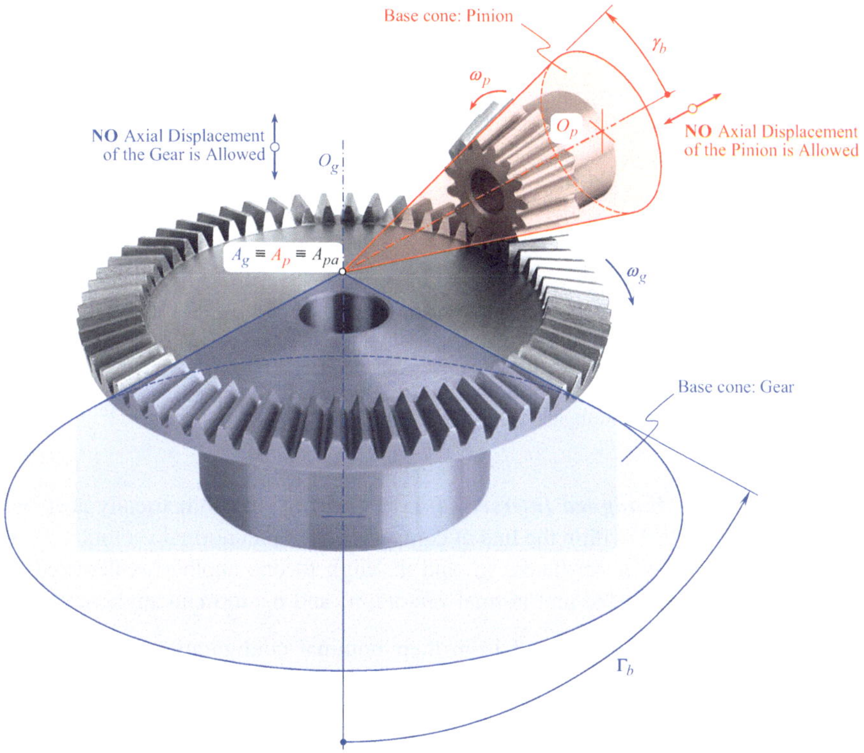

Fig. 1.24 Intersected-axes gearing with associated base cone of a gear and that of a mating pinion

Relative Disposition of Base Cones in Intersected-Axes Gearing A straight bevel gear pair is depicted in Fig. 1.24 as an example. When a bevel gear pair operates, the gear spins, ω_g, about its axis of rotation, O_g. The mating pinion spins, ω_p, about its axis of rotation, O_p. The axes of rotation, O_g and O_p, intersect one another, and the point of intersection coincides with the plane-of-action apex, A_{pa} (the plane of action, *PA*, is not shown in Fig. 1.24) – this is a must for the perfect performance of bevel gearing.

A base cone is associated with the gear. The gear base cone apex, A_g, is located within the gear axis of rotation, O_g. Another base cone is associated with the pinion. The pinion base cone apex, A_p, is located within the pinion axis of rotation axis, O_p. The gear and the pinion must be configured in relation to each other so as to ensure the apexes A_g and A_p snapped together with the plane-of-action apex, A_{pa} – this is a must [7]. Neither the axial displacement of the gear nor the axial displacement of the pinion, from the position shown in Fig. 1.24, is allowed. The plane of action, *PA*, is a plane through the plane-of-action apex, A_{pa}, that is in tangency to the base cones of the gear and the mating pinion.

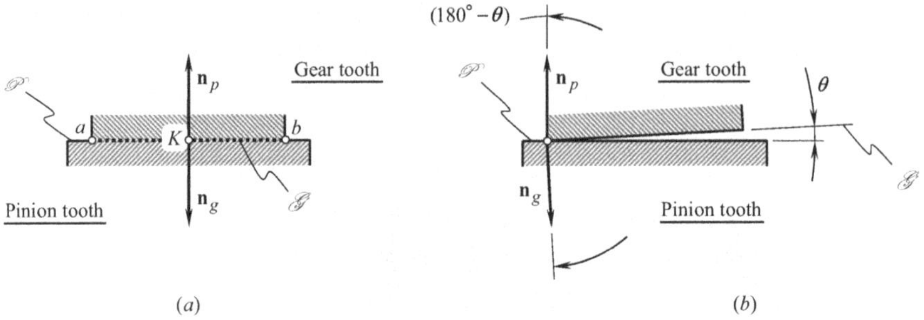

Fig. 1.25 Unit normal vectors, $\mathbf{n}_g$ and $\mathbf{n}_p$, to tooth flanks, $\mathcal{G}$ and $\mathcal{P}$, of a gear and of a mating pinion in cases of correct (**a**), and incorrect (**b**) configuration of the mating gears. The unit normal vectors, $\mathbf{n}_g$ and $\mathbf{n}_p$, are not within the plane of action, *PA*. Instead, these unit vectors are located in a plane of action, constructed for the misaligned gears

Edge Contact in Misaligned Intersected-Axes Gearing In geometrically accurate gears, at every point *K* within the line of contact, *LC*, the unit normal vectors, $\mathbf{n}_g$ and $\mathbf{n}_p$, to the interacting tooth flanks, $\mathcal{G}$ and $\mathcal{P}$, align to one another as depicted in Fig. 1.25a. In this way, the unit normal vectors, $\mathbf{n}_g$ and $\mathbf{n}_p$, form an angle of 180°.

When the gears are displaced from their nominal configuration, as shown in Fig. 1.25b, the unit normal vectors, $\mathbf{n}_g$ and $\mathbf{n}_p$, form an angle that equals $(180° - \theta)$. This schematic (see Fig. 1.25b) is the key to understand how a tolerance for the accuracy of the mounting distance can be calculated. No calculations of the tolerance for the accuracy of the mounting distance can be performed if a tolerance $[\theta]$ for the angle θ is not specified.

Disposition of Base Cone of a Gear in Relation to the Plane of Action in Intersected-Axes Gearing The axis of rotation of a gear, O_g, and that, O_{pa}, of the plane of action, *PA*, in a geometrically accurate intersected-axes gear pair intersect as illustrated in Fig. 1.26a. The base cone apex of the gear, A_g, and the plane-of-action apex, A_{pa}, both coincide with point of intersection of the axes O_g and O_{pa}. A rotation of the gear is designated as ω_g, and rotation of the plane of action is designated as ω_{pa}, correspondingly. When the gears rotate, the base cone of the gear, and the plane of action roll over each other with no sliding.

In reality, an error in mounting distance, ε_g, is always observed. The latter is shown in Fig. 1.26b. In such a scenario, the gear base cone apex, A_g^*, is not coincident with the plane-of-action-apex, A_{pa}, as the apex, A_g^*, is displaced axially in relation to A_{pa} at a distance ε_g. (Here, we are not going into details of analysis of the gap between the gear base cone and the plane of action). In a case, depicted in Fig. 1.26b, the outward displacement, ε_g, is of a positive value. An inward displacement is of a negative value (not shown in Fig. 1.26b).

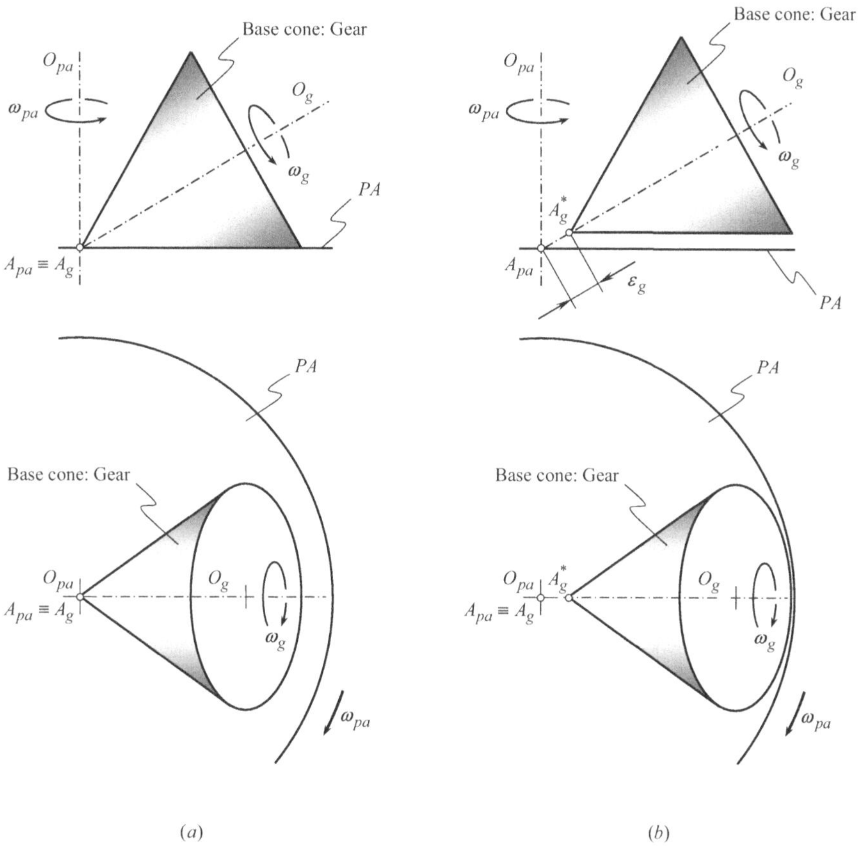

Fig. 1.26 Disposition of base cone of a gear in relation to the plane of action, *PA*, in (**a**) geometrically accurate and in (**b**) misaligned intersected-axes gear pair

The performed analysis in Fig. 1.26 allows to proceed with the derivation of an equation for the calculation of the tolerance for the accuracy of the mounting distance in intersected-axes gear pair.

Mounting of Gears in Intersected-Axes Gear Housing Theoretically, in an intersected-axes gear pair, a gear base cone apex, A_g, a mating pinion base cone apex, A_p, and the plane-of-action apex, A_{pa}, must be snapped together, as schematically illustrated in Fig. 1.27a. In reality, in order to transmit power, gears need to be supported. For this purpose, a gear housing (see Fig. 1.27b) is used. The gear housing features a corresponding number of bores that are used to support gear shafts in the housing. A centerline of a bore for the gear shaft is labeled as O_{hg}. Correspondingly, a centerline of a bore for the pinion shaft is designated by O_{hp}. The centerlines, O_{hg} and O_{hp}, intersect at point A_h. Strong constraints on configuration of the gear and its mating pinion in the gear housing are imposed by this point, A_h. The center-distance, $O_{hg}O_{hp}$, between the centerlines, O_{hg} and O_{hp}, must be as short, as possible ($O_{hg}O_{hp} \rightarrow 0$). Deviation in the angular configuration of the centerlines also must be the smallest possible.

The gear must be configured in the gear housing so as to keep the gear axis of rotation, O_g, aligned with the axis, O_{hg}, in the gear housing (i.e., $O_g \equiv O_{hg}$), as illustrated in Fig. 1.27c. The gear base cone apex, A_g, must be coincident with the point, A_h, of intersection of the centerlines, O_{hg} and O_{hp} (i.e., $A_g \equiv A_h$). Then, the rest of the important components, namely, the bearings, shims, fasteners, and so forth, must be put together so as to keep the gear mounting distance, MD_g, to the blueprint. The gear mounting distance, MD_g is measured between the gear back face, and the corresponding gear housing face, as shown in Fig. 1.27d.

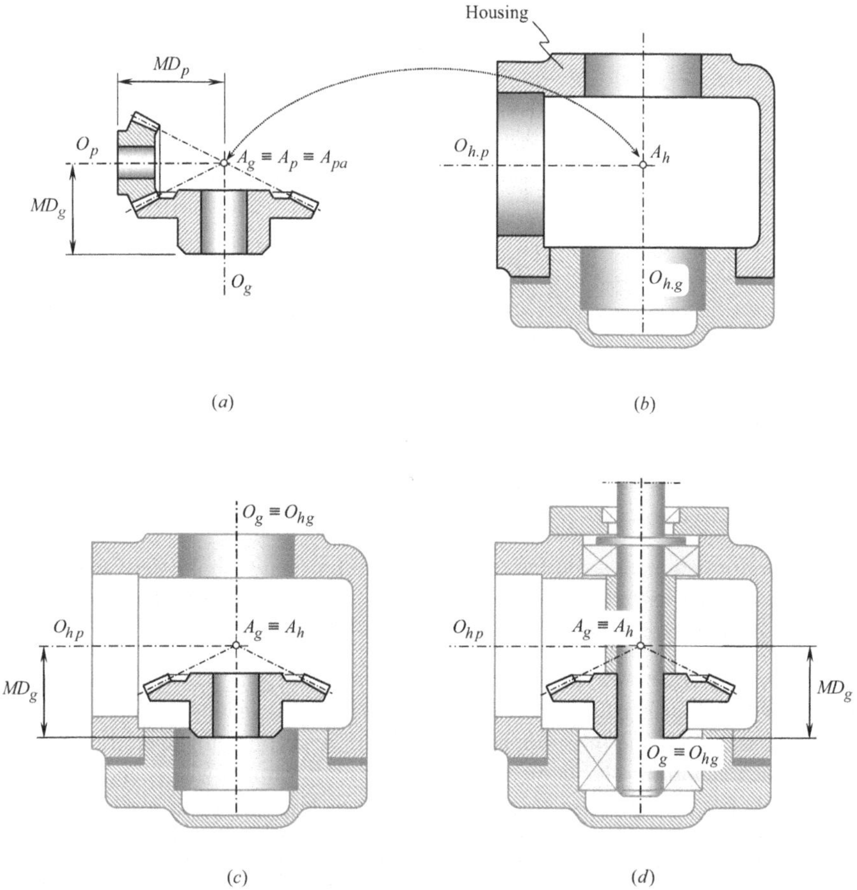

Fig. 1.27 Intersected-axes gearing: mounting of bevel gears in the gear housing. Parts (**a**) through (**g**) are discussed in the text

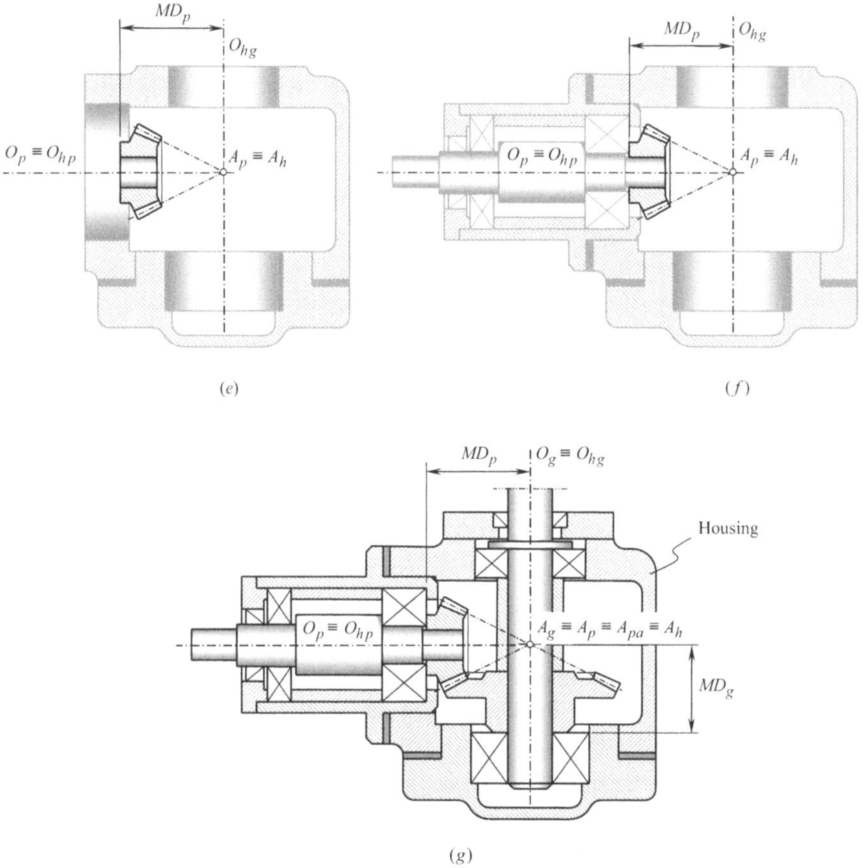

Fig. 1.27 (continued)

Similarly, the pinion must be configured in the gear housing so as to keep the pinion axis of rotation, O_p, aligned with the axis, O_{hp} (i.e., $O_p \equiv O_{hp}$), in the gear housing, as illustrated in Fig. 1.27e. The pinion base cone apex, A_p, must be coincident with the point, A_h, of the intersection of the centerlines, O_{hg} and O_{hp} (i.e., $A_p \equiv A_h$). Then, the rest of the important components, namely, the bearings, shims, fasteners, and so forth, must be put together so as to keep the pinion mounting distance, MD_p, to the blueprint. The pinion mounting distance, MD_p is measured between the pinion back face, and the corresponding gear housing face, as shown in Fig. 1.27f.

Finally, once the gear and the pinion are properly mounted in the gear housing, their axes are aligned to the centerlines of the corresponding bores in the gear housing (i.e., $O_g \equiv O_{hg}$, and $O_p \equiv O_{hp}$), and the apexes, A_g, A_p, and A_{pa}, all of them are snapped together with the point of intersection, A_h, of the centerlines, O_{hg} and O_{hp}, in the gear housing, as schematically illustrated in Fig. 1.27g.

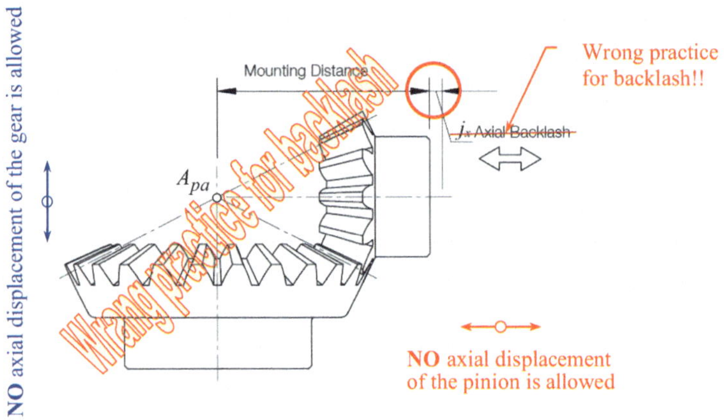

Fig. 1.28 Wrong practice to adjust for the backlash in intersected-axes gearing by means of *inward-outward axial shift of the gears*: NO axial shift of the gears from the point of intersection, A_h, of the centerlines, O_{hg} and O_{hp}, in the gear housing, is allowed in $I_a-gearing$. [The similar is valid with respect to $C_a-gearing$]

An important conclusion can be drawn up from the above discussion: NO axial shift of the gears from the point of intersection, A_h, of the centerlines, O_{hg} and O_{hp}, in the gear housing, is permissible in intersected-axes, gearing (see Fig. 1.28). This is a wrong practice to adjust for the backlash in intersected-axes gearing by means of *inward-outward axial shift of the gears*.

Tolerance for the Accuracy of Mounting Distance in Intersected-Axes Gearing Consider intersection of two teeth in contact by the plane of action, PA, as illustrated in Fig. 1.29a. These teeth contact one another along a straight-line segment ab. (Here, for simplicity, but without loss of generality, a pair of geometrically accurate straight bevel gears is considered). Configuration of the gears in Fig. 1.29a corresponds to a case when the centerline of the gear tooth in a section by the pitch plane is aligned with the axis of instant rotation, P_{ln}, in the gear pair.

Under any circumstances, edge contact between the gear and the pinion teeth has to be avoided. That is why, in order to keep the actual value of the angle θ within a very tight tolerance for the accuracy of this parameter, the tolerances for the accuracy of the rest of the design parameters, on which the axes misalignment depends on, have to be set very tight.

As solely the gear mesh is considered, only the unit normal vectors $\mathbf{n}_g$ and $\mathbf{n}_p$ are taken into account, while the unit normal perpendiculars to the gear, and the pinion faces are ignored.

Moreover, no edge roundness, chamfers, etc. are considered in this analysis.

A case of outward displaced gear is illustrated in Fig. 1.29b. As a result of the displacement, a gap between the teeth is observed. This gap is shown by two straight-line segments, ab and a^*b^*, each of which is entirely located on the tooth flanks, $\mathcal{G}$ and $\mathcal{P}$, of the two interacting teeth.

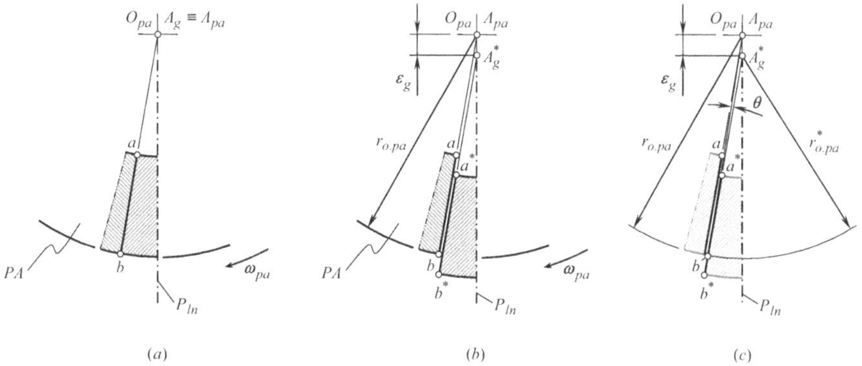

Fig. 1.29 Section of two interacting teeth by the plane of action, *PA*: (**a**) zero mounting distance error, (**b**) mounting distance error of a certain value, ε_g, and (**c**) schematic for the derivation of the formula for the calculation of tolerance for the accuracy of the mounting distance in intersected-axes gearing

A schematic for the derivation of a formula for the calculation of a tolerance for the accuracy of the mounting distance is depicted in Fig. 1.29c. It should be stressed from the very beginning that point b within the straight-line segment ab is the closest point to the straight-line segment a^*b^*. Therefore, the actual value of an angle, θ_g, through which the plane of action, *PA*, has to be turned about the axis of rotation O_{pa} depends on the actual distance of point b to the straight-line segment a^*b^*.

Assume that the gear is motionless, and the plane of action, *PA*, is turned about its axis of rotation, O_{pa}, through an angle at which point, b, touches the straight-line segment, $A_g^*b^*$. Then, consider a triangle, $\Delta A_{pa}A_g^*b$. In this triangle, the following equalities $A_{pa}b = r_{o.\,pa}$, $A_{pa}A_g^* = \varepsilon_g$, and $A_g^*b = r^*_{o.pa}$, are valid.

Law of cosine is used for the determination of the actual value of the angle, θ_g:

$$\varepsilon_g^2 = r_{o.ap}^2 + \left(r^*_{o.pa}\right)^2 - 2 \cdot r_{o.ap} \cdot r^*_{o.pa} \cdot \cos\theta_g \tag{1.37}$$

Then, either the actual value of the angle, θ_g, can be expressed in terms of the mounting distance error, ε_g, and the design parameters of the gear pair, or a maximum permissible value of the displacement, ε_g, can be expressed in terms of the maximum permissible value of the angle θ_g, and the design parameters of the gear pair:

$$\varepsilon_g = \sqrt{r_{o.ap}^2 + \left(r^*_{o.pa}\right)^2 - 2 \cdot r_{o.ap} \cdot r^*_{o.pa} \cdot \cos\theta_g} \leq [\varepsilon_g] \tag{1.38}$$

Here, $[\varepsilon_g]$ is the tolerance for the accuracy of the gear mounting distance, ε_g.

$$\theta_g = \cos^{-1}\left[\frac{r_{o.ap}^2 + \left(r_{o.pa}^*\right)^2 - \varepsilon_g^2}{2 \cdot r_{o.ap} \cdot r_{o.pa}^*}\right] \le [\theta_g] \tag{1.39}$$

Here, $[\theta_g]$ is the tolerance for the accuracy of the gear angle, θ_g.

It can be shown (see Fig. 1.29c) that the equality:

$$r_{o.pa}^* \cong \frac{r_{o.pa} - \varepsilon_g}{\cos \varphi_{pa}} \tag{1.40}$$

is valid.

When the pinion is fully aligned, the equalities $\varepsilon = \varepsilon_g$ and $\theta = \theta_g$, and Eqs. (1.38) and (1.39) can be used for the calculation of tolerance on the mounting distance in the intersected-axes gear pair.

The analysis similar to that above can be performed for a mating bevel pinion:

$$\varepsilon_p = \sqrt{r_{o.ap}^2 + \left(r_{o.pa}^*\right)^2 - 2 \cdot r_{o.ap} \cdot r_{o.pa}^* \cdot \cos\theta_p} \le [\varepsilon_p] \tag{1.41}$$

$$\theta_p = \cos^{-1}\left[\frac{r_{o.ap}^2 + \left(r_{o.pa}^*\right)^2 - \varepsilon_p^2}{2 \cdot r_{o.ap} \cdot r_{o.pa}^*}\right] \le [\theta_p] \tag{1.42}$$

Here, $[\varepsilon_p]$ is the tolerance for the accuracy of the pinion mounting distance, ε_p; and $[\theta_p]$ is the tolerance for the accuracy of the gear angle, θ_p.

Further, when the gear is fully aligned, the equalities $\varepsilon = \varepsilon_p$ and $\theta = \theta_p$, and Eqs. (1.41) and (1.42) can be used for the calculation of tolerance on the mounting distance in the intersected-axes gear pair

Finally, in a more general case, a bevel gear and a mating bevel pinion, both, are displaced. In such a scenario, either the actual value of the angle θ can be expressed in terms of the mounting distance errors, ε_g and ε_p, and the design parameters of the gear pair, or a maximum permissible value of the displacements, ε_g and ε_p, can be expressed in terms of the maximum permissible value of the angle θ, and the design parameters of the gear pair.

The angle θ is the angle that is formed by two perpendiculars, $\mathbf{n}_g$ and $\mathbf{n}_p$ [that is, $\theta = \angle(\mathbf{n}_g, \mathbf{n}_p)$], constructed at point within the edge contact of a gear, $\mathcal{G}$, and a mating pinion, $\mathcal{P}$, tooth flanks, correspondingly: $\mathbf{n}_g$ is the unit normal vector to the gear tooth flank, $\mathcal{G}$, and $\mathbf{n}_p$ the unit normal vector to the pinion tooth flank, $\mathcal{G}$.

For the determination of the tolerances, $[\varepsilon_g]$ and $[\varepsilon_p]$, for the accuracy of the permissible axial displacements, ε_g and ε_p, of the gear and the mating pinion, either one of the tolerances (either the tolerance $[\varepsilon_g]$, or the tolerance $[\varepsilon_p]$), or a ratio of the tolerances, $[\varepsilon_g]/[\varepsilon_p]$, has to be prespecified.

The performed analysis reveals that the actual value of the angle, θ, alters when the gears rotate. The maximum value of the angle, θ, is observed at the very beginning of meshing of two gear teeth. As the rotation proceeds, the angle, θ, reduces to its minimum value. A minimum value of the angle, θ, is observed within a plane through the axis of instant rotation, P_{ln}, oriented perpendicular to the plane of action, *PA*. Further, the angle, θ, increases to its maximum value at the very end of meshing of two gear teeth.

A more in detail analysis is not presented here as the equations are bulky.

In addition to the discussed approach, another approach for the calculation of tolerance for the accuracy of the mounting distance in intersected-axes gearing is developed.

When the displacements ε_g (or the displacement ε_p) are of a negative value, this results in the edge contact occurring at the opposite face of the gear.

1.6.3 Mounting Distance in Crossed-Axes Gearing

In a precision crossed-axes gear pair (see Fig. 1.30), the mounting distance is commonly inspected at nominal operating load, and the preloaded bearings [6].

Relative Disposition of Base Cones in Crossed-Axes Gearing A crossed-axes gear pair that is composed of two spiral bevel gears is shown in Fig. 1.31 as an example. When a crossed-axes gearing operates, the gear spins, ω_g, about its axis of rotation, O_g. The mating pinion spins, ω_p, about its axis of rotation, O_p. The axes of rotation, O_g and O_p, cross one another. The closest distance of approach of the axes of rotation, O_g and O_p, is designated as *C*. The center-distance, *C*, is a straight-line segment of the centerline, ℄. The axes of rotation, O_g and O_p, form a crossed-axes angle, Σ. The angle, Σ, is formed by two rotation vectors, $\boldsymbol{\omega}_g$ and $\boldsymbol{\omega}_p$ [that is, by definition $\Sigma = \angle(\boldsymbol{\omega}_g, \boldsymbol{\omega}_p)$].

The axis of rotation of the gear, O_g, is perpendicular to the centerline, ℄, and intersects the centerline at the gear base cone apex, A_g (see Fig. 1.31). Similarly, the

Fig. 1.30 Crossed-axes gearing in the design of automobile differential

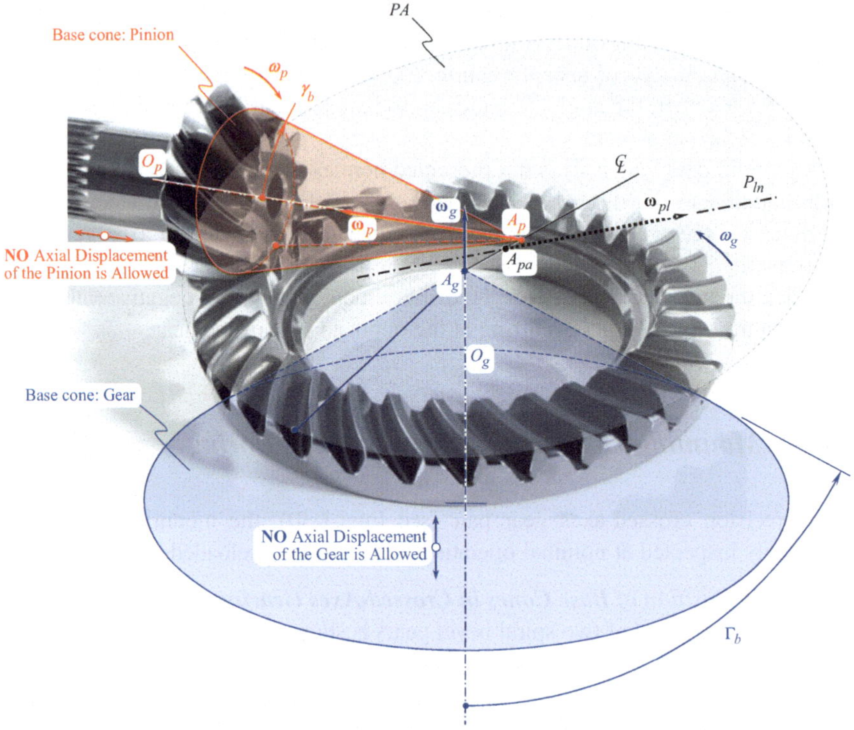

Fig. 1.31 Crossed-axes gearing with associated base cones of a gear and that of a mating pinion

axis of rotation of the pinion, O_p, is also perpendicular to the centerline, ℄, and intersects the centerline at the pinion base cone apex, A_p. The length of the straight-line segment, A_gA_p, equals to the center-distance, C, that is, the equality $A_gA_p = C$ is valid.

The center-distance, C, is subdivided by a point, A_{pa}, onto two straight-line segments, A_gA_{pa} and A_pA_{pa}. The ratio A_gA_{pa}/A_pA_{pa} equals to the gear ratio in the intersected-axes gear pair ($A_gA_{pa}/A_pA_{pa} = u$). Point, A_{pa}, is the plane-of-action-apex, A_{pa}.

The axis of instant rotation, P_{ln}, is a straight line through the plane-of-action-apex, A_{pa}, perpendicular to the centerline, ℄. The axis, P_{ln}, forms the angles, Σ_g and Σ_p, with the axes of rotation, O_g and O_p, correspondingly. The actual value of each of the angles, Σ_g and Σ_p, can be expressed in terms of the gear ratio, u, in the crossed-axes gear pair [4].

A base cone is associated with the gear. The gear base cone apex, A_g, is located within the centerline, ℄. Another base cone is associated with the pinion. The pinion base cone apex, A_p, is also located within the centerline, ℄. The base cones of the gear and the mating pinion are in tangency with the plane of action in the gear pair – this is a must for geometrically accurate crossed-axes gearing [4].

Neither the axial displacement of the gear nor the axial displacement of the pinion, from the position shown in Fig. 1.31, is permissible. Otherwise, the edge contact between the gear and the mating pinion tooth flanks (see Fig. 1.25) becomes inevitable.

Disposition of Base Cone of a Gear in Relation to the Plane of Action in Crossed-Axes Gearing In crossed-axes gearing, the interaction of the tooth flanks, $\mathcal{G}$ and $\mathcal{P}$, of a gear and of a mating pinion occurs within the plane of action, *PA*. Therefore, it makes sense to consider the disposition of base cone of a gear in relation to the plane of action. Later on, the results of the analysis obtained in this way can be applied to the disposition of base cone of a mating pinion in relation to the gear housing.

In a geometrically accurate crossed-axes gear pair the axis of rotation of a gear, O_g, and that, O_{pa}, of the plane of action, *PA*, are crossed, as illustrated in Fig. 1.32a. The base cone apex of the gear, A_g, and the plane-of-action-apex, A_{pa}, are at the gear

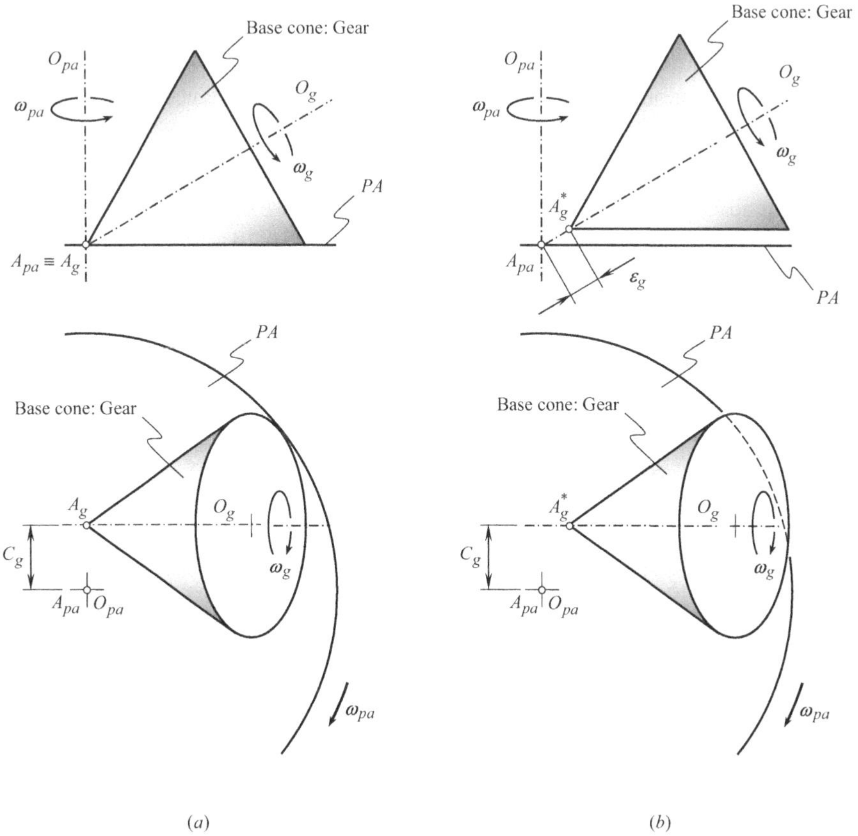

Fig. 1.32 Disposition of base cone of a gear in relation to the plane of action, *PA*, in (**a**) geometrically accurate and (**b**) displaced crossed-axes gear pair

center-distance, C_g, apart from one another. A rotation of the gear is designated as ω_g, and a rotation of the plane of action is designated as ω_{pa}, correspondingly. Sliding between the base cone of the gear, and the plane of action only along the axis, P_{ln}, is observed when the gears rotate – no sliding in a direction perpendicular to the axis, P_{ln}, is observed.

In reality, an error, ε_g, in the mounting distance of the gear is always observed. The latter is shown in Fig. 1.32b. In such a scenario, the gear base cone apex, A_g^*, does not coincide with the point, $A_{h.\ g}$, as A_g^* is displaced axially in relation to $A_{h.\ g}$ at a distance ε_g. In the case, shown in Fig. 1.32b, the outward displacement, ε_g, is of a positive value. An inward displacement is of a negative value (not shown in Fig. 1.32b).

This discussion yields to proceed with the derivation of an equation for the calculation of the tolerance for the accuracy of the mounting distance in a crossed-axes gear pair.

Mounting of Gears in Crossed-Axes Gear Housing To operate properly, gears in crossed-axes gearing have to be properly configured in relation to one another. The mounting distance is a key design parameter that helps keep gears properly configured and is the most important design parameter for ensuring the correct operation of a gear pair. The mounting distance can be specified as the distance from a locating surface on the back of one gear (most commonly a bearing seat) to the gear base cone apex, A_g.

Theoretically, in a crossed-axes gear pair, a gear base cone apex, A_g, a mating pinion base cone apex, A_p, and the plane-of-action-apex, A_{pa}, have to be located within the centerline, ℄, at distances, A_gA_{pa} and A_pA_{pa}, as schematically illustrated in Fig. 1.33.

In reality, in order to transmit power, gears need to be supported. A gear housing is used for this purpose. The gear housing features a corresponding number of bores that are used to support gear shafts in the housing. A centerline of a bore for the gear shaft is labeled as $O_{h.\ g}$. Correspondingly, a centerline of a bore for the pinion shaft is designated by $O_{h.\ p}$. The centerline, $O_{h.\ g}$, intersects the centerline, ℄, at point, $A_{h.\ g}$. Correspondingly, the centerline, $O_{h.\ p}$, intersects the centerline, ℄, at point, $A_{h.\ p}$. Strong constraints on the configuration of the gear and its mating pinion in the gear housing are imposed by the required position of points, $A_{h.\ g}$ and $A_{h.\ p}$.

The gear has to be configured in the gear housing so as to keep the gear axis of rotation, O_g, aligned with the axis, $O_{h.\ g}$, in the gear housing (i.e., $O_g \equiv O_{hg}$). The gear base cone apex, A_g, has to be coincident with the point, $A_{h.\ g}$, of the intersection of the centerlines, $O_{h.\ g}$ and ℄ (i.e., $A_g \equiv A_{h.\ g}$). Then, the rest of the important components, namely, the bearings, shims, fasteners, and so forth, must be put together so as to keep the gear mounting distance, MD_g, to the blueprint. The gear mounting distance, MD_g is measured between the gear back face, and the gear base cone apex, A_g.

Similarly, the pinion must be configured in the gear housing so as to keep the pinion axis of rotation, O_p, aligned with the axis, $O_{h.\ p}$ (i.e., $O_p \equiv O_{h.\ p}$), in the gear housing. The pinion base cone apex, A_p, has to be coincident with the point, $A_{h.\ p}$, of the intersection of the centerlines, $O_{h.\ p}$ and ℄ (i.e., $A_p \equiv A_{h.\ p}$). Then, the rest of the important components, that is, the bearings, shims, fasteners, and so forth, are put together so as to keep the pinion mounting distance, MD_p, to the blueprint. The pinion mounting distance, MD_p is measured between the pinion back face, and the gear base cone apex, A_p.

After the pinion is in position, the tapered roller bearings of the gear assembly are preloaded; at the same time the gear assembly is placed axially in the housing (see Fig. 1.33).

Finally, once the gear and the pinion are properly mounted in the gear housing, their axes are aligned to the centerlines of the corresponding bores in the gear housing (i.e., $O_g \equiv O_{h.\ g}$, and $O_p \equiv O_{h.\ p}$), and the apexes, A_g and A_p, are snapped together with the corresponding points, $A_{h.\ g}$ and $A_{h.\ p}$, as schematically illustrated in Fig. 1.33.

An important conclusion can be drawn up from the above discussion (see Figs. 1.31 and 1.33):

Intermediate Conclusion *In crossed-axes gearing, NO axial displacements of the gears from the position when the base cone apexes, A_g and A_p, of a gear and a mating pinion coincide with the points of intersection, $A_{h.\ g}$ and $A_{h.\ p}$, of the centerlines, $O_{h.\ g}$ and $O_{h.\ p}$, in the gear housing by the centerline ℄, are permissible.*

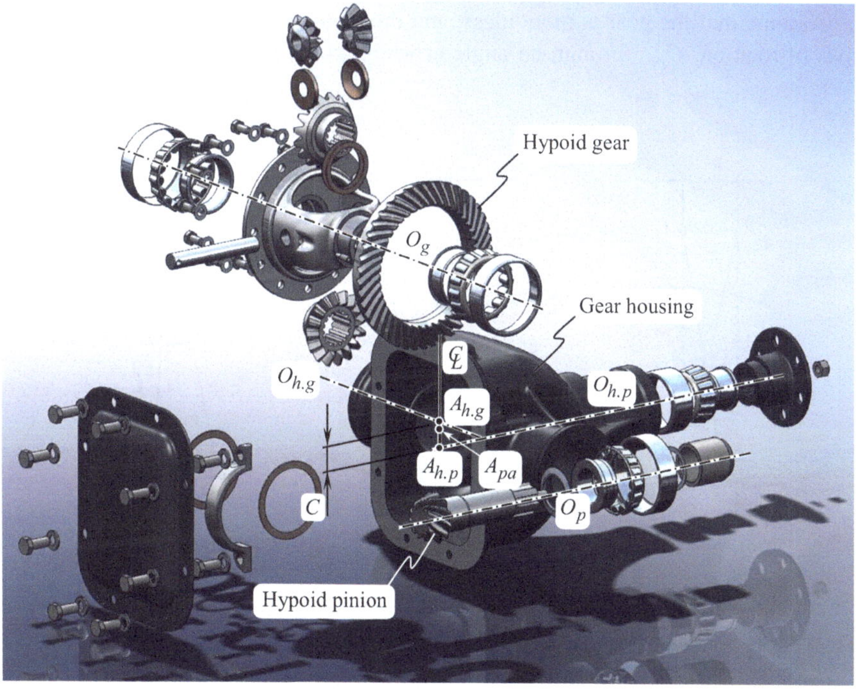

Fig. 1.33 Mounting bevel gears in the gear housing

Therefore, it is wrong practice to adjust for the backlash in crossed-axes gearing by means of inward/outward axial displacement of the gears. NO axial displacements of the gears from their nominal configuration are allowed.

In precision crossed-axes gearing, it is recommended to inspect the mounting distance at nominal operating load and the preloaded bearings. A tolerance for the accuracy of the mounting distance is tight.

Tolerance for the Accuracy of the Mounting Distance in Crossed-Axes Gearing Consider two teeth in contact intersected by the plane of action, *PA*, as illustrated in Fig. 1.34a. The teeth contact one another along a straight-line segment *ab*. (Here, for simplicity, but without loss of generality, a pair of geometrically accurate straight bevel gears is considered).

A case of outward displaced gear is illustrated in Fig. 1.34b. As a result of the displacement, a gap between the teeth is observed. This gap is shown by two straight-line segments, *ab* and a^*b^*, each of which is entirely situated on the tooth flanks, $\mathcal{G}$ and $\mathcal{P}$, of the two interacting teeth.

A schematic for the derivation of a formula for the calculation of a tolerance for the accuracy of the mounting distance is depicted in Fig. 1.34c. It should be stressed here that point *b* within the straight-line segment *ab* is the closest point to the straight-line segment a^*b^*. Therefore, the angle, θ_g, through which the plane of action, *PA*, has to be turned about the axis of rotation O_{pa} depends on the actual distance of point *b* to the straight-line segment a^*b^*.

Assume that the gear is motionless, and the plane of action, *PA*, turns about its axis of rotation, O_{pa}, through an angle at which point, *b*, touches the straight-line

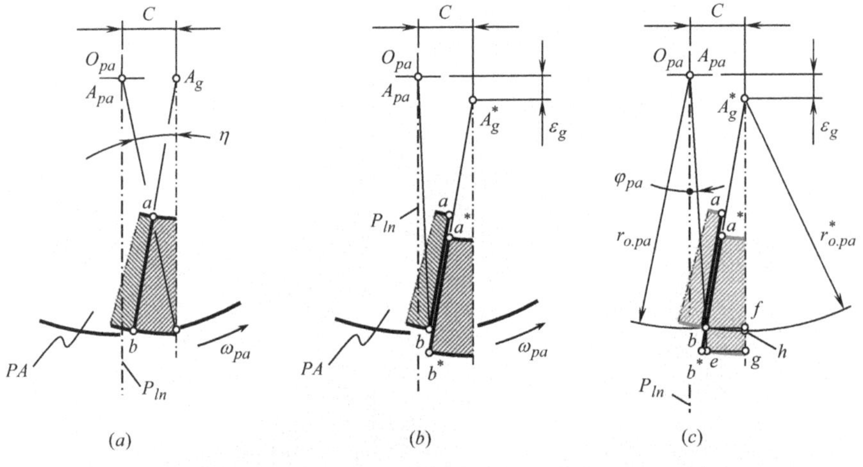

Fig. 1.34 Section of two interacting teeth by the plane of action, *PA*: (**a**) zero mounting distance error, (**b**) mounting distance error of a certain value, ε_g, and (**c**) schematic for the derivation of the formula for the calculation of a tolerance for the accuracy of the mounting distance in crossed-axes gearing

segment, $A_g^* b^*$. Then, consider a triangle, $\Delta A_{pa} A_g^* b$. In this triangle, $A_{pa} b = r_{o.\,pa}$, $A_{pa} A_g^* = \sqrt{C^2 + \varepsilon_g^2}$, and $A_g^* b = r_{o.pa}^*$.

Law of cosine can be used for the determination of the actual value of the angle, θ_g:

$$\varepsilon_g^2 = r_{o.ap}^2 + \left(r_{o.pa}^*\right)^2 - 2 \cdot r_{o.ap} \cdot r_{o.pa}^* \cdot \cos\theta_g - C^2 \tag{1.43}$$

Then, either the actual value of the angle, θ_g, can be expressed in terms of the mounting distance error, ε_g, and the design parameters of the gear pair, or a maximum permissible value of the displacement, ε_g, can be expressed in terms of the maximum permissible value of the angle θ_g, and the design parameters of the gear pair:

$$\varepsilon_g = \sqrt{r_{o.ap}^2 + \left(r_{o.pa}^*\right)^2 - 2 \cdot r_{o.ap} \cdot r_{o.pa}^* \cdot \cos\theta_g - C^2} \leq [\varepsilon_g] \tag{1.44}$$

Here, $[\varepsilon_g]$ is the tolerance for the accuracy of the gear mounting distance, ε_g.

$$\theta_g = \cos^{-1}\left[\frac{r_{o.ap}^2 + \left(r_{o.pa}^*\right)^2 - \varepsilon_g^2 - C^2}{2 \cdot r_{o.ap} \cdot r_{o.pa}^*}\right] \leq [\theta_g] \tag{1.45}$$

Here, $[\theta_g]$ is the tolerance for the accuracy of the gear angle, θ_g.

It can be shown (see Fig. 1.34c) that the equality:

$$r_{o.pa}^* \cong \frac{r_{o.pa} - \varepsilon_g}{\cos\varphi_{pa}} \tag{1.46}$$

is valid.

When the pinion is fully aligned (and, thus, all the displacements are of a zero value), the equalities $\varepsilon = \varepsilon_g$ and $\theta = \theta_g$, and Eqs. (1.44) and (1.45) can be used for the calculation of tolerance on the mounting distance in the crossed-axes gear pair.

An analysis similar to that above can be performed for a mating bevel pinion:

$$\varepsilon_p = \sqrt{r_{o.ap}^2 + \left(r_{o.pa}^*\right)^2 - 2 \cdot r_{o.ap} \cdot r_{o.pa}^* \cdot \cos\theta_p - C^2} \leq [\varepsilon_p] \tag{1.47}$$

$$\theta_p = \cos^{-1}\left[\frac{r_{o.ap}^2 + \left(r_{o.pa}^*\right)^2 - \varepsilon_p^2 - C^2}{2 \cdot r_{o.ap} \cdot r_{o.pa}^*}\right] \leq [\theta_p] \tag{1.48}$$

Here, $[\varepsilon_p]$ is the tolerance for the accuracy of the pinion mounting distance, ε_p; and $[\theta_p]$ is the tolerance for the accuracy of the gear angle, θ_p.

Further, when the gear is aligned, the equalities $\varepsilon = \varepsilon_p$ and $\theta = \theta_p$, and Eqs. (1.47) and (1.48) are used for the calculation of tolerance on the mounting distance in the crossed-axes gear pair.

Finally, in a more general case, a bevel gear and a mating bevel pinion, both, are displaced. Under such a scenario, either the actual value of the angle θ can be expressed in terms of the mounting distance errors, ε_g and ε_p, and the design parameters of the gear pair, or a maximum permissible value of the displacements, ε_g and ε_p, can be expressed in terms of the maximum permissible value of the angle θ, and the design parameters of the gear pair.

The angle θ is formed by two perpendiculars, $\mathbf{n}_g$ and $\mathbf{n}_p$ [that is, $\theta = \angle(\mathbf{n}_g, \ \mathbf{n}_p)$], constructed at point of edge contact of a gear, $\mathcal{G}$, and a mating pinion, $\mathcal{P}$, tooth flanks, correspondingly: $\mathbf{n}_g$ is the unit normal vector to the gear tooth flank, $\mathcal{G}$, and $\mathbf{n}_p$ the unit normal vector to the pinion tooth flank, $\mathcal{P}$.

For the determination of the tolerances, $[\varepsilon_g]$ and $[\varepsilon_p]$, for the accuracy of the permissible axial displacements, ε_g and ε_p, of the gear and the mating pinion, either one of the tolerances (either the tolerance $[\varepsilon_g]$, or the tolerance $[\varepsilon_p]$), or a ratio of the tolerances, $[\varepsilon_g]/[\varepsilon_p]$, has to be prespecified.

The performed analysis reveals that the actual value of the angle, θ, alters when the gears rotate. The maximum value of the angle, θ, is observed at the very beginning of meshing of two gear teeth. As the rotation progresses, the angle, θ, reduces to its minimum value. A minimum value of the angle, θ, is observed within a plane through the axis of instant rotation, P_{ln}, perpendicular to the plane of action, *PA*. Further, the angle, θ, increases to its maximum value at the very end of meshing of two gear teeth.[3]

When the displacement ε_g (or the displacement ε_p) is of a negative value, this results in the edge contact occurring at the opposite face of the gear.

1.7 Contact Pattern

Gears in intersected-axes gear pairs, and in crossed-axes gear pairs, must be assembled in a specific way to ensure smooth running and a favorable load distribution between gears. Even a small axes displacement in intersected-axes gearing (of about three angular minutes or so) results in a severe contact pattern change. The contact area should cover the entire flank surface (without edge contact concentration) if the nominal load rating is reached. An offset error of ±50micrometers leads to a contact pattern change of about the same magnitude.

[3] A more in-detail analysis is not presented here, as the equations are bulky.

Fig. 1.35 An example of acceptable shape and configuration of contact pattern under a low load

To attain the goal (to ensure smooth running and a favorable load distribution between gears), a contact pattern between the interacting tooth flanks of a gear and a mating pinion is commonly used.

It is the right point to mention here that a proper shape along with a desirable location and orientation of the contact pattern in a gear pair is the priority number one from the standpoint of the actual value of the mounting distance.

The contact pattern is a critical attribute of any and all bevel/hypoid gear designs. An example of contact pattern is depicted in Fig. 1.35. For this purpose, the gear teeth are covered by a stain and rolled over one another in tight mesh, as illustrated in Fig. 1.36.

Simply stated, the contact pattern is the area in which the gear teeth come in contact as they engage and disengage during their rotation. When a gear is installed in a gearbox and is powering the designated application, there are varying degrees of pressure – or load – on the gear teeth. These pressures are influenced by box deflections, bearing movement, and temperature changes. When the gear teeth are subjected to these variables, the contact pattern will change. There is a general rule of thumb stating that the heavier the load, the larger the contact pattern and vice versa. For a gear to perform properly under operating load, the contact pattern must be of a certain shape and at a certain location. Typically, an ideal tooth contact pattern under load should encompass the bulk of the tooth surface while avoiding any contact with the edges of the tooth flank. So far, the contact pattern is more a *qualitative* rather than a *quantitative* indicator of a gear pair quality.

Fig. 1.36 Two gears with the teeth covered by stain and rolled over each other in tight mesh to get a favorable shape, size, and configuration of the contact pattern

The permissible variation in contact pattern location, orientation, and shape, can be used to calculate the tolerances for design parameters of a gear pair.[4] Unfortunately, in the present-day gear practice, contact pattern is a qualitative, and not quantitative measure of the gear pair quality.

Instead of the contact pattern analysis, the teeth flank geometry along with the actual value of the mounting distance should be measured. No contact pattern analysis is necessary if the gears for intersected-axes, and/or crossed-axes gear pairs are machined, and assembled according to the properly designed blueprint.

1.8 Permissible Alteration to Bevel Gear Tooth Flank Geometry

As a gear and a mating pinion tooth interact with one another only within the plane of action, *PA*, the geometry of the tooth flanks, $\mathcal{G}$ and $\mathcal{P}$, of a gear and a mating pinion allows for a slight modification (a few examples are illustrated in Fig. 1.37)

[4]In much, the contact pattern is a subjective (and not objective) parameter. Neither the dimensions and orientation, nor the tolerances for the dimensions and orientation of the contact pattern can be assigned (and be directly measured on the shop floor). The contact pattern is an insufficiently engineered parameter of quality of a gear pair. The contact pattern is NOT measurable, that is, it can NOT be reliably measured by means commonly available on the shop floor. It is desirable to substitute the contact pattern with another reliably measurable design parameter of a gear pair. The mounting distance is a promising candidate for this purpose.

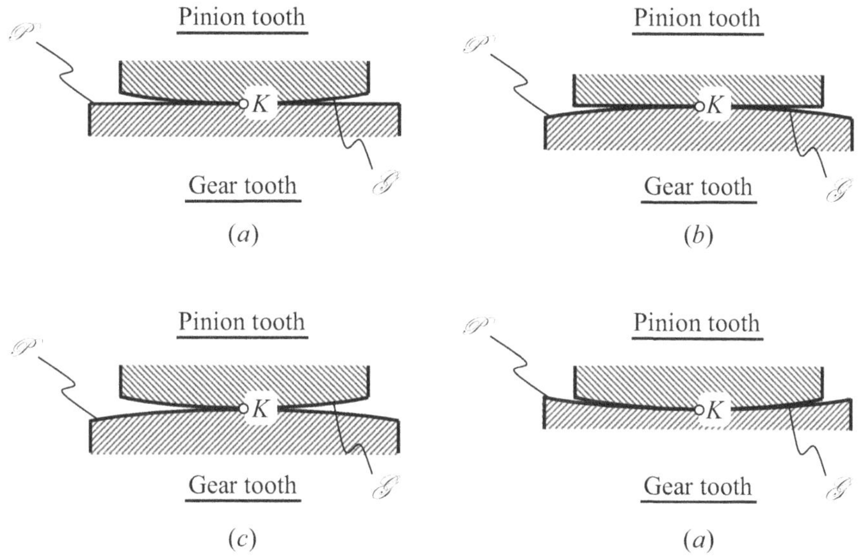

Fig. 1.37 Possible kinds of modification of tooth flanks, $\mathcal{G}$ and $\mathcal{P}$, of a gear and a mating pinion in a section by the plane of action, *PA*: (**a**) straight gear-to-convex pinion, (**b**) convex gear-to-straight pinion, (**c**) convex gear-to-convex pinion, and (**d**) concave gear-to-convex pinion

aiming to avoid the edge contact in gearing. The modification is allowed only for lines of intersection of the tooth flanks, $\mathcal{G}$ and $\mathcal{P}$, by the plane of action. Only under such a scenario angular base pitch, $\varphi_{b.\ op}$, of a gear equals to operating angular base pitch, $\varphi_{b.\ op}$, of the gear pair (i.e., the equality $\varphi_{b.\ g} \equiv \varphi_{b.\ op}$ is observed); and similarly, angular base pitch, $\varphi_{b.\ p}$, of a mating pinion equals to operating angular base pitch, $\varphi_{b.\ op}$, of the gear pair (i.e., the equality $\varphi_{b.\ p} \equiv \varphi_{b.\ op}$ is also observed).

No violation of the equalities $\varphi_{b.\ g} \equiv \varphi_{b.\ op}$ and $\varphi_{b.\ p} \equiv \varphi_{b.\ op}$ is permissible in precision gears for intersected-axes gearing as well as in crossed-axes gearing.

1.9 Gear Noise and Vibration: Root Causes for Vibration Generation and Noise Excitation

Noise and vibration excitation are two annoying problems. These problems have been exacerbated due to the continuous reduction in sounds from other systems' noise exciters, such as engines. All of them are theoretically preventable. Unfortunately, gear noise prevention methods can be costly, in terms of both equipment and labor. The reduction of gear noise/vibration excitation is a complex and challenging engineering problem. It is likely the 2003 book by *J.D. Smith* [8] is the only attempt so far, in which the problem of noise excitation is discussed from the most general point of view.

Many factors, such as transmission error, tooth impacts, mesh stiffness variation, force axial shuttling, friction, air, and lubricant entrapment, cause gear noise and vibration. Only factors that cause noise and vibration in the gear mesh are considered in this chapter.

It is reasonable to begin the analysis of root causes for vibration generation and noise excitation starting from consideration of geometrically accurate gear pairs. Later on, the obtained results of the analysis can be enhanced toward approximate gear pairs.

1.9.1 Root Cause for Vibration Generation and Noise Excitation in Geometrically Accurate Gearing

Geometrically accurate gear pairs of all kinds: parallel-axes, intersected-axes, and crossed-axes gear pairs, met all the three fundamental laws of gearing that gearing of all designs has to fulfill. These laws of gearing are as follows:

- *The law of contact* (or, in other words, the *first fundamental law of gearing*) of the interacting tooth flanks, $\mathcal{G}$ and $\mathcal{P}$, of a gear and a mating pinion. This law of gearing is commonly analytically expressed by *Shishkov equation of contact*, $\mathbf{n}_g \cdot \mathbf{V}_\Sigma = 0$
- *The conjugate action law* (or, in other words, the *second fundamental law of gearing*) for the interacting tooth flanks, $\mathcal{G}$ and $\mathcal{P}$, of a gear and a mating pinion – at every point within the line of contact the unit normal vector, $\mathbf{n}_g$, of a common perpendicular is pointed along a straight line that intersects the axis of instant rotation, P_{ln}, in the gear pair
- *The equal base pitches law* (or, in other words, the *third fundamental law of gearing*): To comply with this law of gearing, angular base pitch of a gear, $\varphi_{b.\ g}$, and angular base pitch of a mating pinion, $\varphi_{b.\ p}$, both, must be equal to operating angular base pitch of the gear pair, $\varphi_{b.\ op}$. This means that the equalities $\varphi_{b.\ g} = \varphi_{b.\ op}$ and $\varphi_{b.\ p} = \varphi_{b.\ op}$ must be valid for any and all configurations of a gear and a mating pinion in relation to each other when the gears rotate.

In geometrically accurate gear pairs, all three fundamental laws of gearing are fulfilled. Therefore, it seems from the first glimpse that there is no reason for vibration generation and noise excitation in geometrically accurate gear pairs. Unfortunately, this is not correct.

In (a) parallel-axes helical gearing, (b) intersected-axes gearing, and in (c) crossed-axes gearing, variation of the *axial component* and the *radial component* of the resultant force of the interaction between a gear and a mating pinion tooth flanks is observed. Therefore, even in the case of geometrically accurate gearing [see items (a) through (b)], noise excitation, and vibration generation become inevitable.

In geometrically accurate intersected-axes, and in crossed-axes gear pairs, the forces that act in the gear mesh vary in time. This is true first of all with respect to the axial component, as well as to the radial component of the resultant force. The tangential component that causes transmission of the rotation also contributes to vibration and noise generation. However, this contribution is not that significant compared to the contribution by the axial, and the radial components.

The variation in time of the axial component of the resultant force and the radial component of the resultant force both are transmitted to the gear housing, and so forth.

1.9.2 Root Cause for Vibration Generation and Noise Excitation in Approximate Gear Pairs

In approximate gearing of all designs, at least one of three fundamental laws of gearing is violated.

Violation of *the Law of Contact* In compliance with the law of contact between tooth flanks, $\mathcal{G}$ and $\mathcal{P}$, of a gear and a mating pinion, a common perpendicular, $\mathbf{n}_g$, to the interacting tooth flanks, $\mathcal{G}$ and $\mathcal{P}$, is always orthogonal to the linear velocity vector, $\mathbf{V}_\Sigma$, of the resultant relative motion of the tooth flanks. Analytically, this law of gearing is described by *Shishkov equation of contact*, $\mathbf{n}_g \cdot \mathbf{V}_\Sigma = 0$.

The law of contact is commonly violated in cases of the so-called *edge contact* of the interacting tooth flanks, $\mathcal{G}$ and $\mathcal{P}$, of a gear and a mating pinion [4].

The law of contact is fulfilled in most cases of approximate gearing, as a gear, $\mathcal{G}$, and a mating pinion, $\mathcal{P}$, tooth flanks in approximate gearing are smooth surfaces having appropriate principal radii of curvature. Commonly, fulfillment of the law of contact causes no problem when designing geometrically accurate gears. However, this could be a challenging problem when designing approximate gears.

Violation of *the Conjugate Action Law* In approximate gearing, the conjugate action law is violated almost always.

Referring to Fig. 1.38, consider the interaction between two local patches of a gear and of a mating pinion tooth flank, $\mathcal{G}$ and $\mathcal{P}$. The local patches are located within the differential vicinity of contact point, K.

In a case of geometrically accurate gearing (see Fig. 38.1a), a gear and a mating pinion rotate, ω_g and ω_p, about their axes of rotation, O_g and O_p. The axes, O_g and O_p, are at a center-distance, C, from one another. The pitch point, P, is located within the centerline, ℄. The ratio, r_g/r_p, of the distances, r_g and r_p, of the pitch point, P, from the axes of rotation, O_g and O_p, is reciprocal to the ratio, ω_p/ω_g, of the rotations of the gear and the mating pinion. Remember, that, by nature, the distances, r_g and r_p, are the gear center-distance, C_g, and the pinion center-distance, C_p, correspondingly (and, thus, the equalities $r_g = C_g$ and $r_p = C_p$ are valid).

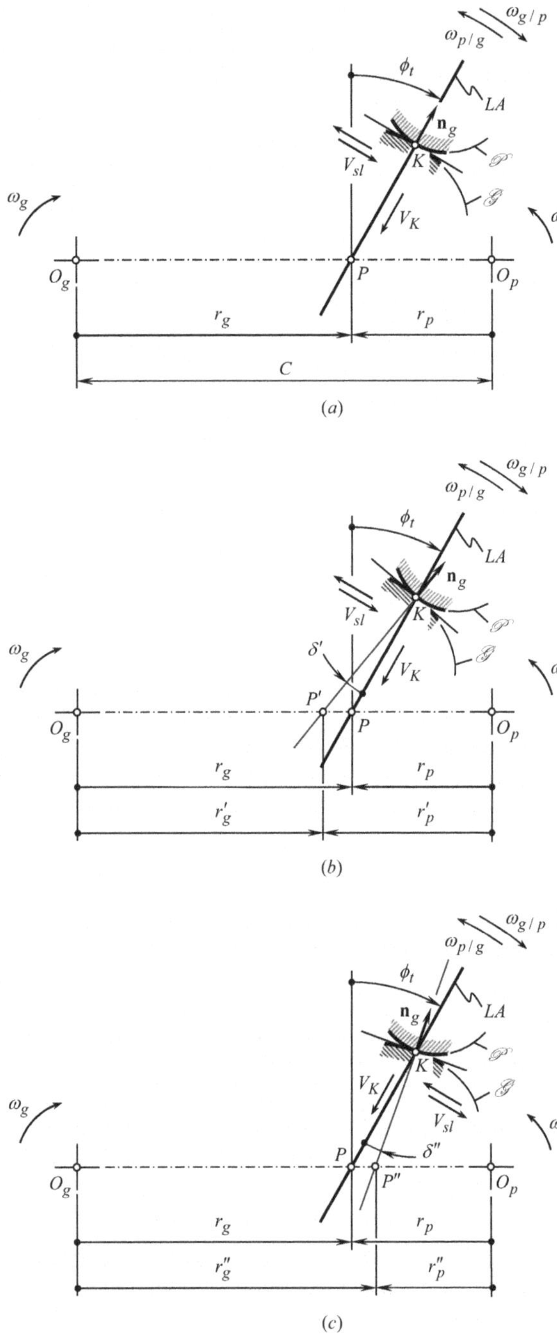

Fig. 1.38 Fulfillment and violation of the *conjugate action law*: (**a**) the *conjugate action law* is fulfilled, (**b**) the *conjugate action law* is violated: the tooth flanks, $\mathcal{G}$ and $\mathcal{P}$, are turned clockwise through an angle, δ', and (**c**) the *conjugate action law* is violated: the tooth flanks, $\mathcal{G}$ and $\mathcal{P}$, are turned counter-clockwise through an angle, δ''

The line of action, *LA*, is a straight line through the pitch point, *P*. The line of action, *LA*, forms a transverse pressure angle, ϕ_t, with a perpendicular to the centerline, ℄, through the pitch point, *P*.

When the gears rotate, the contact point, *K*, travels along the line of action, *LA*, with a linear velocity, V_K. The tooth flanks, $\mathcal{G}$ and $\mathcal{P}$, slide over one another. The linear velocity of the sliding is labeled V_{sl}.

When the gear is considered stationary, the instant motion of a gear and a mating pinion in relation to one another can be viewed as an instant rotation, $\omega_{p/g}$, of the pinion in relation to the gear about the pitch point, *P*. Otherwise, when the pinion is considered stationary, the instant motion of a gear and a mating pinion in relation to one another can be viewed as an instant rotation, $\omega_{g/p}$, of the gear in relation to the pinion about the pitch point, *P*.

It is critical to stress here that in Fig. 1.38a, the common perpendicular vector, $\mathbf{n}_g$, is pointed along the line of action, *LA*. This reveals that the gear and the mating pinion tooth profiles, $\mathcal{G}$ and $\mathcal{P}$, are designed properly, and the conjugate action law for the interacting tooth profiles, $\mathcal{G}$ and $\mathcal{P}$, is fulfilled.

Let's proceed to the analysis of the case shown in Fig. 1.38b. In the case in question, the gear and the mating pinion tooth flanks, $\mathcal{G}$ and $\mathcal{P}$, are designed so that the common perpendicular vector, $\mathbf{n}_g$, is turned about the contact point, *K*, clockwise through an angle, $\delta^{'}$. A straight line along the unit vector, $\mathbf{n}_g$, intersects the centerline, ℄, at point, $P^{'}$ (which, by nature, is an instantaneous pitch point). The ratio, $r_g^{\prime}/r_p^{\prime}$, of the distances, $r_g^{\prime}$ and $r_p^{\prime}$, of the instantaneous pitch point, $P^{'}$, to the axes of rotation, O_g and O_p, is not reciprocal to the ratio, ω_p/ω_g, of the rotations of the gear and the mating pinion. Instead, the ratio, $r_g^{\prime}/r_p^{\prime}$, is reciprocal to the ratio, $\omega_p/\omega_g^{\prime}$, of the actual instant rotations of the gear and the mating pinion. The gears tend to rotate about the instantaneous pitch point, $P^{'}$, and not about the desirable pitch point, *P*. None of the instant rotations (neither $\omega_{p/g}$ nor $\omega_{g/p}$) is feasible. This reveals that the gear and the mating pinion tooth flanks, $\mathcal{G}$ and $\mathcal{P}$, are designed improperly, and the conjugate action law for the interacting tooth profiles, $\mathcal{G}$ and $\mathcal{P}$, is violated.

Finally, let's analyze the case shown in Fig. 1.38c. Here, the gear and the mating pinion tooth flanks, $\mathcal{G}$ and $\mathcal{P}$, are designed so that the common perpendicular vector, $\mathbf{n}_g$, is turned about the contact point, *K*, counter-clockwise through an angle, $\delta^{''}$. A straight line along the unit vector, $\mathbf{n}_g$, intersects the centerline, ℄, at point, $P^{''}$ (which, by nature, is an instantaneous pitch point). The ratio, $r_g^{\prime\prime}/r_p^{\prime\prime}$, of the distances, $r_g^{\prime\prime}$ and $r_p^{\prime\prime}$, of the instantaneous pitch point, $P^{''}$, from the axes of rotation, O_g and O_p, is not reciprocal to the ratio, $\omega_p/\omega_g^{\prime\prime}$, of the rotations of the gear and the mating pinion. Instead, the ratio, $r_g^{\prime\prime}/r_p^{\prime\prime}$, is reciprocal to the ratio, $\omega_p/\omega_g^{\prime\prime}$, of the actual instant rotations of the gear and the mating pinion. The gears tend to rotate about the instantaneous pitch point, $P^{'}$, and not about the desirable pitch point, *P*. None of the instant rotations (neither $\omega_{p/g}$ nor $\omega_{g/p}$) is permissible. This reveals that the gear and the mating pinion tooth flanks, $\mathcal{G}$ and $\mathcal{P}$, are designed improperly, and the conjugate action law for the interacting tooth flanks, $\mathcal{G}$ and $\mathcal{P}$, is violated.

As the line that is pointed along the unit normal vector, $\mathbf{n}_g$, of a common perpendicular does not go through the pitch point, *P*, the actual instantaneous pitch point, P, migrates (P' and P'') up and down along the centerline, ℄. Therefore, the vibration generation, and noise excitation, both, increase when the gears operate.

When friction between a gear tooth flank, $\mathcal{G}$, and a mating pinion tooth flank, $\mathcal{P}$, is taken into account, the friction forces alter the actual direction of a straight line along which the resultant force of the interaction is pointed, as illustrated in Fig. 1.39. The normal forces, $\mathbf{F}_{t.1}$ and $\mathbf{F}_{t.2}$, are pointed along the line of action, LA. The line of action, LA, is a straight line through the pitch point, P. The vectors of friction forces, $\mathbf{F}_{f.1}$ and $\mathbf{F}_{f.2}$, are entirely located within the common tangent plane, and they are perpendicular to the corresponding normal forces, $\mathbf{F}_{t.1}$ and $\mathbf{F}_{t.2}$. As the inequalities $\mathbf{F}_{f.1} \neq 0$ and $\mathbf{F}_{f.2} \neq 0$ are valid, then the resultant force of interaction, $\mathbf{F}_{\Sigma}$, does not align with the line of action, LA, or with the instantaneous line of action, LA_{inst}. As the resultant force, $\mathbf{F}_{\Sigma}$, is no longer pointed along a line through the pitch point, P, an instantaneous pitch point, P_{inst}, is created.

The resultant force is pointed along a straight line that is not through the pitch point, P. The latter along with the variation of the normal forces, $\mathbf{F}_{t.1}$ and $\mathbf{F}_{t.2}$, and friction forces $\mathbf{F}_{f.1}$ and $\mathbf{F}_{f.2}$, depending on an actual configuration of the gear and the pinion in relation to one another (see Fig. 1.39) are potential root causes of an excessive noise excitation. Better lubrication of the gear pair is a potential solution to this problem: the better the lubrication, the lower the noise excitation and vice versa. When performing the analysis, the process of vibration generation, and noise

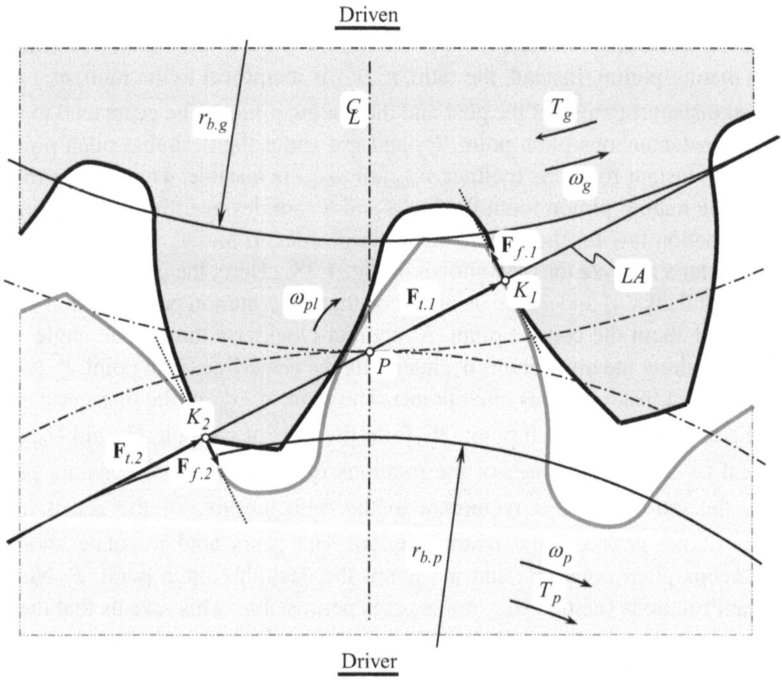

Fig. 1.39 Direction of the resultant force, $\mathbf{F}_{\Sigma}$, in a gear mesh can be affected by the friction forces $\mathbf{F}_{f.1}$ and $\mathbf{F}_{f.2}$

excitation, can be expressed in terms of the torques, $\mathbf{T}_{f.1}$ and $\mathbf{T}_{f.2}$, created by the friction forces, $\mathbf{F}_{f.1}$ and $\mathbf{F}_{f.2}$, as well as in terms of the resultant torque, $\mathbf{F}_{f.1}$ and $\mathbf{F}_{f.2}$, that is created by the resultant friction force, $\mathbf{F}_{f.\Sigma} = \mathbf{F}_{f.1} + \mathbf{F}_{f.2}$.

Violation of the Equality of Base Pitches In approximate gearing, either base pitch of a gear or base pitch of a mating pinion, or both of them, are not equal to operating base pitch in the gear pair. Moreover, often the base pitch either of a gear, or of a mating pinion, or of both, cannot be specified at all. The operating base pitch of a gear pair is not a measurable design parameter – this design parameter can either be calculated or be measured indirectly.

The base pitch variation is a root cause for the vibration generation and for an excessive noise excitation. The parameters of noise excitation can be expressed in terms of the variation of the base pitch when the gears rotate. Thus, the noise frequency must correlate to the base pitch variation.

The base pitch variation is caused by one of the following factors (or by a combination of these factors):

(a) The deviation of the gear tooth flank geometry in relation to the desirable geometry
(b) A displacement (both linear displacement and angular displacement) of the gear tooth flank in relation to the axis of rotation of the gear/pinion (of the gear/pinion body)
(c) The axes misalignment

The actual value of the base pitch variation can be expressed in terms of these three elementary components.

The deviations of the tooth flank geometry from its desirable geometry as well as the displacements of the tooth flanks from their desirable location and orientation are measured from the nominal tooth flank geometry, the location, and orientation specified by the gear blueprint. In this scenario, base pitches of the gear, of the mating pinion, and the operating base pitch are equal to that specified by the gear pair blueprint.

Without loss of generality, consider a simplified example to illustrate the approach employed for the analysis of the base pitch deviations, and their impact on the noise excitation in a gear pair. A parallel-axes helical gear pair is diagrammatically depicted in Fig. 1.40. The gear pair features a zero axes misalignment. The gear and the pinion are rotated about their axes of rotation, O_g and O_p, correspondingly. The angular velocities, ω_g and ω_p, are synchronized with one another so as to keep the ratio valid:

$$\frac{\omega_p}{\omega_g} = \frac{N_g}{N_p} \tag{1.49}$$

Here, in Eq. (1.49), the tooth numbers of the gear and the pinion are designated as N_g and N_p, correspondingly.

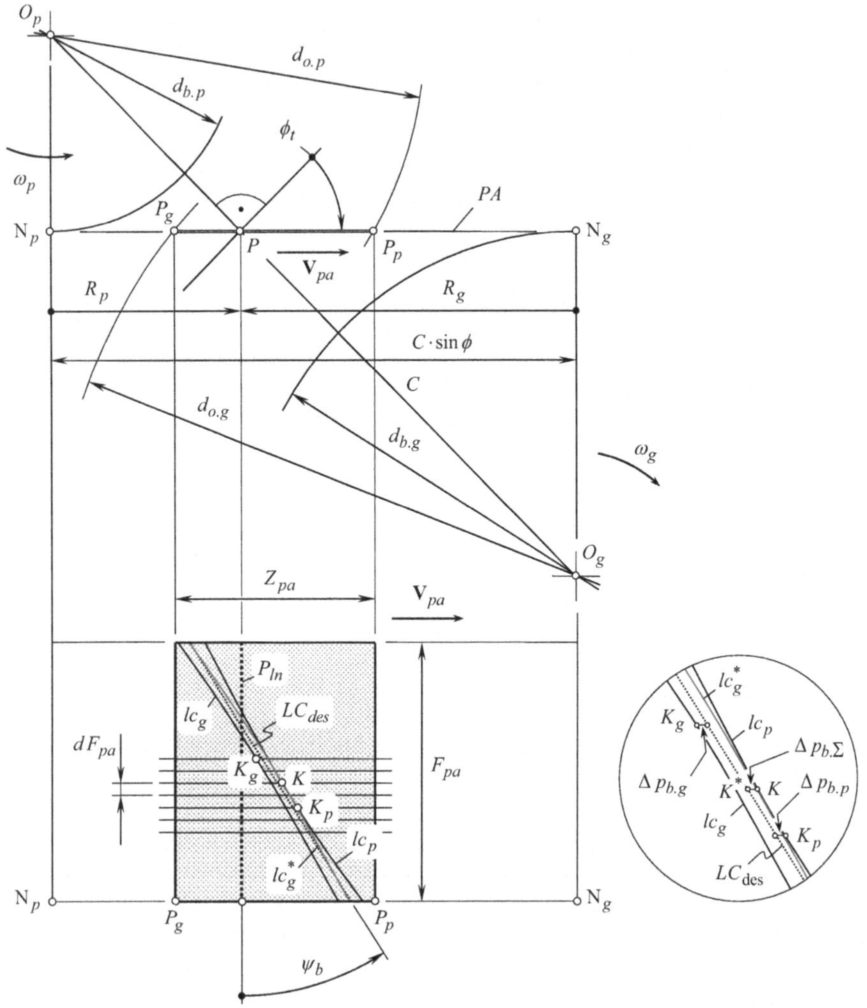

Fig. 1.40 Details of the concept used for the analysis of the deviations: of the base pitch, $\Delta p_{b.\ g}$, in a gear, of the base pitch, $\Delta p_{b.\ p}$, in a mating pinion, and the operating base pitch, $\Delta p_{b.\ \Sigma}$, in a gear pair

The axes of rotation, O_g and O_p, of the gear and the pinion are at a center-distance, C, apart from one another. The transverse pressure angle is denoted by ϕ_t.

The plane of action, PA, is tangent from opposite sides to the base cylinders of diameters, $d_{b.\ g}$ and $d_{b.\ p}$, of the gear and the pinion, correspondingly. The width of the zone of action, ZA, equals to the functional face width, F_{pa}, and the length is equal to the length of the active portion of the zone of action, Z_{pa}. The desirable line of contact, LC_{des}, forms a base helix angle, ψ_b, with the axis of instant rotation, P_{ln}.

The tooth flanks, $\mathcal{G}$ and $\mathcal{P}$, of the geometrically accurate gears are intersected by the plane of action, *PA*, along the desirable line of contact, LC_{des}.

The tooth flanks, $\mathcal{G}_a$ and $\mathcal{P}_a$, of the actual gear and its mating pinion deviate from their desirable geometries, $\mathcal{G}$ and $\mathcal{P}$. Therefore, the gear tooth flank, $\mathcal{G}_a$, is intersected by the plane of action, *PA*, along the *gear line of contact*, lc_g. The line of intersection, lc_g, does not align with the desirable line of contact, LC_{des}. Similarly, the pinion tooth flank, $\mathcal{P}_a$, is intersected by the plane of action, *PA*, along the *pinion line of contact*, lc_p. The line of intersection, lc_p, also does not align with the desirable line of contact, LC_{des}. Evidently, the lines of intersection, lc_g and lc_p, do not align with the desirable line of contact, LC_{des}.

The plane of action, *PA*, is sliced by an infinite number of planes, all of which are perpendicular to the axis of instant rotation, P_{ln}. The planes are at an infinitesimally short distance, $d F_{pa}$, from one another. Within every slice, the *gear line of contact*, lc_g, deviates from the desirable line of contact, LC_{des}, at a distance, $\Delta p_{b.\ g}$. At point, K_g, the deviation, $\Delta p_{b.\ g}$, is of a minimal value. Similarly, within every slice, the *pinion line of contact*, lc_p, deviates from the desirable line of contact, LC_{des}, at a distance, $\Delta p_{b.\ p}$. At point, K_p, the deviation, $\Delta p_{b.\ p}$, is of a minimal value. In fact, the deviations, $\Delta p_{b.\ g}$ and $\Delta p_{b.\ p}$, are equal to the deviations of the linear base pitches, $p_{b.\ g}$ and $p_{b.\ p}$, of the gear and the mating pinion from the operating linear base pitch, $p_{b.\ op}$, of the gear pair. As it is clear from the above discussion, the base pitch deviations are caused by the deviations in tooth flank geometries, $\mathcal{G}_a$ and $\mathcal{P}_a$, of the gear and the mating pinion from their desirable geometries, $\mathcal{G}$ and $\mathcal{P}$.

Theoretically, the tooth flanks, $\mathcal{G}$ and $\mathcal{P}$, are at certain distance from one another, but in reality, they contact each other at point, as both the driving pinion and the driven gear are loaded by the operating torque. As the gears are loaded, there is no gap between the *gear line of contact*, lc_g, and the *pinion line of contact*, lc_p. The gear line of contact, lc_g, travels toward the pinion line of contact, lc_p. In a shifted position, the line, lc_g, is labeled as lc_g^*. These two lines of contact have a common point, that is, they share contact point, *K*. The distance covered by the pinion line of contact, lc_p, to get in contact with the gear line of contact, lc_g, depends on the actual values of the deviations, $\Delta p_{b.\ g}$ and $\Delta p_{b.\ p}$. By nature, this distance is the operating base pitch error. In a case when minimal deviations are observed in a common slice, the total deviation of the base pitches, $\Delta p_{b.\ \Sigma}$, is equal to the sum of the instantaneous deviations,[5] $\Delta p_{b.\ g}$ and $\Delta p_{b.\ p}$:

$$\Delta p_{b.\Sigma} = \Delta p_{b.g} + \Delta p_{b.p} \tag{1.50}$$

Commonly, the minimal deviations, $\Delta p_{b.\ g}$ and $\Delta p_{b.\ p}$, are observed in different slices, and not in a common slice. Therefore, the total deviation of the base pitches, $\Delta p_{b.\ \Sigma}$, is calculated as:

[5]The instantaneous values of the deviations, $\Delta p_{b.\ g}$ and $\Delta p_{b.\ p}$, commonly are not known. However, the calculations can be performed for the tolerances, $[\Delta p_{b.\ g}]$ and $[\Delta p_{b.\ p}]$, for the accuracy of the corresponding deviations.

$$\Delta p_{b.\Sigma} = a\left(\varphi_p\right) \cdot \Delta p_{b.g} + b\left(\varphi_p\right) \cdot \Delta p_{b.p} \tag{1.51}$$

Here, in Eq. (1.51), the multipliers, $a\,(\varphi_p)$ and $b\,(\varphi_p)$, are functions of the angular configuration of the input shaft, namely, of the angular configuration of the pinion.

The instantaneous value of the total deviation of the base pitch, $\Delta\, p_{b.\ \Sigma}$, falls into the interval:

$$\Delta p_{b.g} - b\Delta p_{b.p} \leq \Delta p_{b.\Sigma} \leq \Delta p_{b.g} + b\Delta p_{b.p} \tag{1.52}$$

This idea of relative displacement (in micrometers) is the cause of a force variation, and hence vibration is unusual since traditionally an external force such as an out-of-balance rotating component, or vibration of the supporting ground is excited to produce a vibration. In gearing, the relative displacement between the mating gears generates forces between the teeth and the subsequent vibrations through the system.

Once the deviations of the base pitch are measured both in the gear and in the mating pinion, then each of the components of the total deviation of the base pitch, $\Delta\, p_{b.\ \Sigma}$, can be corrected so, as to keep the base pitches, $p_{b.\ g}$ and $p_{b.\ p}$, equal to the operating base pitch, $p_{b.\ op}$. This is important, as it makes clear how many deviations in the tooth flank geometry in a gear, and that in a mating pinion, contribute to the total deviation in the base pitch. Moreover, it becomes clear which one of the components, either the gear or the pinion, needs more improvement to reduce the operating base pitch variation.

Here and below, it is assumed that the total deviation of the base pitches, $\Delta\, p_{b.\ \Sigma}$, is specified at every instant of time.[6] Moreover, for convenience of the analysis below, it is also assumed that the deviation, $\Delta\, p_{b.\ \Sigma}$, is always observed in that same slice, and the contact point, *K*, does not migrate within the face of the gear pair.

The model of parallel-axes gear pair simplified in such a way is used below for the purpose to demonstrate one of the root causes of vibration generation and noise excitation in approximate gear pairs.

The variation in the current value of the base pitch causes a corresponding transmission error, $\Delta\varphi = \Delta\varphi\,(\varphi_p)$.

When approximate gears of any and all designs are used, no gear noise problem can be solved by means of sources based on the theory of gearing in its current stage of development (gear noise and gear vibration are challenging problems that motivate the further development of the scientific theory of gearing). Noisy gears have to be *isolated* from the environment, but they will still produce noise.

[6]It is realized here that the actual values of the deviations, $\delta p_{b.\ g}$ and $\delta p_{b.\ p}$, are never known. However, corresponding tolerances, $[\delta p_{b.\ g}]$ and $[\delta p_{b.\ p}]$, can be set for the accuracy of each of the deviations. Often, it is more convenient solving problems using the tolerances for the accuracy of the deviations, rather than the actual values of the deviations.

1.10 Transmission Error

Gear noise is caused, to a great extent, by the dynamic phenomena in tooth meshing. It can be characterized by transmission error, which is the root cause for vibration generation and for noise excitation.

A brief discussion immediately below is helpful for better understanding of the nature of transmission error.

On the Nature of Transmission Error For simplicity but without loss of generality, transmission error in a gear pair may be illustrated by means of a wheel that rolls over a surface, as illustrated in Fig. 1.41 [3].

When a perfect wheel rolls over a perfect flat surface (see Fig. 1.41a) with no slippage, the relative motion of the wheel may be viewed as a superposition of a rotation, ω, and a translation, V. This case corresponds to meshing of two geometrically accurate gears. No sources for noise excitation are observed under such a scenario, as this is a trivial case of rolling a circle over a plane.

In another scenario, a perfect wheel rolls over a rough surface as diagrammatically shown in Fig 1.41b. The relative motion of the wheel may be viewed as a superposition of a rotation, ω, and a translation, V, and a certain additional motion. This case corresponds to meshing of an approximate gear with a geometrically accurate pinion. As the gear is imperfect, an additional motion, V_{noise}, that causes the noise excitation, becomes inevitable. The transmission error is designated here as $\Delta\varphi_g$, because it is caused by the imperfect gear. It is evident that the gear (and NOT the pinion) needs to be improved in order to reduce/eliminate the transmission error and to reduce noise excitation in the gear pair.

In another scenario, a rough wheel rolls over a perfect plane surface as schematically shown in Figure 28.4c. The relative motion of the wheel may be viewed as a superposition of a rotation, ω, and a translation, V, and a certain additional motion. This case corresponds to meshing of a geometrically accurate gear with an approximate pinion. As the pinion is imperfect, an additional motion, V_{noise}, that causes the noise excitation, becomes inevitable. The transmission error is designated here as $\Delta\varphi_p$, because it is caused by the imperfect pinion. It is evident that the pinion (and NOT the gear) needs to be improved in order to reduce/eliminate the transmission error and to reduce noise excitation in the gear pair.

Finally, a rough wheel rolls over a rough surface as schematically shown in Fig. 1.41d. The relative motion of the wheel may be viewed as a superposition of a rotation, ω, and a translation, V, and a certain additional motion. This case corresponds to the meshing of an approximate gear with an approximate pinion. As both the gear and the pinion are imperfect, an additional motion, V_{noise}, that causes the noise excitation, becomes inevitable. The transmission error is designated here as $\Delta\varphi$, because it is caused by the imperfect gear and an imperfect pinion together. The transmission error, $\Delta\varphi$, in this case, is a function of the components $\Delta\varphi_g$ and $\Delta\varphi_p$, that is, $\Delta\varphi = \Delta\varphi(\Delta\varphi_g, \Delta\varphi_p)$. In order to reduce/eliminate the noise excitation when the gear pair operates, it is necessary to know how much each of the components,

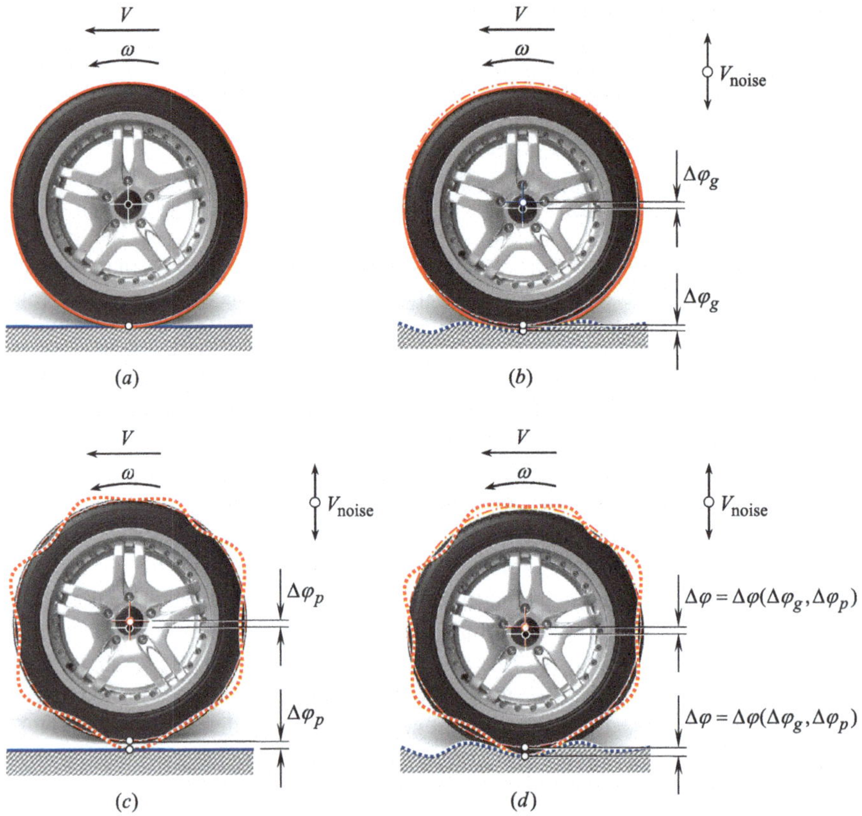

Fig. 1.41 Analogs of the contributions to the transmission error, $\Delta\varphi$, in cases of meshing: (**a**) of perfect gear-to perfect pinion, (**b**) of perfect gear-to real pinion, (**c**) of real gear-to perfect pinion, and (**d**) of real gear-to real pinion

$\Delta\varphi_g$ and $\Delta\varphi_p$, contributes to the transmission error, $\Delta\varphi$. Unfortunately, the actual values of the components, $\Delta\varphi_g$ and $\Delta\varphi_p$, commonly are not known. Therefore, the problem of reduction of gear noise in gear pairs of this kind turns to an indefinite one. Thus, the measured values of the transmission error cannot be used for the improvement neither of the gear nor of the pinion.

Fortunately, the actual values of the components, $\Delta\varphi_g$ and $\Delta\varphi_p$, as well as of the transmission error, $\Delta\varphi$, can be expressed in terms of variation of base pitches both of a gear and of a mating pinion.

Another approach can be utilized for the explanation of transmission error in gearing.

Consider a gear pair composed of a geometrically accurate gear that is engaged in mesh with an inaccurate gear. The driving gear is rotated uniformly. Regardless of which of the two gears is a driving one (either the geometrically accurate gear or the inaccurate gear), the driven gear is rotated with certain fluctuation, that is, the driven gear is rotated unsteadily.

Assume, a rotation vector, $\boldsymbol{\omega}_g$, of the inaccurate gear in the gear pair is specified for a case that this gear also is geometrically accurate. In reality (for the imperfect gear), the rotation vector of the gear is designated as, $\boldsymbol{\omega}_{g\Delta}$. The transmission error in gearing can be expressed in terms of the rotation vector $\Delta\boldsymbol{\omega}_g = \boldsymbol{\omega}_g - \boldsymbol{\omega}_{g\Delta}$.

Determination of Transmission Error Transmission error is the most important factor in the generation of gear noise. Transmission error is the error between the gear teeth. It can be defined in the following way: Imagine that the input gear is being driven at an absolutely uniform angular velocity. It is hoped that the output gear is rotating at a uniform angular velocity as well. Any variation from this uniform velocity gives a variation from the correct position of the output, and this is the *transmission error*, which will subsequently generate vibration and noise excitation. More formally:

Definition *Transmission error is the difference between the angular position that the output shaft of a drive would occupy if the drive were perfect and the actual position of the output.*

In practical terms, the successive angular positions of the output, where the output should be, can be taken. These can be subtracted from the measured output positions to give the error in position. Measurements are made by measuring angular displacements, so the answers appear initially in units of seconds of arc [8].

By definition:

$$TE(t) = r_{b.g}\theta_g(t) - r_{b.p}\theta_p(t) \tag{1.53}$$

where $r_{b.\,g}$ and $r_{b.\,p}$ are the base pitch radii, and the angles, $\theta_g(t)$ and $\theta_p(t)$ are the angles of rotation of the gear and the mating pinion, correspondingly.

When no transmission error occurs, and the input shaft rotates steadily $[\,\omega_p(t) = \text{const}\,]$, the output shaft also rotates steadily $\left[\omega_g^{des}(t) = \text{const}\right]$, as schematically illustrated in Fig. 1.42a. Under the influence of transmission error, the output shaft does not rotate steadily $[\,\omega_g(t) \neq \text{const}\,]$. The period of oscillation of the output shaft is designated as C_m.

Transmission error for an ideal gear train is almost a linear function of time. It can be represented in the form of a function $\varphi_g = (N_g/N_p) \cdot \varphi_p$. Here, N_g and N_p are the number of teeth of the gear and the pinion, and φ_g and φ_p are the rotation angles of the gear and the pinion. Due to the axis displacement, the transmission function becomes piecewise almost linear with the period of the cycle of meshing, C_m, of a pair of teeth. Due to the jump of the angular velocity at the junction of cycles (see Fig. 1.42a) [5], the acceleration approaches an infinitely large value. This causes vibration generation and noise excitation.

A desirable gear transmission function, $\varphi_g^{des}(\varphi_p) = a \cdot \varphi_p$, is a linear function of the angle of rotation, φ_p, of the input shaft. Here is the factor $a = \tan\ (u_t)$. The actual gear transmission function, $\varphi_g = \varphi_g(\varphi_p)$, can be represented in the form of a piecewise function, as shown in Fig. 1.42b.

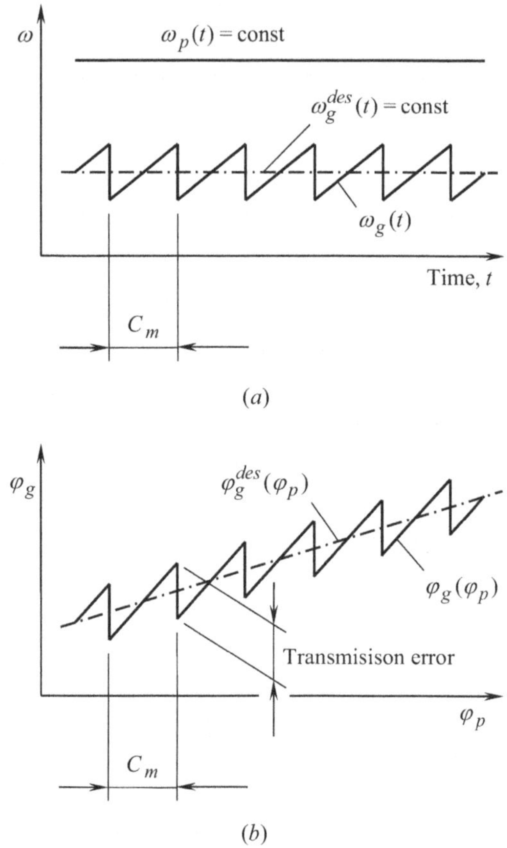

Fig. 1.42 Rotation of a driver, $\omega_p(t)$, and desirable, $\omega_g^{des}(t)$, and real, $\omega_g(t)$, rotations of the driven functions vs. time, t (**a**), and desirable, $\varphi_g^{des}(\varphi_p)$, and real, $\varphi_g(\varphi_p)$, transmission functions (**b**). (After Prof. *S. P. Radzevich* [5]

A uniform rotation of the driving pinion, cannot be transmitted by the gear pair smoothly to the driven gear if the total deviation of the base pitch, $\Delta p_{b.\ \Sigma}$, is not of a zero value ($\Delta p_{b.\ \Sigma} \neq 0$). The base pitch error, $\Delta p_{b.\ \Sigma}$, causes a transmission error, $\Delta\varphi$. The transmission error, $\Delta\varphi$, can be expressed in terms of the total deviation of the base pitch, $\Delta p_{b.\ \Sigma}$.

Refer to Fig. 1.43 to express an instantaneous value of the transmission error, $\Delta\varphi$, in terms of the total deviation of the base pitch, $\Delta p_{b.\ \Sigma}$.

In the case of geometrically accurate parallel-axes gear pair, contact point, *K*, occupies a certain position within the line of action, *LA*. For a specified value of the angle of rotation of the gear, φ_g, the distance, *PK*, that is traveled by the contact point together with the plane of action, *PA*, can be determined from the $\triangle PKO_g$:

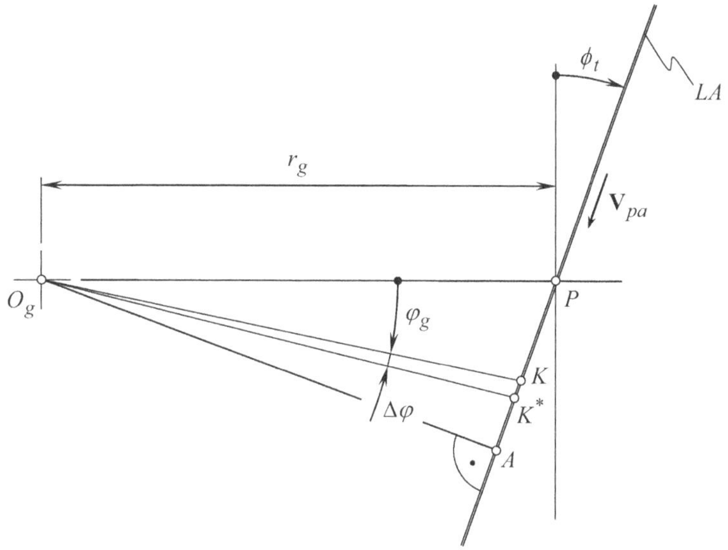

Fig. 1.43 A schematic for the calculation of the instantaneous value of the transmission error, $\Delta\varphi$, in parallel-axes gear pair

$$PK = r_g \frac{\sin \varphi_g}{\cos\left(\varphi_g - \phi_t\right)} \tag{1.54}$$

where:

r_g – is the radius of the pitch circle of the gear
ϕ_t – is the transverse pressure angle in the gear pair

To derive Eq. (1.54), the sine rule was applied to the $\triangle PKO_g$.

In reality, because the base pitch error is not equal to zero, the gears make contact not at point, K, but they contact one another at point, K^*, instead. The distance between the points, K and K^*, equals to the total deviation of the base pitch, $\Delta p_{b.\ \Sigma}$, that is, the equality $KK^* = \Delta p_{b.\ \Sigma}$ is valid. The deviation, $\Delta p_{b.\ \Sigma}$, is calculated from Eq. (1.51).

An additional rotation through an angle, $\Delta\varphi$, is caused by the deviation of $\Delta p_{b.\ \Sigma}$. By nature, this additional rotation is equal to the instantaneous value of the transmission error.

In order to determine the angle, $\Delta\varphi$, it is necessary to determine the distance, K^*O_g. This distance can be determined from the $\triangle PK^*O_g$:

$$K^{*}O_{g} = r_{g}\sqrt{1 + \frac{\sin^{2}\varphi_{g}}{\cos^{2}(\varphi_{g} - \phi_{t})} - 2\frac{\sin\varphi_{g}}{\cos(\varphi_{g} - \phi_{t})}\sin\phi_{t}} \tag{1.55}$$

To derive the Eq. (1.55), the cosine rule was applied to the $\triangle PK^{*}O_{g}$. Then, the rule of sine is applied to the $\triangle PK^{*}O_{g}$:

$$\sin(\varphi_{g} + \Delta\varphi) = \underbrace{\frac{r_{g}\sin\varphi_{g} + \Delta p_{b.\Sigma}\cos(\varphi_{g} - \phi_{t})}{K^{*}O_{g}\cos(\varphi_{g} - \phi_{t})}\cos\phi_{t}}_{A} \tag{1.56}$$

Ultimately, the angle, $\Delta\varphi$, equals to:

$$\Delta\varphi = \sin^{-1}(a) - \varphi_{g} \tag{1.57}$$

For the determination of the distance, KO_{g}, another approach can be used. It can be written from the $\triangle APO_{g}$ that:

$$O_{g}A = r_{g}\cos\phi_{t} \tag{1.58}$$

$$AP = r_{g}\sin\phi_{t} \tag{1.59}$$

and then from the $\triangle AKO_{g}$:

$$AK = r_{g}\left(\sin\phi_{t} - \frac{\sin\varphi_{g}}{\sin(\varphi_{g} - \phi_{t})}\right) \tag{1.60}$$

Equations (1.58)–(1.60) yield an expression for the calculation of the distance, KO_{g}:

$$KO_{g} = r_{g}\frac{\sqrt{[\cos(\varphi_{g} - \phi_{t}) - 2\sin\phi_{t}\sin\varphi_{g}]\cos(\varphi_{g} - \phi_{t})}}{\cos(\varphi_{g} - \phi_{t})} \tag{1.61}$$

The determined distance, KO_{g} [see Eq. (1.61)], is used for the derivation of Eq. (1.57).

The approach, applied for the derivation of Eq. (1.57), can also be used for the calculation of the components, $\Delta\varphi_{g}$ and $\Delta\varphi_{p}$, of the transmission error, $\Delta\varphi$. The components, $\Delta\varphi_{g}$ and $\Delta\varphi_{p}$, are contributed by the gear and the pinion, correspondingly.

Another approach can be used for the derivation of Eq. (1.57). The transmission error, $\Delta\varphi$, can be expressed in terms of the total deviation of the base pitch, $\Delta p_{b.\ \Sigma}$, and of base diameters of a gear and a mating pinion in a gear pair.

As the driving pinion rotates uniformly, and the distance, KK^*, is specified, then this distance is equivalent to an additional rotation of the driving pinion through an angle, $\Delta\varphi_p$:

$$\Delta\varphi_p = \frac{KK^*}{r_{b.p}} \tag{1.62}$$

Taking into account that $\Delta\varphi_g = \Delta\varphi_p/u$ (here u is the gear ratio of the gear pair), Eq. (1.62) immediately returns an expression for the angular deviation, $\Delta\varphi$:

$$\Delta\varphi = \Delta\varphi_p = \frac{\Delta p_{b.\Sigma}}{u \cdot r_{b.p}} \tag{1.63}$$

In this case, there is no need to determine the deviation, $\Delta\varphi$, using the rule of sine for this purpose.

The rotation of the gear through an angle, $\Delta\varphi$, causes an additional rotation of the gear. This additional rotation can be represented by the rotation vector, $\boldsymbol{\omega}_{bpv}(\varphi_g)$. The rotation vector, $\boldsymbol{\omega}_{bpv}(\varphi_g)$, is referred to as the *base pitch variation rotation,* $\boldsymbol{\omega}_{bpv}(\varphi_g)$.

The resultant rotation vector of the gear:

$$\boldsymbol{\omega}_\Sigma(\varphi_g) = \boldsymbol{\omega}_g + \boldsymbol{\omega}_{bpv}(\varphi_g) \tag{1.64}$$

The transmission error is an integral parameter of the accuracy of the gear pair. When only transmission error is specified, it is not clear what has to be done with the gears in order to keep the base pitches equal, that is, whether a gear or a mating pinion has to be corrected, and to what extent.

It seems ridiculous that 1mm module (25DP) gear less than an inch in diameter will have roughly the same transmission error as a 25mm module (1DP) gear of 3m diameter of the same quality, but this is surprisingly close to what happens in practice. This unexpected constant size of errors is liable to cause problems in the future with the current trend toward micromechanics. If a gear tooth is only 20 μm tall, the base pitch is about 20 μm, but errors of 2 μm in pitch or profile are still likely with corresponding transmission errors so that a speed variation of 10% becomes possible.

Rotation Vector of Base Pitch Variation When a geometrically accurate gear is rotated in mesh with an inaccurate gear, the rotation vector, $\boldsymbol{\omega}_{bpv}(\varphi_g)$, is considered. The rotation vector, $\boldsymbol{\omega}_{bpv}(\varphi_g)$, is referred to as the *base pitch variation of rotation,* $\boldsymbol{\omega}_{bpv}(\varphi_g)$. This rotation vector influences both the magnitude and the configuration of the vector of instant rotation, $\boldsymbol{\omega}_{pl}$. Ultimately, the operating base pitch, $\varphi_{b.\ op}$, varies, that is, the equality $\varphi_{b.\ op} = \varphi_{b.\ op}(t)$ is valid. Consequently, the geometrically accurate gear in mesh becomes also inaccurate, as the gear base pitch, $\varphi_{b.\ g}$

(a constant value), cannot be equal to a variable value $\varphi_{b.\ op} = \varphi_{b.\ op}(t)$: namely, an inequality $\varphi_{b.\ op} \neq \varphi_{b.\ op}(t)$ is observed. In such a mesh, the geometrically accurate gear is no longer a perfect one.

The variation in the configuration of the vector of instant rotation, $\boldsymbol{\omega}_{pl}$, can be illustrated by a corresponding vector diagram. In the vector diagram, the vector, $\boldsymbol{\omega}_{\Sigma}(\varphi_g)$, of total rotation of a gear equals to:

$$\boldsymbol{\omega}_{\Sigma}\left(\varphi_g\right) = \boldsymbol{\omega}_g + \boldsymbol{\omega}_{bpv}\left(\varphi_g\right) \tag{1.65}$$

This is also a good illustration of usefulness of gear vector diagrams for the research and analysis of gears and gearing.

1.11 Variation of Axial and Radial Forces

Transmission error is not the only cause for vibration generation and noise excitation in gear transmission systems. A variation of the instantaneous (equivalent) force, namely, of its magnitude, coordinates of a point where the force is applied (especially in gears that feature a low tooth count), and the orientation, also cause the vibration generation and the noise excitation. The variation of the loading is transmitted to the housing that works like panels, and so forth.

Helical gears, for example, are subjected to variation of the force between the gears. They are sensitive to the force variation in particular.

Consider a gear pair that has a nominal contact ratio of $\overline{m}$. At a particular instant of time, the lengths of the theoretical lines of contact get the two extreme positions schematically depicted in Fig. 1.44.

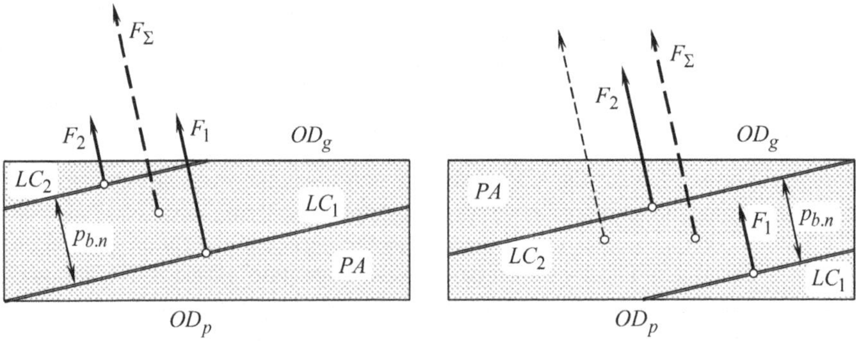

Fig. 1.44 Extreme positions of the lines of contact, *LC*, in the plane of action, *PA*, showing how the forces at the centers of each section of the line of contact give a resultant force whose position varies

These show the plane of action for the worst case with a correct face width of an axial pitch and a small helix angle. In the case under consideration, any end relief effects or tip relief effects are ignored. A constant loading along the contact line is assumed. Practical teeth tend to give slightly larger effects.

The extreme position of the center of action of the resultant force is determined by taking moments about one end and is approximately $\left[(\overline{m}-1)^2+1\right]/2\overline{m}_t$ from one end. This has a minimum when the total contact ratio, $\overline{m}$, is equal to $\sqrt{2}$ and the center of force oscillates about 0.086 of the face widths on either side of the center of the face.

There is a corresponding radial force variation at the bearing housings of the order of 8% of the mean value when the gears are well supported by close shafts or less if the supporting shafts are long [8].

1.12 Alternative Causes of Noise Excitation in Gear Pair

Transmission error is not the only cause for vibration generation and noise excitation. Contact ratio and variation of point at which the resultant load is applied (especially in gears with a low tooth count) are the alternative causes for vibration generation and noise excitation. The said result in variation of load that is transmitted from the gear mesh to the housing. In the case under consideration elements of the housing work like panels generating noise.

1.12.1 Influence of Contact Ratio

The influence of contact ratio onto the gear transmission function, $\varphi_g = \varphi_g(\varphi_p)$, is more, or less clear as long, as the consideration relates to an elementary gear drive (that is, to a single *gear-to-pinion mesh*) with a contact ratio exactly equal to $\overline{m}=1$. However, in reality, the contact ratio always is greater than one ($\overline{m}>1$). Because of this, during certain periods of meshing, not one but more pairs of teeth are engaged in mesh simultaneously. In general, the piecewise linear functions of transmission errors for distinct pairs of contacting teeth do not coincide with each other. The cycle of meshing, $c_m^{(i)}$, for the i–th pair of teeth is shifted in relation to the previous/consequent cycle of meshing, $c_m^{(i\pm1)}$, in the direction of φ_g–axis at a certain distance, ΔC_m. The actual value of the shift, ΔC_m, depends on the actual value of contact ratio, $\overline{m}$. The transmission functions, $Tr_f^{(i)}$ and $Tr_f^{(i\pm1)}$, make it possible to compose the resultant transmission function, $Tr_f^{(\Sigma)}$, for the elementary gear drive that has a total contact ratio $\overline{m}>1$. In the same way, the resultant transmission function, $Tr_f^{(\Sigma)}$, can be composed for any actual value of contact ratio that exceeds $\overline{m}>2$ [5].

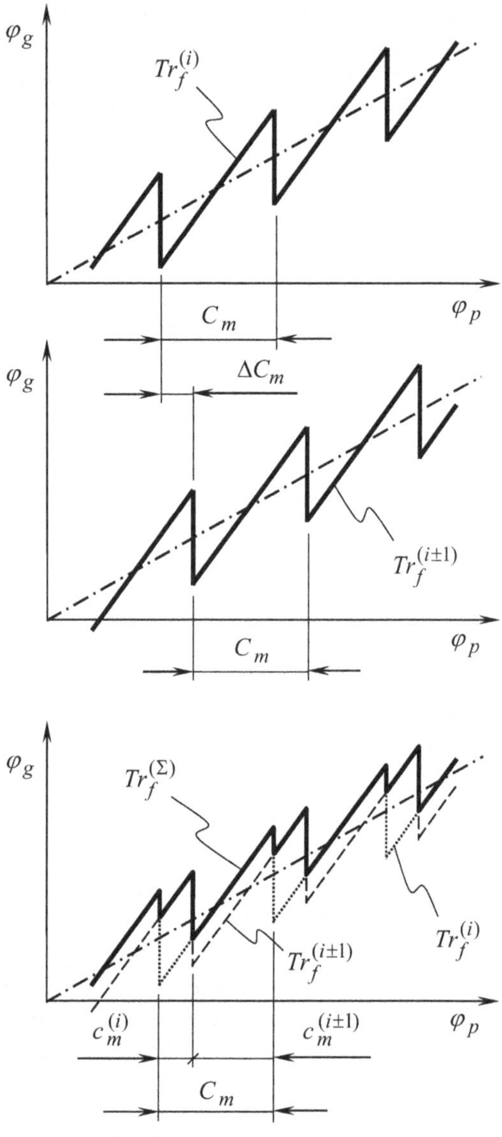

Fig. 1.45 Transmission function as a superposition of a linear function, $\varphi_g^{des}(\varphi_p)$, and piecewise linear functions for the case when total contact ratio $m_t > 1$. (Adapted from Radzevich [5])

Figure 1.45 shows that for the case when the contact ratio is greater than one, $\overline{m} > 1$, the resultant transmission function, $Tr_f^{(\Sigma)}$, significantly differs from that for the case when the equality, $\overline{m} = 1$, is valid. The situation gets more severe when there are no common multipliers in the gear and the pinion tooth number, N_g and N_p [5].

The impact of transmission error onto noise excitation is commonly considered from the geometrical and kinematic points of view. No tooth flank wear is incorporated into the analysis.

The most reliable way to reduce noise excitation is to ensure equality of the operating base pitch of a gear to the operating base pitch of a mating pinion, and both of them keep equal to operating base pitch of the gear pair. Once the base pitches are equal to one another, the gear mesh generates no (or almost no) vibration and produces no noise excitation.

1.12.2 Influence of Location of the Point at Which the Resultant Load Is Applied

There is no variation neither of the equivalent (resultant) tangential force nor of the equivalent (resultant) axial force in a gear pair. Only variation of coordinates of point (at which the resultant force is applied) is observed. This variation is an additional root cause of noise excitation in gearing.

Migration of point at which the resultant load is applied can cause an excessive noise excitation even in geometrically accurate gearing.

When a gear rotates, ω_g, the plane of action, *PA*, travels straight, $\mathbf{V}_{pa}$ (see Fig. 1.46) [2]. The leverage, $\mathbf{R}_{cg}$, of the resultant (or the equivalent) force, $\mathbf{F}_t$, with respect to the bearing support is remained the same and equals:

$$|\,\mathbf{R}_{cg}\,| = \sqrt{r_{b.g}^2 + a^2} \tag{1.66}$$

where:

$r_{b.\,g}$ – is the base radius of the gear
a – is the distance of the gear from the bearing support (see Fig. 1.46)

Center-of-gravity point, c_g [4], does not migrate in the axial direction of the gear. Therefore, the force, $\mathbf{F}_t$, creates a bending moment of a constant value with respect to the bearing support. Thus, no variation of the deflection of the shafts, of the gear housing, and so forth, is observed in the case under consideration. Spur gears create no additional source for noise excitation.

A more general case of contact between tooth flanks, $\mathcal{G}$ and $\mathcal{P}$, of a gear and a mating pinion is observed in geometrically accurate *helical* P_a*–axes gearing*.

When the gears rotate, commonly either two, or three lines of contact, LC_i, are observed in geometrically accurate *helical* P_a*–axes gearing*. It is convenient to distinguish in two different cases helical gearing, namely:

(a) when all the lines of contact, LC_i, are of a full length, and
(b) when one or more line(s) of contact is of a reduced length (is truncated).

An equivalent load is applied at *center-of-gravity point*, c_g. When the gears rotate, point cg migrates within the plane of action, *PA*. The migration of *center-of-gravity*

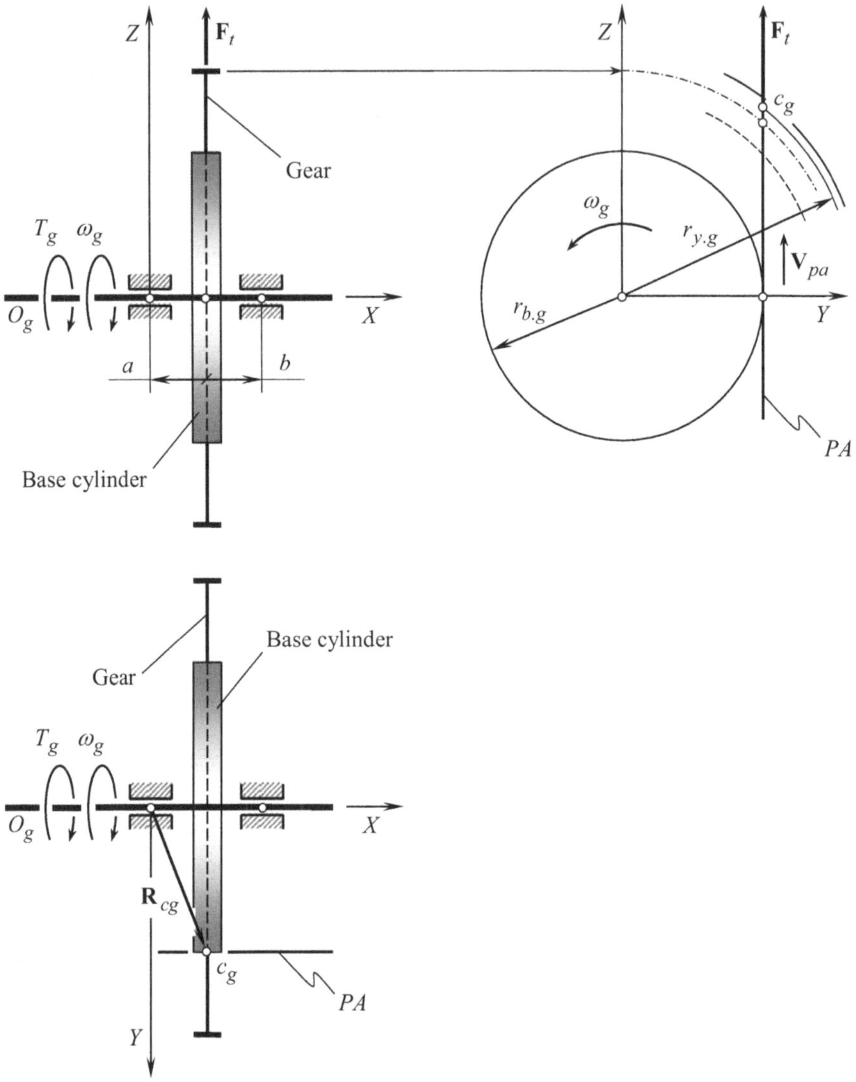

Fig. 1.46 Bending torque in geometrically accurate spur parallel-axes gearing

point, c_g, with a linear velocity, $\mathbf{V}_{cg}$, in the axial direction of the gear pair results in a corresponding alteration of the leverage, $\mathbf{R}_{cg}$, of the resultant force, $\mathbf{F}_t$, in relation to the bearing support of the gear shaft as schematically illustrated in Fig. 1.47. When *center-of-gravity point*, c_g, migrates between points c_g^* and c_g^{**}, the leverage $|\mathbf{R}_{cg}|$ of the resultant force, $\mathbf{F}_t$, changes in the range of:

Fig. 1.47 Variation of bending torque in helical parallel-axes gearing

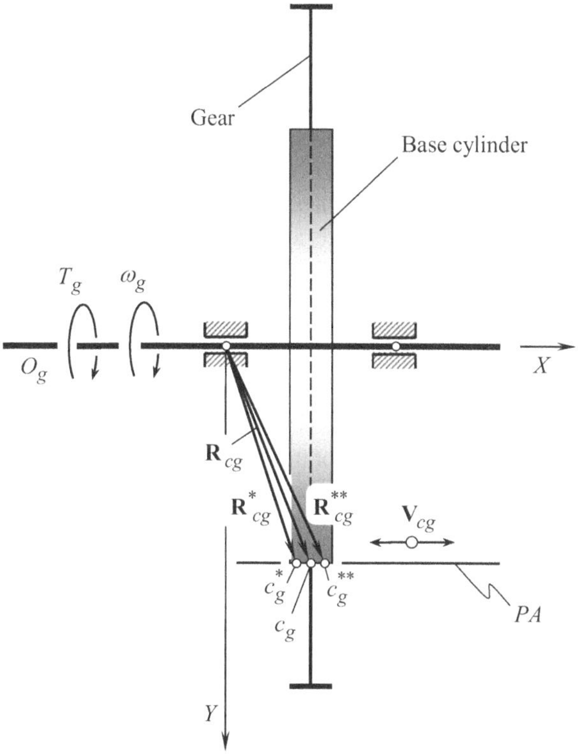

$$|\mathbf{R}^*_{cg}| \leq |\mathbf{R}_{cg}| \leq |\mathbf{R}^{**}_{cg}| \tag{1.67}$$

A variation of magnitude, $|\Delta\mathbf{R}_{cg}|$, of the arm, $\mathbf{R}_{cg}$, results in a variable in time torque that bends the gear shaft, and deforms the gear housing, and can cause an extensive unfavorable vibration of the gear housing, and an extensive noise excitation.

When *center-of-gravity point*, c_g, migrates between points c^*_g and c^{**}_g, the leverage $|\mathbf{R}_{cg}|$ of the resultant force, $\mathbf{F}_t$, alters in the range of $|\mathbf{R}^*_{cg}| \leq |\mathbf{R}_{cg}| \leq |\mathbf{R}^{**}_{cg}|$ (see Fig. 1.47).

A variation of magnitude, $|\Delta\mathbf{R}_{cg}|$, of the leverage, $\mathbf{R}_{cg}$, results in a variable in time torque that bends the gear shaft, and deforms the gear housing, and can cause an extensive unfavorable vibration of the gear housing, and an extensive noise excitation.

The disclosed approach for the determining of loads acting in every pair of teeth in geometrically accurate spur and helical parallel-axes gearing can be enhanced to gear pairs with lines of contact of arbitrary geometry, that is, to gear pairs with lines of contact in the form of arcs of a circle, of a spiral curve, and so forth. Generally speaking, for this purpose the plane-of-action face width, F_{pa}, is subdivided onto several segments within each of which either zero, or one, or two, and so forth, lines of contact are observed. In a case of line of contact in a form of a planar curve, the corresponding portion of plane-of-action face width, F_{pa}, is sliced onto an infinite number of infinitesimally narrow slices. Portion of a line(s) of contact within each infinitesimally narrow slice is considered a straight-line segment. In such a manner, loads in geometrically accurate parallel-axes gear pairs with the lines of contact of an arbitrary geometry can be determined.

Variation of the gear loading along with a variation of *center-of-gravity point*, at which the resultant load is applied is the root cause for an excessive noise excitation in geometrically accurate parallel-axes gearing. This results in a corresponding variation of the deformation of the gear shafts, of the gear housing, and so forth. Ultimately, geometrically accurate parallel-axes gearing produces noise. Noise of this kind is also referred to as the *gear noise*; however, noise of this kind is produced not by the gear mesh, but by other components of the gear mechanism instead: by the shafts, by bearings, as well as by the gear housing, and so forth (all of this stuff vibrate due to variation of the resultant force in gear mesh). This source of noise has to be considered as a *gear noise*, as control over the level of the noise can be achieved by proper selection of design parameters of the gears: profile angle, contact (total) ratio, helix angle, and so forth.

1.13 On a Possibility of Prediction of Gear Noise Excitation

The problem of analytical prediction of noise excitation in a gear pair falls into three levels/iterations (parallel-axes gearing is considered below solely as an example for all three iterations):

- ***The first level of iteration*** – The geometry and configuration of a desirable line of contact, LC_{des}, is specified within the plane of action, PA. When the gears rotate, the transverse pressure angle, ϕ_t, in the gear pair is retained of a constant value. Because of variation of the tooth flank geometry of a gear, $\mathcal{G}$, and that of a mating pinion, $\mathcal{P}$, the lines of intersection, LC_g and LC_p, of the tooth flanks, $\mathcal{G}$ and $\mathcal{P}$, by the plane of action, PA, differ from the desirable line of contact, LC_{des}. The lines, LC_g and LC_{des}, and the linesLC_p and LC_{des}, make contact at points, K_g and K_p, correspondingly. As considered in this section of the book, a gap exists between the lines of intersection, LC_g and LC_p. The gap equals to the closest distance of approach between the lines of intersection, LC_g and LC_p, and, by nature, it is equal to the variation of the base pitch in a gear, $\Delta p_{b.\ g}$, and in the mating pinion, $\Delta p_{b.\ p}$, that is, the gap equals to the algebraic summa $\Delta p_{b.\ g} + \Delta p_{b.\ p}$.

Physically, the gap does not exist as the gears transmit a load. Zero gap is observed between the gear, $\mathcal{G}$, and the mating pinion, $\mathcal{P}$, tooth flanks when the gears transmit power.

As the input shaft rotates uniformly (with a constant angular velocity, ω_p), the driven shaft must make an additional turn through an angle, $\Delta \varphi_g$, to bridge the gap. The angle, $\Delta \varphi_g$, can be expressed in terms of the distance, $(\Delta p_{b.\ g} + \Delta p_{b.\ p})$, and in terms of the parameters of configuration of the lines of intersection, LC_g and LC_p, by the plane of action, PA.

The variation, $(\Delta p_{b.\ g} + \Delta p_{b.\ p})$, of the base pitch is the root cause for an excessive noise excitation in a gear pair. An analysis like that above can be performed for intersected-axes gear pairs as well as for crossed-axes gear pairs.

Conclusion 1 *The deviations, those measured within the plane of action, PA, contribute the most to the variation of the base pitch as well as to the transmission error.*

- ***The second level of iteration*** *– It is assumed in the first level of iteration that the condition of conjugacy of the transverse sections of a gear and a mating pinion is fulfilled for all points within the line of intersection,* LC_g, *and for all the corresponding points within the line of intersection,* LC_p. *For more accurate calculations, points at which the conjugate action law is fulfilled for the gear and for the mating pinion transverse tooth profiles should be considered within spatial curves, both for the gear and for the pinion.*

As the input shaft rotates uniformly (with a constant angular velocity, ω_p), the driven shaft must make an additional turn through an angle, $\Delta\,\varphi_g$, to bridge the gap. The angle, $\Delta\,\varphi_g$, can be expressed in terms of the distance, $(\Delta p_{b.\ g} + \Delta p_{b.\ p})$, and in terms of the parameters of configuration of the lines of intersection, LC_g and LC_p, by the plane of action, PA.

The variation, $(\Delta p_{b.\ g} + \Delta p_{b.\ p})$, of the base pitch is the root cause for an excessive noise excitation in a gear pair. An analysis like that above can be performed for intersected-axes gear pairs as well as for crossed-axes gear pairs.

Conclusion 2 *The deviations, measured perpendicular to the plane of action, PA, contribute less to the variation of the base pitch as well as to the transmission error (projections onto the plane of action, PA).*

- ***The third level of iteration*** – Finally, as the desirable line of contact, LC_{des}, travels together with the plane of action, PA, the actual value of the transverse pressure angle, ϕ_t, could vary depending on the actual angular configuration of the gear and the pinion in relation to one another.

As the input shaft rotates uniformly (with a constant angular velocity, ω_p), the driven shaft must make an additional turn through an angle, $\Delta\,\varphi_g$, to bridge the gap. The angle, $\Delta\,\varphi_g$, can be expressed in terms of the distance, $(\Delta p_{b.\ g} + \Delta p_{b.\ p})$, and in terms of the parameters of configuration of the lines of intersection, LC_g and LC_p, by the plane of action, PA.

The variation, $(\Delta p_{b.\ g} + \Delta p_{b.\ p})$, of the base pitch is the root cause for an excessive noise excitation in a gear pair. An analysis like that above can be performed for intersected-axes gear pairs as well as for crossed-axes gear pairs.

Conclusion 3 *The deviations, measured perpendicular to the plane of action, PA, contribute less to the variation of the base pitch as well as to the transmission error (projections onto the plane of action, PA).*

Once the importance of equality of the base pitches onto the vibration generation and noise excitation is realized, then other sources of gear noise can be taken into account.

References

1. Radzevich, S. P. (Ed.). (2021). *Dudley's handbook of practical gear design and manufacture* (4th ed.). CRC Press. 1170 pages, 718 B/W Illustrations.
2. Radzevich, S. P. (2019, October). Root cause for noise excitation in geometrically-accurate (perfect) parallel-axes gearing. *Gear Solutions Magazine*, 36–44.
3. Radzevich, S. P. (2018, December). The root cause for gear noise excitation. *Gear Solutions Magazine*, 24–29.
4. Radzevich, S. P. (2022). *Theory of gearing: Kinematics, geometry, and synthesis* (3rd ed.). CRC Press. 1192 pages. ISBN 9781032318578.
5. Radzevich, S. P. (2006). Technological methods for noise/vibration reduction in driveline/transmission of truck and all-wheel-drive vehicles. *International Journal of Vehicle Noise and Vibration, 2*(4), 283–291.
6. Radzevich, S. P. (2020, February). Understanding the mounting distance: Crossed-axes gearing (hypoid gearing). *Gear Solutions Magazine*, 38–43.
7. Radzevich, S. P. (2019, December). Understanding the mounting distance: Intersected-axes gearing (bevel gearing). *Gear Solutions Magazine*, 42–47.
8. Smith, J. D. (2003). *Gear noise and vibration* (2nd ed., 318 pages). Marcel Dekker Inc..

Bibliography

Radzevich, S. P. (Ed.). (2019). *Advances in gear design and manufacture* (549 pages). CRC Press.
Radzevich, S. P. (2018, February). An examination of high-conformal gearing. *Gear Solutions*, 31–39.
Radzevich, S. P. (2019, February). Design features of perfect gears for crossed-axes gear pairs. *Gear Solutions Magazine*, 36–43.
Radzevich, S. P. (2020). *High-conformal gearing: Kinematics and geometry* (2nd ed., 506 pages). Elsevier.
Radzevich, S. P. (2020, August). Knowledge (of gear theory) is power in the design, production, and application of gears. *Gear Solutions Magazine*, 38–44. [Upon request, a .pdf of this article can be ordered from the author (at no charge)].
Radzevich, S. P. 2022, November 15. *Novikov/conformal gearing: Scientific theory and practice*. Springer. 33+493 (526 pages).
Radzevich, S. P. (2021, June 25). *Recent advances in gearing: Scientific theory and applications* (1st ed., 2022 edition). Springer. 569 pages
Radzevich, S. P. (2021, February). The commonalities and differences between helical "low-tooth-count gears" and "multiple-start worms". *Gear Solutions Magazine*, 34–39.
Radzevich, S. P., & Storchak, M. G. (Eds.). (2022). *Advances in gear theory and gear cutting tool design* (500 pages). Springer.

Chapter 2
Tooth Contact Analysis (TCA) of Cylindrical Gears with Tooth Surface Deviations

Michał Batsch

2.1 Introduction

Gear noise and vibration excitation are influenced mostly by gear teeth accuracy and resulting pitch [13] and transmission errors [9] as well as by overall gearbox accuracy (e.g. shafts misalignment) and rigidity (e.g. bearing, shafts and mesh stiffness) [5]. Many authors have correlated gear vibrations and noise with transmission error [6, 9, 12, 15] thus exposing its importance within the design of low-noise gearboxes. There are two types of transmission errors (TEs): no-load transmission error resulting from the geometry of mated gears and loaded transmission error resulting from no-load TE and deflection of teeth under load. Both of them affect gearbox operation and contribute to vibration generation. For the calculation of loaded TE, one can use the finite element method [10, 11] or the analytical and semi-analytical approach [14]. For obtaining no-load TE the numerical methods may be used similarly as in the case of spiral bevel gears [8], which takes into account manufacturing deviations of gear teeth like pitch and surface errors or runout. The importance of the influence of manufacturing deviations (in particular surface deviations) on gear operation was also recognized within the industry, which has led to the inclusion of such TCA calculation procedures in commercial software like KissSoft [7]. The present chapter discussed the method of no-load TE calculation resulting from surface deviations and its influence on gearbox vibration excitation.

M. Batsch (✉)
Faculty of Mechanical Engineering and Aeronautics, Department of Mechanical Engineering, Rzeszow University of Technology, Rzeszów, Poland
e-mail: mbatsch@prz.edu.pl

S. P. Radzevich (ed.), *Gear Accuracy*, Mechanical Engineering Series,
https://doi.org/10.1007/978-3-031-74798-4_2

2.2 Tooth Contact Analysis

2.2.1 Coordinate Systems

This chapter analyses the cylindrical external gear pairs with parallel axes. To simulate meshing, the coordinate systems 1 and 2 are introduced and attached to the pinion and gear, respectively. Gears rotate around a z_1 and z_2 axis about angles φ_1 and φ_2 according to the directions of angular velocities ω_1 and ω_2 (Fig. 2.1).

Teeth surfaces are represented at each coordinate system with the aid of vectors:

$$\mathbf{r}_1^{(1)}(\theta_1, \zeta_1) = \begin{bmatrix} x_1^{(1)} & y_1^{(1)} & z_1^{(1)} & 1 \end{bmatrix}^{\mathrm{T}}$$

and

$$\mathbf{r}_2^{(2)}(\theta_2, \zeta_2) = \begin{bmatrix} x_2^{(2)} & y_2^{(2)} & z_2^{(2)} & 1 \end{bmatrix}^{\mathrm{T}}$$

where θ_1, ζ_1 are the parameters of the pinion tooth surface, and θ_2, ζ_2—are parameters of the gear tooth surface. Both of the above vectors might be transformed into a fixed coordinate system *f*:

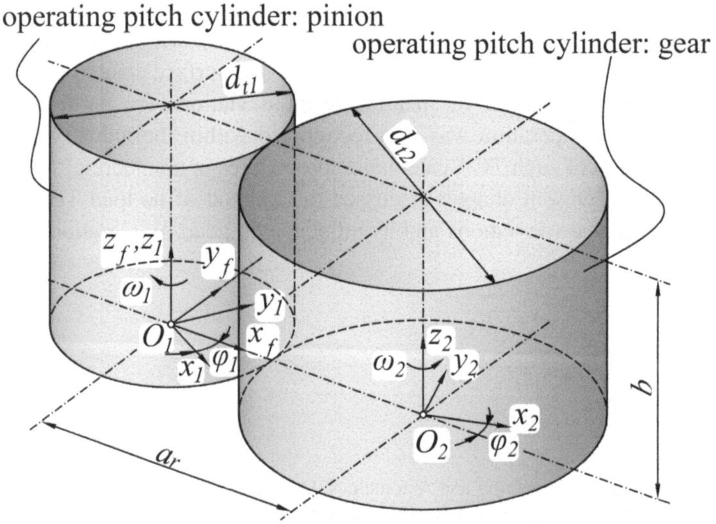

Fig. 2.1 Coordinate systems in cylindrical gear analysis

$$\mathbf{r}_1^{(f)} = \begin{bmatrix} x_1^{(f)} \\ y_1^{(f)} \\ z_1^{(f)} \\ 1 \end{bmatrix} = \mathbf{Rs}(1 \mapsto f) \cdot \mathbf{r}_1^{(1)} \tag{2.1}$$

$$\mathbf{r}_2^{(f)} = \begin{bmatrix} x_2^{(f)} \\ y_2^{(f)} \\ z_2^{(f)} \\ 1 \end{bmatrix} = \mathbf{Rs}(2 \mapsto f) \cdot \mathbf{r}_2^{(2)} \tag{2.2}$$

where:

$$\mathbf{Rs}(1 \mapsto f) = \mathbf{Rt}(-\varphi_1, Z) \tag{2.3}$$

$$\mathbf{Rs}(2 \mapsto f) = \mathbf{Tr}(a_r, X) \cdot \mathbf{Rt}(\varphi_2, Z) \tag{2.4}$$

Homogenous matrices of transformation occurring in Eqs. (2.3) and (2.4) were defined in Appendix 4 [3]. Assuming that $\mathbf{n}_1^{(1)}$ is the unit normal to the pinion tooth surface in coordinate system 1 and $\mathbf{n}_2^{(2)}$ is the unit normal to the gear tooth surface in coordinate system 2, it can be written that:

$$\mathbf{n}_1^{(f)} = \begin{bmatrix} n_{x1}^{(f)} \\ n_{y1}^{(f)} \\ n_{z1}^{(f)} \end{bmatrix} = \mathbf{L}_{1 \mapsto f} \cdot \mathbf{n}_1^{(1)} \tag{2.5}$$

$$\mathbf{n}_2^{(f)} = \begin{bmatrix} n_{x2}^{(f)} \\ n_{y2}^{(f)} \\ n_{z2}^{(f)} \end{bmatrix} = \mathbf{L}_{2 \mapsto f} \cdot \mathbf{n}_2^{(2)} \tag{2.6}$$

where $\mathbf{L}_{1 \mapsto f}$ is the matrix of transformation from coordinate system 1 to f, and $\mathbf{L}_{2 \mapsto f}$- is the matrix of transformation from coordinate system 2 to *f*. Matrices of transformation $\mathbf{L}_{1 \mapsto f}$ and $\mathbf{L}_{2 \mapsto f}$ are obtained by crossing out the last row and the last column in homogenous matrices of transformation (2.3) and (2.4).

2.2.2 Tooth Surfaces

Let us consider the meshing of two screw involute surfaces that represent the teeth surfaces of cylindrical gear pair. These surfaces in coordinate systems defined as in Fig. 2.1 are given by vectors:

$$\mathbf{r}_1^{(1)} = \begin{bmatrix} r_{b1}\left(\cos\left(\pm\zeta_1 + t_1 - \varphi_{\text{sym1}}\right) + t_1 \sin\left(\pm\zeta_1 + t_1 - \varphi_{\text{sym1}}\right)\right) \\ r_{b1}\left(\sin\left(\pm\zeta_1 + t_1 - \varphi_{\text{sym1}}\right) - t_1 \cos\left(\pm\zeta_1 + t_1 - \varphi_{\text{sym1}}\right)\right) \\ r_1\zeta_1 \cot\beta \\ 1 \end{bmatrix} \quad (2.7)$$

$$\mathbf{r}_2^{(2)} = \begin{bmatrix} r_{b2}\left(-\cos\left(\pm\zeta_2 - t_2 + \varphi_{\text{sym2}}\right) + t_2 \sin\left(\pm\zeta_2 - t_2 + \varphi_{\text{sym2}}\right)\right) \\ r_{b2}\left(\sin\left(\pm\zeta_2 - t_2 + \varphi_{\text{sym2}}\right) + t_2 \cos\left(\pm\zeta_2 - t_2 + \varphi_{\text{sym2}}\right)\right) \\ r_2\zeta_2 \cot\beta \\ 1 \end{bmatrix} \quad (2.8)$$

where the upper sign stands for the pinion with the right-hand helix mated with a gear with left-hand helix, while the lower sign for the left-hand helix of pinion and right-hand helix of gear, r_{b1}, r_{b2} are base radii for pinion and gear respectively, t_1, t_2—are the tooth profile parameters for pinion and gear (roll angles), ζ_1, ζ_2—are the screw tooth line parameters for pinion and gear, φ_{sym1}—is the involute of transverse pressure angle decreased by the central angle corresponding to half of the tooth space on pitch diameter in transverse section, φ_{sym2}—is the involute of transverse pressure angle increased by the central angle corresponding to half of the tooth width on pitch diameter in transverse, β—is the helix angle. The units normal to teeth surfaces are therefore given as:

$$\mathbf{n}_1^{(1)} = \frac{\frac{\partial \mathbf{r}_1^{(1)}}{\partial \zeta_1} \times \frac{\partial \mathbf{r}_1^{(1)}}{\partial t_1}}{\left|\frac{\partial \mathbf{r}_1^{(1)}}{\partial \zeta_1} \times \frac{\partial \mathbf{r}_1^{(1)}}{\partial t_1}\right|} = \frac{1}{\sqrt{r_1^2 \cot^2\beta + r_{b1}^2}} \cdot \begin{bmatrix} r_1 \cot\beta \cos\left(\pm\zeta_1 + t_1 - \varphi_{\text{sym1}}\right) \\ -r_1 \cot\beta \sin\left(\pm\zeta_1 + t_1 - \varphi_{\text{sym1}}\right) \\ r_{b1} \end{bmatrix} \quad (2.9)$$

$$\mathbf{n}_2^{(2)} = \frac{\frac{\partial \mathbf{r}_2^{(2)}}{\partial \zeta_2} \times \frac{\partial \mathbf{r}_2^{(2)}}{\partial t_2}}{\left|\frac{\partial \mathbf{r}_2^{(2)}}{\partial \zeta_2} \times \frac{\partial \mathbf{r}_2^{(2)}}{\partial t_2}\right|} = \frac{1}{\sqrt{r_2^2 \cot^2\beta + r_{b2}^2}} \cdot \begin{bmatrix} -r_2 \cot\beta \sin\left(\pm\zeta_2 - t_2 + \varphi_{\text{sym2}}\right) \\ r_2 \cot\beta \cos\left(\pm\zeta_2 - t_2 + \varphi_{\text{sym2}}\right) \\ -r_{b2} \end{bmatrix} \quad (2.10)$$

2.2.3 Tooth Surface Deviations

Tooth surface deviations resulting as an error in topography may be represented over the grid of coordinates $z_i^{(i)}$:

$$r_{yi} = \sqrt{\left(x_i^{(i)}\right)^2 + \left(y_i^{(i)}\right)^2}$$

as ΔS_i (Fig. 2.2) where $i = 1{,}2$ refers to pinion and gear respectively.

This surface may be obtained from real gear measurement, e.g. by interpolation over the measured tooth profiles. Tooth surface with surface deviation is defined as:

$$\mathbf{r}_i^{*(i)} = \mathbf{r}_i^{(i)} - \Delta S_i \cdot \mathbf{n}_i^{(i)} \tag{2.11}$$

In analyses instead of the ideal surface described by vectors (2.7) and (2.8) the surfaces with topography errors (2.11) should be taken into account.

2.2.4 Transmission Error

This subsection briefly discusses the author's implementation of the method for calculating the transmission error which was described in previously published works [3, 4]. It assumes a perfect mesh between pinion tooth Σ_1 surface and gear tooth Σ_2 surface, i.e. one in which the gears are rotated by angles resulting from the theoretical gear ratio $i_{12} = z_2/z_1 = \varphi_1/\varphi_2$. Surfaces in the nth position are shown in Fig. 2.3.

It is assumed that pinion rotation axis z_1 is aligned with axis z_f of fixed coordinate system f, and the surfaces in that system are given by vectors (2.11). In each position n resulting from the discretization of gear rotation angles, between the surfaces, there is a certain distance measured along the arc with a radius:

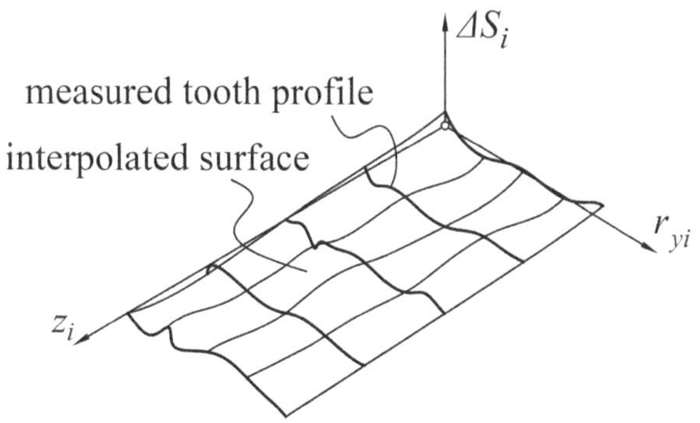

Fig. 2.2 Tooth surface topography deviations

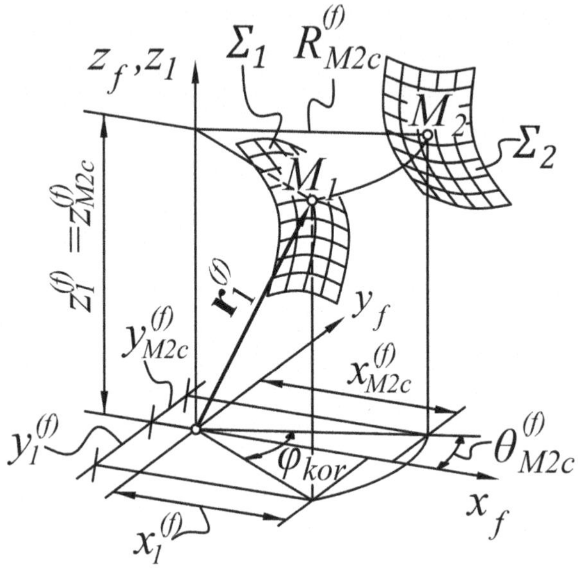

Fig. 2.3 Tooth flanks in the n-th position [3, 4]

$$R_{M2c}^{(f)} = \sqrt{\left(x_1^{(f)}\right)^2 + \left(y_1^{(f)}\right)^2},$$

corresponding to the length of arc $k_{\mathrm{kor}} = \widehat{M_1 M_2}$, where M_1 is a point on surface Σ_1 with coordinates $z_1^{(f)}$, and M_2 a point on surface Σ_2 with coordinates $z_1^{(f)} = z_{M2c}^{(f)}$. Angle φ_{kor} by which surface Σ_1 would have to be rotated for points M_1 and M_2 to overlap is the central angle corresponding to the length of the arc $k_{\mathrm{kor}} = \widehat{M_1 M_2}$ and can be determined based on coordinates as:

$$\varphi_{\mathrm{kor}} = 2 \arcsin\left(\frac{\sqrt{\left(x_1^{(f)} - x_{M2c}^{(f)}\right)^2 + \left(y_1^{(f)} - y_{M2c}^{(f)}\right)^2}}{2R_{M2c}^{(f)}}\right). \tag{2.12}$$

The distance between flank surfaces is expressed as:

$$k_{\mathrm{kor}} = R_{M2c}^{(f)} \varphi_{\mathrm{kor}}. \tag{2.13}$$

In each n-th position, there is a certain minimum distance $k_{\mathrm{kor}}^{\min}$ with its corresponding angle $\varphi_{\mathrm{kor}}^{\min}$. The relationship between minimum angle $\varphi_{\mathrm{kor}}^{\min}$ and gear rotation angle φ_2 determined discretely for all n positions constitutes the transmission error. The greatest challenge of the method is to specify the location of point M_2 on surface Σ_2 that corresponds to the coordinate $z_1^{(f)} = z_{M2c}^{(f)}$ and radius $R_{M2c}^{(f)}$. The issue may be easily solved by interpolation in a cylindrical coordinate system $\theta_{2c}^{(f)}, R_{2c}^{(f)}, z_2^{(f)}$.

2.3 Results

2.3.1 Analysed Gear Pair

The analysed gear pair is the helical cylindrical gear set manufactured from 42CrMo4 steel, quenched and tempered to 28-30HRC and gas-nitrided (650–700 HV) with CNC hobbed teeth (Fig. 2.4), whose data are listed in Table 2.1.

After measurements taken on *Klingelnberg* P40 coordinate measuring machine the gears were classified according to ISO accuracy grade (Table 2.1). The results of topography measurements are presented in Fig. 2.5.

Red lines indicate the measured tooth profiles, while the coloured surface was interpolated at measured profiles over the grid of $z_i^{(i)}$, r_{yi} coordinates with the aid of cubic *Hermite interpolation* [1]. Within the experiment, the gear pair was operating on a power circulation test stand which is described in the paper [2]. The pinion rotation velocity was 2280 rpm and the applied torque 455 Nm. This results in a frequency of meshing $f_m = 1139$ Hz.

2.3.2 Transmission Error Excitation

The interpolated values of deviations (Fig. 2.5) were put into Eq. (2.11) to obtain the transmission error according to the method described in Sect. 2.2. The resulting transmission error ($x_{te} = k_{\text{kor}}^{\text{min}}$) and its normalized amplitude spectrums are presented in Fig. 2.6.

Fig. 2.4 Tested gear pair

Table 2.1 Data of the analysed gear pair

Parameters	Pinion	Gear
Geometry		
Normal module m, mm	3	
Number of teeth	$z_1 = 30$	$z_2 = 47$
Face width b, mm	30	
Helix angle β, °	22.428	
Normal pressure angle α_n, °	20	
Axis distance a_r, mm	125	
Profile shift X, mm	0	
Base diameter, mm	$d_{b1} = 90.62$	$d_{b2} = 141.97$
Pitch diameter, mm	$d_1 = 97.40$	$d_2 = 152.59$
Addendum diameter, mm	$d_{a1} = 103.40$	$d_{a2} = 158.59$
Dedendum diameter, mm	$d_{f1} = 89.90$	$d_{f2} = 145.09$
Accuracy		
Pitch accuracy grade	7	
Topography (teeth line and profile) accuracy grade	7	8
Total profile deviation F_α, µm	13.5	15.1
Profile form deviation $f_{f\alpha}$, µm	11.6	8.4
Profile slope deviation $f_{H\alpha}$, µm	5.3	−12.3
Total helix deviation F_β, µm	12.3	14.1
Helix form deviation $f_{f\beta}$, µm	7.0	12.7
Helix slope deviation $f_{H\beta}$, µm	9.5	17.9

The amplitude spectrum was obtained with the aid of a fast Fourier transform (FFT) of the time signal of transmission error x_{te}, and its acceleration $a_{\mathrm{te}} = \frac{\mathrm{d}^2 x_{\mathrm{te}}}{\mathrm{d}t^2}$:

$$A(x) = \frac{1}{N}\sqrt{\mathrm{Re}\,(\mathrm{FFT}(x))^2 + \mathrm{Im}(\mathrm{FFT}(x))^2} \tag{2.14}$$

where FFT(x) is the fast *Fourier transform* of signal x, x is a time signal (x_{te} or a_{te}), Re denotes the real while Im is the imaginary part of a complex number and N is the number of signal samples. Next, the amplitude spectrum was normalized according to the equation:

$$A_{\mathrm{norm}}(x) = \frac{A(x) - \max(A(x))}{\max(A(x)) - \min(A(x))} \tag{2.15}$$

The maximum value of transmission error was about 100 µm (Fig. 2.6a). It should be noted that this transmission error results only from topography errors (profile and helix deviations). Other factors influencing its value such as pitch errors and deflections are not included. Increased amplitudes of transmission error are observed below the frequency of 2000 Hz (Fig. 2.6b), while in the case of acceleration, they are present at a higher order of harmonics (Fig. 2.6c).

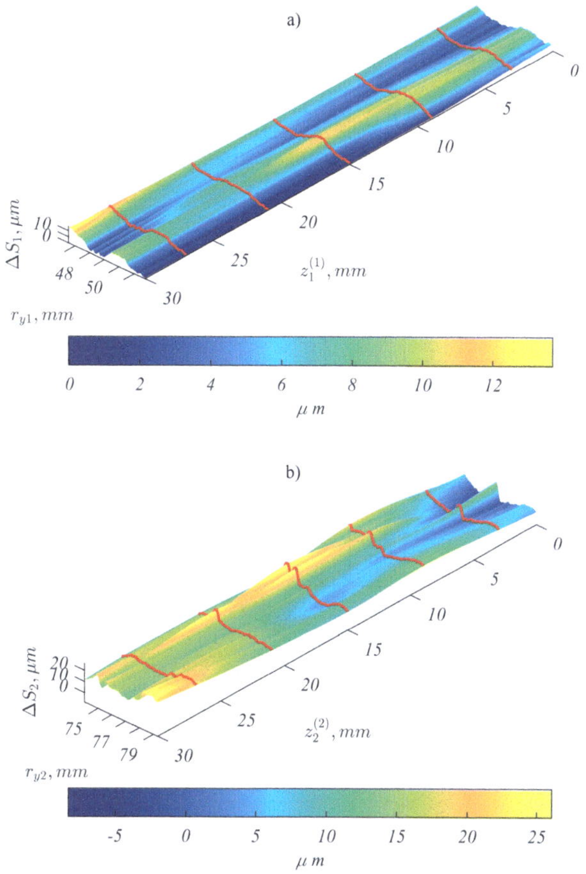

Fig. 2.5 Topography measurements: (**a**) pinion flank, (**b**) gear flank

2.3.3 *Correlation with Measurement Data*

The vibrations were measured by a piezoelectric sensor mounted on the gear case, in a vertical direction (perpendicular to the shafts' plane). Measurements were taken several times through the whole operating time of 80 h (which corresponds to $1.25 \cdot 10^7$ number of cycles of pinion) with a sampling frequency of 25 kHz. This results in an acquired acceleration time signal with 225,000 samples a part of which is presented in Fig. 2.7a. No signs of failure were detected during gear operation (2280 rpm, 455 Nm—on pinion). The amplitude spectrum of the gear case acceleration signal (Fig. 2.7b) was calculated with the formula (2.14).

In order to compare the results of tooth contact analysis (Fig. 2.6) with actual vibrations (Fig. 2.7) as a measure of similarity the cross-power spectral density was chosen:

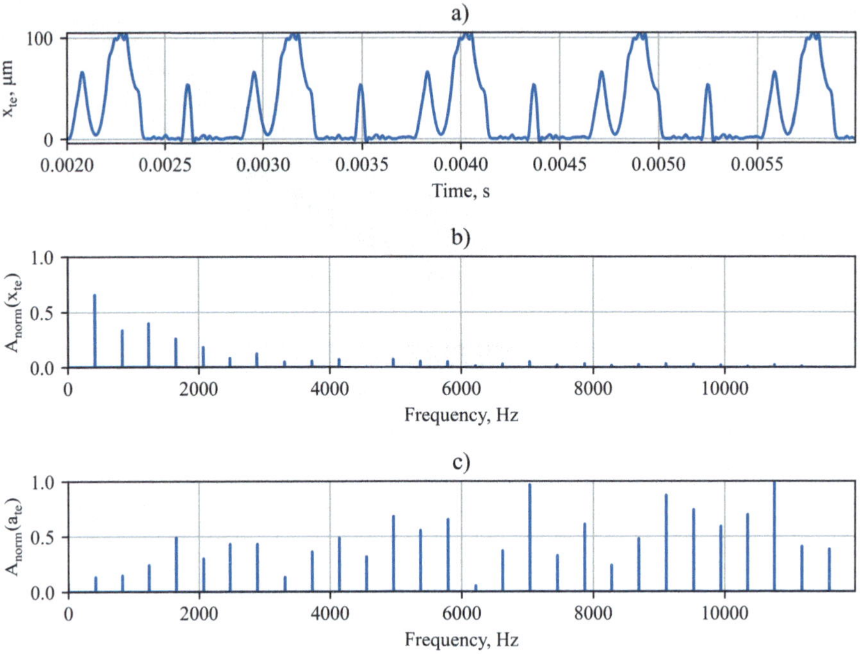

Fig. 2.6 Transmission error excitation: (**a**) transmission error time signal, (**b**) normalized amplitude spectrum of transmission error, (**c**) normalized amplitude spectrum of transmission error acceleration

$$\mathrm{CSD}(x, y) = \mathrm{W}(x * y) \tag{2.16}$$

where W denotes the power spectral density estimate according to *Welch*'s *approach* [16] with the Hann window function, *x* and *y* are normalized time signals and x^*y is a cross-correlation of time signals *x* and *y*. The results of the calculations are shown in Fig. 2.8.

High peaks can be observed within a cross-spectral density which means that the transmission error signal and its acceleration are periodically correlated with the gear case acceleration signal. The greatest power of transmission error is shared with gear case acceleration within the first three orders of harmonics—f_m, $2f_m$, $3f_m$ (Fig. 2.8a). The strongest correlation in the case of transmission error acceleration is observed within $2f_\mathrm{m}$ frequency. High peaks occurred also at f_m, and $5f_m$ (Fig. 2.8b).

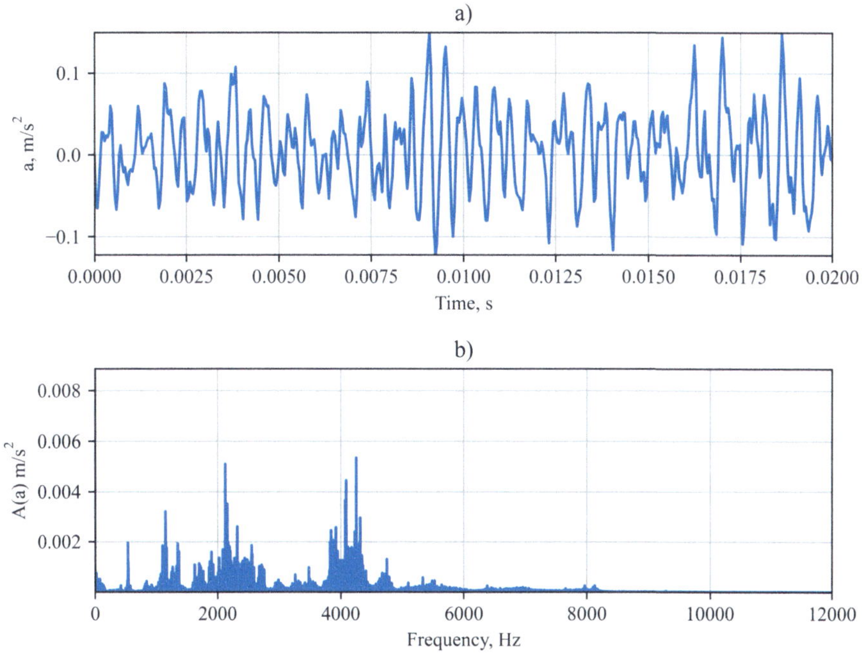

Fig. 2.7 Gear case vibrations: (**a**) acceleration time signal, (**b**) amplitude spectrum

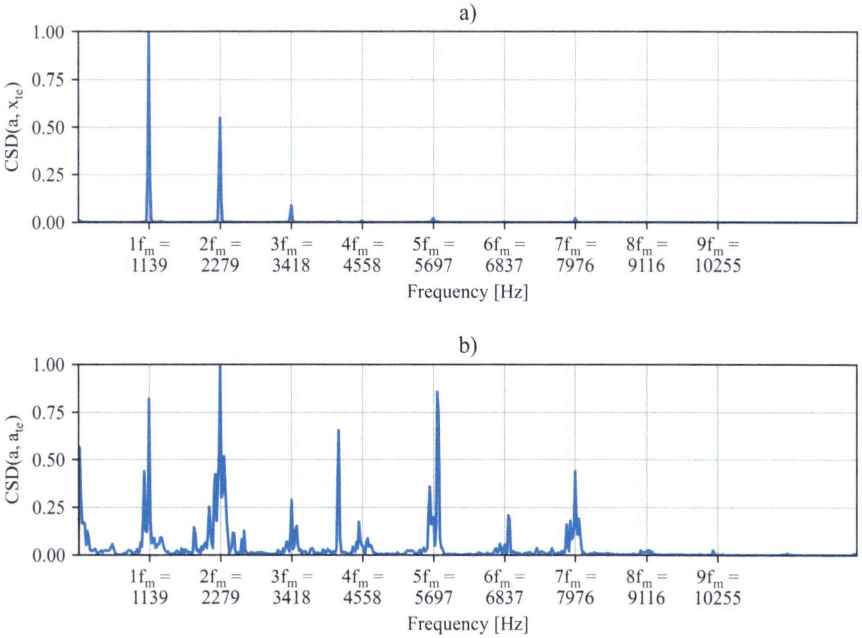

Fig. 2.8 Cross power spectral density of gear case acceleration signal and: (**a**) transmission error, (**b**) acceleration of transmission error

2.4 Conclusion

Based on the obtained results, the following conclusions can be drawn:

- Surface deviations have a significant influence on gearbox vibrations.
- In order to minimize the vibration excitation, the transmission error resulting from surface deviations should be minimized.
- Particular care must be taken on transmission error acceleration if higher-order harmonics are under consideration.

References

1. Akima, H. (1974). A method of bivariate interpolation and smooth surface fitting based on local procedures. https://doi.org/10.1145/360767.360779
2. Batsch, M. (2016). Comparative fatigue testing of gears with involute and convexo-concave teeth profiles. *Advances in Manufacturing Science and Technology, 40*(2), 5–25. https://doi.org/10.2478/amst-2016-0007
3. Batsch, M. (2023). Helical bevel Novikov gears. In S. P. Radzevich (Ed.), *Novikov/conformal gearing: Scientific theory and practice*. Springer. https://doi.org/10.1007/978-3-031-10019-2_3
4. Batsch, M., et al. (2022). Tooth contact analysis of cylindrical gears with an unconventional tooth profile. *Advances in Science and Technology Research Journal., 16*(4), 119–129. https://doi.org/10.12913/22998624/152172
5. Bruzzone, F., & Rosso, C. (2020). Sources of excitation and models for cylindrical gear dynamics: A review. *Machines, 8*(3), 37. https://doi.org/10.3390/machines8030037
6. Henriksson, M., & Pang, Y.-Y. (2009). Transmission error as gear noise excitation. In *Volume 6: ASME power transmission and gearing conference; 3rd international conference on micro- and nanosystems; 11th international conference on advanced vehicle and tire technologies* (pp. 197–203). ASMEDC. https://doi.org/10.1115/DETC2009-86695
7. Kissling, U. (2022, September/October). Taking into account the production methodology and estimating the influence of manufacturing quality on NVH performance when sizing gears. *Gear Technology*, 36–45.
8. Klingelnberg, J. (2016). *Bevel gear: Fundamentals and applications*. Springer.
9. Lahoti, S., et al. (2021). *Gear whine noise reduction by transmission error and contact pattern optimization: Simulation and testing correlation*. Presented at the noise and vibration conference & exhibition. https://www.sae.org/content/2021-01-1103/. https://doi.org/10.4271/2021-01-1103.
10. Mao, K. (2006). An approach for powertrain gear transmission error prediction using the non-linear finite element method. *Proceedings of the Institution of Mechanical Engineers, Part D: Journal of Automobile Engineering., 220*(10), 1455–1463. https://doi.org/10.1243/09544070JAUTO251
11. Marciniec, A., et al. (2018). Comparative analysis of numerical methods for the determination of contact pattern of spiral bevel gears. *Aircraft Engineering and Aerospace Technology, 90*(2), 359–367. https://doi.org/10.1108/AEAT-08-2016-0133
12. Radu, M. C., et al. (2018). A perspective on gear meshing quality based on transmission error analysis. *IOP Conference Series: Materials Science and Engineering, 444*, 052011. https://doi.org/10.1088/1757-899X/444/5/052011
13. Radzevich, S. P. (2018, December). The root cause for gear noise excitation. Gear. *Solutions*, 24–29.

14. Velex, P., & Ajmi, M. (2006). On the modelling of excitations in geared systems by transmission errors. *Journal of Sound and Vibration, 290*(3–5), 882–909. https://doi.org/10.1016/j.jsv.2005.04.033
15. Wang, C., et al. (2016). Modified optimization and experimental investigation of transmission error, vibration and noise for double helical gears. *Journal of Vibration and Control, 22*(1), 108–120. https://doi.org/10.1177/1077546314525216
16. Welch, P. (1967). The use of fast Fourier transform for the estimation of power spectra: A method based on time averaging over short, modified periodograms. *IEEE Transactions on Audio and Electroacoustics, 15*(2), 70–73.

Chapter 3
Influence of Tooth Profile Deviation Due to Wear on Fluctuation of Gear Ratio

Pavlo M. Tkach

3.1 Introduction

In the design of machine drives used in any industry and on all types of transportation, the correlation between the transmitted load and the size and weight parameters is relevant. When increasing the requirements for the load capacity of the designed drives, an increase in transmission dimensions may not always be permissible. For example, in on-board transmissions of agricultural and road machinery, traction transmissions of locomotives, and other vehicles, the dimensions of the drive are limited by the dimensions of the space where it is located. In aircraft, in addition to possible layout constraints on the drive, there are requirements for its overall weight and, consequently, its dimensions. As a result, drives of modern machines operate at the limit of their load capacity. In such conditions, some types of failure that are typical for the transmission may appear long before the end of the design service life.

Despite the fact that the type of machinery (aviation engines, marine gearboxes, automotive and tractor transmissions, traction transmissions of locomotives, etc.) determines its specific operating conditions for gears, the main types of gear tooth failures and their reasons are similar everywhere. An analysis of the damage to gear teeth, conducted by several researchers and the author, has shown that typical failures in heavy loaded transmissions include wear, cracks at the tooth root, chipping and breaking of teeth and pitting of the flanks. There are various factors that influence their performance: the steel grade and structure, design and operational features, etc. One or several factors may be decisive in different types of failures. In case of abrasive wear, the main factors are as follows: presence of abrasive particles in the oil, surface roughness, hardness, and circumferential speed. The presence of

P. M. Tkach (✉)
E.O. Paton Electric Welding Institute of the National Academy of Sciences of Ukraine, Kyiv, Ukraine

S. P. Radzevich (ed.), *Gear Accuracy*, Mechanical Engineering Series,
https://doi.org/10.1007/978-3-031-74798-4_3

metallic particles from wear products does not affect the wear of the tooth profile. For tooth breakage, the main factors are hardness, module, tooth length and the diameter of the base circle.

For all open and encased transmissions operating in aggressive conditions, where the entry of abrasive particles and water into the lubricant is possible, the predominant mode of failure is wear of the tooth flanks. It is often that mating gears of such transmissions operate under heavy conditions due to several more reasons: presence of significant variable misalignments of engaged gears and deviations in the center distance due to shaft deformations; dynamic forces in the transmission caused by forced oscillations of the machine executive element; poor lubrication of mating tooth surfaces etc. The combined effect of the aforementioned factors predisposes the wear of the mating teeth.

Among the listed factors, dynamic loading in the transmission is highly influential. In addition to external dynamic forces, internal dynamic forces also act in the meshing. Speaking about the nature of their occurrence, it should be noted that during the operation of the power drive, the entire system becomes elastically deformed by three external torques:

(i) the driving torque from the engine
(ii) the driven torque from the executive element, and
(iii) the balancing torque applied to the drive elements from the stationary base

In the best-case scenario, the moments are constant, and a constant gear ratio is achieved in the meshing. However, even uniform rotation of shafts and gears leads to the excitation of elastic vibrations and variable stresses in the system, because: gear teeth are sequentially loaded and unloaded as they pass through the meshing zone; the plane of shaft bending changes; inevitable contact and bending deformations occur as teeth start one-pair or two-pair meshing. All of this leads to periodic disruptions in the mutual position of the engaging elements, which inevitably causes fluctuations in the gear ratio, i.e., leads to kinematic meshing errors. The consequence of this is the occurrence of torque fluctuations, generating torsional vibrations, and increasing dynamic loads in the transmission.

In real transmissions, tooth profiles deviate from the ideal form, causing kinematic errors that lead to uneven rotation of gears and torque fluctuations. However, based on the experience, the described fluctuations in gear transmissions, as well as fluctuations caused by inaccuracies in gears manufacturing, are much less intense than fluctuations generated by variable gear ratio in meshing of teeth with worn-out flanks (i.e., with non-involute flanks). The mentioned fluctuations, acting together, induce vibrations and significant dynamic forces in the meshing. These are detrimental phenomena that affect other components and assemblies connected to the transmissions and further intensify teeth wear. Therefore, the research of impact of the wear value on the gear ratio is a relevant task.

3.2 Literature Review

In most cases, gear transmissions in power drives operate until the limit state is reached. Typically, this state is regulated by normative documentation for any specific type of transmission and is associated with appearance of defects or their maximum sizes. Accordingly, the limit state is determined either by qualitative or quantitative indicators. For example, the appearance of cracks or the breaking of a tooth, or its part, does not have a quantitative measure, and the rejection of gears occurs based on the presence of cracks or breakage. At the same time, these defects may not be detected at the moment of their appearance, so the transmissions may continue to be operating with defects until the next inspection. Unlike the cracks, wear and pitting have a quantitative measure, and transmissions with such defects can continue to be operating for the most of their lives. This is confirmed by numerous studies and operational experience. For example, the authors of [1], while monitoring the performance of mine reducers for 12,000 h of operation, found that after only 2000 h, the wear exceeds the tolerances for gear tooth manufacturing. A significant amount of work is devoted to the study of the operability of transmissions with defects. For example, in [2], a model was developed to assess the condition of the contact of a gear pair with damage in the form of partial tooth breakage across the entire width at different heights. Such damage generates corner contact of the teeth, which causes the appearance of a group of sidebands in the amplitude spectrum during the diagnosis of transmission operation. The obtained result indicates the necessity of considering the corner contact, which will allow to increase the accuracy of tooth breakage detection. In contrast to [2], a greater variety of fracture scenarios, including partial tooth breakage, were considered in the work [3]. The authors proposed a two-stage strategy for defect significance assessment based on artificial intelligence. The first stage involves defect detection, while the second stage involves a three-level classification of fracture severity. Subsequent fault severity classification can be achieved with zero-shot learning based on fault geometry. The proposed approach has a great potential not only for diagnosing gear transmission failures but also for the future failure prediction. Even minor tooth damage, such as chipping fault, can significantly affect meshing characteristics. The authors of [4] investigated the chipping fault along various curves and assessed the influence of this factor on mesh stiffness. The stiffness is an important parameter for the mesh dynamics. A model for evaluation of this influence for both external and internal meshing was developed and confirmed by finite element method (FEM). It is known that the presence of not only broken teeth but also cracks at the tooth root affects meshing characteristics. This influence was studied in [5] for two types of spatial cracks in the tooth root and beyond. An analytical method for calculating the stiffness of meshing teeth with cracks of various geometries was presented in the study. The method was verified by the authors using FEM, and it can be applied to analyze tooth damage. The study [6] is also dedicated to determining the influence of profile deviations due to (i) profile modification and (ii) wear of active tooth surfaces on gear mesh stiffness and quasi-static transmission error. To assess such influence, the authors proposed an enhanced model that takes into account the effects of corner contact and structure coupling that are considered together. The model was verified using FEM and allows to evaluate the influence of both profile modification and profile wear during transmission operation up to $100 \cdot 10^6$ load cycles. In [7], the same authors considered the effect of profile deviations due to wear on load distribution. It should be noted that in [6, 7], the profiles of worn teeth are represented theoretically, based on Archard's wear equation [8], i.e., with no wear at the pitch circle. The researchers in [9] investigated the influence of wear on the stiffness variation of teeth by analyzing the vibrational characteristics of transmissions under accelerated wear testing conditions. The developed methodology is applicable for quantitative assessing the wear of spur gears. The numerical model presented in [10] is intended for predicting the occurrence and progression of another important type of failure—micropitting. This irreversible surface damage is caused by contact fatigue and occurs through the initiation and progression of microcracks. According to [11], this can occur after 20,000 load cycles. Externally, the phenomenon may appear similar to wear, i.e., in some time the distortions of the involute profile can appear. The authors of the research [10] incorporated into the model a crack initiation criterion that involves six factors. The modeling results were experimentally validated, and the model can be used for predicting micropitting on tooth flanks with profile modification. An algorithm for quantitative assessment of micropitting and pitting at different stages of transmission operation was proposed by the authors of [12]. The results were also experimentally validated by obtaining the real tooth profiles with micro- and macro-pitted surfaces, and they can be used to predict those types of tooth failures. Thus, research on transmissions with various damages at different stages of their operation is carried out very actively.

A lot of studies are dedicated to researching the factors influencing wear as a phenomenon. Among important factors, the presence of dynamic loading should be highlighted, which increases with wear increase, and the intensification of dynamic loads leads to increased wear intensity [13]. The absence of a model that could accurately describe this feedback loop makes predicting the wear of gear transmissions very costly and unrealistic. Therefore, the authors of [13] developed such a model based on the analysis of vibration signals. The feature of the model is that the worn profile at different stages of operation is returned into the model as a new profile error, i.e., a deviation from the involute curve. Due to this update, the model allows to predict the remaining service life of the gear transmission. Study [14] is a logical continuation of the research presented in [13]. The model developed in [13] was refined and integrated with two tribological models to assess wear depth and pitting density. The approach proposed by the authors allows interaction between the dynamic model and wear models, enabling continuous modeling of tooth profile changes. It should be noted that real profiles of worn teeth were obtained by the authors after wear tests without lubrication. These profiles significantly differ from the theoretical ones presented in [6, 7]. Thus, real profiles show nonzero wear at the pitch circle, and their shape corresponds to that presented in [15]. This conclusion is also corroborated in [1]. In [16], the models were developed and experimentally validated to predict the wear volume of a polymer gear working in gearing with a steel pinion. The researchers in [17] proposed a progressive methodology for identifying wear products, the application of which helps better understand the nature of wear. Authors of [18] investigated wear and pitting at different stages of transmission operation in the presence of lubrication using optical microscopy and laser scanning confocal microscopy of teeth molds for qualitative and quantitative analysis, respectively. The phenomenon was investigated both at the macro-level (profile) and at the micro-level (surface roughness), providing new insights into the evolution of wear. A comprehensive analysis of factors determining wear in an open gear transmission was conducted in [19]. Based on the analysis, recommendations are provided for extending the service life of the pinion. In particular, it is recommended to replace the worn pinion not individually, but only together with the gear. In [20], the influence of lubricant viscosity and the parameters of its delivery into meshing zone on tooth wear under mixed lubrication conditions, considering surface roughness, was investigated. Authors of the article [21] developed a nonlinear tribo-dynamic model of spur gear drives for investigation of the dynamic behavior of high-speed gear drives under loss-of-lubrication condition and extended to other gear transmissions and rotating mechanisms. The authors of [22] investigated the influence of materials on tooth wear during the operation of a spur gear pair under conditions of lubrication absence. Authors applied a wear prediction model that allows the construction of profiles after wear. The model indicates maximum wear amount at the tooth dedendum and zero wear at the pitch circle. The absence of lubrication can lead not only to abrasive wear but also to the occurrence of scuffing. Articles [23, 24] are dedicated to the investigation of such cases, where wear models are proposed, taking into account the instantaneous temperature in the contact zone [24] and an enhanced model using the *Block theory* [23], as well as the eccentricity of pitch circumference of gears and the edge contact of teeth [24]. The author is modeling the continuous interaction of the profiles of mining machines gear teeth in boundary lubrication conditions depending on contact parameters, as well as the influence of these parameters on the shape of tooth profiles. The profiles obtained using the predictive model were compared to real profiles obtained with a specially developed device for this purpose in [23] and by the method of replicas in [24]. The comparison proves the validity of the created model that can be used to predict gear parameters during operation. It should be noted that the profiles of worn teeth in [23, 24] do not also correspond to the theoretical profiles in [6, 7], but they are close to the profiles presented in [1, 14]. Additionally, it is shown in [24] that the worn profile at a certain stage gets its main shape and retains it thereafter. Only its position changes as it moves away from the original involute profile during the operation. Not only the presence of lubrication and its characteristics but also its condition plays an important role. In [25] the evolution of wear at the presence of quartz sand in the lubricant was investigated. The study was conducted using two approaches: offline (replicas and measurement methods) and online (vibration diagnostics and transmission error method). The molds were examined using an optical microscope, allowing for the assessment of various wear mechanisms from a tribological perspective. Additionally, the replicas were measured in cross-section to evaluate changes in the wear profile. The wear profile in [25] is similar to that in [1, 14, 23, 24]. Furthermore, the preservation of the main wear profile shape, as mentioned in [24], is corroborated by the results presented in [25]. In addition to lubricants, other factors are also actively researched. In [26], an important factor influencing wear, namely tooth deformation during meshing, was investigated. Reliable methods for assessing tooth inclination were proposed. The work [27] is dedicated to assessing the influence of changes in tooth stiffness due to the presence of a crack in the tooth root on the tribological and dynamic characteristics determining wear of a gear pair. In [28], the influence of combined (radial-axial-angular direction) misalignments of gears on wear was investigated, and theoretical values of wear similar to those in [6, 7] were also obtained in [28]. Thus, it can be concluded that there are two approaches to modeling worn profiles: (i) theoretical, as reflected in [6, 7, 22, 28], and (ii) experiment-based, as seen in [1, 14, 15, 23–25].

Research aimed at decreasing wear are highly relevant. Both tribological aspects of wear and the influence of teeth geometry enhancements on wear are being investigated. For the purposes of this study, the geometry enhancements are of particular interest. They variate from traditional profile modification to the application of nonstandard solutions. One of the most widely used ways to reduce wear is the tooth shape modification. The most common are tip, root, and end relieving. Such modifications reduce the sensitivity of meshing to manufacturing and assembly errors. An example of profile modification aimed at reducing vibrations in meshing is presented in [29]. The authors selected modification parameters optimal for reducing dynamic loads. The effectiveness of the chosen parameters was experimentally confirmed. In [30], a comprehensive approach to modification is demonstrated. To eliminate corner contact in meshing, the authors proposed tip, root, and end relieving of spur gears by the pulsed electrochemical finishing (PECF) process. The authors found that PECF also significantly reduces roughness of the processed surfaces. Roughness is an important tribological parameter, and its reduction increases the wear resistance of the teeth. An effective way to reduce roughness during machining is the use of lubricating-cooling fluids with an improved composition [31]. In [32], a method of modification considering gear wear is proposed. For gears with high (HCR) and low (LCR) contact ratios, the authors obtained optimal modification parameters that minimize transmission error for both new gears and gears after one million load cycles. Authors in [33] suggested a nonstandard approach to modifying tooth width, making it different for one-pair and two-pair contact zones. As a result, uniform contact pressures along the tooth height were achieved, and wear amount along the meshing area became almost uniform. Additionally, addendum modification also positively contributes to wear reduction. In the study published in two parts [34, 35], a mathematical model of a profile shifted cylindrical arc-tooth-trace gear with involute teeth is presented. Using the model, meshing characteristics in the mesh-in zone where the gear addendum starts contact with the pinion dedendum were obtained depending on the profile shift coefficient and the angle of tooth trace. The range of the angle was established for which the addendum modification is the most effective. In the study [36], a new type of gearing is proposed for HCR gears. The shape of teeth is completely different from the involute. The work introduces a generalized mathematical model for synthesis of tooth profile by the given path of contact. The model was validated using FEM. Meshing of non-involute profiles [36] provides better meshing characteristics compared to involute profiles. Among the characteristics influencing wear resistance are the following: lower specific sliding, lower contact pressures, and higher thickness of oil film. Additionally, a non-involute tooth shape provides lower bending stresses at the tooth root. In the study presented in [37], a model for synthesis of the basic rack profile directly based on the wear criterion by *I.V. Kragelsky* [15] is proposed. Authors of [38] developed an asymmetric logarithmic spiral gear drive with pure rolling gearing. Another nonstandard gearing for HCR gears is proposed in [39]. While the tooth profiles are involute, the gear and pinion have different tooth thickness at the pitch circle. This allows to achieve a balanced transmission in terms of tooth bending and surface wear. Article [40] by the same authors, focuses on HCR gears. They investigate the influence of gear ratio on wear for nonstandard gears proposed in [39]. The authors of [41] proposed to increase the wear resistance of the standard involute spur gear by creating longitudinal grooves on the flanks using laser processing technology. This significantly reduced the coefficient of friction between the tooth surfaces under fluid lubrication and increased their wear resistance. In [42], another microtexture, in the form of pits with diameters of 30–120 μm, was applied for the same purpose. Such improvement allows for increased wear resistance of the gear pair under dry friction, especially at low speeds and high torque. In [43], a comprehensive modification, i.e., tip and end relieving, is proposed to improve meshing of gears with asymmetric construction and thin rims. The proposed modification provides more uniform load distribution, thus slowing down the wear process and leading to a significant reduction of vibration.

Numerous studies are dedicated to investigation of the negative impact of tooth wear on the performance of gear transmissions as a whole and the increase in dynamic forces in particular. Authors of [44] developed and validated a numerical method designed for simulating double-flank roll testing. Using this method, the variation in the center distance during double-flank roll testing of gears with worn tooth profiles was investigated, and an attempt was made to develop an algorithm for solving the inverse problem. The goal of the latter was to predict the geometric parameters of teeth based on the presence of random fluctuations in the center distance. In the study [1], the influence of profile distortions on the dynamics of spur gears due to wear was investigated. Profiles corresponding to operating times of 3000 and 12,000 h were obtained for gears operating under similar conditions, i.e., significant dynamic loads and the presence of abrasive particles in the lubricant. Real profiles of worn teeth of these gears were obtained using the method of replicas. Based on the developed dynamic model, dynamic coefficient values within one angular pitch were determined. Additionally, the amplitudes of gear ratio fluctuations were identified. It was justified that the changes in profiles due to wear need to be taken into account when calculating the durability of gears. In a later work by *V. Onishchenko* [24], the influence of profile wear on the performance of the transmission at various stages of its operation, from the beginning of operation to 12,000 h, was investigated. The evolution of gear ratio fluctuations within the angular pitch was determined, and the sliding and rolling velocities in meshing, contact pressures, and dynamic coefficient were calculated. The study confirmed the findings of [1] that teeth wear is nonuniform along their height and exceeds manufacturing errors by the order of magnitude, and the intensity of the wear process varies throughout the service life. The study presented in [45] provides an assessment of the influence of spur gears teeth wear on the mesh stiffness, considering progressive wear. The authors found that as the teeth wear increase, the time-varying mesh stiffness significantly decreases, especially in two-pair meshing. Moreover, the reduction in stiffness is nearly proportional to the maximum tooth wear within the range of up to 500,000 load cycles. In [46], a dynamic model was developed that considers not only tooth wear but also the temperature in the tooth contact zone. The model enables the consideration of the influence of these factors on time-varying mesh stiffness. It was found that temperature, along with wear, contributes to the reduction in mesh stiffness. Typically, gears experience more than one type of damage. That is why, in the studies like [23, 24, 46] and other, temperature in the contact zone is considered along with wear as a factor capable of affecting wear. Similarly, in the work [47], the combined action of wear and fatigue surface damage is studied. The authors evaluated the influence of wear on the fatigue crack growth rate from its initiation to the detachment of a metal fragment from the surface, i.e., to the formation of a pit. The service life of the contact point on the pinion dedendum is reduced by 98% under the combined effects of wear and fatigue compared to fatigue alone.

As the presented literature review shows, there are numerous models that take into account wear and enable the assessment of its impact on the performance of gear transmissions. A significant obstacle hindering the development of a unified model for gear wear is the need to take into account the real operating conditions of specific machines. In most cases, these conditions are quite different. It leads to significant diversity in models that take into account different operating conditions: speed, load, short-term overloads, lubrication (presence or absence of lubricant, its cleanliness, and method of lubrication), etc.

3.3 Wear of Tooth Flanks

3.3.1 Experimental Gears

As it was noted in Sect. 3.2, the shape of the worn profile is formed in transmissions at a certain stage of operation and remains unchanged until the onset of the limit state. Additionally, the profile shape does not depend on the module; therefore, all subsequent parameters are related to the module of $m = 1$ mm.

The transmissions were investigated with the following parameters: profile angle is 20°; the number of teeth of the pinion is $z_1 = 15$; the number of teeth of the gear is $z_2 = 68$; the profile shift coefficient of the pinion is $x_1 = 0.8$; the profile shift coefficient of the gear is $x_2 = 0.424$; the tooth width is $b = 12.73$.

Table 3.1 Chemical composition (%) of steels used in the manufacturing of the investigated gears

	Chemical composition, % by mass according to [48]							
Steel grade	C	Si	Mn	Cr	Ni	P	S	Cu
20KhN3A (pinion)	0.17–0.24	0.17–0.37	0.3–0.6	0.6–0.9	2.75–3.15	≤0.025	≤0.025	≤0.03
45KhN (gear)	0.41–0.49	0.17–0.37	0.5–0.8	0.45–0.75	1,0–1,4	≤0.035	≤0.035	≤0.03

In the studied transmissions, the circumferential speed and the load are close to their limit values, the overloading, i.e., short-term torque variations, reach 100% of the nominal value. The operating mode is reversible. The service life of the gear pair until the onset of the limit state is approximately 12,000 h, which corresponds to about 5 years of operation or about $N_{c1} = 8 \cdot 10^8$ load cycles for the pinion or about $N_{c2} = 1.77 \cdot 10^8$ load cycles for the gear. The transmissions belong to the encased class, but due to hard operating conditions, the lubricant of the experimental transmissions at different stages of operation contains up to 7% of abrasive particles with an average size of more than 0.1 mm (sand, coal dust, clay), and about 20–35% of water. As a result, the main type of failure is wear (appears in 70–85% of pinions and 55–75% of gears). A common type of failure is fatigue fracture of teeth (appears as cracks and chips in 15–20% of pinions and 20–40% of gears), while pitting of flanks is almost not common.

The teeth have a high surface hardness. The pinions of the experimental transmissions were made of chromium-nickel steel 20KhN3A with subsequent carburizing and full contour quenching of the teeth. Carburizing was performed at the temperature of 950 °C with air cooling to 600 °C. This was followed by a high tempering at 600–650 °C in two stages, each is lasting 6 hours, after which quenching was carried out at 800 °C with oil cooling and low tempering at 160 °C for 2 h. The surface hardness of the pinion teeth was at least HRC 59, and the core hardness was HRC 30–49. Gear blanks were forged, then machined, teeth were cut, the aforementioned heat treatment was performed, and finally, the teeth were finished by grinding process.

The gears were made of steel 45KhN with contour quenching of the teeth. The surface hardness of the teeth was HRC 47–53, and the core hardness was HRC 19–36. Gear blanks were manufactured by stamping, then normalized at the temperature of 860–880 °C, and subjected to rough machining. Subsequently, quenching (840–870 °C, oil cooling) with tempering (450–480 °C, 5 h) were performed. After this final machining and teeth cutting, induction hardening of the teeth surface (880 °C, water cooling), low tempering (220 °C, 3 h) and final teeth grinding were done.

In Table 3.1, the chemical composition of the steels is presented, while in Table 3.2, their mechanical properties according to GOST 4543-71 [48] are given.

GOST 4543-71 is a former Soviet standard applicable at the time of the tests. The gears tolerance corresponded to the class 8-A accuracy according to the standards of kinematic accuracy and smoothness of operation according to GOST 1643-72 [49]

Table 3.2 The minimum guaranteed mechanical properties of the steels used in the manufacturing of the investigated gears

	Heat treatment condition								
	Quenching		Tempering		Mechanical properties at 20 °C				
	Quenching temperature		Tempering temperature		Proof strength	Tensile strength	Elongation after fracture	Reduction in area	Impact energy
Steel grade	T, °C	Type of cooling	T, °C	Type of cooling	$R_{p0.2}$, MPa	R_m, MPa	A, %	Z, %	KCU, J/cm^2
20KhN3A (pinion)	820	Oil	500	Water or oil	735	930	12	55	108
45KhN (gear)	820	Water or oil	530	Water or oil	835	1030	10	45	69

(also a former Soviet standard, which was valid in Ukraine when the tests were carried out).

3.3.2 Research Methods

To assess the influence of wear on the gear ratio, data received from measuring the teeth (micrometry) and obtaining their profiles using the method of replicas (profilography) was used.

According to [50], micrometry belongs to the differential methods of size determination, its required accuracy is 0.01–0.001 mm. The measuring tool (gear tooth caliper) has a measurement accuracy of 0.02 mm. Therefore, two measurement methods as mentioned above were used.

Both procedures were carried out during maintenance at different stages of the gear transmission operation. Tests conducted under the normal operating conditions allow comprehensive conclusions about the operational reliability of the transmissions, accurately determining the main factors influencing their performance, and obtaining data that does not require further verification or clarification. To obtain reliable results and eliminate the influence of random factors, the duration of operation testing was set to be at least a half of the designated service life of 12,000 h.

To determine the number of objects to be tested, it was necessary to establish the distribution law of the controlled variables. To determine the distribution law of the tooth thickness, measurements of four diametrically arranged teeth were taken on 20–28 pairs of gears and pinions, assuming a hypothesis of normal distribution. Then, the statistical relative frequencies were determined [51]:

$$h_1 = \frac{n_1}{n};\ \ h_2 = \frac{n_1}{n} + \frac{n_2}{n};\ \ h_3 = \frac{n_1}{n} + \frac{n_2}{n} + \frac{n_3}{n} \tag{3.1}$$

where:

n_i—is the number of measurements in the i-th interval;
h_i—is the statistical relative frequencies in the i-th interval;
n—is the total number of measurements.

The values of the statistical relative frequencies, calculated according to the Eq. (3.1), were plotted on a probability paper in accordance with the requirements of DSTU ISO 5479:2009 [52] (Ukrainian standard, which is identical to ISO 5479: 1997). To illustrate the approach, the statistical relative frequencies obtained during the measurement of new pinions are shown in Fig. 3.1a, and the statistical relative frequencies obtained during the measurement of new gears are shown in Fig. 3.1b (the measured thickness δ_0 is related to the module $m = 1$ mm).

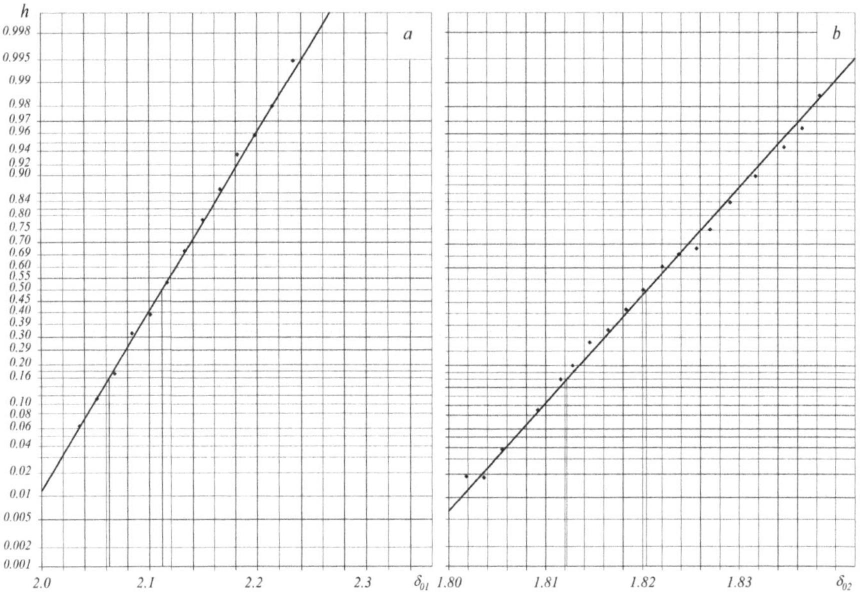

Fig. 3.1 The statistical relative frequencies based on the measurement results of pinions (**a**) and gears (**b**)

The points are located on a straight line (with a small dispersion), indicating a normal distribution of the tooth thickness values and confirming the correctness of the hypothesis.

At the normal distribution law, when the probability density function is given by:

$$\varphi(\delta) = \frac{1}{\sqrt{2\pi}\sigma}\exp\left\{-\frac{\left(\delta-\overline{\delta}\right)^2}{2\sigma^2}\right\},$$

where:

δ—is the independent variable (in our case, δ is the thickness of the tooth);
$\overline{\delta}$—is the mean value of the normal distribution;
σ—is the standard deviation,

the number of objects that need to be tested, with an unknown σ, is determined according to [53] by the formula:

$$n = \left(\frac{u_q}{\xi}\right)^2 \ln\left(\nu^2+1\right)\left[1+\frac{\ln\left(\nu^2+1\right)}{2}\right], \tag{3.2}$$

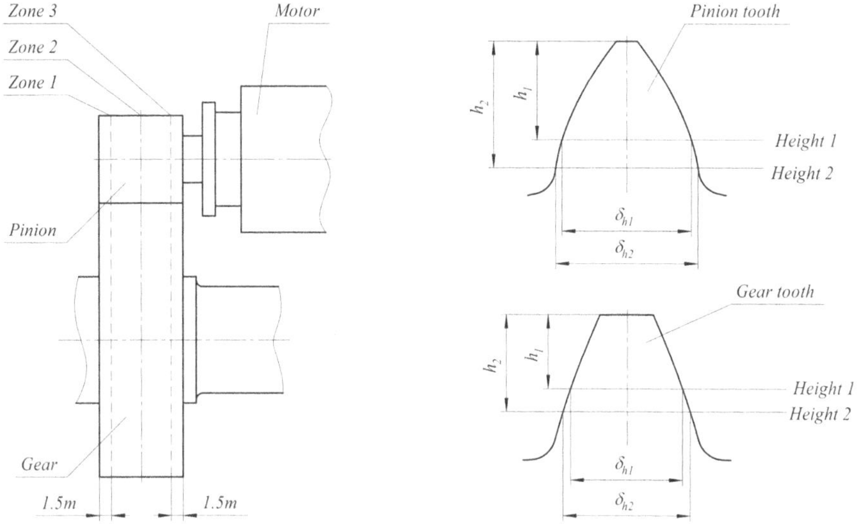

Fig. 3.2 Diagram of the pinion and gear teeth measurement

where:

ξ—is the relative error of the mean value $\overline{\delta}$;
ν—is the coefficient of variation;
u_q—is the quantile of the normal distribution for a confidence probability q.

Let us determine the number n of gear pairs that need to be tested at $\xi = 0.1$ (for basic components in serial production according to [53]) and $\nu = 0.3$ (according to [53], $\nu = 0.2$–0.4 for gear transmissions), using the Eq. (3.2). The quantile $u_q = 1.28$ is chosen from the table [53] for the confidence probability $q = 0.9$ (for basic components in serial production according to [53]). Then, $n = 15$, i.e., it is necessary to measure at least 15 pinions or gears with the same operating time. An operational testing map was developed for information accumulation, generalization and comparison of obtained data, as well as for the assessment of the influence of various factors on the reliability of the studied gears. The map includes the tooth thickness measurement of pinions and gears in three zones at two heights, when a certain operating time is reached (see Fig. 3.2), as well as the measurement of the profile deviation from the involute.

For example, Fig. 3.1 illustrates the processing of measurement results for new pinions and gears in zone 1 at the height of h_2.

The assessment of deviation from the involute profile and measurements that are necessary for obtaining the actual profiles of worn teeth were carried out by the replicas taken from the teeth of pinions and gears. AST-T plastic (self-curing thermal-resisting acrylic plastic for technical purposes) was used to make the replicas. The plastic was produced from powder (emulsion polymethyl methacrylate with the addition of pigment and benzoyl peroxide in an amount of 1%) and liquid (a solution of dimethylaniline in methyl methacrylate). Before its use, the powder

Fig. 3.3 Measurement of profile deviations on the BMI-1 microscope

(2 parts by weight) was mixed with the liquid (1 part by weight). Initially, the mixture turns into a cream-like mass, then into a dough-like one, and then it solidifies. The mixture was used before it became solid. For this purpose, after 15–20 min, it was placed in special steel molds installed on the teeth, kept for 20–30 min at the temperature of 20 ± 5 °C. After that the solidified replica of the tooth was removed. To facilitate the removal of the hardened replica from the gear tooth and mold surface, a thin layer of machine oil was applied before filling with the mixture. The finished replica had a hardness of HB 13-19, tensile strength of 45–50 MPa, and bending strength of 80–120 MPa.

The accuracy of the measurements, taken with a gear tooth caliper, was verified by the replicas using a big tool microscope BMI-1 (see Fig. 3.3a).

Additional measurements were conducted at various heights of the tooth. For measuring and profiling, a 30× magnification with a standard eyepiece and an additional objective lens was used (see Fig. 3.3b). The accuracy of measurement conducted using the standard micrometric screw pairs of the microscope (see Fig. 3.3c) is ±0.003 mm.

As a result of measurements carried out at real operating conditions, series of tooth thickness values δ were obtained. They corresponded to the values of the argument, i.e., the operating time of the gears L, which was converted to millions of cycles, starting from production $L_0 = 0$ to an operating time $L = N_{c1}$, which is not less than a half of the design life $L_{\max} = N_{c1\max} \approx 800$ millions of cycles, i.e., $L_{\max} = 12,000$ h related to the pinion.

To obtain, a theoretical relationship (linear or nonlinear) that would closely approximate the true functional dependency of the tooth thickness on the operating time $\delta = f(L)$, based on the experimental data points, the least-squares method [54] was used.

Assuming that the tooth thickness has a linear dependence on the operating time, based on the arrangement of experimental points, we have:

$$\delta(L) = \delta_0 + aL, \tag{3.3}$$

and applying the least-squares method, we determine the parameters of Eq. (3.3) as follows:

$$a = \frac{n\sum_{i=1}^{n}\delta_i L_i - \sum_{i=1}^{n}\delta_i \cdot \sum_{i=1}^{n} L_i}{n\sum_{i=1}^{n} L_i^2 - \left(\sum_{i=1}^{n} L_i\right)^2} \quad \delta_0 = \frac{1}{n}\left(\sum_{i=1}^{n}\delta_i - a\sum_{i=1}^{n} L_i\right). \tag{3.4}$$

The measure of significance of the influence of a particular factor on the effective feature is determined by the indicators of the strength of relationship. To assess the strength of relationship between two dependent series of δ_i and L_i, let us define the linear correlation coefficient. For the relationship of Eq. (3.3), it is calculated using the formula [55]:

$$r = \frac{n\sum_{i=1}^{n}\delta_i L_i - \sum_{i=1}^{n}\delta_i \cdot \sum_{i=1}^{n} L_i}{\sqrt{\left[n\sum_{i=1}^{n}\delta_i^2 - \left(\sum_{i=1}^{n}\delta_i\right)^2\right]\left[n\sum_{i=1}^{n} L_i^2 - \left(\sum_{i=1}^{n} L_i\right)^2\right]}}. \tag{3.5}$$

Assuming that the thickness of the tooth has a quadratic dependence on the operating time, represented as:

$$\delta(L) = \delta_0 + bL + cL^2 \tag{3.6}$$

and applying the least-squares method, we determine the parameters b and c of the Eq. (3.6), taking into account that the constant term in equation δ_0 is equal to the initial tooth thickness measured after manufacturing:

$$c = \frac{\sum_{i=1}^{n} L_i \cdot \sum_{i=1}^{n} \delta_i L_i - \delta_0 \left(\sum_{i=1}^{n} L_i \right)^2 - \sum_{i=1}^{n} L_i^2 \cdot \sum_{i=1}^{n} \delta_i + \sum_{i=1}^{n} L_i^2 \cdot n\delta_0}{\left(\sum_{i=1}^{n} L_i^2 \right)^2 - \sum_{i=1}^{n} L_i \cdot \sum_{i=1}^{n} L_i^3};$$
$$b = \frac{\sum_{i=1}^{n} \delta_i - n\delta_0 - c \sum_{i=1}^{n} L_i^2}{\sum_{i=1}^{n} L_i} \tag{3.7}$$

In the case of nonlinear relationship, the usage of the linear correlation coefficient as a measure of the strength of relationship is not always applicable. Therefore, let us use the relationship [56] to determine the correlation ratio η, which is used when grouping the data:

$$\eta = \sqrt{\frac{\sigma^2}{\sigma_\delta^2}}, \tag{3.8}$$

where:

σ^2—is the external variance;
σ_δ^2—is the total variance.

The external variance is defined as follows:

$$\sigma^2 = \sigma_\delta^2 - \sigma_{\delta(L)}^2,$$

where $\sigma_{\delta(L)}^2$ is the within-group variance:

$$\sigma_{\delta(L)}^2 = \frac{\sum_{i=1}^{n} [\delta_i - \delta_i(L)]^2}{n},$$

and the total variance is as follows:

$$\sigma_\delta^2 = \frac{\sum_{i=1}^{n} \delta_i^2}{n} - \bar{\delta}^2.$$

3.3.3 Results of Wear Study

Based on the statistical data taken during operational testing, functional dependencies of tooth thickness on operating time were determined, using the calculation methodology outlined in Sect. 3.3.2. The actual tooth thickness values of new pinions and gears were used as initial data for the calculations made at two heights and in three zones (Fig. 3.2). Measurements were taken on 20–28 manufactured pinions and gears. The number of measurements significantly exceeded the minimum required quantity $n = 15$ determined in Sect. 3.3.2, indicating sufficiency of the sample size. Measurements were carried out on four teeth evenly distributed across the diameter. It was established earlier, that the distribution law of tooth thickness values is normal. The empirical parameters of the normal distribution law are the expectation $M(\delta)$ of the random variable, representing tooth thickness, and its standard deviation $\sigma(\delta)$, which are determined by the following formulas:

$$M(\delta) = \bar{\delta} = \frac{1}{n}\sum_{i=1}^{n}\delta_i; \quad \sigma(\delta) = \sigma = \sqrt{\frac{1}{n-1}\sum_{i=1}^{n}\left(\delta_i - \bar{\delta}\right)^2}. \tag{3.9}$$

The results of the main parameters calculation, i.e., the mean and standard deviation, based on the Eq. (3.9), are summarized in Table 3.3 (values are related to the module $m = 1$ mm).

Measurements of the same pinions and gears were taken during various stages of operation for the gear transmissions, also on four teeth evenly distributed across the diameter. The regression coefficients and correlation indicators, which characterize the significance of the influence of operation on the wear value were determined:

- according to Eqs. (3.4) and (3.5) for the linear dependency (3.3), and
- according to Eqs. (3.7) and (3.8) for the parabolic dependency (3.6).

The results of calculation of the regression coefficients according to Eqs. (3.3) and (3.6) are presented in Tables 3.4 and 3.5, respectively.

Negative values of correlation coefficients in Table 3.4 indicate a reverse relationship between the studied parameters, i.e., the tooth thickness decrease with increasing of the operating time. A comparison of the correlation coefficients in Tables 3.4 and 3.5 shows that the strength of relationship when using regression (3.6) is higher than when using regression (3.3). Linear regression provides for a

Table 3.3 The values of the measured parameters for new gears

	Height		$M(\delta)$ in the zone			$\sigma(\delta)$ in the zone		
Parameters for	Designation	Value	1	2	3	1	2	3
Pinion	h_1	1.345	1.882	1.877	1.869	0.00727	0.00691	0.00655
	h_2	1.727	2.103	2.099	2.086	0.00564	0.00518	0.00709
Gear	h_1	1.015	1.605	1.613	1.615	0.01	0.01091	0.01182
	h_2	1.327	1.818	1.818	1.832	0.00909	0.01	0.01045

Table 3.4 The results of calculation of the parameters of linear regression

	δ_0 in the zone			a in the zone			η in the zone		
Height	1	2	3	1	2	3	1	2	3
Pinion $h_1 = 1.345$	1.917	1.913	1.907	−0.0002145	−0.0002303	−0.00026875	−0.876	−0.86	−0.909
Pinion $h_2 = 1.727$	2.122	2.117	2.107	−0.00016258	−0.00017655	−0.00018494	−0.912	−0.922	−0.93
Gear $h_1 = 1.015$	1.631	1.634	1.634	−0.0009388	−0.0010758	−0.00119117	−0.975	−0.982	−0.988
Gear $h_2 = 1.327$	1.84	1.838	1.845	−0.00061635	−0.00067524	−0.00083193	−0.905	−0.937	−0.943

Table 3.5 The results of calculation of the parameters of quadratic regression

	b in the zone			c in the zone			η in the zone		
Height	1	2	3	1	2	3	1	2	3
Pinion $h_1 = 1.345$	0.00012886	0.00012527	0.00010589	−0.00000053	−0.00000055	−0.00000058	0.991	0.984	0.995
Pinion $h_2 = 1.727$	0.00002527	0.0000044	0.00001501	−0.00000029	−0.00000028	−0.00000031	0.99	0.99	0.996
Gear $h_1 = 1.015$	−0.00020362	−0.00046543	−0.00061639	−0.00000335	−0.00000278	−0.00000262	0.992	0.99	0.993
Gear $h_2 = 1.327$	0.00002264	−0.00010023	−0.00045423	−0.00000291	−0.00000262	−0.00000172	0.953	0.971	0.963

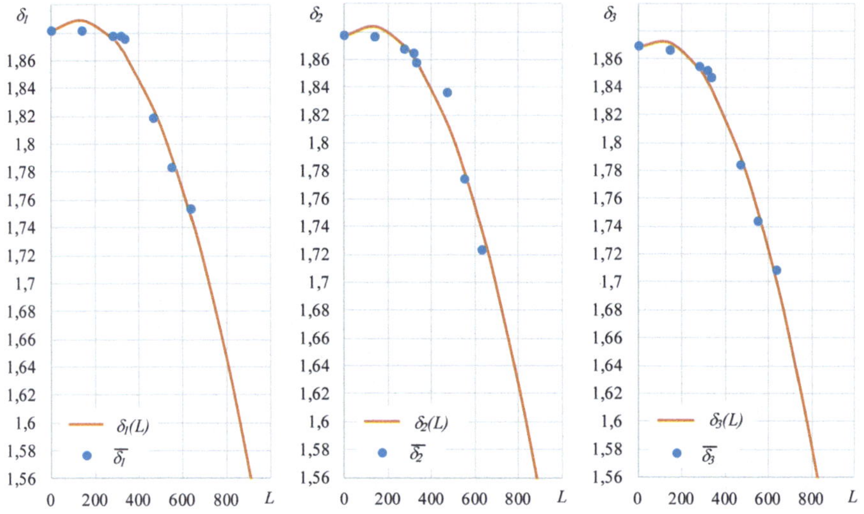

Fig. 3.4 Pinion tooth thickness at the height h_1 depending on its operating time

correlation coefficient range of −0.86 to −0.988, while quadratic regression provides for a range of 0.953 to 0.996. Obviously, for describing the change in thickness during operation up to the value of $L_{\max} = N_{c1\max} \approx 800 \cdot 10^6$ load cycles, the linear model is less suitable, hence the quadratic model was applied thereafter.

The results obtained by regression (3.6) are more visually represented by the graphs showing the dependency of tooth thickness on the operating time, as presented in Figs. 3.4, 3.5, 3.6 and 3.7.

The notations in Figs. 3.4, 3.5, 3.6 and 3.7 are as follows: $\overline{\delta}_1$, $\overline{\delta}_2$, and $\overline{\delta}_3$ represent the mean thicknesses of the teeth in zones 1, 2, and 3, respectively, while $\delta_1(L)$, $\delta_2(L)$, and $\delta_3(L)$ represent the functional dependencies of the tooth thicknesses in zones 1, 2, and 3 on the operating time, calculated using Eq. (3.6) with coefficients from Table 3.5. The operating time in millions of load cycles for the pinion $L = N_{c1}$ and the gear $L = N_{c2}$, respectively, is plotted on the x-axis. Obviously, the connection between the operating times of the pinion and the gear is as follows:

$$N_{c1} = iN_{c2},$$

where i is the gear ratio: $i = z_2/z_1 = 68/15 = 4.53$.

Thus, as a result of the calculations for the investigated gears, functional dependencies of the tooth thickness on the operating time and the correlation coefficients were obtained. Based on the conducted calculations and Figs. 3.4, 3.5, 3.6 and 3.7, it can be concluded that the best theoretical approximations to the true functional dependence of the tooth thickness on the operating time are provided by the parabolic dependence expressed by Eq. (3.6), where δ_0 represents the initial thickness of the tooth; L represents the operating time in millions of cycles for both the

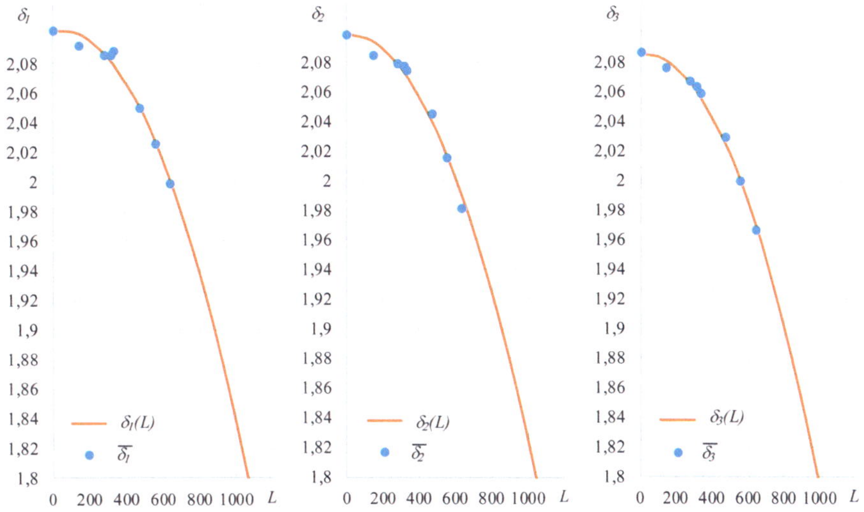

Fig. 3.5 Pinion tooth thickness at the height h_2 depending on its operating time

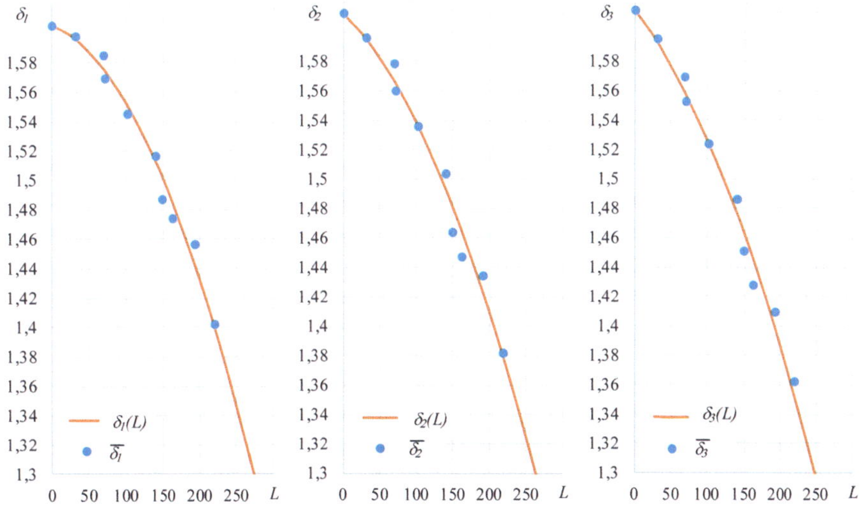

Fig. 3.6 Gear tooth thickness at the height h_1 depending on its operating time

pinion and the gear; *b* represents the specific wear intensity of the teeth, per million of cycles; *c* represents the coefficient accounting the increasing of wear due to the increase of dynamic forces in the transmission. Moreover, the correlation coefficient, ranging from 0.953 to 0.996 (Table 3.5), indicates the significance of the functional dependence of the gear teeth thickness on their operating time. Obviously, that linear regression can only provide a close relationship at the initial stage of gear operation (operating time is approximately $300 \cdot 10^6$ load cycles for the pinion or about $70 \cdot 10^6$ load cycles for the gear).

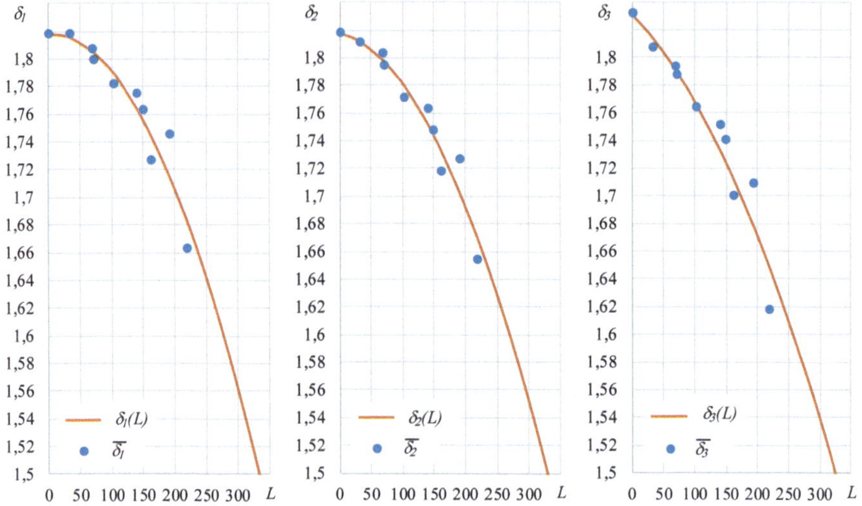

Fig. 3.7 Gear tooth thickness at the height h_2 depending on its operating time

Furthermore, the results presented in Table 3.5 and Figs. 3.4, 3.5, 3.6 and 3.7 show that the greatest wear is observed on the teeth of the pinion at height 1 (corresponds to the measurement height of the constant chord) in zone 3 (according to Fig. 3.2). The maximum wear in zone 3 can be explained by the concentration of load within a small segment of the tooth length during the operation at misalignment of axes. By relying on the data from zone 3, the maximum allowable operating time can be estimated. For the gears under consideration, according to the current regulatory documentation, the wear of the teeth should not exceed 0.3 m. By solving the regression Eq. (3.6) for the operating time at the minimally acceptable tooth thickness $\delta_{\min} = \delta_0 - 0, 3$, the predictive allowable number of cycles for the pinion and gear can be determined for the transmissions described in Sect. 3.3.1.

3.4 Influence of Wear on Gear Ratio

According to the methodology outlined in Sect. 3.3.2, the replicas of the pinions and gears teeth with different operating times were made in three zones (according to Fig. 3.2). Measurements were taken using the measuring microscope BMI-1, and actual tooth profiles of the gears and pinions were constructed for the number of load cycles as follows: $N_{c1} = 172 \cdot 10^6$ ($N_{c2} = 38 \cdot 10^6$); $N_{c1} = 322 \cdot 10^6$ ($N_{c2} = 71 \cdot 10^6$); $N_{c1} = 635 \cdot 10^6$ ($N_{c2} = 140 \cdot 10^6$).

To study the nature of the tooth profiles distortion of the pinions and gears during operational testing, the actual tooth profiles were compared to the theoretical ones for the pinions (Fig. 3.8) and the gears (Fig. 3.9).

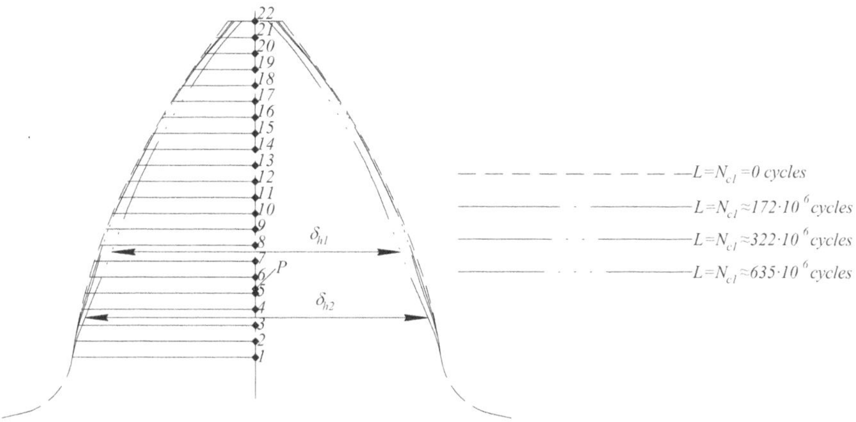

Fig. 3.8 The actual pinions tooth profiles at different stages of operation

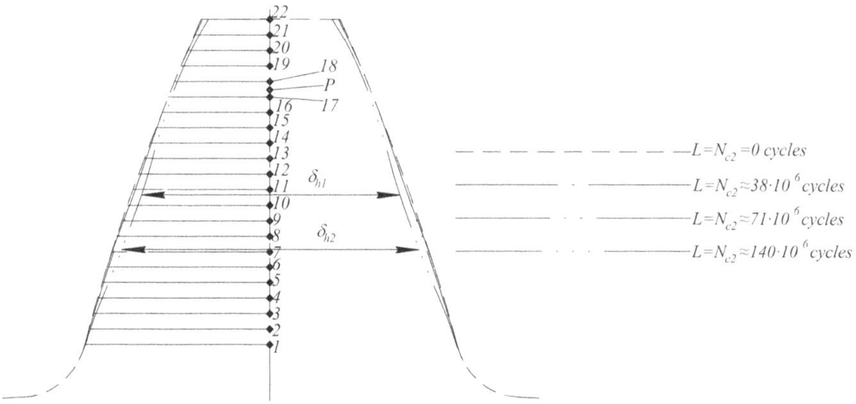

Fig. 3.9 The actual gears tooth profiles at different stages of operation

Comparisons demonstrate that after operating for $N_{c1} = 172 \cdot 10^6$ ($N_{c2} = 38 \cdot 10^6$), a slight wear appears on the tooth addendum and dedendum of both gears and pinions. As the operation continues up to $N_{c1} = 322 \cdot 10^6$ ($N_{c2} = 71 \cdot 10^6$), the profile wear occurs relatively uniformly. For the pinions, it closely resembles the theoretical wear, while for the gears, a concave profile with a larger radius appears at the tooth dedendum. With further operation up to $N_{c1} = 635 \cdot 10^6$ ($N_{c2} = 140 \cdot 10^6$), a sharp increase in wear is observed. The pinions tooth addendum is worn intensively, approaching the tip pointed shape. The maximum profile wear occurs at the pinions tooth dedendum near the constant chord, where a concave profile appears, smoothly transitioning into a convex profile on the worn tooth addendum. The gears tooth addendum experience minimal wear, with the maximum wear occurring near the constant chord, resulting also in a concave profile at the tooth dedendum that smoothly transitions into a convex profile on the tooth addendum.

The wear profile shape at later stages of gear operation corresponds to that described in the works [1, 14, 15, 23–25]. Furthermore, as noted above, although wear increases at later stages of operation, the shape of the worn tooth profile remains unchanged thereafter. Therefore, based on this experience, it can be assumed that on the pinions and gears, as the operation exceeds $N_{c1} = 635 \cdot 10^6$ ($N_{c2} = 140 \cdot 10^6$), the profile shape will also remain unchanged. A constant profile shape like this can be considered as the main wear profile.

For modeling the meshing of an extremely worn pinion with a gear at the corresponding stage of operation, actual tooth profiles are required. Under the absence of such profiles, they can be predicted. Two previously mentioned facts were considered for obtaining the predicted profile. First, after the operating time exceeds $N_{c1} = 635 \cdot 10^6$ ($N_{c2} = 140 \cdot 10^6$), the curves outlining the profiles of the pinion and gear teeth shift towards the axis of tooth symmetry as the operating time increases. Second, for the gears under consideration, the maximum allowable deviation of the main wear profile at the measurement height $h_1 = 1.345$m is known. It is determined based on the fact that for the examined gears, according to the regulatory documentation, the criterion for the limit state is a reduction in the pinion tooth thickness by 0.3 m, i.e., the maximum total wear on both sides of the tooth is $\Delta^{\Sigma}_{h1p} = 0.3$. Thus, realizing the shape of the main wear profile (Figs. 3.8 and 3.9) and its position relative to the axis of tooth symmetry, it is possible to construct the predicted shape of an extremely worn pinion tooth.

The operating time $N_{c1\max}$ corresponding to maximum allowed wear can be determined from the regression Eq. (3.6) by substituting the regression coefficients from Table 3.5, corresponding to the height $h_1 = 1.345$m in zone 3:

$$\delta_{\min} = \delta_0 + bN_{c1\max} + cN^2_{c1\max}$$

Realizing that for maximum wear $\delta_0 - \delta_{\min} = 0.3$, we have the equation

$$cN^2_{c1\max} + bN_{c1\max} + 0.3 = 0.$$

Or, taking into account the values of the regression coefficients, we have:

$$-0.00000058N^2_{c1\max} + 0.00010589N_{c1\max} + 0.3 = 0.$$

This quadratic equation has one positive solution: $N_{c1\max} = 816$ millions of cycles. Then, the thickness of the gear tooth at height $h_1 = 1.015$m at the same stage of operation at $N_{c2\max} = N_{c1\max}/i = 816/4.53 = 180$ millions of cycles will be:

$$\delta = 1.615 - 0.00061639 \cdot 180 - 0.00000262 \cdot 180^2 = 1.419$$

Thus, the value of wear of the gear tooth at the measurement height $h_1 = 1.015$m is $\Delta^{\Sigma}_{h1g} = 1.615 - 1.419 = 0.196$.

Similarly, solving the equations for the height h_2 at the maximum operating time of the pinion $N_{c1\max} = 816 \cdot 10^6$ and the gear $N_{c2\max} = 180 \cdot 10^6$, we obtain the maximum wear for the pinion $\Delta_{h2p}^{\Sigma} = 0.18$ and gear $\Delta_{h2g}^{\Sigma} = 0.138$ respectively.

As mentioned above, the gear operation mode is reversible, this is why the wear of the tooth flanks is not equal (Figs. 3.8 and 3.9). The maximum wear of the flanks on the more worn sides of the pinion and gear teeth (in Figs. 3.8 and 3.9, these are the left sides of the teeth) at $N_{c1} = 635 \cdot 10^6$ ($N_{c2} = 140 \cdot 10^6$) is about 57% of the maximum total wear on both sides. By distributing $\Delta_{\Sigma 1} = 0.3$ and $\Delta_{\Sigma 2} = 0.196$ between the tooth sides according to the specified ratio, we obtain the predicted tooth profile of the pinion and gear for the transmission in the limit state. And, the maximum wear on the left sides at the height h_1 will be $\Delta_{h1p}^{left} = 0.171$ and $\Delta_{h1g}^{left} = 0.112$ for the pinion and gear, respectively. By performing similar actions for the height h_2, we obtain $\Delta_{h2p}^{left} = 0.1$ and $\Delta_{h2g}^{left} = 0.079$.

Based on the results of profilography, the experimental wear distribution of the pinion and gear teeth along the tooth height was obtained. From practical considerations, it makes sense to model the meshing of the most worn (left) profiles. Therefore, Figs. 3.10 and 3.11 show the deviations of these profiles from the involute.

It should be noted that the wear dependence on the position of the point on the profile, shown in Figs. 3.10 and 3.11, generally corresponds to those presented in works [1, 14, 15, 23–25]. The predicted wear distribution of gear teeth along the height is also shown in Figs. 3.10 and 3.11. It was obtained at every profile point 1–22, using quadratic extrapolation of experimental data for the points, taking into account the previously obtained values of tooth thickness at heights h_1 and h_2. Quadratic extrapolation was chosen based on the fact that the dependence of tooth thickness on the operating time is effectively approximated by a second-degree

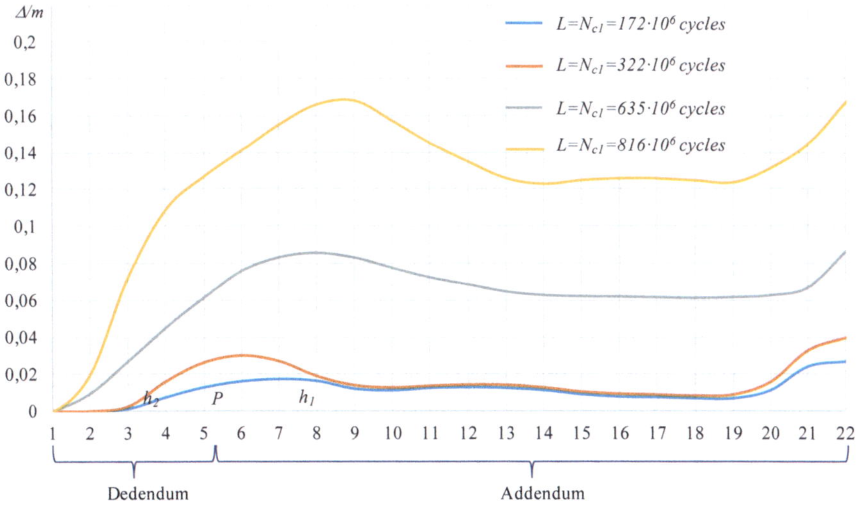

Fig. 3.10 Wear values of pinion teeth along the height (* is predicted)

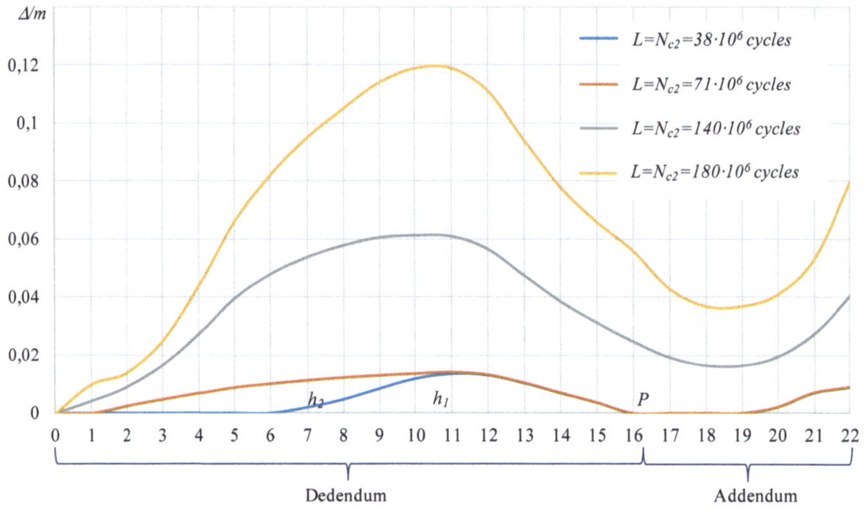

Fig. 3.11 Wear values of gear teeth along the height (* is predicted)

equation, as it was demonstrated in Sect. 3.3.3. For the predicted profile on the addendum near the top and dedendum near the root a correction was performed based on empirical data and the nature of wear distribution along the tooth height, as described in works [1, 14, 15, 23–25]. The profile points were approximated using nonuniform rational *B*-splines.

Worn pinions and gears at various points along the profile height have different radii of curvature. This is why the path of contact of one pair of teeth within the meshing phase (Fig. 3.12) is not already a theoretical straight line (shown as dashed line in Fig. 3.12), but a curve (shown as solid line in Fig. 3.12).

Consequently, the pitch point changes its position at the center line. The unstable position of the pitch point results in uneven rotation of one gear when another one is rotating uniformly, i.e., the transmission of motion occurs with a variable gear ratio. The nature of the gear ratio fluctuations was determined based on the positions of intersection of the common normal to the profiles with the center line. The meshing of one pair of teeth within meshing phase (Fig. 3.12) was considered. The boundary points of the path of contact were determined by the condition of the existence of a common normal to the worn profiles.

It was mentioned in Sect. 3.3.1, that the operating conditions of the transmission are severe. The described conditions, such as overloads up to 100% of the nominal, water and abrasive contamination into the housing, is also common for the shaft support units. The wear of these units lead to the increasing of the center distance. Therefore, the meshing of worn teeth with an increase in the center distance at three levels of $\Delta a_w = +0.1$m; $\Delta a_w = +0.2$m and $\Delta a_w = +0.27$m were studied. The value of $\Delta a_w = +0.27$m is the allowable one according to the regulatory documentation for the transmission. To illustrate the influence of Δa_w on the nature of gear ratio fluctuations, a schematic representation of the meshing of worn teeth at $\Delta a_w = +0.1$m is presented in Fig. 3.13.

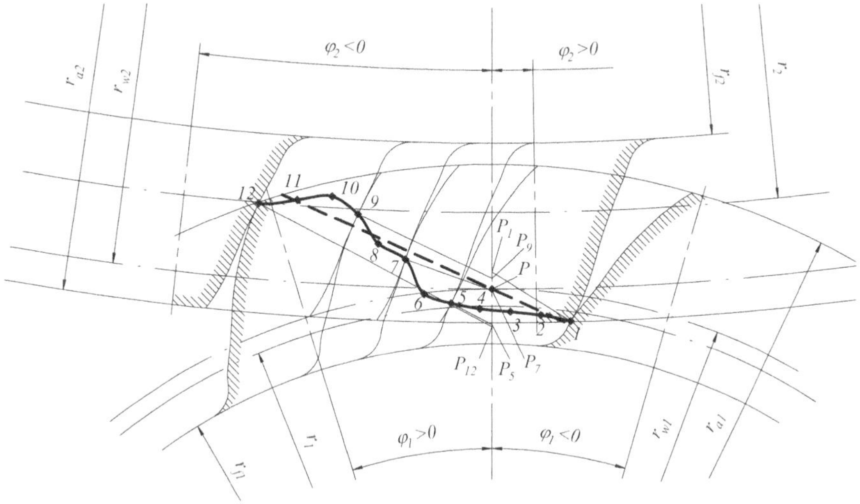

Fig. 3.12 Schematic representation of the meshing of worn gears after $N_{c1} = 635 \cdot 10^6$ ($N_{c2} = 140 \cdot 10^6$) cycles

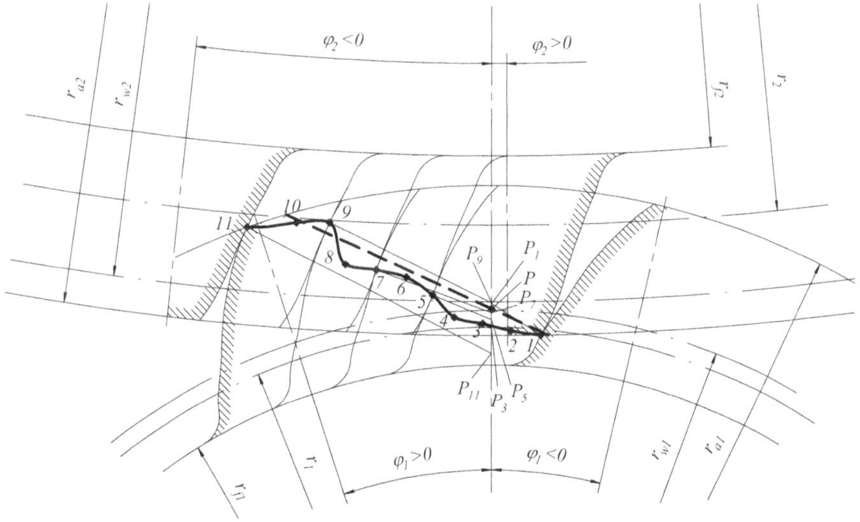

Fig. 3.13 Schematic representation of the meshing of worn gears after $N_{c1} = 635 \cdot 10^6$ ($N_{c2} = 140 \cdot 10^6$) cycles at $\Delta a_w = +0.1\text{m}$

As shown in Fig. 3.13, the greatest deviation of the actual path of contact (solid line) from the theoretical one (dotted line) occurs at the end of the meshing phase, but not in the middle one. This trend persists with further increases in Δa_w. It is reflected in a significant increase in the deviation of the gear ratio Δi at large positive values of the pinion rotation angle ϕ_1. The dependencies of the gear ratio on the gear

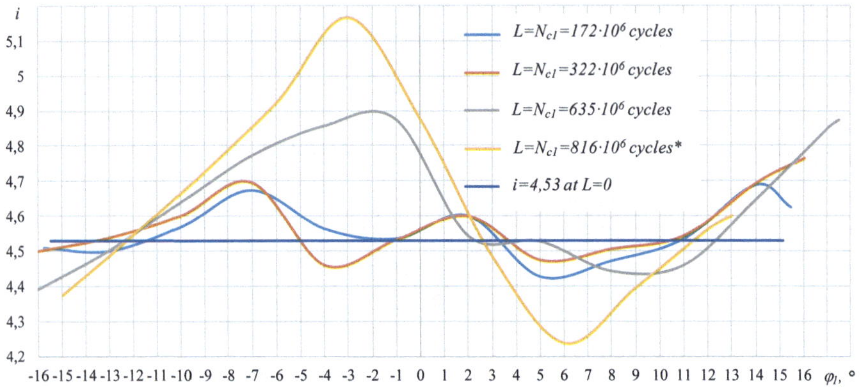

Fig. 3.14 Gear ratio fluctuation in meshing at $\Delta a_w = 0$ (* is predicted values)

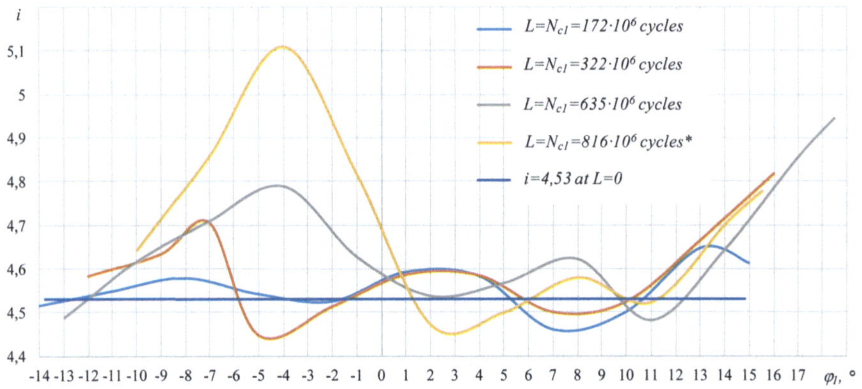

Fig. 3.15 Gear ratio fluctuation in meshing at $\Delta a_w = +0.1$m (* is predicted values)

rotation angle $i(\phi_1)$ were plotted (Figs. 3.14, 3.5, 3.6 and 3.17) to assess the gear ratio fluctuations and the influence of Δa_w on the nature of the fluctuations.

The analysis of Figs. 3.14, 3.15, 3.16, and 3.17 indicates that at the early stage of operation of the experimental transmissions ($N_{c1} = 172 \cdot 10^6$ or $N_{c2} = 38 \cdot 10^6$), the gear ratio during the meshing of a pair of teeth changes insignificantly. This is due to small deviations of the profiles of worn pinion and gear teeth from the original involute profile. It should come as no surprise that along with increasing of the operating time up to $N_{c1} = 635 \cdot 10^6$ ($N_{c2} = 140 \cdot 10^6$), these fluctuations became more significant. However, the case of maximum allowed wear is of practical interest, which occurs on the pinion much earlier than on the gear after $N_{c1} = 816 \cdot 10^6$ ($N_{c2} = 180 \cdot 10^6$). In this case, the gear ratio deviates from the nominal value of $i = 4.53$ by 6.5% on the lower side and by 14% on the higher side.

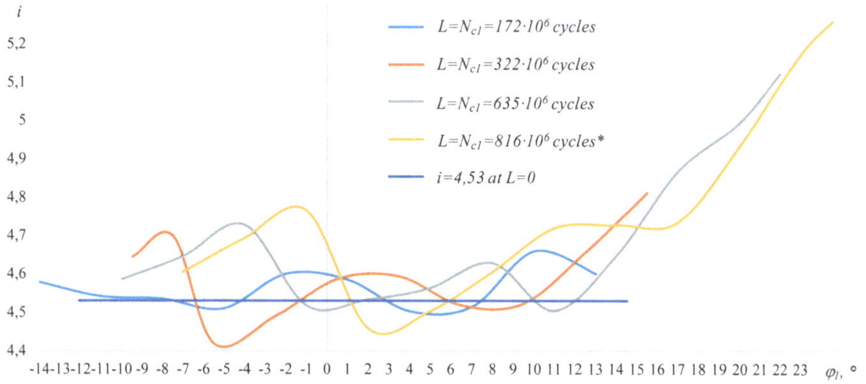

Fig. 3.16 Gear ratio fluctuation in meshing at $\Delta a_w = +0.2$m (* is predicted value)

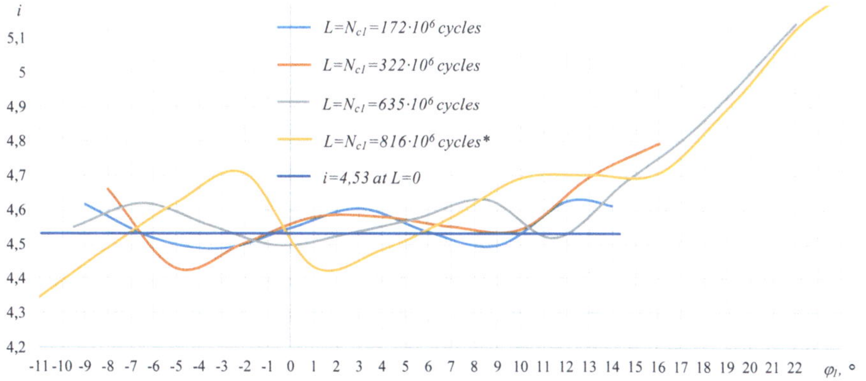

Fig. 3.17 Gear ratio fluctuation in meshing at $\Delta a_w = +0.27$m (* is predicted value)

The change of Δa_w does not have an unequivocal impact on the fluctuations of the gear ratio. At low teeth wear values, an increase of the center distance leads to the displacement of the oscillation phases of $i(\phi_1)$ function. As wear progresses, the slope of the curves $i(\phi_1)$ changes. The main effect of increasing the center distance is the displacement of the maximum value of instantaneous gear ratio towards the boundaries of the meshing phase. At $N_{c1} = 816 \cdot 10^6$ ($N_{c2} = 180 \cdot 10^6$), its maximum deviation increases to $\Delta i = +14.8\%$, and there is a sharper change in the gear ratio at the beginning and end of the meshing phase. More detailed information on the ranges of Δi depending on Δa_w is given in Table 3.6.

Table 3.6 Deviation of gear ratio depending on operating time and increasing of center distance

	Deviation of gear ratio at operating time, millions of cycles							
	$N_{c1} = 172$, $N_{c2} = 38$		$N_{c1} = 322$, $N_{c2} = 71$		$N_{c1} = 635$, $N_{c2} = 140$		$N_{c1} = 816$, $N_{c2} = 180$	
Δa_w	$-\Delta i$, %	$+\Delta i$, %	$-\Delta i$, %	$+\Delta i$, %	$-\Delta i$, %	$+\Delta i$, %	$-\Delta i$, %	$+\Delta i$, %
0	2.3	3.4	1.6	5.1	3.1	7.5	6.5	14
+0.1m	1.6	2.5	1.9	6.3	1.1	9.1	1.3	12.7
+0.2m	0.7	2.7	2.6	6.1	0.7	12.9	1.7	16
+0.27m	1	2	2.3	5.7	0.8	13.5	4.1	14.8

3.5 Conclusion

1. The dynamic load in common determines the intensity of wear, and the wear along the profile is nonuniform, which leads to additional dynamic loads in the meshing. Besides the mentioned feedback loop, the materials of the gears, the hardness of the teeth surfaces, the accuracy of their manufacture, and many other features associated with the actual operating conditions need to be considered. These factors restrain the development of a universal model of tooth wear.
2. The tooth thickness has a quadratic dependence on the operating time, and it can be approximated by an equation of the form $\delta(L) = \delta_0 + bL + cL^2$, with a correlation parameter of 0.953 to 0.996. The obtained equation allows to predict the value of the pinion and gear teeth wear until their limit state.
3. The measurement of the teeth of heavy-duty experimental transmissions within the period of approximately 2/3 of their designated service life showed that the maximum value of wear occurs at the height of the constant chord of the tooth. This characteristic is typical for both the pinion teeth and the gear teeth. Concave profile segments on the tooth dedendum of the pinion and gear teeth, smoothly transitioning into convex segments on the tooth addendum. The profile shape closely resembles the theoretical one (with wear at the pitch circle equal to zero) only at the early stages of transmission operation.
4. Distortions of the profiles, i.e., deviations from the involute, leads to the variation of pitch point position on the center line due to the variable position of common normal to the profiles at different contact points. This causes deviations of the instantaneous values of the gear ratio from the nominal value. As the transmission operates, fluctuations of the gear ratio in the meshing of one pair of teeth increase, reaching the value of $\Delta i = +14\%$ at the number of load cycles of $N_{c1\max} = 816 \cdot 10^6$ for the pinion and $N_{c2\max} = 180 \cdot 10^6$ for the gear.
5. The increasing of the center distance, due to the wear of shaft support units, does not lead to an unequivocal increase of Δi value at different stages of transmission operation. However, it changes the character of the gear ratio fluctuation. In particular, when the value of the tooth wear is low, the increasing of the center distance results in a shift in the phases of the fluctuation. At later stages of

operation, the slopes of the curves $i(\phi_1)$ change, and the maximum value of the instantaneous gear ratio shifts towards the boundaries of the meshing phase. With the number of load cycles of $N_{c1\max} = 816 \cdot 10^6$ for the pinion and $N_{c2\max} = 180 \cdot 10^6$ for the gear a sharper change of the gear ratio occurs in mesh-in and mesh-out zones of the path of contact, and the fluctuation of gear ratio reaches its maximum value of $\Delta i = +14.8\%$.

References

1. Wojnarowski, J., & Onishchenko, V. (2003). Tooth wear effects on spur gear dynamics. *Mechanism and Machine Theory, 38*(2), 161–178. https://doi.org/10.1016/S0094-114X(02)00091-5. ISSN 0094-114X.
2. Han, H., Yuan, K., Ma, H., Peng, Z., Li, Z., Zhao, S., & Wen, B. (2023). Mesh characteristic analysis and dynamic simulation of spur gear pair considering corner contact and tooth broken fault, 106883. *Engineering Failure Analysis, 143*(Part A). https://doi.org/10.1016/j.engfailanal.2022.106883. ISSN 1350-6307.
3. Lior Bachar, Omri Matania, Roee Cohen, Renata Klein, Michael G. Lipsett, Jacob Bortman, A novel hybrid physical AI-based strategy for fault severity estimation in spur gears with zero-shot learning, Mechanical Systems and Signal Processing, Volume 204, 2023, 110748. doi: https://doi.org/10.1016/j.ymssp.2023.110748. ISSN 0888-3270.
4. Liu, Y., Shi, Z., Shen, G., Dong, Z., Wang, F., & Fengshou, G. (2021). Evaluation model of mesh stiffness for spur gear with tooth tip chipping fault. *Mechanism and Machine Theory, 158*, 104238. https://doi.org/10.1016/j.mechmachtheory.2020.104238. ISSN 0094-114X.
5. Yang, Y., Tang, J., Hu, N., Shen, G., Li, Y., & Zhang, L. (2023). Research on the time-varying mesh stiffness method and dynamic analysis of cracked spur gear system considering the crack position. *Journal of Sound and Vibration, 548*, 117505. https://doi.org/10.1016/j.jsv.2022.117505. ISSN 0022-460X.
6. Chen, Z., & Ji, P. (2021). Research on the variation of mesh stiffness and transmission error for spur gear with tooth profile modification and wear fault, 105184. *Engineering Failure Analysis, 122*. https://doi.org/10.1016/j.engfailanal.2020.105184. ISSN 1350-6307.
7. Chen, Z., & Ji, P. (2020). Study on wear in spur gears based on an improved load distribution model considering the effects of corner contact, 104605. *Engineering Failure Analysis, 115*. https://doi.org/10.1016/j.engfailanal.2020.104605. ISSN 1350-6307.
8. Archard, J. F. (1953). Contact and rubbing of flat surfaces. *Journal of Applied Physics, 24*, 981–988.
9. Yesilyurt, I., Fengshou, G., & Ball, A. D. (2003). Gear tooth stiffness reduction measurement using modal analysis and its use in wear fault severity assessment of spur gears. *NDT & E International, 36*(5), 357–372. https://doi.org/10.1016/S0963-8695(03)00011-2. ISSN 0963-8695.
10. Xiangyang, X., Lai, J., Lohmann, C., Tenberge, P., Weibring, M., & Dong, P. (2019). A model to predict initiation and propagation of micro-pitting on tooth flanks of spur gears. *International Journal of Fatigue, 122*, 106–115. https://doi.org/10.1016/j.ijfatigue.2019.01.004. ISSN 0142-1123.
11. Mallipeddi, D., Norell, M., Sosa, M., & Nyborg, L. (2017). Influence of running-in on surface characteristics of efficiency tested ground gears. *Tribology International, 115*. https://doi.org/10.1016/j.triboint.2017.05.018
12. Lin, J., Teng, C., Bergstedt, E., Li, H., Shi, Z., & Olofsson, U. (2021). A quantitatively distributed wear-measurement method for spur gears during micro-pitting and pitting tests. *Tribology International, 157*, 106839. https://doi.org/10.1016/j.triboint.2020.106839. ISSN 0301-679X.

13. Feng, K., Borghesani, P., Smith, W. A., Randall, R. B., Chin, Z. Y., Ren, J., & Peng, Z. (2019). Vibration-based updating of wear prediction for spur gears. *Wear, 426–427*(Part B), 1410–1415. https://doi.org/10.1016/j.wear.2019.01.017. ISSN 0043-1648.
14. Feng, K., Smith, W. A., Randall, R. B., Hongkun, W., & Peng, Z. (2022). Vibration-based monitoring and prediction of surface profile change and pitting density in a spur gear wear process, 108319. *Mechanical Systems and Signal Processing, 165*. https://doi.org/10.1016/j.ymssp.2021.108319. ISSN 0888-3270.
15. Kragelsky, I. V., & Alisin, V. V. (Eds.). (1981). *Friction wear lubrication: Tribology handbook* (Vol. 2, 280p.). Mir Publ.
16. Evans, S. M., & Keogh, P. S. (2019). Wear mechanisms in polyoxymethylene spur gears. *Wear, 428–429*, 356–365. https://doi.org/10.1016/j.wear.2019.03.027. ISSN 0043-1648.
17. Balitskii, O., & Kolesnikov, V. (2019). Identification of Wear products in the automotive tribotechnical system using computer vision methods, artificial intelligence and big data. In *2019 XIth international scientific and practical conference on electronics and information technologies (ELIT), Lviv, Ukraine* (pp. 24–27). https://doi.org/10.1109/ELIT.2019.8892275
18. Chang, H., Borghesani, P., Smith, W. A., & Peng, Z. (2019). Application of surface replication combined with image analysis to investigate wear evolution on gear teeth—A case study. *Wear, 430–431*, 355–368. https://doi.org/10.1016/j.wear.2019.05.024. ISSN 0043-1648.
19. Satter, M. A., Ebadi, A. R., Aghaie, K. K., Fotovati, M. R., & Piroozan, A. (1993). Wear of gears: A case study at shiraz cement plant. *Wear, 162–164*(Part B), 1054–1058. https://doi.org/10.1016/0043-1648(93)90122-3. ISSN 0043-1648.
20. Zhao, J., Sheng, W., Li, Z., Zhang, H., & Zhu, R. (2022). Effect of lubricant selection on the wear characteristics of spur gear under oil-air mixed lubrication. *Tribology International, 167*, 107382. https://doi.org/10.1016/j.triboint.2021.107382. ISSN 0301-679X.
21. Bo, H., Zhou, C., Wang, H., & Chen, S. (2021). Nonlinear tribo-dynamic model and experimental verification of a spur gear drive under loss-of-lubrication condition, 107509. *Mechanical Systems and Signal Processing, 153*. https://doi.org/10.1016/j.ymssp.2020.107509. ISSN 0888-3270.
22. Dhanasekaran, S., & Gnanamoorthy, R. (2008). Gear tooth wear in sintered spur gears under dry running conditions. *Wear, 265*(1–2), 81–87. https://doi.org/10.1016/j.wear.2007.08.025. ISSN 0043-1648.
23. Onishchenko, V. (2015). Investigation of tooth wears from scuffing of heavy-duty machine spur gears. *Mechanism and Machine Theory, 83*, 38–55. https://doi.org/10.1016/j.mechmachtheory.2014.08.016. ISSN 0094-114X.
24. Onishchenko, V. (2008). Tooth wear modeling and prognostication parameters of engagement of spur gear power transmissions. *Mechanism and Machine Theory, 43*(12), 1639–1664. https://doi.org/10.1016/j.mechmachtheory.2007.12.005. ISSN 0094-114X.
25. Mao, Y., Tong, J., Chin, Z. Y., Borghesani, P., & Peng, Z. (2023). Transmission-error- and vibration-based condition monitoring of gear wear with contaminated lubricant, 204760. *Wear, 523*. https://doi.org/10.1016/j.wear.2023.204760. ISSN 0043-1648.
26. Černe, B., & Petkovšek, M. (2022). High-speed camera-based optical measurement methods for in-mesh tooth deflection analysis of thermoplastic spur gears. *Materials & Design, 223*, 111184. https://doi.org/10.1016/j.matdes.2022.111184. ISSN 0264-1275.
27. Zhao, J., Hou, L., Li, Z., Zhang, H., & Zhu, R. (2022). Prediction of tribological and dynamical behaviors of spur gear pair considering tooth root crack, 106145. *Engineering Failure Analysis, 135*. https://doi.org/10.1016/j.engfailanal.2022.106145. ISSN 1350-6307.
28. Jangra, D., Hirani, H., & Darpe, A. K. (2023). Effect of combined (radial-axial-angular direction) misalignment on sliding wear of spur gears: A comprehensive study. *Tribology International, 189*, 108908. https://doi.org/10.1016/j.triboint.2023.108908. ISSN 0301-679X.
29. Gao, P., Liu, H., Yan, P., Xie, Y., Xiang, C., & Wang, C. (2022). Research on application of dynamic optimization modification for an involute spur gear in a fixed-shaft gear transmission system, 109530. *Mechanical Systems and Signal Processing, 181*. https://doi.org/10.1016/j.ymssp.2022.109530. ISSN 0888-3270.

30. Rana, V., Jain, N. K., & Pathak, S. (2023). Development of analytical models for tip, root, and end relieving of spur gears by pulsed electrochemical finishing process. *CIRP Journal of Manufacturing Science and Technology, 47*, 132–157. https://doi.org/10.1016/j.cirpj.2023.09.004. ISSN 1755-5817.
31. Balyts'kyi, O. I., Kolesnikov, V. O., & Havrylyuk, M. R. (2019). Influence of lubricating liquid on the formation of the products of cutting of 38KhN3MFA steel. *Materials Science, 54*, 722–727. https://doi.org/10.1007/s11003-019-00238-7
32. Sánchez, M. B., Pleguezuelos, M., & Pedrero, J. I. (2024). Influence of profile modification on the transmission error of spur gears under surface wear. *Mechanism and Machine Theory, 191*, 105473. https://doi.org/10.1016/j.mechmachtheory.2023.105473. ISSN 0094-114X.
33. İmrek, H., & Düzcükoğlu, H. (2007). Relation between wear and tooth width modification in spur gears. *Wear, 262*(3–4), 390–394. https://doi.org/10.1016/j.wear.2006.06.004. ISSN 0043-1648.
34. Tkach, P., Reviakina, O., Kryvosheia, A., Melnyk, V., Ustynenko, O., & Protasov, R. (2022). Meshing characteristics of profile shifted cylindrical quasi-involute arc-tooth-trace gears. Part 1. Theoretical base. *Strojnícky časopis—Journal of Mechanical Engineering, 72*(2), 201–210. https://doi.org/10.2478/scjme-2022-0029
35. Tkach, P., Reviakina, O., Kryvosheia, A., Melnyk, V., Ustynenko, O., & Protasov, R. (2023). Meshing characteristics of profile shifted cylindrical quasi-involute arc-tooth-trace gears. Part 2. Calculation results. *Strojnícky časopis—Journal of Mechanical Engineering, 74*(2), 181–192. https://doi.org/10.2478/scjme-2023-0032
36. Hussein, A. W., & Abdullah, M. Q. (2022). High-contact ratio spur gears with conformal contact and reduced sliding, 100412. *Results in Engineering, 14*. https://doi.org/10.1016/j.rineng.2022.100412. ISSN 2590-1230.
37. Shishov, V. P., Nosko, P. L., Tkach, P. M., Sklyar, Y. O., & Tkach, I. H. (2009). *Fundamentals of synthesis of reference profile of cylindrical gears with increased teeth wear resistance* (Bulletin of the National Technical University "KhPI". Coll. of scientific papers on thematic issue "Problems of mechanical drive", No. 20) (pp. 171–176). NTU "KhPI" Publ. [In Ukrainian].
38. Zenghuang He, X., Gong, S. F., Luo, S., & Mo, J. (2022). Design and investigation of an asymmetric logarithmic spiral gear drive. *Applied Bionics and Biomechanics*, 1176–2322. https://doi.org/10.1155/2022/8550869
39. Ravivarman, R., Palaniradja, K., & Prabhu Sekar, R. (2018). Evolution of balanced root stress and tribological properties in high contact ratio spur gear drive. *Mechanism and Machine Theory, 126*, 491–513. https://doi.org/10.1016/j.mechmachtheory.2018.04.025. ISSN 0094-114X.
40. Ravivarman, R., Palaniradja, K., & Prabhu Sekar, R. (2018). Influence of gear ratio on wear depth of nonstandard HCR spur gear drive with balanced fillet stress. *Materials Today: Proceedings, 5*(9, Part 1), 17350–17359. https://doi.org/10.1016/j.matpr.2018.04.148. ISSN 2214-7853.
41. Chang, X., Renqing, D., Liao, L., Zhu, P., Lin, B., Huang, Y., & Luo, S. (2023). Study on hydrodynamic lubrication and friction reduction performance of spur gear with groove texture. *Tribology International, 177*, 107978. https://doi.org/10.1016/j.triboint.2022.107978. ISSN 0301-679X.
42. Chen, Z., Zhou, J., Bao, L., Hao, F., Meng, X., Ji, J., Zhang, Y., Hua, X., Xu, X., & Yonghong, F. (2023). Meshing frictional characteristics of spur gears under dry friction and heavy loads: Effects of the preset pitting-like micro-textures array. *Tribology International, 180*, 108296. https://doi.org/10.1016/j.triboint.2023.108296. ISSN 0301-679X.
43. Huangfu, Y., Zhao, Z., Ma, H., Han, H., & Chen, K. (2020). Effects of tooth modifications on the dynamic characteristics of thin-rimmed gears under surface wear. *Mechanism and Machine Theory, 150*, 103870. https://doi.org/10.1016/j.mechmachtheory.2020.103870. ISSN 0094-114X.

44. Kalligeros, C., Papalexis, C., Vasileiou, G., Tzouganakis, P., Spitas, C., & Spitas, V. (2023). Exploiting double-flank roll testing spur gear measurements to determine gear parameter deviations through numerical simulation of free-form meshing gears. *Simulation Modelling Practice and Theory, 126*, 102757. https://doi.org/10.1016/j.simpat.2023.102757. ISSN 1569-190X.
45. Chen, W., Lei, Y., Yao, F., & Hou, L. (2021). A study of effects of tooth surface wear on time-varying mesh stiffness of external spur gear considering wear evolution process. *Mechanism and Machine Theory, 155*, 104055. https://doi.org/10.1016/j.mechmachtheory.2020.104055. ISSN 0094-114X.
46. Xu, Z., Zhong, J., Li, W., & Bocian, M. (2022). Nonlinear dynamic analysis of high-speed gear pair with wear fault and tooth contact temperature for a wind turbine gearbox. *Mechanism and Machine Theory, 173*, 104840. https://doi.org/10.1016/j.mechmachtheory.2022.104840. ISSN 0094-114X.
47. Yan, Y., Jiang, C., & Li, W. (2022). Simulation on coupling effects between surface wear and fatigue in spur gear, 106055. *Engineering Failure Analysis, 134*. https://doi.org/10.1016/j.engfailanal.2022.106055. ISSN 1350-6307.
48. GOST 4543-71 Structural alloy steel bars. Specifications [In Russian].
49. GOST 1643-81 Basic requirements for interchangeability. Cylindrical gearings. Tolerances [In Russian].
50. Pronikov, A. S. (1978). *Durability of machines* (p. 592). Mashinostroenie Publ. pp [In Russian].
51. Korn, G. A., & Korn, T. M. (2000). *Mathematical handbook for scientists and engineers. Definitions, theorems, and formulas for reference and review* (1151pp.). Dover Publications
52. DSTU ISO 5479:2009 Statistical interpretation of data. Tests for departure from the normal distribution (ISO 5479:1997, IDT)—ISO 5479:1997 Statistical interpretation of data. Tests for departure from the normal distribution.
53. DSTU 3004-95 Reliability of equipment. Experimental data based methods of estimation of reliability indicators [In Ukrainian].
54. Kassandrova, O. N., & Lebedev, V. V. (1970). *Processing of observed results* (104p.). Nauka Publ. [In Russian].
55. Lukomsky, Ya.I. (1958). *Correlation theory and its application to analysis of manufacturing* (388p.). Gosstatizdat Publ. [In Russian].
56. Venetsky, I. G., Kildyshev, G. E. (1968). *Fundamentals of probability theory and mathematical statistics* (360p.) Statistika Publ. [In Russian].

Chapter 4
Elements of Dynamics of Geometrically Accurate Crossed-Axes Gear Pair with Line Contact Between the Tooth Flanks: *R-Gearing*

Stephen P. Radzevich

4.1 Fundamentals of *R-Gearing*

R–gearing is a novel kind of gearing. Gearing of this design was invented by Prof. *S.P. Radzevich* at around 2008. Since that time, the crossed-axes gearing of the proposed design was discussed in numerous sources that are available in the public domain [1, 2, 3, 4, 5], and others.

The key accomplishments in the field of *R– gearing* are also extensively used in design of gear cutting tools for machining gears for *R– gear* pairs [6], and others.

4.1.1 Principal Kinematics of R-Gearing

In a crossed-axes gear pair, a rotation, ω_p, from a driving shaft is transformed and transmitted to a rotation, ω_g, of a driven shaft by means of a driving pinion and a driven gear, as illustrated in Fig. 4.1. Three rotation vectors, namely, a rotation vector of the gear, $\boldsymbol{\omega}_g$, a rotation vector of a mating pinion, $\boldsymbol{\omega}_p$, and a vector of instantaneous rotation, $\boldsymbol{\omega}_{pl}$, of the pinion in relation to the gear, are used to construct a gear vector diagram.

Here:

(a) the rotation vector $\boldsymbol{\omega}_g$ is pointed along the gear axis of rotation, O_g,
(b) the rotation vector $\boldsymbol{\omega}_p$ is pointed along the pinion axis of rotation, O_p, and
(c) the rotation vector $\boldsymbol{\omega}_{pl}$ is pointed along the axis of instantaneous rotation, P_{ln}.

S. P. Radzevich (✉)
Machine Design Department, National Technical University of Ukraine "Igor Sikorsky Kyiv Polytechnic Institute", Sterling Heights, MI, USA

S. P. Radzevich (ed.), *Gear Accuracy*, Mechanical Engineering Series,
https://doi.org/10.1007/978-3-031-74798-4_4

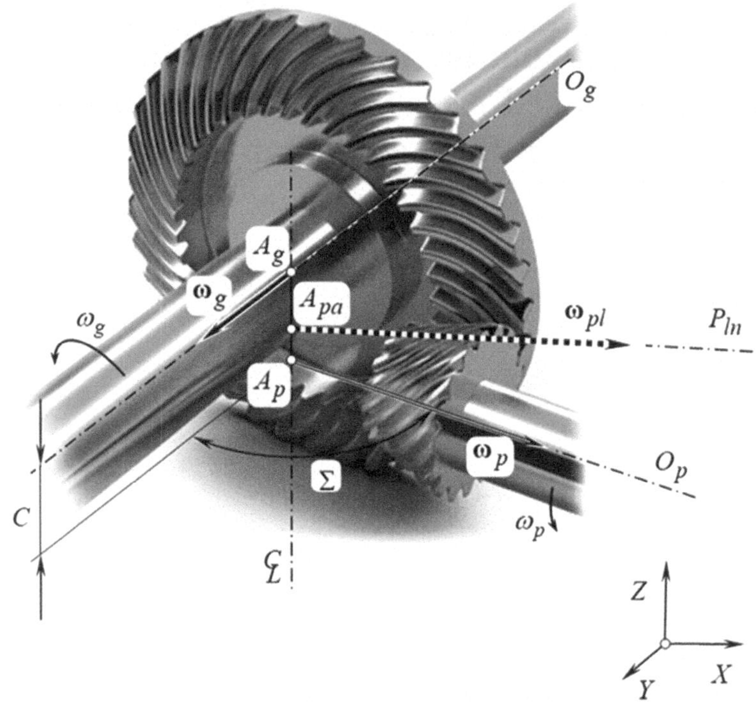

Fig. 4.1 Gear vector diagram of crossed-axes gear pair

The axes, O_g and O_p, are apart from one another at a center-distance, C, and form a shaft angle, Σ. Here, the shaft angle, Σ, is specified as $\Sigma = \angle\,(\boldsymbol{\omega}_g,\ \boldsymbol{\omega}_p)$. The gear base-cone-apex, the pinion base-cone-apex, and the plane-of-action apex are designated as A_g, A_p, and A_{pa}, correspondingly. All of the apexes are situated within the centerline, ℄.

4.1.2 Tooth Flank Geometry in R-Gearing

To generate tooth flanks, $\mathcal{G}$ and $\mathcal{P}$, of a gear, and of a mating pinion, in a crossed-axes gear pair, a desirable line of contact, LC_{des}, is used. The desirable line of contact, LC_{des}, is a planar curve that is entirely situated within the plane of action, PA. This curve is commonly specified in the reference system, $X_{pa}Y_{pa}Z_{pa}$, associated with the plane of action, PA. The desirable lines of contact, LC_{des}, of various geometries are used in present day practice: this could be a straight-line segment, a circular arc, LC_{circ} (see Fig. 4.2), an arc of cycloid, and so forth.

In a particular case of desirable line of contact in the form of straight-line segment, a line LC_{des}, of one of the following geometries can be employed:

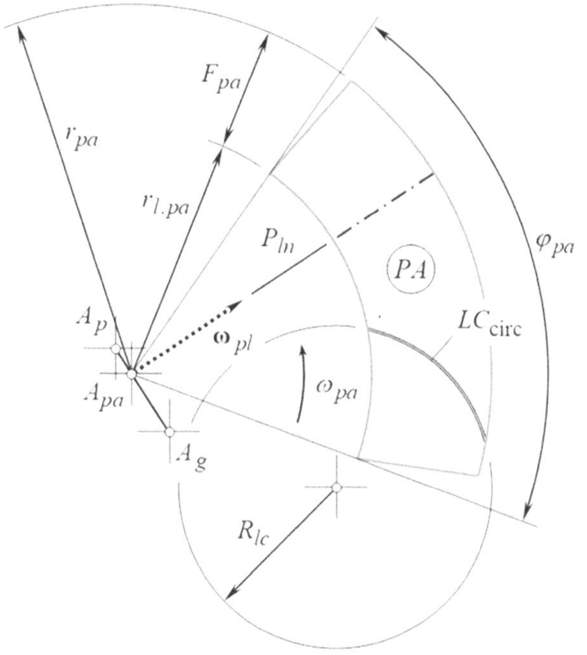

Fig. 4.2 Configuration of the desirable line of contact, LC_{circ}, within the plane of action, PA, in geometrically accurate crossed-axes gear pair

- the line of contact of the base cone of the gear and the plane of action (straight tooth gear is generated in this case); gear pair of this design is referred to as "*straight-gear* C_a- *gearing*"
- the line contact of the base cone of the pinion and the plane of action (straight tooth pinion is generated in this case); gear pair of this design is referred to as "*straight-pinion* C_a- *gearing*"
- the axis of instantaneous rotation, P_{ln} ; (neither straight tooth gear, nor straight tooth pinion is generated in this case); gear pair of this design is referred to as "*straight* C_a*–gearing*"

To generate the tooth flanks, $\mathcal{G}$ and $\mathcal{P}$, three coordinate systems are employed. They are: (a) a *Cartesian coordinate system*, $X_gY_gZ_g$, associated with the gear, (b) a *Cartesian coordinate system*, $X_pY_pZ_p$, associated with the pinion, and (c) a *Cartesian coordinate system*, $X_{pa}Y_{pa}Z_{pa}$, associated with the plane of action, PA.

In the reference system, $X_{pa}Y_{pa}Z_{pa}$, position vector of point, $\mathbf{r}_{\text{des}}$, of the desirable line of contact, LC_{des}, is given by an equation in vector representation $\mathbf{r}_{\text{des}} = \mathbf{r}_{\text{des}}(u)$, where u is the parameter of the planar curve $\mathbf{r}_{\text{des}}$.

When the gears rotate, the desirable line of contact, LC_{des}, forms a family of consecutive positions in both the reference systems, $X_gY_gZ_g$ and $X_pY_pZ_p$.

The position vector of point, $\mathbf{r}_g$, of the gear tooth flank, $\mathcal{G}$, is specified by a family of consecutive positions of the desirable line of contact, LC_{des}, given in the *Cartesian coordinate system*, $X_gY_gZ_g$:

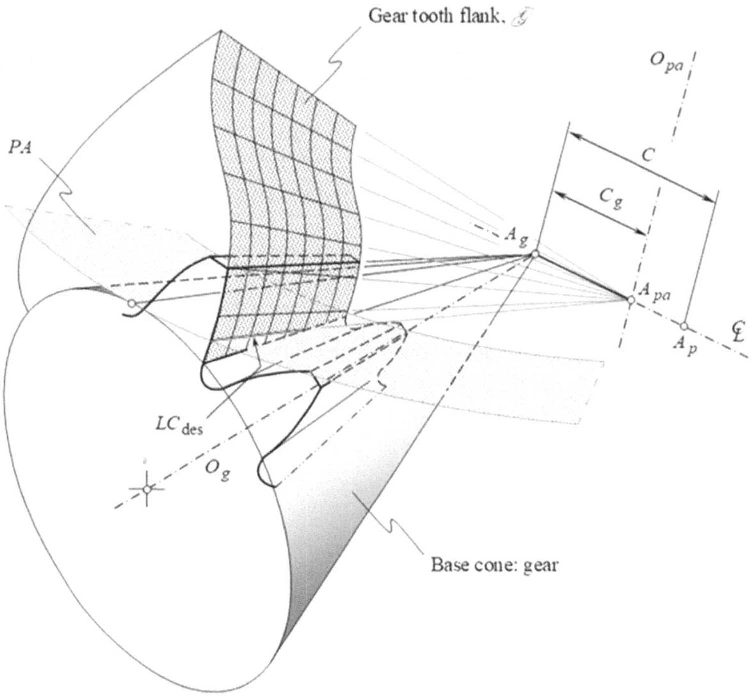

Fig. 4.3 Tooth flank, $\mathcal{G}$ in geometrically accurate gear for C_a- axes gear pair

$$\mathcal{G} \;\Rightarrow\; \mathbf{r}_g(u,\;\varphi_{pa}) = \mathbf{Rs}(pa \mapsto \mathcal{G}) \cdot \mathbf{r}_{\text{des}} \tag{4.1}$$

Here, φ_{pa} is the angle of rotation of the plane of action, PA; $\mathbf{Rs}(pa \mapsto \mathcal{G})$ is the operator/matrix of the transition from the coordinate system, $X_{pa}Y_{pa}Z_{pa}$, to the gear coordinate system, $X_gY_gZ_g$.

The actual geometry of a geometrically accurate gear for crossed-axes gear pair is illustrated in Fig. 4.3.

The position vector of point, $\mathbf{r}_p$, of the pinion tooth flank, $\mathcal{P}$, is specified by a family of consecutive positions of the desirable line of contact, LC_{des}, given in the *Cartesian coordinate system*, $X_pY_pZ_p$:

$$\mathcal{P} \;\Rightarrow\; \mathbf{r}_p(u,\;\varphi_{pa}) = \mathbf{Rs}(pa \mapsto \mathcal{P}) \cdot \mathbf{r}_{\text{des}} \tag{4.2}$$

Here, $\mathbf{Rs}(pa \mapsto \mathcal{P})$ is the operator/matrix of transition from the coordinate system; $X_{pa}Y_{pa}Z_{pa}$, to the gear coordinate system, $X_pY_pZ_p$.

All three fundamental laws of gearing are met [3] if the tooth flanks fulfill Eq. (4.1) and Eq. (4.2).

4.2 Contact Motion Characteristics in *R-Gearing*

Rotation of the input shaft and rotation of the output shaft in a crossed-axes gear pair cause several other motions. Rolling and sliding of the tooth flanks are among these motions. Sliding is inevitable in crossed-axes gearing.

For better understanding of the nature of sliding, an in-detail investigation into teeth sliding in geometrically accurate crossed-axes gearing is required to be undertaken.

4.2.1 Tooth Flank Sliding in R-Gearing

The sliding is inevitable in crossed-axes gearing. The velocity of sliding of tooth flanks in crossed-axes gearing can be viewed as a component of the velocity of the relative motion of tooth flanks when the gears rotate.

In the particular case under consideration, it is convenient to interpret the plane of action, *PA* (see Fig. 4.4), as driving component, and to consider both the gear and the pinion as driven components.

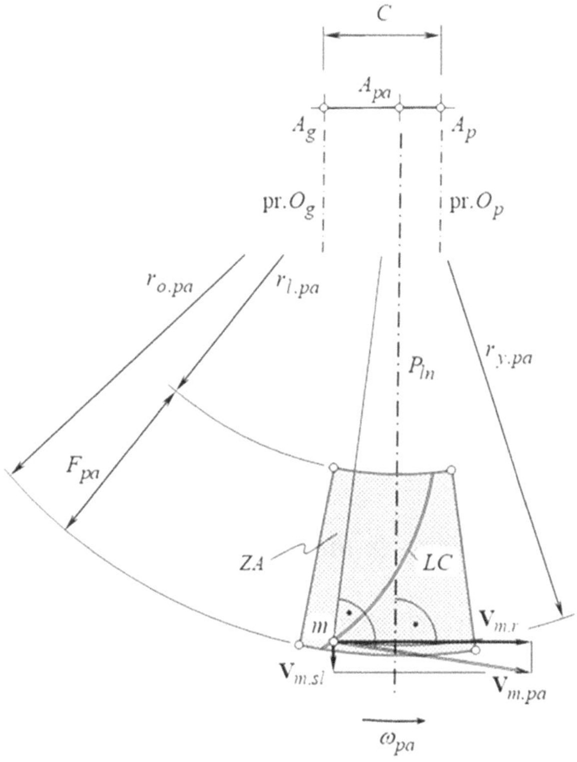

Fig. 4.4 Origination of axial sliding in geometrically accurate crossed-axes gear pair

Point m is an arbitrary point of the desirable line of contact, LC, between the tooth flanks, $\mathcal{G}$ and $\mathcal{P}$. The linear velocity vector of instantaneous motion of point m is designated as $\mathbf{V}_{m.\ pa}$. Considered in a reference system, associated with the gear, the projection of the linear velocity vector, $\mathbf{V}_{m.\ pa}$, onto the straight line that is tangent to LC at point m, equals to the linear velocity vector of the tooth flanks sliding in their lengthwise direction. This projection is entirely situated within the plane of action. Similarly, considered in that same reference system, the projection of the linear velocity vector, $\mathbf{V}_{m.\ pa}$, onto the straight line that is tangent to the tooth profile at point m, equals to the linear velocity vector of the tooth flanks sliding in their transverse section. This projection is entirely situated within the plane that is perpendicular to LC at point m.

In order to determine the sliding velocity in geometrically accurate C_a– gearing, equations for the linear velocity vectors, $\mathbf{V}_{m.\ g}$ and $\mathbf{V}_{m.\ p}$, of point of interest, m, are required to be derived, and represented in a common reference system. A stationary *Cartesian coordinate system*, $X_{pa.\ s}Y_{pa.\ s}Z_{pa.\ s}$, associated with the plane of action, PA, is especially handy for solving the problem under consideration.

Operators of linear transformations are extensively used below for the derivation of analytical solution to the problem under consideration. The chief reason for that is as follows.

Commonly, it is convenient to derive an analytical expression for each of the linear velocity vectors, $\mathbf{V}_{m.\ g}$ and $\mathbf{V}_{m.\ p}$, in a specific reference system, and then to represent the derived equations in a common reference system. For instance, it is convenient to determine the linear velocity vector, $\mathbf{V}_{m.\ g}$, in a coordinate system associated with the gear. Similarly, it is convenient to determine the linear velocity vector, $\mathbf{V}_{m.\ p}$, in a coordinate system associated with the pinion. Later on, the derived equations for both the vectors, $\mathbf{V}_{m.\ g}$ and $\mathbf{V}_{m.\ p}$, can be represented in a common reference system. Use of operators of linear transformations, namely, the operators of translation, and of the operators of rotation, along and about the coordinate axes is proven to be very helpful for this purpose.

Relative motion of a driving member, of a driven member in a gear pair, as well, as of the plane of action, form the foundation, on which the corresponding operators of the coordinate system transformations, $\mathbf{Rs}_{Ca}(g \rightarrow pa)$ and $\mathbf{Rs}_{Ca}(p \rightarrow pa)$, can be derived. When a gear pair operates, the driving member, the driven member, and the plane of action, all of them are rotated in a timely manner, that is, the rotations, ω_g, ω_p, and ω_{pa}, are synchronized with one another.

4.2.2 *Sliding in the Lengthwise Direction of Gear Teeth*

Sliding of two different kinds are distinguished when crossed-axes gearing operates. The tooth flank sliding in the lengthwise direction of the teeth is the first kind of sliding. The tooth flank sliding in the transverse direction of the teeth is the second kind of sliding.

Coordinate system transformations are extensively used in the analysis of sliding of a gear, $\mathcal{G}$, and of a mating pinion, $\mathcal{P}$, tooth flanks.

Analytical Solution to the Problem

For the analytical description of the linear transformation, namely, for the transition from the reference system, $X_g Y_g Z_g$, associated with the rotated gear, to the reference system, $X_{pa} Y_{pa} Z_{pa}$, associated with the rotated plane of action, *PA*, the operator, $\mathbf{Rs}_{Ca}\,(g \rightarrow pa)$, of the resultant coordinate system is used [3]. A similar operator of linear transformation, $\mathbf{Rs}_{Ca}\,(p \rightarrow pa)$, is derive for the pinion [3].

The operators of linear transformations, $\mathbf{Rs}_{Ca}\,(g \rightarrow pa)$ and $\mathbf{Rs}_{Ca}\,(p \rightarrow pa)$, are not presented here in expanded form, as both of them are bulky.

In the reference system, $X_g Y_g Z_g$, associated with the rotated gear, the linear velocity vector, $\mathbf{V}_{m.\ g}$, of point of interest, m, is analytically described by a column matrix:

$$\mathbf{V}_{m.g}\big|_g = \begin{bmatrix} -r_{b.g}\sin\varphi_g \\ r_{b.g}\cos\varphi_g \\ 0 \\ 1 \end{bmatrix} \tag{4.3}$$

To represent that same linear velocity vector, $\mathbf{V}_{m.\ g}$, in the coordinate system, $X_{pa} Y_{pa} Z_{pa}$, associated with the rotated plane of action, *PA*, the operator of linear transformation, $\mathbf{Rs}_{Ca}\,(g \rightarrow pa)$, is employed:

$$\mathbf{V}_{m.g}\big|_{pa} = \mathbf{Rs}_{Ca}\,(g \rightarrow pa)\cdot\mathbf{V}_{m.g}\big|_g \tag{4.4}$$

An equation similar to Eq. (4.3), can be derived for the linear velocity vector, $\mathbf{V}_{m.\ p}$, of the rotated pinion.

$$\mathbf{V}_{m.p}\big|_{pa} = \mathbf{Rs}_{Ca}\,(p \rightarrow pa)\cdot\mathbf{V}_{m.p}\big|_p \tag{4.5}$$

Here, in Eq. (4.5), the operator, $\mathbf{Rs}_{Ca}\,(p \rightarrow pa)$, of the linear transformation is employed.

Consider the relative motion of point within the line of contact of the tooth flanks, $\mathcal{G}$ and $\mathcal{P}$, of a gear, and of a mating pinion in a crossed-axes gear pair.

The vector, $\mathbf{V}_{rel}$, of relative motion of point of interest, m, equals to the difference:

$$\mathbf{V}_{rel} = \mathbf{V}_{m.p}\big|_{pa} - \mathbf{V}_{m.g}\big|_{pa} \tag{4.6}$$

where both the linear velocity vectors, $\mathbf{V}_{m.\ g}$ and $\mathbf{V}_{m.\ p}$, are represented in a common reference system, $X_{pa} Y_{pa} Z_{pa}$, associated with the rotated plane of action, *PA*.

These velocity vectors, $\mathbf{V}_{m.\ g}$ and $\mathbf{V}_{m.\ p}$, decisively affect the lubrication and friction parameters on the mating flanks and, hence, influence the load capacity and efficiency of the gear set.

In expanded form, the vector, $\mathbf{V}_{rel}$, can be represented in matrix form as:

$$\mathbf{V}_{rel}\big|_{pa} = \begin{bmatrix} V_{x.rel} \\ V_{y.rel} \\ V_{z.rel} \\ 1 \end{bmatrix} \tag{4.7}$$

The projection, $V_{y.\ rel}$, of the linear velocity vector, $\mathbf{V}_{rel}$, onto the axis, Y_{pa}, causes pure rolling of the gear and the mating pinion tooth flanks, $\mathcal{G}$ and $\mathcal{P}$, over each other. Therefore, this component of the linear velocity vector, $\mathbf{V}_{rel}$, is labeled below as $\mathbf{V}_{rol}$.

The projection, $V_{z.\ rel}$, of the linear velocity vector, $\mathbf{V}_{rel}$, onto the axis, Z_{pa}, causes profile sliding of the gear and the mating pinion tooth flanks, $\mathcal{G}$ and $\mathcal{P}$. Therefore, this component of the linear velocity vector, $\mathbf{V}_{rel}$, is labeled below as $\mathbf{V}_{sl.\ p}$.

The projection, $V_{x.\ rel}$, of the linear velocity vector, $\mathbf{V}_{rel}$, onto the axis, X_{pa}, causes *drag* sliding of the gear, and of the mating pinion tooth flanks, $\mathcal{G}$ and $\mathcal{P}$. Because of this, this component of the linear velocity vector, $\mathbf{V}_{rel}$, is labeled below as $\mathbf{V}_{sl.\ d}$.

Total sliding, $\mathbf{V}_{sl}$, in crossed-axes gearing can be calculated as the sum:

$$\mathbf{V}_{sl} = \mathbf{V}_{sl.p} + \mathbf{V}_{sl.d} \tag{4.8}$$

At different points within the zone of action, *ZA*, each of the components of the linear velocity vectors, namely, the vectors $\mathbf{V}_{sl}$, $\mathbf{V}_{sl.\ p}$, and $\mathbf{V}_{sl.\ d}$, vary.

A computer code for the calculation of all the components, $\mathbf{V}_{rol}$, $\mathbf{V}_{sl.\ p}$, and $\mathbf{V}_{sl.\ d}$, of the linear velocity vector, $\mathbf{V}_{rel}$, can be developed on the premise of the equations derived above in this subsection of the chapter. Ultimately, the distribution of the components, $\mathbf{V}_{rol}$, $\mathbf{V}_{sl.\ p}$, and $\mathbf{V}_{sl.\ d}$, within the zone of action, *ZA*, in a crossed-axes gear pair can be interpreted graphically.

Specific Sliding in Geometrically Accurate Crossed-Axes Gearing

For the specification of sliding between tooth flanks, $\mathcal{G}$ and $\mathcal{P}$, of a gear and of a mating pinion in crossed-axes gear pair, a dimensionless parameter can be used. This sliding parameter can be also referred to as *specific sliding*. The *specific sliding* is denoted by γ_{Σ}. An actual value of *specific sliding* does not depend on rotation of the input/output shafts, and depends only on the design parameters of a gear, and of a mating pinion. The latter is especially important when optimizing the design parameters of gears in crossed-axes gear pairs.

Two different parameters, γ_{Σ}, are distinguished.

First, the slide/roll ratio for tooth flank, $\mathcal{G}$, of a gear that is calculated from a formula:

$$\gamma_{\Sigma.g} = \frac{V^{m}_{sl.g} - V^{m}_{sl.p}}{V^{m}_{sl.g}} \tag{4.9}$$

Second, the slide/roll ratio for tooth flank, $\mathcal{P}$, of a pinion that is calculated from a formula:

$$\gamma_{\Sigma.p} = \frac{V^{m}_{sl.p} - V^{m}_{sl.g}}{V^{m}_{sl.p}} \tag{4.10}$$

The calculation of sliding velocities, $V^{m}_{sl.g}$ and $V^{m}_{sl.p}$, in Eq. (4.9) and Eq. (4.10) is discussed immediately below.

In the reference system, $X_{pa}Y_{pa}Z_{pa}$, associated with the rotated plane of action, PA, the linear velocity vector, $\mathbf{V}_{m.\ g}|_{pa}$, can be represented in matrix form:

$$\mathbf{V}_{m.g}\big|_{pa} = \begin{bmatrix} V_{x.m.g} \\ V_{y.m.g} \\ V_{z.m.g} \\ 1 \end{bmatrix} \tag{4.11}$$

In a crossed-axes gear pair, the component, $V_{z.\ m.\ g}$, contributes to the profile sliding, and the component, $V_{x.\ m.\ g}$, contributes to the drag sliding (the component, $V_{y.\ m.\ g}$, contributes to pure rolling of the tooth flanks, $\mathcal{G}$ and $\mathcal{P}$). The sliding velocity, $V^{m}_{sl.g}$ [see Eqs. (4.9) and (4.10)], can be expressed in terms of the velocities, $V_{x.\ m.\ g}$ and $V_{z.\ m.\ g}$, as:

$$V^{m}_{sl.g} = \sqrt{V^{2}_{x.m.g} + V^{2}_{z.m.g}} \tag{4.12}$$

Similarly, in the reference system, $X_{pa}Y_{pa}Z_{pa}$, associated with the rotated plane of action, PA, the linear velocity vector, $\mathbf{V}_{m.\ p}|_{pa}$, can be represented in the form of column mantrix:

$$\mathbf{V}_{m.p}\big|_{pa} = \begin{bmatrix} V_{x.m.p} \\ V_{y.m.p} \\ V_{z.m.p} \\ 1 \end{bmatrix} \tag{4.13}$$

In a crossed-axes gear pair, the component, $V_{z.\ m.\ p}$, contributes to the profile sliding, and the component, $V_{x.\ m.\ p}$, contributes to sliding in the lengthwise direction of the gear tooth (the component, $V_{y.\ m.\ p}$, contributes to pure rolling of the tooth flanks, $\mathcal{G}$ and $\mathcal{P}$). The sliding velocity, $V_{sl.p}^{m}$, can be expressed in terms of the velocities, $V_{x.\ m.\ p}$ and $V_{z.\ m.\ p}$, as:

$$V_{sl.p}^{m} = \sqrt{V_{x.m.p}^{2} + V_{z.m.p}^{2}} \tag{4.14}$$

The sliding velocities, $V_{sl.g}^{m}$ and $V_{sl.p}^{m}$ [see Eqs. (4.12) and (4.14)], can be entered into the expressions:

$$\gamma_{\Sigma.g} = \frac{V_{sl.g}^{m} - V_{sl.p}^{m}}{V_{sl.g}^{m}} \tag{4.15}$$

$$\gamma_{\Sigma.p} = \frac{V_{sl.p}^{m} - V_{sl.g}^{m}}{V_{sl.p}^{m}} \tag{4.16}$$

to calculate the specific roll/slide ratios, $\gamma_{\Sigma.\ g}$ and $\gamma_{\Sigma.\ p}$, in crossed-axes gearing.

The specific sliding, γ_{Σ}, is of a positive value on the addendum portions of the tooth flanks. The parameter, γ_{Σ}, does not exceed 1. At points within the axis of instantaneous rotation, P_{ln}, specific sliding, γ_{Σ}, is equal to zero, and it is equal to 1 at the base cone of the mating gear.

The specific sliding on the dedendum portion of the tooth flanks is of a negative value. It is equal to zero at points within the axis of instantaneous rotation, P_{ln}, and it approaches a minus infinity at the base cone.

The specific rolling/sliding ratios, $\gamma_{\Sigma.\ g}$ and $\gamma_{\Sigma.\ p}$, can be plotted within the zone of action, *ZA*, as only the region, *ZA*, of the plane of action comes into effect when investigating the engagement of the gear teeth.

Features of Specific Sliding in Geometrically Accurate Crossed-Axes Gearing

The interaction of the tooth flanks, $\mathcal{G}$ and $\mathcal{P}$, in geometrically accurate crossed-axes gearing features sliding in the lengthwise direction of gear teeth (see Fig. 4.4). Therefore, in addition to specific roll/sliding ratios, $\gamma_{\Sigma.\ g}$ and $\gamma_{\Sigma.\ p}$ [see Eqs. (4.9) and (4.10)], two more characteristics are required to be introduced to specify specific sliding in crossed-axes gear pairs.

Specific profile roll/slide ratios, $\gamma_{p.\ g}$ and $\gamma_{p.\ p}$, are the first additional characteristics. These two ratios are specified as:

$$\gamma_{p.g} = \frac{V_{z.m.g} - V_{z.m.p}}{V_{z.m.g}} \tag{4.17}$$

$$\gamma_{p.p} = \frac{V_{z.m.p} - V_{z.m.g}}{V_{z.m.p}} \tag{4.18}$$

Specific roll/slide ratios, $\gamma_{d.\,g}$ and $\gamma_{d.\,p}$, for the sliding in the lengthwise direction of the gear teeth are the second additional characteristics. These two ratios are specified as:

$$\gamma_{d.g} = \frac{V_{x.m.g} - V_{x.m.p}}{V_{x.m.g}} \tag{4.19}$$

$$\gamma_{d.p} = \frac{V_{x.m.p} - V_{x.m.g}}{V_{x.m.p}} \tag{4.20}$$

Specific profile roll/slide ratios, $\gamma_{p.\,g}$ and $\gamma_{p.\,p}$, and specific drag roll/slide ratios, $\gamma_{d.\,g}$ and $\gamma_{d.\,p}$, are helpful for more in-detail analysis of sliding conditions in crossed-axes gearing.

It is the right point to turn the readers' attention here to the fact that gears with a low tooth count are more vulnerable to sliding between the tooth flanks, $\mathcal{G}$ and $\mathcal{P}$. They are also more sensitive to the variation of the roll/slide conditions within the zone of action, *ZA*.

4.3 Elements of Dynamics of Geometrically Accurate Crossed-Axes Gearing

In a crossed-axes gear pair, the input shaft and the output shaft are loaded by an input torque and output torque, correspondingly. As the gears interact with one another, a force of the interaction is exerted from the driving member of the gear pair. An actual value of the force of interaction, as well, as the components of this force, depend on the input torque, and on the design parameters of the gear, and of the mating pinion. Considering an input torque and an input rotation of constant values (namely, no acceleration/deceleration is taken into account in the performed below analysis), it is required to determine the forces that act between a gear and a mating pinion in a crossed-axes gear pair

4.3.1 The Principal Assumption Adopted in the Load Analysis of R-Gearing

When a crossed-axes gear pair operates, the tooth flanks, $\mathcal{G}$ and $\mathcal{P}$, of the gear and of the mating pinion interact with one another at points within the line(s) of contact, *LC*. The line(s) of contact is entirely situated within the plane of action, *PA*. This makes possible a conclusion that the force of interaction between the tooth flanks, $\mathcal{G}$ and $\mathcal{P}$, acts along a straight line that is also situated within the plane of action, *PA*.

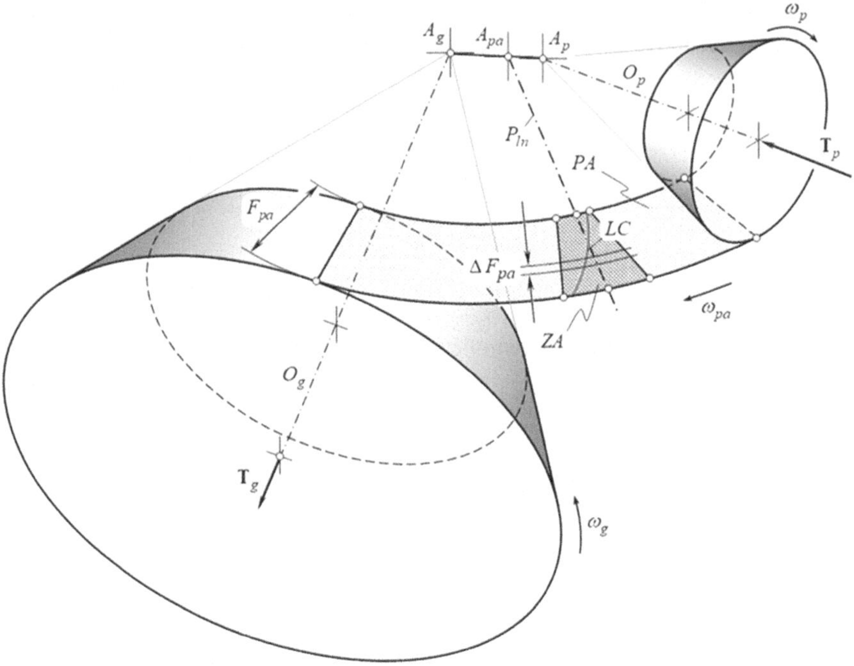

Fig. 4.5 Schematic for calculation of forces that act in geometrically accurate crossed-axes gear pair

As the line of action of the force is entirely situated within the plane of action, *PA*, then the following assumption seems to be reasonable in the load analysis of geometrically accurate crossed-axes gearing.

Referring to Fig. 4.5, consider base cones of a gear, and of a mating pinion, along with the plane of action. The gear is rotated, ω_g, about its axis of rotation, O_g, and the pinion is rotated, ω_p, about its axis of rotation, O_p, as schematically illustrated in Fig. 4.5. The plane of action, *PA*, is in tangency to both the base cones. The plane of action, *PA*, rotates, ω_{pa}, about its axis of rotation (not shown in Fig. 4.5) that passes through the *plane-of-action apex*, A_{pa}, and is pointed perpendicular to the plane of action. It is instructive to point out here that in contrast to the *equivalent pulley-and-belt transmission*, where the belt ***pools*** the driven member, in reality, in a gear pair, the driving gear ***pushes*** the driven member. In this way, the direction of the rotation, ω_{pa}, is specified in Fig. 4.5.

The rotations, ω_g, ω_p, and ω_{pa}, are synchronized with one another so, as to fulfill the equations:

$$\frac{\omega_g}{\omega_{pa}} = \sin\Gamma_b; \ \frac{\omega_p}{\omega_{pa}} = \sin\gamma_b; \ \frac{\omega_g}{\omega_p} = \frac{\sin\Gamma_b}{\sin\gamma_b} \tag{4.21}$$

The input torque, $\mathbf{T}_p$, is applied to the pinion shaft, and the output torque, $\mathbf{T}_g$, is applied to the gear shaft. The torque, $\mathbf{T}_{pa}$, is a *virtual* parameter. This torque is applied to the rotated plane of action, *PA*. The ratios of the magnitudes, T_g, T_p, and T_{pa}, of the torques, $\mathbf{T}_g$, $\mathbf{T}_p$, and $\mathbf{T}_{pa}$, are inverse to the corresponding ratios of the rotations, ω_g, ω_p, and ω_{pa}:

$$\frac{T_g}{T_{pa}} = \frac{1}{\sin \Gamma_b}; \ \frac{T_p}{T_{pa}} = \frac{1}{\sin \gamma_b}; \ \frac{T_g}{T_p} = \frac{\sin \gamma_b}{\sin \Gamma_b} \tag{4.22}$$

Interaction between a gear tooth flank, $\mathcal{G}$, and of a mating pinion tooth flank, $\mathcal{P}$, occurs only within the zone of action, *ZA*. Width of the zone of action is designated as F_{pa}. Width of the zone of action, F_{pa}, is calculated as the difference $F_{pa} = r_{o.\,pa} - r_{l.\,pa}$ between the outer, $r_{o.\,pa}$, and the inner, $r_{l.\,pa}$, radii of the active portion of the plane of action, *PA* (see Fig. 4.6).

The interaction between the tooth flanks, $\mathcal{G}$ and $\mathcal{P}$, is observed only along a line (s) of contact, *LC* (or portions of the lines of contact), those situated within the zone of action, *ZA*. There are one or two (and not more than three) lines of contact (portions of the lines of contact) within the zone of action, *ZA*, at that same time.

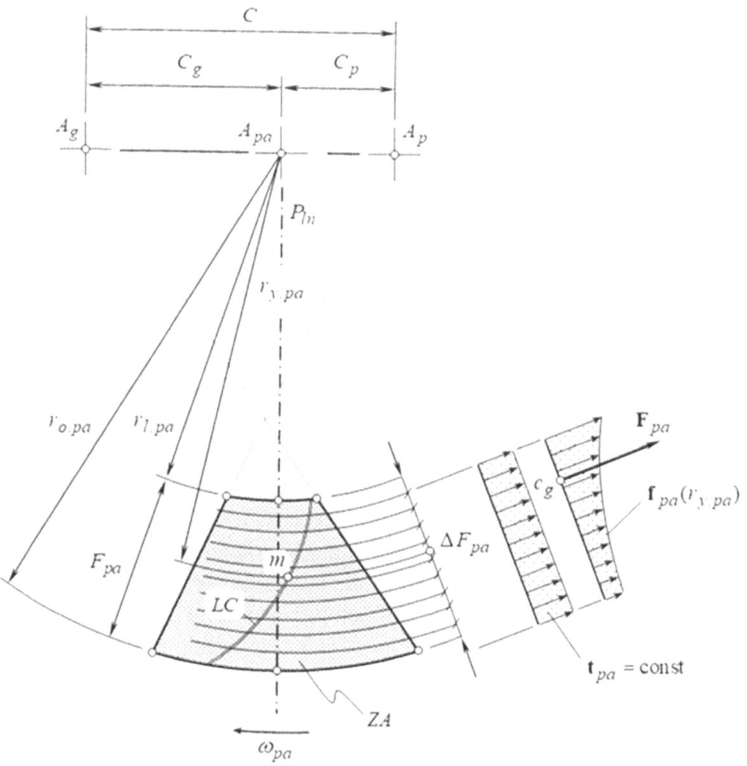

Fig. 4.6 Forces acting in geometrically accurate crossed-axes gear pair

In crossed-axes gear pairs of a special design the total number of the lines of contact can be greater than three.

A gear and a mating pinion in a crossed-axes gear pair can be sliced onto a large enough number, N_{sl}, of slices, each of which is perpendicular to the gear axis of rotation, O_g, and to the pinion axis of rotation, O_p, correspondingly. It is handy to illustrate the slicing of a gear pair using the plane of action, *PA*, for this purpose. The thickness of the slices is designated as ΔF_{pa}. It can be calculated as:

$$\Delta F_{pa} = \frac{F_{pa}}{N_{sl}} \tag{4.23}$$

In the best-case scenario, the slice thickness approaches to a zero ($\Delta F_{pa} \rightarrow 0$), and the number of the slices approaches an infinity ($N_{sl} \rightarrow \infty$). In this latter case the slice thickness is denoted by $d F_{pa}$. It is assumed here and below that the torque being transmitted by each of the slices is equal to one another. Considering the plane-of-action torque, $\mathbf{T}_{pa}$, the following expression:

$$\mathbf{t}_{pa} = \frac{\mathbf{T}_{pa}}{F_{pa}} = \text{const} \tag{4.24}$$

can be composed for a torque per unit length, $\mathbf{t}_{pa}$.

Similar equations:

$$\mathbf{t}_g = \frac{\mathbf{T}_g}{F_{pa}} = \text{const}; \mathbf{t}_p = \frac{\mathbf{T}_p}{F_{pa}} = \text{const} \tag{4.25}$$

Are valid with respect to the gear, as well, as to the mating pinion.

As the torque per unit length, $\mathbf{t}_{pa}$, is of a constant value, and the equality, $\mathbf{t}_{pa}F_{pa} = \mathbf{T}_{pa}$, is valid, then the actual value of the plane-of-action torque, $\mathbf{T}_{pa}$ (lumped torque), is proportional to the shadowed area in Fig. 4.6.

Equal torque share among the slices is graphically illustrated in Fig. 4.6.

In order to transmit a given power, it is always desirable to design and to implement gearboxes of the smallest possible size. From this perspective, the active portion of the line of contact, *LC*, should begin from the *plane-of-action-apex*, A_{pa}. Evidently, the line of contact of such a geometry is far from to be practical, as the maximum contact and bending strength of the gear teeth is restricted by physical properties of a material the gear and the mating pinion are made of.

Calculation of the design parameters of the favorable portion of the line of contact is based on the assumption that the power being transmitting by a gear pair is equally shared within the active portion of the gear pair face width. Under the torque of a constant value the smaller diameter of a gear/pinion, the large force and vice versa. Therefore, a practical value of the smallest possible diameter of the gear/pinion is limited by the yield contact and bending stress in the gear tooth.

The above discussion makes reasonable the following assumption that is referred to as the *first fundamental assumption* in dynamics of crossed-axes gearing:

Assumption 4.1 *In a crossed-axes gear pair, the torque per unit length,* $\mathbf{t}_{pa}$, *is equally shared in the radial direction among an infinite number of infinitesimally narrow slices, each of which is perpendicular to the axis of rotation either of the plane of action, or of the gear, or of the pinion.*

This conclusion is derived on the premise of mandatory equilibrium of the adjacent infinitesimally narrow strips of the plane of action, *PA*, which must be stationary in relation to one another (and, thus, are not allowed traveling with respect to each other).

A similar assumption has been made with respect to parallel-axes gear pairs, as well, as with respect to intersected-axes gearing.

4.3.2 *Forces of Interaction in Geometrically Accurate Crossed-Axes Gearing*

The forces that act in a geometrically-accurate crossed-axes gear pair are considered below in different reference systems. The analysis begins with the load applied within the plane of action, *PA*. Then, this analysis enhanced to the forces that act on the bearings, on the gear housing, and so forth.

Total Force Acting in Geometrically Accurate Crossed-Axes Gearing

In the plane-of-action, *PA*, the torque, $\mathbf{T}_{pa}$, creates the plane-of-action tangential force, $\mathbf{F}_{pa}$. The tangential force per unite length, $\mathbf{f}_{pa}$, can be expressed in terms of the torque per unit length, $\mathbf{t}_{pa}$, and the distance, $r_{y.\,pa}$, of a particular slice from the plane-of-action point, A_{pa}, as:

$$\mathbf{f}_{pa} = \frac{\mathbf{t}_{pa}}{r_{y.pa}} \tag{4.26}$$

It is instructive to point out here that the tangential force per unite length, $\mathbf{f}_{pa}$, is a function of the distance, $r_{y.\,pa}$, of point of interest, m, from the plane-of-action point, A_{pa}, that is: $\mathbf{f}_{pa} = \mathbf{f}_{pa}(r_{y.\,pa})$.

The distribution of the tangential force per unite length, $\mathbf{f}_{pa}(r_{y.\,pa})$, in the radial direction of the plane of action is illustrated in Fig. 4.6.

Having the tangential force per unite length, $\mathbf{f}_{pa}$, determined, the total tangential forth, $\mathbf{F}_{pa}$, is calculated from the expression:

$$\mathbf{F}_{pa} = F_{pa} \cdot \mathbf{f}_{pa} + \mathbf{t}_{pa} \int\limits_{r_{l.pa}}^{r_{o.pa}} \frac{1}{r_{y.pa}} dr_{y.pa} = F_{pa} \cdot \mathbf{f}_{pa} + \mathbf{t}_{pa} \cdot \left(\ln|r_{o.pa}| - \ln|r_{l.pa}|\right) \quad (4.27)$$

$$\mathbf{F}_{pa} = F_{pa} \cdot \mathbf{f}_{pa} + \mathbf{t}_{pa} \cdot \ln\left|\frac{r_{o.pa}}{r_{l.pa}}\right| \quad (4.28)$$

The similar equations:

$$\mathbf{F}_{g} = F_{pa} \cdot \mathbf{f}_{g} + \mathbf{t}_{g} \cdot \ln\left|\frac{r_{o.pa}}{r_{l.pa}}\right| \quad (4.29)$$

$$\mathbf{F}_{p} = F_{pa} \cdot \mathbf{f}_{p} + \mathbf{t}_{p} \cdot \ln\left|\frac{r_{o.pa}}{r_{l.pa}}\right| \quad (4.30)$$

are valid with respect to the gear, as well, as to the mating pinion.

The total (lumped) forth, $\mathbf{F}_{pa}$, is applied at point, c_g, within the line of contact, LC, that is remote from the plane-of-action apex, A_{pa}, at distance, r_{cg} (see Fig. 4.6).

Transmission of a rotation from a driving shaft to a driven shaft is due to the tangential forth, $\mathbf{F}_{pa}$, that acts between the tooth flanks, $\mathcal{G}$ and $\mathcal{P}$, of the mating gears.

When the gears rotate, friction is observed between the gear tooth flank, $\mathcal{G}$, and the mating pinion tooth flank, $\mathcal{P}$. The friction between the compressed gear and pinion tooth flanks, $\mathcal{G}$ and $\mathcal{P}$, is due to the sliding that occurs between the tooth flanks, $\mathcal{G}$ and $\mathcal{P}$. As it is already shown earlier, sliding of two types are distinguished in crossed-axes gear pairs. The profile sliding is the first kind, and the sliding in the lengthwise direction of the gear teeth is the second kind of sliding in crossed-axes gearings. Therefore, friction forces of two types have to be recognized, namely, the profile friction force, $\mathbf{F}_{pr}$, and the friction force in the lengthwise direction, $\mathbf{F}_{dr}$. These friction forces contribute to the total force of the interaction, $\mathbf{F}_{\Sigma}$, between the gear tooth flank, $\mathcal{G}$, and the mating pinion tooth flank, $\mathcal{P}$:

$$\mathbf{F}_{\Sigma} = \mathbf{F}_{pa} + \mathbf{F}_{pr} + \mathbf{F}_{dr} \quad (4.31)$$

The component, $\mathbf{F}_{pa}$, is specified by Eq. (4.28).

In the reference system, $X_{pa}Y_{pa}Z_{pa}$, associated with the rotated plane of action, PA, the profile (transverse) friction force, $\mathbf{F}_{pr}$, is pointed along the axis Z_{pa}, as illustrated in Fig. 4.7. The actual value of this force equals $\mathbf{F}_{pr} = \mu_{pr} \cdot \mathbf{F}_{pa}$. Here, μ_{pr} designates the coefficient of friction in profile sliding of the gear, $\mathcal{G}$, and of the pinion, $\mathcal{P}$, tooth flanks.

In that same reference system, $X_{pa}Y_{pa}Z_{pa}$, the friction force in the lengthwise direction, $\mathbf{F}_{dr}$, is pointed along the axis X_{pa}. The actual value of this component equals (see Fig. 4.7) $\mathbf{F}_{dr} = \mu_{dr} \cdot \mathbf{F}_{pa}$. The coefficient of friction in "*drag*" sliding of the gear, $\mathcal{G}$, and of the pinion, $\mathcal{P}$, tooth flanks is denoted by μ_{dr}.

The coefficients of friction, μ_{pr} and μ_{dr}, can be of different values, as the normal curvatures of the gear and of the pinion tooth flanks, $\mathcal{G}$ and $\mathcal{P}$, at point of interest are

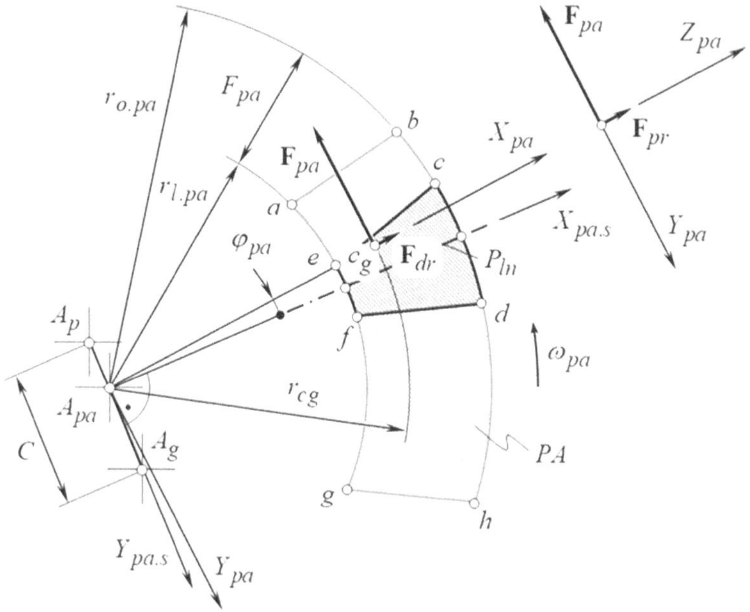

Fig. 4.7 Tangential force, $\mathbf{F}_{pa}$, and friction forces, $\mathbf{F}_{pr}$ and $\mathbf{F}_{dr}$, in crossed-axes gear pair

significantly different in the transverse, and in the lengthwise directions of the tooth flanks $\mathcal{G}$ and $\mathcal{P}$. Moreover, the conditions of profile (transverse) sliding, and that in the lengthwise direction of the gear teeth, are also different. However, when performing a preliminary analysis, the coefficients of friction, μ_{pr} and μ_{dr}, can be considered equal to one another ($\mu_{pr} = \mu_{dr}$).

The total friction force (lumped force), $\mathbf{F}_{fr.\ \Sigma}$, in geometrically accurate crossed-axes gearing is calculated as:

$$\mathbf{F}_{fr.\Sigma} = \mathbf{F}_{pr} + \mathbf{F}_{dr} \tag{4.32}$$

The friction forces, $\mathbf{F}_{pr}$ and $\mathbf{F}_{dr}$, in a crossed-axes gear pair are negligibly small, and, thus, commonly they are not taken into account when calculating the design parameters of the gear transmission. Moreover, the friction forces, $\mathbf{F}_{pr}$ and $\mathbf{F}_{dr}$, do not affect the uniformity of the output rotation in a gear pair, as both, $\mathbf{F}_{pr}$ and $\mathbf{F}_{dr}$, act along the straight lines that are entirely situated within a plane that is perpendicular to the plane of action, *PA* (and that is tangent to the gear, $\mathcal{G}$, and the pinion, $\mathcal{P}$, tooth flanks at point of interest, m). Only component of the total force of interaction, $\mathbf{F}_{\Sigma}$, that is pointed along the common perpendicular to the tooth flanks, $\mathcal{G}$ and $\mathcal{P}$, causes a rotation of the driven gear in a crossed-axes gear pair.

Different components of the total tangential force, $\mathbf{F}_{pa}$, of interaction between the tooth flanks, $\mathcal{G}$ and $\mathcal{P}$, of a gear and of a mating pinion, along with the forces $\mathbf{F}_g$ and $\mathbf{F}_p$ that act on the gear and the pinion, are entered into equations for the calculation of the bending, and of the contact strength of the gear and of the pinion teeth, the bearings, the housing, the shafts, and so forth.

The forces that act on the gear, have to be expressed in a stationary coordinate system, $X_{g.\,s}Y_{g.\,s}Z_{g.\,s}$, associated with the motionless gear (that is, associated with the gear pair housing).

After the total tangential force, $\mathbf{F}_{pa}$, of interaction between the tooth flanks, $\mathcal{G}$ and $\mathcal{P}$, of a gear and of a mating pinion has been pre-multiplied by the operator of the linear transformation, $\mathbf{Rs}_{Ca}\,(pa \rightarrow g_s)$, the total force, $\mathbf{F}_{g.\,s}$, that acts over the gear can be represented in the form of a column matrix:

$$\mathbf{F}_{g.s} = \mathbf{Rs}\left(pa \rightarrow g_s\right) \cdot \mathbf{F}_{pa} = \begin{bmatrix} F_{x.g.s} \\ F_{y.g.s} \\ F_{z.g.s} \\ 1 \end{bmatrix} \tag{4.33}$$

The components $F_{x.\,g.\,s}$, $F_{y.\,g.\,s}$, and $F_{z.\,g.\,s}$, of the total force, $\mathbf{F}_{g.\,s}$, in Eq. (4.33) are equal to the separating force, $F_{g.\,sep}$, the tangential force, $F_{g.\,tg}$, and the axial thrust, $F_{g.\,ax}$, correspondingly, that is, the equalities $F_{x.\,g.\,s} = F_{g.\,sep}$, $F_{y.\,g.\,s} = F_{g.\,tg}$, and $F_{z.\,g.\,s} = F_{g.\,ax}$, are valid. The forces, $\mathbf{F}_{g.\,sep}$, $\mathbf{F}_{g.\,tg}$, and $\mathbf{F}_{g.\,ax}$, are entered into the equations for the calculation of the design parameters of a gearbox.

Forces Acting on the Gear and the Pinion in Geometrically Accurate Crossed-Axes Gearing

In a reference system, $X_{pa}Y_{pa}Z_{pa}$, associated with the rotated plane of action, PA, orientation of the force, $\mathbf{F}_{pa}$ [namely, of the force specified by Eq. (4.28)], or of the total force, $\mathbf{F}_{\Sigma}$ [that is specified by Eq. (4.31)], does not alter when the gears rotate. However, actual configuration of the coordinate system, $X_{pa}Y_{pa}Z_{pa}$, in relation to a motionless reference system, associated with the gear pair housing, alters when the gears rotate. Therefore, in a stationary coordinate system, $X_{g.\,s}Y_{g.\,s}Z_{g.\,s}$, associated with the motionless gear, Eq. (4.33) has to be rewritten in a form:

$$\mathbf{F}_{g.s}\left(\varphi_g\right) = \begin{bmatrix} F_{g.sep}\left(\varphi_g\right) \\ F_{g.tg}\left(\varphi_g\right) \\ F_{g.ax}\left(\varphi_g\right) \\ 1 \end{bmatrix} \tag{4.34}$$

As it follows from the analysis of Eq. (4.34), the total lumped force, $\mathbf{F}_{g.\,s}$, as well, as all three components, $\mathbf{F}_{g.\,sep}$, $\mathbf{F}_{g.\,tg}$, and $\mathbf{F}_{g.\,ax}$, that act over the gear, depend on the angle of rotation of the gear, that is, all of them are functions the angle, φ_g, of rotation of the gear:

$$\begin{array}{ll} \mathbf{F}_{g.s} = \mathbf{F}_{g.s}(\varphi_g) & \mathbf{F}_{g.tg} = \mathbf{F}_{g.tg}(\varphi_g) \\ \mathbf{F}_{g.sep} = \mathbf{F}_{g.sep}(\varphi_g) & \mathbf{F}_{g.ax} = \mathbf{F}_{g.ax}(\varphi_g) \end{array} \tag{4.35}$$

The lumped force, $\mathbf{F}_{g.\ s}(\varphi_g)$, and the components, $\mathbf{F}_{g.\ sep}(\varphi_g)$, $\mathbf{F}_{g.\ tg}(\varphi_g)$, and $\mathbf{F}_{g.\ ax}(\varphi_g)$, that depend on the angular parameter, φ_g, are an additional source of vibration generation, and noise excitation in crossed-axes gearing. When the gears rotate, point of intersection of the axis of instantaneous rotation, P_{ln}, and the instantaneous line of action, LA_{inst}, are not allowed to migrate along the axis of instantaneous rotation, P_{ln}, or, at least, the distance covered in this migration has to be the shortest possible. It is desirable to design gears so, as to minimize the travel distance along the axis of instantaneous rotation.

The variation of the direction of the (total) lumped force, $\mathbf{F}_{g.\ s}(\varphi_g)$, of interaction of tooth flanks, $\mathcal{G}$ and $\mathcal{P}$, of a gear and of a mating pinion is illustrated in Fig. 4.8. For an arbitrary configuration of the line of contact, LC, between the tooth flanks, $\mathcal{G}$ and

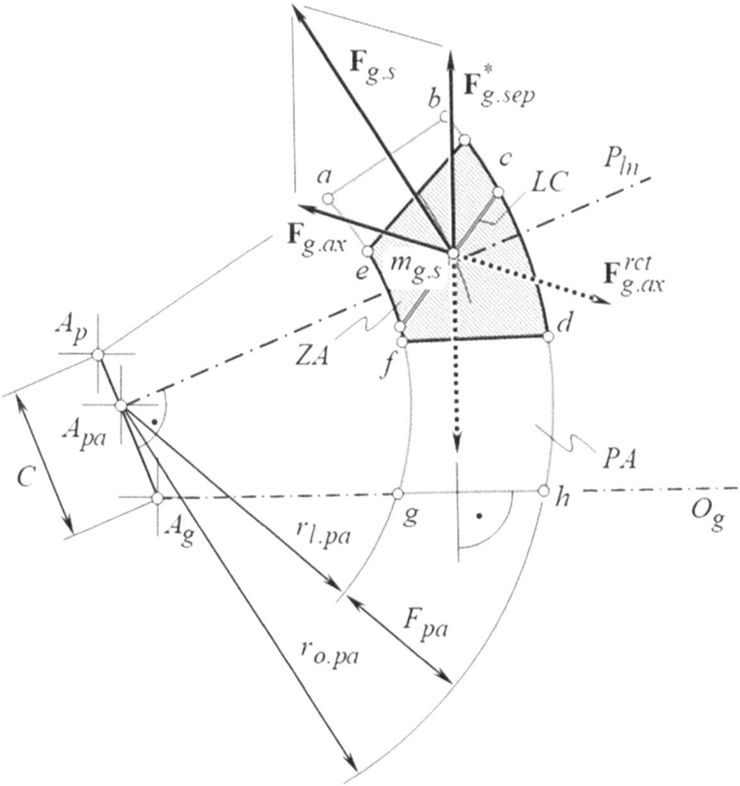

Fig. 4.8 The axial thrust, $\mathbf{F}_{g.\ ax}(\varphi_g)$, and the separating force, $\mathbf{F}_{g.\ sep}(\varphi_g)$, as the components of the resultant (lumped) force, $\mathbf{F}_{g.\ s}(\varphi_g)$, of interaction of the tooth flanks, $\mathcal{G}$ and $\mathcal{P}$, of a gear, and of a mating pinion in geometrically accurate crossed-axes gear pair

$\mathcal{P}$, the coordinates of point, c_g, at which the total force, $\mathbf{F}_{g.\ s}$, is applied, can be determined. The component, $\mathbf{F}_{g.\ ax}$, is parallel to the gear axis of rotation, O_g. The component, $\mathbf{F}_{g.\ sep}$, is perpendicular to the gear axis of rotation, O_g. The magnitude, $F^*_{g.sep}$, of the shown in Fig. 4.9 the force vector, $\mathbf{F}^*_{g.sep}$, equals to $F^*_{g.sep} = F_{g.sep}/\sin\phi_{t.\omega}$, where $\phi_{t.\ \omega}$ is the transverse pressure angle in the crossed-axes gear pair. (The separating force vector, $\mathbf{F}_{g.\ sep}$, is entirely situated in the normal $N_{ln}-$ plane. The normal plane (the $N_{ln}-$ *plane*) is the plane through the plane-of-action-apex, A_{pa}, perpendicular to the centerline, ℄, in the gear pair. As it follows from the analysis of the schematic shown in Fig. 4.9, at any configuration of the line of contact, *LC*, for which the point, c_g, is situated within the zone of action, *ZA*, the forces, $\mathbf{F}_{g.\ ax}$ and $\mathbf{F}_{g.\ sep}$, never equals to zero. Thus, the reaction forces, $\mathbf{F}^{rct}_{g.ax}$ and $\mathbf{F}^{rct}_{g.sep}$, also are of nonzero values. As the forces, $\mathbf{F}^{rct}_{g.ax}$ and $\mathbf{F}^{rct}_{g.sep}$, fluctuate when the gears rotate [see Eqs. (4.35)], an excessive vibration generation, and noise excitation can be observed when a crossed-axes gear pair operates. The variation of the forces, $\mathbf{F}^{rct}_{g.ax}$ and $\mathbf{F}^{rct}_{g.sep}$, does not depend on the geometry of the line of contact, *LC*. This discussion is also valid in cases, when multiple lines of contact, as well, as multiple portion of the lines of contact, are observed in a gear pair of a particular design, as the forces, $\mathbf{F}^{rct}_{g.ax}$ and $\mathbf{F}^{rct}_{g.sep}$, fluctuate when the gears rotate. Fluctuation of the forces, $\mathbf{F}^{rct}_{g.ax}$ and $\mathbf{F}^{rct}_{g.sep}$, is inevitable in geometrically accurate crossed-axes gear pairs.

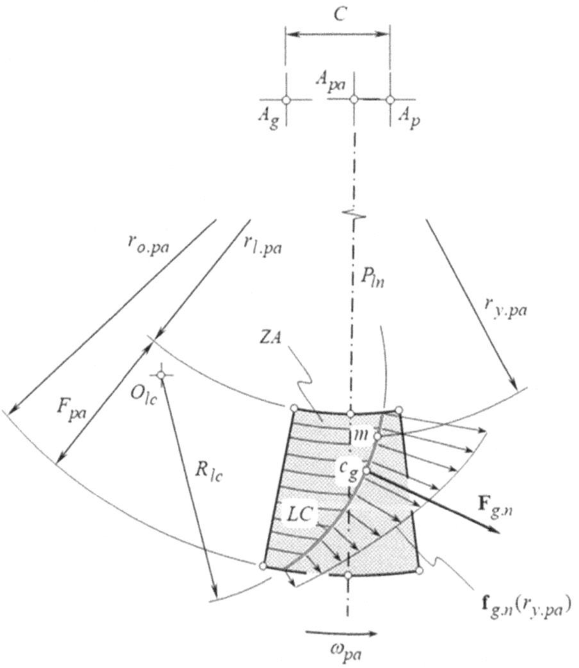

Fig. 4.9 Lumped normal force, $\mathbf{F}_{g.\ n}$, and normal force per unit length, $\mathbf{f}_{g.\ n}$, acting in geometrically accurate crossed-axes gear pair

An analysis, similar to the aforementioned one, can be performed with respect to loading of the pinion in a geometrically accurate crossed-axes gear pair.

As it follows from the analysis of Eq. (4.34), the resultant lumped force, $\mathbf{F}_{p.\ s}$, as well, as all three components, $\mathbf{F}_{p.\ sep}$, $\mathbf{F}_{p.\ tg}$, and $\mathbf{F}_{p.\ ax}$, that act over the pinion, depend on the angle of rotation of the pinion, that is, all of them are functions the angle, φ_p, of rotation of the pinion:

$$\begin{array}{ll} \mathbf{F}_{p.s} = \mathbf{F}_{p.s}(\varphi_p) & \mathbf{F}_{p.tg} = \mathbf{F}_{p.tg}(\varphi_p) \\ \mathbf{F}_{p.sep} = \mathbf{F}_{p.sep}(\varphi_p) & \mathbf{F}_{p.ax} = \mathbf{F}_{p.ax}(\varphi_p) \end{array} \tag{4.36}$$

The lumped force, $\mathbf{F}_{p.\ s}(\varphi_p)$, and the components, $\mathbf{F}_{p.\ sep}(\varphi_p)$, $\mathbf{F}_{p.\ tg}(\varphi_p)$, and $\mathbf{F}_{p.\ ax}(\varphi_p)$, that depend on the angular parameter, φ_p, are an additional source of vibration generation, and noise excitation in crossed-axes gearing. When the gears rotate, no migration of point of intersection of the axis of instantaneous rotation, P_{ln}, as well, as of the instantaneous line of action, LA_{inst}, along the axis of instantaneous rotation, P_{ln}, is allowed (or, at least, the distance covered in this migration has to be the shortest possible). It is desirable to design gears so, as to minimize the travel distance along the axis of instantaneous rotation.

For any configuration of the line of contact, *LC*, when the point, c_g, is situated within the zone of action, *ZA*, the forces, $\mathbf{F}_{p.\ ax}$ and $\mathbf{F}_{p.\ sep}$, never equal to zero. Thus, the reaction forces, $\mathbf{F}^{rct}_{p.ax}$ and $\mathbf{F}^{rct}_{p.sep}$, also are of nonzero values. As the forces, $\mathbf{F}^{rct}_{p.ax}$ and $\mathbf{F}^{rct}_{p.sep}$, fluctuate when the gears rotate, an excessive vibration generation, and noise excitation are commonly observed when a crossed-axes gear pair operates. The variation of the forces, $\mathbf{F}^{rct}_{p.ax}$ and $\mathbf{F}^{rct}_{p.sep}$, does not depend on the geometry of the line of contact, *LC*. This discussion is also valid with respect to the cases when multiple lines of contact, as well, as multiple portion of the lines of contact, are observed in a gear pair of a particular design, as the fluctuation of the forces, $\mathbf{F}^{rct}_{p.ax}$ and $\mathbf{F}^{rct}_{p.sep}$, is inevitable in geometrically accurate crossed-axes gear pairs.

Normal Force Acting on the Gear in Geometrically Accurate Crossed-Axes Gearing

For the calculations of gear teeth for contact strength, as well, as for contact strength analysis, a component, $\mathbf{F}_{g.\ n}$, of the total forth, $\mathbf{F}_{g.\ s}$, is entered into the corresponding equations for the calculations. This component is commonly referred to as the *normal force*, $\mathbf{F}_{g.\ n}$. The vector of normal load, $\mathbf{F}_{g.\ n}$, is situated within the plane of action, *PA*, and is perpendicular to the line of contact, *LC*, at point, c_g, at which the normal load is applied. The magnitude, $F_{g.\ n}$, of the normal force, $\mathbf{F}_{g.\ n}$, is calculated from the expression $F_{g.\ n} = F_{g.\ s} \cos \psi_g$, where ψ_g is the actual value of the spiral angle of the gear teeth at *center-of-gravity point*, c_g.

The normal force per unit length, $\mathbf{f}_{g.\ n}$, is also entirely situated within the plane of action, *PA*. The normal force per unit length, $\mathbf{f}_{g.\ n}$, depends on the geometry of a particular line of contact. At every point of the line of contact, *LC*, the applied normal load per unit length, $\mathbf{f}_{g.\ n}$, is perpendicular to the line of contact at that point.

In straight bevel gearing, the normal force per unit length, $\mathbf{f}_{g.\ n}$, is distributed much similar to that shown in Fig. 4.6 for the force per unit length, $\mathbf{f}_{pa}(r_{y.\ pa})$. In skew crossed-axes gearing the normal force per unit length, $\mathbf{f}_{g.\ n}$, is distributed similar to that in straight bevel gearing – all the elementary forces are perpendicular to the line of contact, *LC*. The distribution of the normal force per unit length, $\mathbf{f}_{g.\ n}$, in spiral bevel gearing with a circular-arc line of contact, *LC*, of a radius, R_{lc}, is illustrated in Fig. 4.9. In this particular case, all the elementary forces are pointed along the radius, R_{lc}, to the corresponding point, *m*, within the line of contact, *LC*.

In a more general case of geometrically accurate crossed-axes gearing, in face-hobbed bevel gears in particular, the following approach can be used to determine the distribution of the normal force per unit length along the line of contact, *LC*.

First, the unit normal vector, $\mathbf{n}_{lc}$, is constructed at every point of the line of contact, *LC*.

Second, the distribution of the normal load per unit length, $\mathbf{f}_{g.\ n}$, is constructed as a product of the unit vector, $\mathbf{n}_{lc}$, by the function of the distribution of the magnitude, $f_{g.\ n}$, of the normal load per unit length, that is, $\mathbf{f}_{g.\ n} = f_{g.\ n} \cdot \mathbf{n}_{lc}$.

When two or more lines of contact (or portions of the line of contact), LC_i, are observed, then positon vector, $\mathbf{r}_{cg.\ \Sigma}$, of the point, $c_{g.\ \Sigma}$, at which the resultant load, $\mathbf{F}_{g.\ n.\ \Sigma}$, is applied, can be calculated from the formula:

$$\mathbf{r}_{cg.\Sigma} = \frac{\sum_i l_{lc.i} \cdot \mathbf{r}_{cg.i}}{\sum_i l_{lc.i}} \tag{4.37}$$

where:

i– is the number of a lines of contact (or portions of line of contact), LC_i

$l_{lc.\ i}$– is the length of a line of contact (or a portions of a line of contact), LC_i

$\mathbf{r}_{cg.\ i}$– is the position vector of point, $c_{g.\ i}$, of a line of contact (or a portion of a line of contact), LC_i

$\sum_i l_{lc.i}$– is the total length, *TLLC*, of all the lines of contact, $l_{lc.\ i}$

The variation of the direction of the total lumped force of the interaction of the tooth flanks, $\mathcal{G}$ and $\mathcal{P}$, can be minimized by means of selection of the desirable line of contact, LC_{des}, of proper geometry. When the gears rotate, point of intersection of the axis of instantaneous rotation, P_{ln}, and the line along the total force of interaction of the tooth flanks, $\mathcal{G}$ and $\mathcal{P}$, are not allowed to migrate along the P_{ln}, or, at least, the distance of this migration has to be minimized. In this way, the vibration generation, and noise excitation can be minimized.

Cases of Multiple Lines of Contact

Gear pairs of all designs feature the contact ratio greater than one ($\overline{m} > 1$). The contact ratio in the range of $\overline{m} = 1.4 \ldots 1.8$ is common in present-day parctice.

As the inequality, $\overline{m} > 1$, is always valid, periods of gear meshing when either one or two (or more) portions of the line of contact are occur.

When a single line of contact is observed, the entire load is transmitted through this line of conact between the interacting tooth flanks, $\mathcal{G}$ and $\mathcal{P}$.

When two (or more) portions of the line of contact are observed, the active lines of contact share the load being transmitted by the gear teeth.

The load carrying capacity of a gear pair depends on the manner the load is shared between the lines of contact.

Consider an active portion of the plane of action, *PA*, in a geometrically accurate crossed-axes gear pair, as illustrated in Fig. 4.10. In this schematic, an arbitrary point of interest, m_i, is chosen within an arbitrary line of contact, LC_i. When the gears rotate, point m_i traces the path of contact, P_c. The path of contact, P_c, intersects the adjacent line of contact, $LC_{i \pm 1}$, at point, $m_{i \pm 1}$. At a certain instant of time, contact in the gear pair occurs in both points, m_i and $m_{i \pm 1}$, at that same time. In such a scenario, the load to be transmitted is equally shared between the loacal arcs of the lines of contact in differential vicinity of contact points, m_i and $m_{i \pm 1}$. In cases of multiple lines of contact, the load to be transmitted by a gear pair is shared equally among all the active pairs of teeth (proportionally to the actual number of tooth pairs involved in mesh).

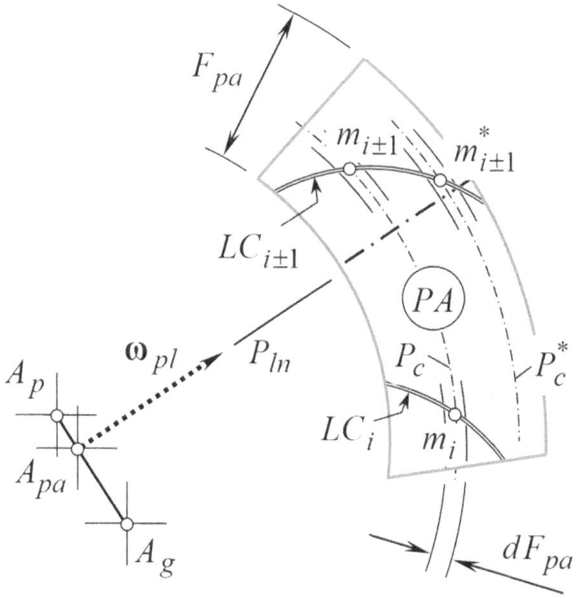

Fig. 4.10 Case of multiple lines of contact

In contrast to the path of contact, P_c, when the gears rotate, another point, $m^*_{i\pm1}$, of the line of contact traces another path of contact, P^*_c. The path of contact, P^*_c, does not intersect any other line of contact between the tooth flanks, $\mathcal{G}$ and $\mathcal{P}$. In this scenario, the entire load to be transmitted is transmitted by a local portion of the arc of the line of contact within the differential vicinity of the contact point, $m^*_{i\pm1}$. This load is twice as big compared to the load, shared in the previous case.

The performed analysis allows the following assumption that is commonly referred to as the *second fundamental assumption* in dynamics of crossed-axes gearing:

Assumption 4.2 *The load to be transmitted by a gear pair is shared equally among the all active pairs of teeth (proportionally to the actual number of tooth pairs involved in mesh).*

This conclusion is derived on the premise of equilibrium of the adjacent infinitesimally narrow strips of the plane of action, *PA*, which must be stationary in relation to one another (and, thus, are not allowed traveling with respect to each other).

The greater the contact ratio in a gear pair, the better the load share among the tooth pairs, and vice versa.

4.4 Conclusion

The book chapter deals with geometrically accurate crossed-axes gearing with line contact between the tooth flanks of the gear, and of the mating pinion. Gearing of this particular design is commonly referred to " $R-$ *gearing*."

In the first subsection of the chapter, fundamentals of $R-$ *gearing* is briefly outlined. Actually, the principal kinematics of $R-$ *gear pairs* and the tooth flank geometry of the mating gears are covered in this subsection of the chapter.

Contact motion characteristics in $R-$ *gearing* are discussed in the second subsection of the chapter. A gear, and a mating pinion tooth flanks sliding are considered. This section of the book chapter ends with the analysis of specific sliding and features of the specific sliding in $R-$ *gearing*.

Elements of dynamics of geometrically accurate crossed-axes gearing are disclosed in the third subsection of the chapter. This section begins with the first principal assumption adopted in the load analysis of $R-$ *gearing*, and is followed by the analysis of forces of interaction in geometrically accurate crossed-axes gearing. This includes, but is not limited to, the following: (a) Total force acting in geometrically accurate crossed-axes gearing; (b) Forces acting on the gear and the pinion in geometrically accurate crossed-axes gearing; and (c) Normal forces acting on the gear and the pinion in geometrically accurate crossed-axes gearing. The case of multiple lines of contact in $R-$ *gear pairs* get special consideration. Here, in this subsection, the second principal assumption adopted in the load analysis of $R-$ *gearing* is introduced.

References

1. Radzevich, S. P. (2013). On fundamentals of the scientific theory of gearing. In *Power transmissions*, Proceedings of the 4th international conference, held at Sinaia, Romania, June 20–23, 2012, Editor: George Dobre, Assisted by Mihai Robert Vladu, Mechanism and Machine Science, Vol. 13, Springer, 2013, pp. 165–172.
2. Radzevich, S. P. (2013). *R*-gearing: A novel kind of crossed-axis gearing having line contact between the tooth flanks of the gear and of the pinion. In *Proceedings of international conference on gears*, October 7–9, 2013, Technical University of Munich (TUM), Garching (Near Munich), 2013, pp. 1303–1314. (Vol. 2).
3. Radzevich, S. P. (2018). *Theory of gearing: Kinematics, geometry, and synthesis* (2nd ed., revised and expanded). CRC Press. 934 pages.
4. Radzevich, S. P. (2019, February). Design features of perfect gears for crossed-axes gear pairs. *Gear Solutions*, pp. 36–43.
5. Radzevich, S. P. (2020, June), Conjugate action law in intersected-axes gear pairs and in crossed-axes gear pairs, *Gear Solutions Magazine* (pp. 42-48).
6. Radzevich, S. P. (2017). *Gear cutting tools: Science and engineering* (2nd ed., 606 pages). CRC Press.

Bibliography

Radzevich, S. P. (Ed.). (2019). *Advances in gear design and manufacture* (549 pages). CRC Press.

Radzevich, S. P. (2018, February). An examination of high-conformal gearing. *Gear Solutions*, 31–39.

Radzevich, S. P. (Ed.). (2021). *Dudley's handbook of practical gear design and manufacture* (4th ed.). CRC Press. 1170 pages, 718 B/W Illustrations.

Radzevich, S. P. (2020). *High-conformal gearing: Kinematics and geometry* (2nd ed., 506 pages). Elsevier.

Radzevich, S. P. (2020), "Knowledge (of gear theory) is power in the design, production, and application of gears", Gear Solutions Magazine, August 2020, 38-44. [Upon request, a.pdf of this article can be ordered (for free) from the author].

Radzevich, S. P. (2022). *Theory of gearing: Kinematics, geometry, and synthesis* (3rd ed., 1128 pages). CRC Press.

Radzevich, S. P. (Ed.). (2023). *Novikov/conformal gearing: Scientific theory and practice* (1st ed., 505 pages). Springer.

Radzevich, S. P. (Ed.). (2022). *Recent advances in gearing: Scientific theory and applications* (1st ed., 569 pages). Springer.

Radzevich, S. P., & Storchak, M. G. (Eds.). (2022). *Advances in gear theory and gear cutting tool design* (500 pages). Springer.

Chapter 5
On the Lack of Understanding of the *Scientific Theory of Gearing* by the Majority of Gear Experts

Stephen P. Radzevich

The *scientific theory of gearing* outlined in the monograph [1] is a new theory in the realm of gear science and arts. Despite the origin of the scientific theory of gearing can be traced back to the fundamental research carried out by *Leonhard Euler* in the eighteenth century, the rapid growth of this theory began only in the early twenty-first century. Through the first two decades of the twenty-first century most (but not all!!) of the key problems in the proposed theory of gearing are solved. It is the right time now to study what is already achieved in this area of mechanical engineering, and to implement these results in the everyday gear practice. Unfortunately, the latest achievements in the scientific theory of gearing are ignored by the majority of the gear scientists and gear engineers. It is likely that the lack of understanding of the scientific theory of gearing by the majority of the gear scientists, engineers, and practitioners is the root cause for that.

It takes time for a considerable amount of gear experts to get familiar with the theory, study it, and begin using it in everyday practice. Evidently, this cannot be done instantly.

To properly evaluate the advantages and benefits of usage of the proposed *scientific theory of gearing*, in-depth knowledge of the theory and a reasonably extensive experience in implementation of it in practice are necessary. Otherwise, there is no chance to evaluate the theory properly and to be benefited from the theory. Moreover, toxic environment can be formed in the gear community due to the lack of knowledge about the proposed *scientific theory of gearing*.

Nowadays, some well-known gear experts, those who pose themselves as gear theoreticians, commit huge mistakes when trying to evaluate the theory, not being qualified to do such an evaluation.

S. P. Radzevich (✉)
Machine Design Department, National Technical University of Ukraine "Igor Sikorsky Kyiv Polytechnic Institute", Sterling Heights, MI, USA

S. P. Radzevich (ed.), *Gear Accuracy*, Mechanical Engineering Series,
https://doi.org/10.1007/978-3-031-74798-4_5

As an illustrative example, consider a recently published article [2] in the *Gear Solutions* magazine. The author of this article has made a weak attempt to review some key accomplishments in the *scientific theory of gearing* and has demonstrated the absence of understanding of the cornerstone ideas of gear science (see Fig. 5.1).

The article [2] is instructive, as it makes clear the poor knowledge of the gear experts when they make an attempt to participate in a discussion on the theoretical aspects of the gear art in a general sense of this term.

Over a year later (in August 2020), another article was published in *Gear Solutions* magazine [3]. The author's attention in the article [3] is focused mainly on the analysis of the principal mistakes committed in the article [2].

For a very short period, this article has appeared in the August 2020 issue of the *Gear Solutions* magazine. The so-called ***established theory*** is not a *theory of gearing at all*, rather than a collection of recipes. *Trial-and-error method* (and NOT a gear science!!) is the key tool to derive such recipes. Shortly this article was removed from the magazine. The August 2020 issue of the *Gear Solutions* magazine no longer contains the earlier published article [3]. This is nonsense, but the already published article was withdrawn from the issue of the magazine[1]!! (Evidence of that can be found on page 6 of this issue of the *Gear Solutions* magazine—for details, see Fig. 5.2, where the name of the authors of this article is mistakenly not removed). Scientific problems cannot be solved this way.

The article [3] is important, as numerous common mistakes committed by other gear experts are responded there. However, because the article [3] is removed from August 2020 issue of the *Gear Solutions* magazine, and, thus, it is no longer available in the public domain, facsimile of this article can be found in Fig. 5.3.

In this second article [3], the author's comments on all the mistakes and inconsistencies committed in the article [2], those pertaining to the gear science, are concisely outlined.

Conclusion *The outlined in this text scientific theory of gearing is an efficient means for solving sophisticated problems in gear design, gear production, gear inspection, and gear application.*

Only proficient and knowledgeable gear experts are allowed to use the proposed theory of gearing. This theory is developed not for amateurs and not for inexperienced gear experts.

[1] The original version of the entire August 2020 issue of the *Gear Solutions* magazine (that contains the article [3]) is available (at no charge) from the author upon request.

Fig. 5.1 Facsimile of the article: [2]

Fig. 5.1 (continued)

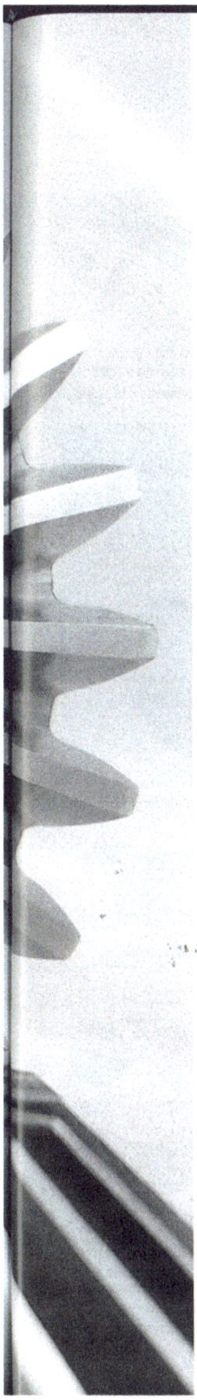

Gear scientists, gear engineers, and gear manufacturers have worked successfully for many decades on finding the optimal flank forms and the optimal non-conjugate flank surface interaction.

By Dr. HERMANN J. STADTFELD

In 1924, Ernest Wildhaber, a well-known gear scientist, invented hypoid gearing. Compared to spiral bevel gears, hypoid gears provide an offset that allows lowering of the body of rear wheel drive vehicles by 50mm or more. This is possible because the propeller shaft between engine/transmission and the driving axle is not positioned at the center of the drive axle but is lowered by the offset amount (Figure 1). This allows the vehicle designer to lower the floor of the vehicle and subsequently the entire body by the same amount. Lowering the center of gravity of a passenger car by 50mm reduces the inertia responsible for sideways rolling by more than 10 percent, which provides better vehicle handling and more active safety. The lower body also reduces the CV coefficient for air resistance, providing higher gas mileage. Less than five years after the invention of hypoid gearing, all large automotive manufacturers around the world had converted their passenger cars and trucks to hypoid drive axles with a lower vehicle body.

Ernest Wildhaber emigrated from Switzerland to the U.S. in 1919 and was hired by The Gleason Works as a gear theoretician. Wildhaber received 279 patents, many of which changed the world of gearing. The cylindrical gear tooth profile that is today called *Wildhaber-Novikov* gearing was invented by Ernest Wildhaber in 1926. Mikhail Novikov, a Russian scientist with no access to western publications, invented the same tooth profile independent from Wildhaber in 1956. The contributions of both scientists are honored today by calling this system *Wildhaber-Novikov* gears.

Ernest Wildhaber is the father of modern gear theory. His pioneering contributions have been invaluable for the development of today's gear calculation and manufacturing processes.

NEWS ABOUT HYPOID GEARS?

A recently published article by S.P. Radzevich [1] attempts to explain that hypoid gearsets designed and manufactured today lack good performance and are not perfect but approximate. The article seemingly questions the credibility of many scientists and gear engineers who worked on the theory and its improvement over approximately 100 years. However, reading the article carefully quickly reveals some key misperceptions and false assumptions. This write-up is meant to be a response on behalf of established gear theory and manufacturing.

In the following sections, certain statements and mathematical relationships of the published article are discussed and countering responses are presented and explained. Also, some assumptions about manufacturing and application procedures are discussed and compared to established practices today. The topics of this paper are structured accordingly in:

- The three fundamental laws of gearing.
- Perfect conjugacy.
- Real world applications.
- Transmission design.
- Heat treatment, lapping and grinding.

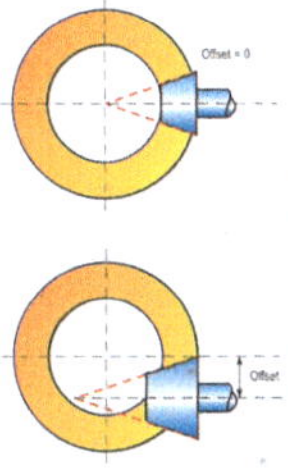

Hypoid Offset

- favorable design freedom
- results in tooth length sliding
- increases the pinion diameter
- delivers good hydrodynamic properties
- has good damping properties
- delivers increased face contact ratio
- results in higher pinion strength
- increased efficiency in combination with correct lubrication

Figure 1: Features of pinion offset.

THE THREE FUNDAMENTAL LAWS OF GEARING

The first fundamental law of gearing quoted in [1] $n_g \bullet V_\Sigma = 0$ also implies $|N_1 \times R_1| = i \bullet |N_2 \times R_2|$, where i is the constant transmission ratio. The three cases in Figure 2 visualize the problem of non-constant ratio and are noncompliant with the first gearing law because of a ratio change from case to case. This problem led Leonard Euler to discover the involute tooth profile. A simplistic mathematical approach teaches that the effective radius vector **R** remains unchanged while the contacting point between two mating flanks moves from R_{b2} to R_{b1} as shown to the

Fig. 5.1 (continued)

right in Figure 3 (movement along the line of action).

The line of action in parallel axes cylindrical gearing is a straight line, connecting the two base circles. If the surface normal vectors N1 and N2 are within the line of action, then the vector product $|\mathbf{N_i} \times \mathbf{R_i}|$ remains constant during a complete mesh cycle. The consequent application of this principle leads to the construction of an involute, as shown to the right in Figure 3.

The line that forms the tooth surface elements while traveling from position "a" to "f" along the line of engagement (line of action) is always perpendicular to the line of action. This principle implies that a tool, simply with straight cutting edges as shown in Figure 4, can be used to form the complex involute profile. Figure 4 also demonstrates the principle of profile shift while maintaining the first fundamental law of gearing.

The second fundamental law of gearing, which was proposed in 2017, $p_{In} \times V_m + n_g = 0$ [1], is a redundant relationship to the first gearing law and it is limited to cylindrical gears with parallel axes and straight bevel gears without hypoid offset. In this case, it adds nothing to the first gearing law; conjugacy is already given by the relationships required in the first gearing law.

The third fundamental gearing law is quoted in [1] in two different notations. The first notation covers only the special case of ratio = 1:

$$\varphi_{b.g} = \varphi_{b.p} = \varphi_{b.op}$$

However, the first notation is incorrect for all cases with different base pitch diameters because the circular pitch and not the angular pitch has to be equal between pinion, gear, and the operating base pitch. The second notation:

$$p_{b.g} = p_{b.p} = p_{b.op}$$

is consistent with the requirement of equal circular pitch, but the article makes no mention of the different notations.

PERFECT CONJUGACY IN STRAIGHT BEVEL GEARSETS

Bevel gears with intersecting axes is the topic of a series of three papers published between October 2014 and January 2015 [2]. A straight bevel gearset with skew teeth was modeled, and a sample was manufactured. This publication addressed two points: the design of a gearset with low tooth count and the solution for perfect conjugacy. There is no mention in the article that the Coniflex® straight bevel gear system has been used since the 1940s for low tooth count gearsets such as differential gears.

Straight bevel gears such as Gleason Coniflex have tapered depth teeth, where the pitch cones roll on each other and the pitch apexes of pinion and gear match the crossing point of the axes. In the standard case, the face and root cone apexes match the crossing point. In such a standard Coniflex design, the base elements are also cones with cone apexes that match the crossing point of the axes. The involute development in Figure 3 can be applied to an infinite number of normal sections along the face width of a straight bevel gear which allows an involute development similar to that for cylindrical gears. The conical base elements of both members can be connected with a straight line (the line of action) in each section along the face width, whereas the plurality of all lines of action forms a plane (the plane of action).

This principle is shown in Figure 5. The two cones in Figure 5 are base cones of a straight bevel or a spiral bevel gearset. In the right two graphics, the view is directed such that the plane of action appears as a line that is tangential to the two cone-enveloping surfaces. The left side graphic in Figure 5 shows the plane of action three-dimensionally and how it connects the two base elements.

The plane of action cannot be extended beyond its tangential contacting line with the base element as shown incorrectly in Figure 1 in [2] and Figure 6 in [1]. The plane of action only exists where tooth engagement is possible and it is different than the generating gear plane (more specifically explained later).

There is, however, one difference from the true involute of cylin-

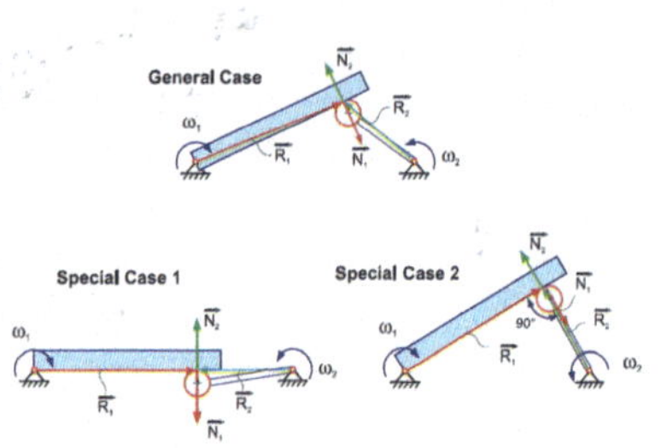

Figure 2: Three general cases of motion transmission with a bar and a crank.

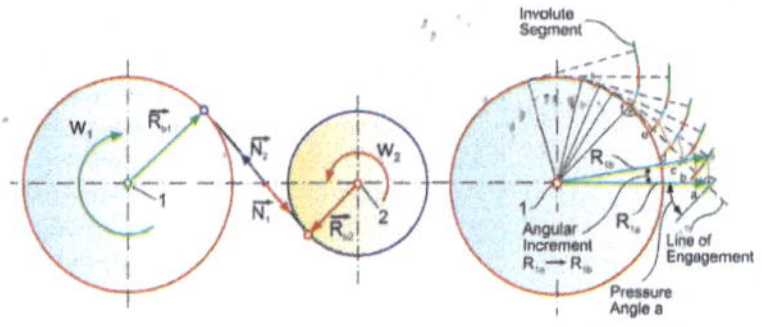

Figure 3: Line of action between base circles (left) and subsequent involute development.

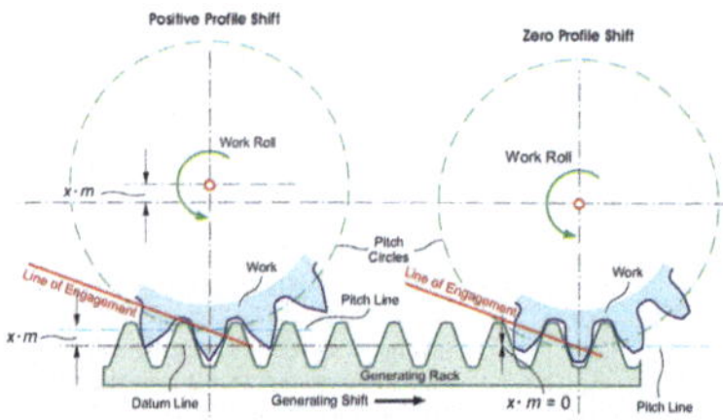

Figure 4: Generating rack with straight profile forming involute profiles.

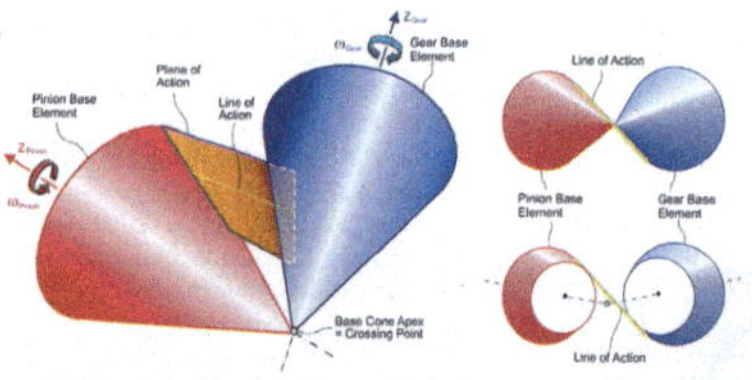

Figure 5: Conical base elements and plane of engagement.

Fig. 5.1 (continued)

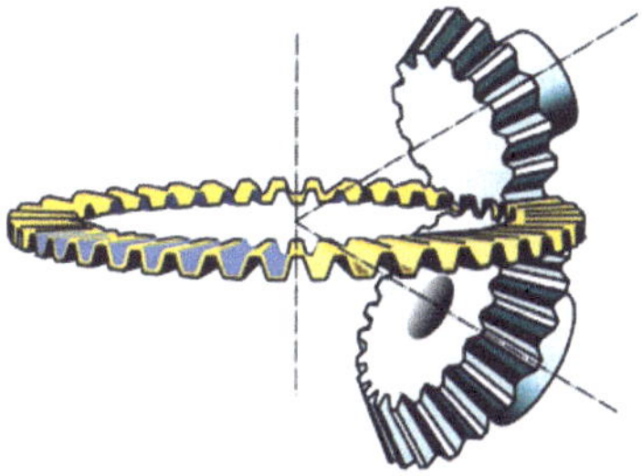

Figure 6: Ring shaped generating rack with trapezoidal profile.

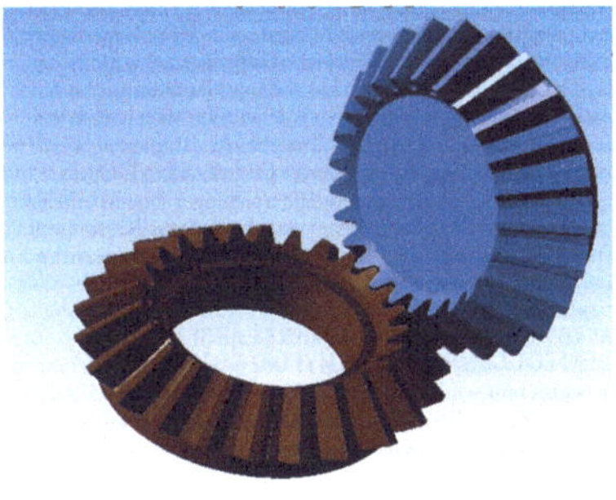

Figure 7: Solid model computer graphic of a perfectly conjugate straight bevel gearset.

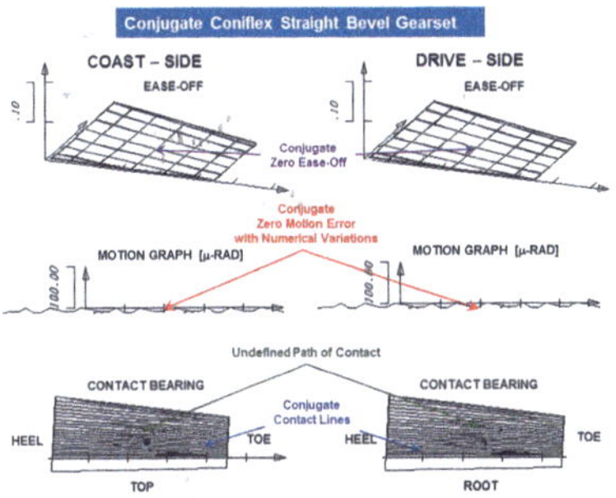

Figure 8: Tooth contact analysis of the perfectly conjugate straight bevel gearset.

drical gears. The rotation of pinion and gear does not occur in the normal plane but in the transverse plane. Because of this difference, the flank profile of straight bevel gears (and all other bevel gear types) is called Octoide. The Octoide is the analog function of an involute, and it provides to bevel gears the same advantages as an involute provides to cylindrical gears. Those advantages are constant ratio, center distance insensitivity, and ease of manufacturing.

Like cylindrical gears, bevel gears also have a trapezoidal generating profile. The straight rack of cylindrical gears becomes a ring, as shown in Figure 6. It is required to establish certain conditions in order to make the ring rack the generating gear for a pinion and a ring gear that will mesh perfectly together with zero transmission error and line contact identical to cylindrical gears. Those conditions are postulated in the kinematic coupling requirements:

- The flank surfaces of the generating gears of the two mating bevel gears are congruent (same shape but mirror images, as given in the example of Figure 6).
- The generating gears of the two mating bevel gears require identical axes of rotation (Top and bottom side of the generating gear in Figure 6 form the same generating gear, which rotates in both cases around the same axis and therefore satisfies condition 2).
- The surface of engagement of pinion and ring gear must be identical to the surface of engagement between pinion and generating gear and also to the one between ring gear and generating gear (without detailed knowledge of the surfaces of engagement, the global condition in Figure 6 seems to satisfy this requirement).

The generating gear principle has to be understood as the ultimate vehicle to form the teeth of two mating gears. The first fundamental law of gearing is fully executed by choosing trapezoidal profiles and by applying the kinematic coupling requirements. Gears are designed and manufactured in order to mesh with each other. What better way to manufacture them by way of a generating gear? The generating gear is represented by the manufacturing machine; it forms the teeth of the gear (at the bottom in Figure 6) while meshing with this gear perfectly. If the pinion is manufactured with the same generating gear but on the opposite side (at the top in Figure 6) and if the generating gear is imagined infinitely thin, then the result is a pinion that perfectly meshes with the gear having zero motion error. It is also given, in such a case, that line contact between the pinion and gear flank surfaces exists along the entire face width. The extraordinary effort shown in [2] to calculate straight bevel gear tooth surfaces that are skew with asymmetric teeth is based on the falsely assumed conjugate requirement. The sample gear pair shown in [2] was manufactured on a machining center with a ball nose endmill in a milling operation likely taking many hours to complete. It is rather simple and straightforward to manufacture a significantly better, perfectly conjugate straight bevel gearset in only a few minutes by using the generating gear method and a standard bevel gear manufacturing machine.

An example straight bevel gearset computer model is shown in Figure 7. The solid model in Figure 7 has been generated by using standard Gleason basic settings, based on the generating gear approach and by applying a standard Coniflex® Plus cutter head as used on Phoenix bevel gear manufacturing machines. The Coniflex straight bevel gear calculation for conjugate contact has to be conducted with a dish angle of 0° and no root angle correction (ΔGammaM = 0.). The dish angle is creating the length crowning and a ΔGammaM creates the profile crowning. With a dish angle of 0° and a ΔGammaM of 0° flank lead lines are straight lines and the profile is a true Octoide (involute equivalent).

A contact analysis of the gearset in Figure 7 is shown in Figure 8. The top of the figure shows the Ease-Offs of left and right flank (called coast and drive side in the graphic). The center of the figure shows the motion transmission errors of the pinion and gear flank pairs. The two bottom graphics are the representation of the tooth contact pattern. The contact pattern graphics are axial

May 2019 45

Fig. 5.1 (continued)

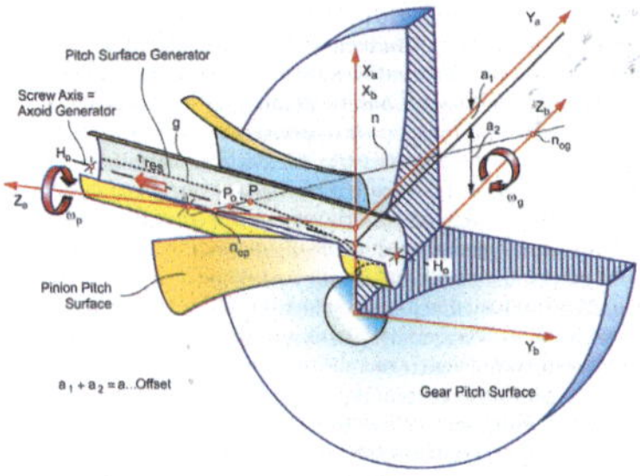

Figure 9: The correct pinion and gear pitch surfaces.

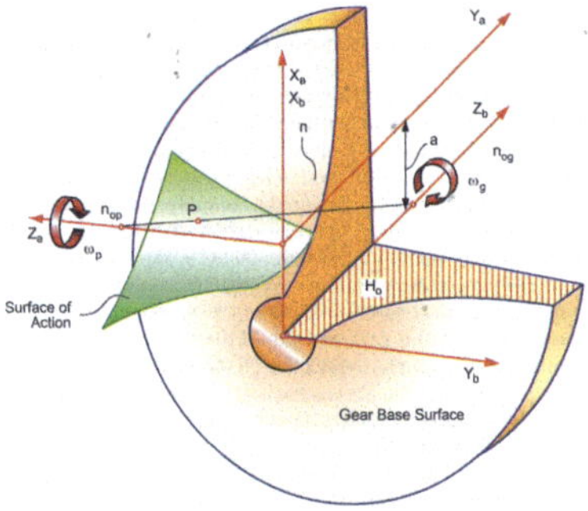

Figure 10: Surface of action connecting the gear and pinion base surfaces.

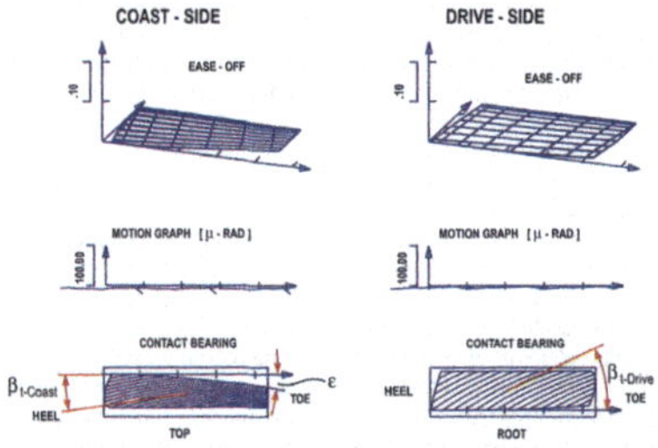

Figure 11: Ease-Offs, motion graphs, and contact lines of a real conjugate hypoid gearset.

projections of the flank surfaces and the contact lines in the same plane a two-dimensional part print would show the tooth area.

The contact analysis in Figure 8 confirms the full line contact in each roll position (lower graphics) as well as the zero motion error (only numerical static) in the center graphic which makes this example the perfectly conjugate straight bevel gearset. In the lower graphic the path of contact was calculated as a zig-zag line which indicates an undefined contact path. This means that due to the conjugacy, every point along each contact line is a path of contact point which makes the analysis program pick random points.

The Ease-Off base plane (top graphics in Figure 8) defines the conjugate state of a flank surface pair. Because the Ease-Off graph of the calculated flank pairs matches the presentation plane (base plane) precisely, that is the proof that a conjugate and precisely rolling gearset was the input of this contact analysis calculation.

The above experiment creating a conjugate straight bevel gearset is strictly academic. Conjugacy is the basis of all gearsets manufactured in high volume on dedicated manufacturing machines. A conjugate bevel gearset cannot be used for a power transmission because manufacturing tolerances and load affected deflections as well as material expansions and deformations under high operating temperatures will result in edge contact and high load concentrations. The load concentrations already start with moderate load and cause material damage and considerable noise emission. Although conjugacy is used as a reference for each design, predetermined amounts of length and profile crowning are applied. The right amount of crowning makes a gearset quiet and gives it the required load carrying capacity. The crowning is shown in the Ease-Off graphics with the conjugate reference always being present as the Ease-Off base plane. Several Ease-Off examples of a gearset with length and profile crowning are shown in the proceedings of this paper.

PERFECT CONJUGACY IN HYPOID GEARSETS

It begins to become more problematic for hypoid gears. In [1] the pitch elements of crossed axes hypoid gears are drawn as cones. Even though the face cones of hypoid gears and pinions are machined conical, the pitch elements are hyperboloids.

Ernest Wildhaber and Arthur Stewart described their invention of hypoid gearing in 1926 [3]. Boris Shtipelman published in 1978 the relationships and derivations required to understand hypoid gears and their hyperbolic pitch elements [4]. Figure 9 offers a graphical interpretation of the hyperbolic pitch elements and their generator. The pitch surface generator is a line which winds on the surface of a cylinder beginning at the crossing point of the axes, equal to the first contact point of the pinion and gear pitch surfaces. The pitch surface generator is developed by the connecting line between the pinion and gear pitch surfaces (n_{op}- n_{og}) by shifting the connecting line along the pinion and gear axes. The connecting line (n_{op}- n_{og}) is normal to the pitch elements. Point P is one point of the pitch surface generator. If the division of the vector products of the distances between the axes and point P and their respective axis direction equals the ratio i of the hypoid gearset then one point of the pitch surface generator is found with:

$$\{(n_{op} - P) \times Z_a\} / \{(P - n_{og}) \times Z_b\} = i$$

Although the pitch elements are hyperbolic and not conical, it is possible to use conical faces for the blanks of pinion and gear. If point P in Figure 9 was chosen at the center of the face width, then line (n_{op}- n_{og}) can be used as a normal vector to define the face angle a blank with straight face cones if the hypoid set was manufactured by face hobbing, which implies parallel depth teeth.

Straight face cones will merely influence the top root clearance of the gearset in the range of 30 to 60 microns. Using straight face cones will not change the form of the pitch surface nor will it influence the base surface. Those functional surfaces are given by the

Fig. 5.1 (continued)

kinematical relationships and have to be considered when thinking about the shape of the surface of action. No plane of action can exist between two hyperbolic base elements. The correct surface of action is curved and warped, as shown in Figure 10.

The conclusion is that the second fundamental law of gearing according to [1] does not exist and is not related in any way to the conjugacy of gears (as explained and demonstrated in the following section).

CONJUGACY BETWEEN MESHING FLANKS

The term conjugate is used in mathematics for two or more surfaces that contact each other along a line. Since the 1970s, the term conjugate has also been employed in gear technology literature to define the "exact" gear pair which presents a triple plurality of line contact between two gear flanks during the meshing process [5].

DEFINITION OF THE CONJUGATE GEAR PAIR

1: The flanks contact along a line (contact line), which is only limited by the boundaries of the teeth i.e. the working area.

2: The line contact between the flanks exists within the entire area of engagement in every mesh position.

3: Line contact is maintained in the entire area of engagement if pinion and ring gear are rotated by angular increments where: (angular pinion increment) / (angular ring gear increment) = transmission ratio.

The Ease-Off is a three-dimensional graphic of the flank deviations from a conjugate pair. It is calculated by transformation of a pinion flank "into" the gear coordinate system according to the first gearing law, resulting in a virtual gear flank that is conjugate to the actual pinion flank. This conjugate gear flank will then be compared to the present gear flank, where all differences in arc length are plotted point by point in ordinate direction into the Ease-Off graphic.

If both mating bevel gears have conjugate manufacturing data, then the Ease-Off graphic has no deviations in ordinate direction. Also, if the pinion flanks and the gear flanks have spiral angle and pressure angle errors of equal amounts, the Ease-Off graphic will not show any deviation. Although the individual gears are considered incorrect in this case, they will roll conjugate with each other, which subsequently leads to an Ease-Off without any ordinate values. Figure 11 shows the analysis results of a typical conjugate hypoid gearset. The Ease-Off graphics have zero crowning magnitudes in the ordinate direction. The motion graph has, next to some numerical entrance and exit variation, zero motion error. The contact bearings show line contact within the entire working area. The coast side contact ends at a toe root undercut (section ε).

Each spiral bevel and hypoid gearset with uniform tooth depth has a conjugate base design. This applies to all face hobbed and some face milled gearsets. A simple explanation of hypoid gearset conjugacy is possible with a non-generated gear that meshes with a generated pinion. For the calculation and manufacturing process, the hyperbolic pitch elements are calculated for the gear first. Then a suitable blade profile (gear cutter in Figure 12) is chosen and positioned in a face cutter head. The cutter head is positioned to create the desired spiral angle. With this procedure, a non-generated gear can be created by computer simulation and it can be manufactured with a bevel gear cutting machine [6].

A pinion cutter (see Figure 12) is positioned in a mathematical model or in a bevel gear cutting machine such that it represents one tooth of the non-generated gear by rotating around its axis. An additional simultaneous rotation around the pinion generating gear axis results in this pinion cutter becoming the generat-

Fig. 5.1 (continued)

ing gear of a conjugate pinion. If the pinion is positioned with the same offset that was used to determine the pitch surfaces (Figure 9), then the cutter rotation around the pinion generating gear axis will form a pinion that is perfectly conjugate to the non-generated gear. The tooth contact analysis in Figure 11 has been obtained from such a non-generated hypoid gearset and therefore shows perfect conjugacy.

More complicated is the generation of a conjugate bevel or hypoid gearset with tapered depth teeth (see Figure 13). If the generating gear axes have identical axes of rotation that are perpendicular to the pitch line, the rotating cutter heads and their blades will not follow the root line of a tapered depth tooth. Tilting the cutter head in order to follow the root line would violate the first kinematic coupling requirement for teeth that are congruent to the slots of the mating member. The solution was developed in the 1940s [7].

If the cutting edges are adjusted in the cutting machine such that tooth reference profile and depth is matched at midface, and if an axial motion of the cradle is introduced that guides the blades along the tapered root line while the generating roll progresses along the face width, then the requirements of congruent teeth and slots are fulfilled with the result of perfect conjugacy.

The process configuration and kinematics in Figure 13 is called *Duplex Completing*. Today, all face-milled and ground spiral bevel and hypoid gears are manufactured with the Duplex Completing process. The axial cradle movement in this process is called Helical Motion and was first introduced with mechanical bevel gear machines in the late 1940s. The *Helical Motion* of the days of mechanical machines required an additional change gear box which actuated a cam that moved the sliding base during the generation process. Today's CNC controlled Phoenix® free form machines have the Helical Motion capability automatically by their interpolated axes movement.

WHY IS CONJUGACY NOT DESIRABLE FOR REAL WORLD APPLICATIONS?

In 1926, Ernest Wildhaber [3, 8] was the first to propose applying surface crowning on hypoid gears. Wildhaber acknowledged that the slightest deviations in the gear housing and in the building position, as well as deflections affected by load and heat, will cause edge contact with peak stress levels of a multiple of the allowable values the gearset had been designed for. The conjugate gearset used for the TCA in Figure 11 was repeated with realistic displacement values of 50μm offset, 50μm pinion cone and 30' of shaft angle change. The results in Figure 14 show warped and tilted Ease-Offs and severe edge contact on heel and top. This edge contact will cause noisy operation followed by pittings and tooth fracture.

The theoretically conjugate hypoid set will not fulfill any of the fundamental gearing laws in case of the smallest gearbox inaccuracies or deflections. As mentioned above, already small deflections at moderate loads lead to load concentrations on the edges of conjugate flank pairs and can cause material damage and considerable noise emission. As such, the conjugate gear pair is not suitable for any task in a power transmission.

The introduction of 80μm length crowning and 15μm profile crowning to the conjugate hypoid design delivers the analysis results shown in Figure 15. The crowning makes the gearset insensitive to expected inaccuracies in the gear housing and load and heat affected deflections. Applying the same amounts of shaft displacements then used for the TCA of the conjugate hypoid set in Figure 14 moves the mean point slightly out of the initial position (see Figure 16), but a large contact area within the tooth boundaries is still maintained.

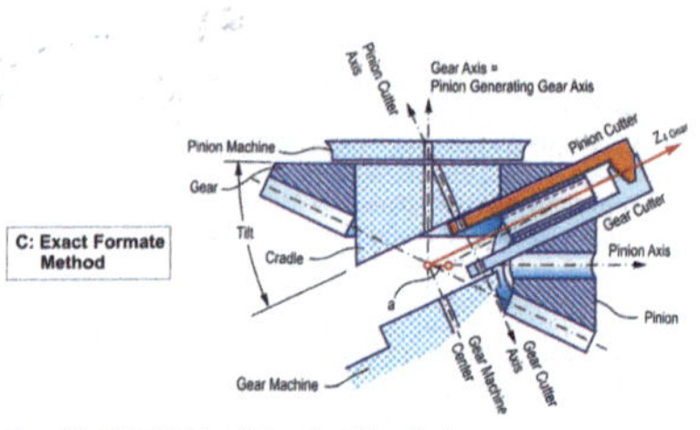

Figure 12: Generating of perfect conjugate hypoid pair.

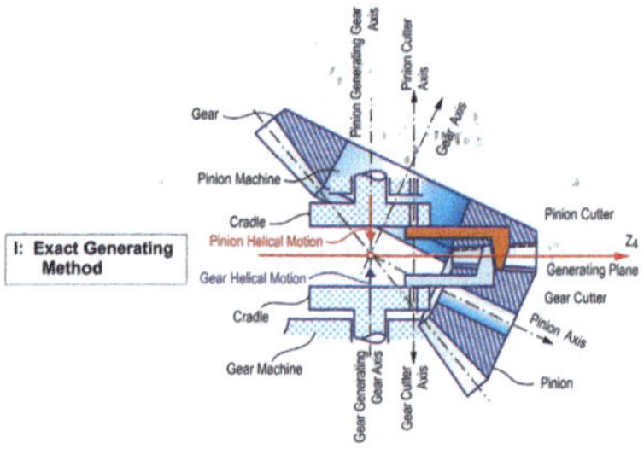

Figure 13: Generating of a conjugate tapered depth bevel gearset.

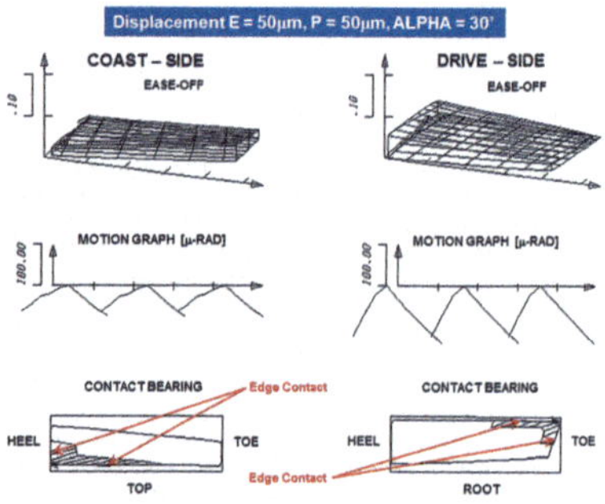

Figure 14: TCA of conjugate hypoid set from Figure 11 with shaft displacements.

It was demonstrated that the hypoid gearset with length and profile crowning in Figure 15 was developed based on a conjugate design. The first and third fundamental gearing law, mentioned in this paper, applies to the hypoid set in Figure 15 at the mean point roll position, if the load is zero. The first and third fundamental gearing law will apply in the area of contact as the load increases and the Hertzian contact spreads in contact line direction as well as in path of contact direction. This ideal condition can only be

Fig. 5.1 (continued)

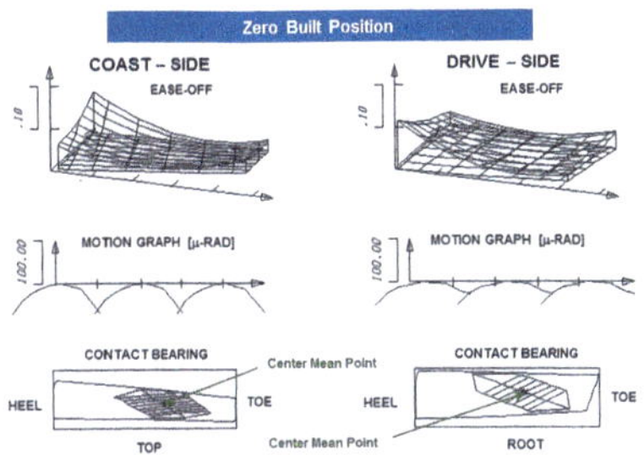

Figure 15: TCA of hypoid set from Figure 11 without any displacements.

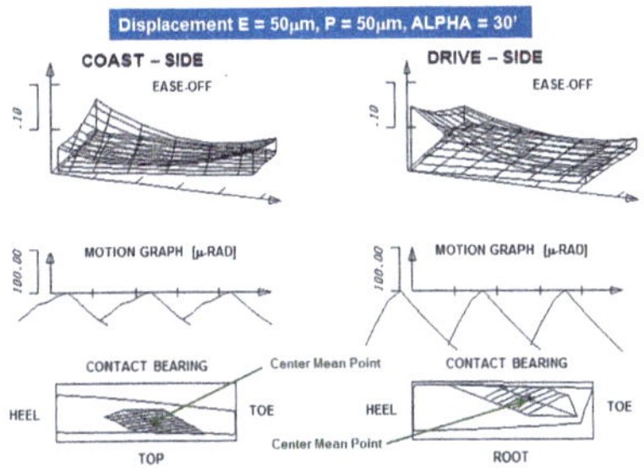

Figure 16: TCA of hypoid set from Figure 11 with added crowning and shaft displacement.

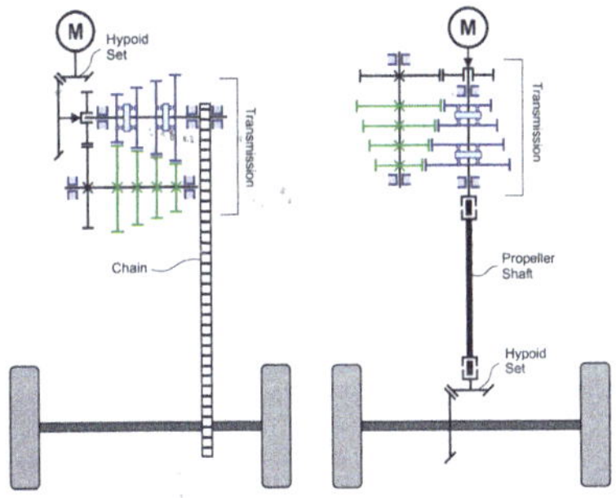

Figure 17: Hypoid gearset locations in vehicles with rear wheel drive.

achieved with correct amounts of crowning, adjusted to the operating displacements. It is interesting to mention that a hypoid gearset with crowning will fulfill the first and third fundamental gearing laws even in case of gearbox inaccuracies.

TRANSMISSION ORIENTATIONS

The proposal in [1] to place the hypoid gearset between motor and transmission is not practical. Hypoid gears are not used as simple reducers, but their purpose is to redirect rotation and torque by a certain angle, commonly 90°. In case of hypoids, the second purpose of lowering the center of gravity of a vehicle body has become very important in the automotive and truck industry. The redirection of rotation and torque has to be done at the driving axle of a vehicle. The engine orientation of vehicles with a rear wheel drive is longitudinal. Because engines of cars and trucks are commonly in the front, a longitudinal oriented propeller shaft transmits the engine rotation to the rear drive axle (Figure 17, right side). The transmission, which is located between engine and propeller shaft, needs to have a gear shaft orientation identical to the direction of the engine crankshaft. The hypoid gearset redirects rotation and torque and provides the final reduction at the drive axle. The advantage of this concept is the lower torque in the complex shift or automatic transmission and the high torque only at the ring gear at the drive axle.

A concept as shown in the left side graphic in Figure 17 would require, for example, a chain connection from the transmission output to the drive axle. This solution has low efficiency, causes high noise, provides an unacceptable packaging, and is not very reliable.

IS LAPPING AN ATTEMPT TO MAKE HYPOID GEARS CONJUGATE?

In [1, 2] it is mentioned that lapping of bevel and hypoid gearsets can be omitted if the gear geometry consists of a conjugate design. This reveals a misperception about the reason for lapping (and grinding). Lapping and grinding are hard-finishing operations. The soft manufactured bevel and hypoid gearsets have to be heat-treated, which in the most common case begins with a case carburizing of standard gear steels steel such as AISI 8620 or 16MnCr5. In order to give the low carbon steel a surface hardness in the 60 HRc range, a layer of carbon enrichment below the surface of 0.8 to 1,5mm depth is placed by a diffusion process. After the carburizing a quenching in oil and an additional tempering takes place. The result is a surface hardness that is commonly close to 60HRc and a core hardness in the 30HRc range. Case hardening provides an ideal transition between surface and core hardness that makes gears on the surface hard and wear resistant and in the core ductile. This makes shock loads and certain small plastic deformations tolerable without failure of the gearset. One major side effect of the heat-treatment process is the distortion of the gears that is caused by the carburizing, the re-crystallization of the steel, and the quenching. In order to make a gearset after heat-treatment suitable for power transmissions, for example in cars and trucks, a hard-finishing operation is required. The hard-finishing operating eliminates the heat-treatment distortions hereby providing the flank surfaces with the correct geometry from before the heat-treatment back. In addition, hard-finishing improves the surface finish to a low roughness and waviness, which enhances the hydrodynamic lubrication and reduces noise. Grinding and skiving are the preferred hard-finishing methods, creating a defined surface form that duplicates the original designed surfaces in the single micron range.

In case of face-hobbed bevel and hypoid gearsets, grinding is not possible, because of the epicyclical flank lead function. Skiving can generate epicyclical lead functions, but is not yet accepted for the high production volumes in the automotive and truck industry. This leaves only the lapping process for the hard-finishing of face-hobbed angular gearsets. However, the face-hobbed surface texture and the relative sliding between the flanks of hypoid gears make lapping an ideal alternative. Lapping can remove the surface scale

May 2019 49

Fig. 5.1 (continued)

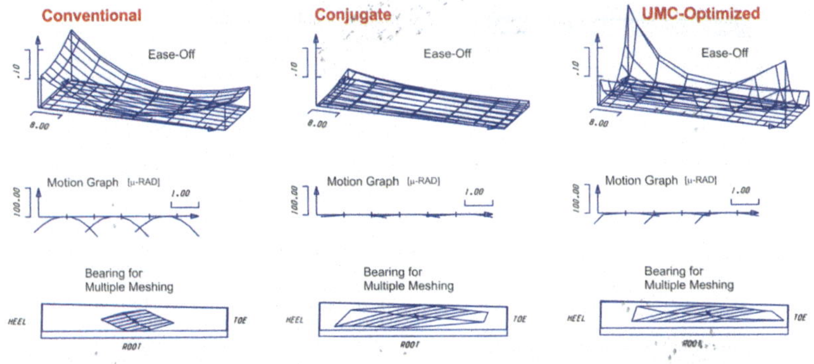

Figure 18: From conventional crowning via conjugate to UMC-optimized.

left from heat-treatment, and it re-matches two mating members by removing some runout and flank form distortions. Lapping can reduce the transmission error in many cases due to the fact that the major material removal is in the center region of the teeth where the tooth contact under light load is expected. In order for the lapping to work well, more crowning than required in the hard-finished gearset is used in the gearset design for the soft cutting. Lapping removes about 30 percent of this crowning, such that the length and profile crowning is just right after the lapping. Soft cutting of parts that are lapped after heat treatment considers a stock allowance of 0.03mm in the pinion and 0.01mm in the gear. If grinding is the hard finishing process (for face milled gearsets) then the design crowning is identical to the desired crowning after hard finishing. Between soft cutting and grinding, a uniform stock allowance of 0.10mm to 0.15mm is applied to the pinion and gear flanks.

SUMMARY

Conjugacy between the members of straight bevel, spiral bevel, and hypoid gears was only the first step and goes back more than 100 years. Quickly, the early scientists and engineers found out that conjugacy only gives us an important basis, but not a solution for power transmissions. Angular gearsets under load experience deflections that move them away from their theoretical position by half a millimeter and more. Well-designed and manufactured bevel and hypoid gearsets today can live up to those requirements and still maintain a power density that is four times higher than it was 50 years ago.

Transmission errors of 50 to 150 micro-radiant that were normal in the 1970s are in today's high-power density gearsets only between 5 and 15 micro-radiant. All this was achieved by converting a global length and profile crowning (Figure 18, left) first back to conjugacy (Figure 18, center) and then into a UMC™-optimized selective crowning, which is limited to particular regions of the teeth as shown in the right graphic in Figure 18. It is notable that the flank center of the UMC-optimized Ease-Off is conjugate, and the transmission error is next to zero. In lapping, similar effects as in grinding are achieved by using low inertia spindles with rotational compliance and high-speed torque control (SmartLap™).

The dream of conjugate angular gearsets turned out to be a false objective. Gear scientists, gear engineers, and gear manufacturers worked successfully for many decades on finding the optimal flank forms and the optimal non-conjugate flank surface interaction. The conjugate tooth design is today considered simple compared to a sophisticated higher order surface modulation. There is still room for improvement, but this cannot be achieved by going back to antiquated conjugate designs.

REFERENCES

[1] Radzevich, S.P.; Design Features of Perfect Gears for Crossed-Axes Gear Pairs; Gear Solutions Magazine, February 2019, Pages 36 to 43.

[2] Radzevich, S.P.; Preliminary Results of Testing of Low Tooth-Count Bevel Gears of a Novel Design, Part I, II and III, Gear Solutions Magazine, October 2014, December 2014 and January 2015.

[3] Wildhaber, E.; Stewart, A.J.; Design, production and application of the hypoid rear axle gear; The journal of the Society of Automotive Engineers Vol. 18, Issue 6, June 1926, Pages 575 to 589.

[4] Shtipelman, B.; Design and Manufacture of Hypoid Gears; John Wiley & Sons, Inc., New York, 1978.

[5] Stadtfeld, H.J.; Gleason Bevel Gear Technology; Expert Publishing, Renningen, Germany, May 2017.

[6] Schriefer, H.; Verzahnungsgeometrie und Laufverhalten bogenverzahnter; Kegelradgetriebe Dissertation RWTH Aachen, 1983.

[7] Wildhaber, E.; Relationships of Bevel Gears – Design for Duplex Cutting; American Machinist, October 1946.

[8] Wildhaber, E.; Relationships of Bevel Gears -Tooth Contact; American Machinist, February 1946.

ABOUT THE AUTHOR Dr. Hermann J. Stadtfeld is vice president of Bevel Gear Technology at The Gleason Works.

Fig. 5.1 (continued)

FROM THE EDITOR

Industry keeps adapting to the new normal

The craziness that seems to be the driving force of 2020 has affected almost every part of our lives – from the tiniest things we may take for granted to the larger health issues that keep us up at night.

As we try to make sense and adapt to what has become a new normal, *Gear Solutions* has tried to adapt as well.

Our August issue and our September issue were originally scheduled to be a springboard for the International Manufacturing Technology Show. But like so many other 2020 tradeshows, the organizers were forced to shift gears due to the coronavirus pandemic.

I must applaud IMTS for working toward making the best lemonade out of the lemons COVID-19 has thrown at us. Even though the Chicago show is officially canceled, IMTS organizers are working to take much of the tradeshow virtual. A virtual experience means no face-to-face camaraderie as with past shows, but attendees will still get the benefit of IMTS' experts and presenters. And I suspect no one is going to really miss their annual soft pretzel food run. (OK, full confession: I do love a soft pretzel.)

It is somewhat of a relief to know that technology has been an awesome tool in keeping a lot of our business going in these weird times.

With physical tradeshows on hold, please take time to see how *Gear Solutions* can also be your ally in getting your message to your customers. We offer many ways in which to remind the industry that your products and services are available.

That's good news for your audience in search of the very services and products that you can provide every day. And with the world trying to cope with economic and medical hardships, the deep reach *Gear Solutions* can provide is more important than ever.

With that in mind, I hope you find the involute gearing and gear design articles in our August issue of interest.

In a fascinating article from Christian Weber, Thomas Tobie, and Karsten Stahl – all with the University of Munich's Gear Research Centre – the authors share their insights on the rapid and precise manufacturing of special involute gears for prototype testing.

Next up, frequent contributor Stephen Radzevich discusses the most critical accomplishments in the scientific theory of gearing.

And, as always, you'll find some expert and innovative advice from our columnists. I always appreciate the wisdom and time they volunteer for the magazine.

With all that in mind, a final reminder that *Gear Solutions* is here to serve you. If you have any suggestions or would like to contribute, please contact me. I'm always looking for exciting articles to share.

Stay safe and healthy out there, and, as always, thanks for reading!

KENNETH CARTER, editor
editor@gearsolutions.com
(800) 366-2185 x204

CALL FOR ARTICLES Have a technical paper or other work with an educational angle? Let *Gear Solutions* publish it. Contact the editor, Kenneth Carter, at editor@gearsolutions.com for how you can share your expertise with our readers.

6 gearsolutions.com

GEARSolutions

David C. Cooper
PUBLISHER

EDITORIAL
Kenneth Carter
EDITOR
Jennifer Jacobson
ASSOCIATE EDITOR
Joe Crowe
CONTRIBUTING EDITOR

SALES
Dave Gomez
NATIONAL SALES MANAGER
Tom McNulty
REGIONAL SALES MANAGER

CIRCULATION
Teresa Cooper
MANAGER
Jamie Willett
ASSISTANT

DESIGN
Rick Frennea
CREATIVE DIRECTOR
Michele Hall
GRAPHIC DESIGNER

CONTRIBUTING WRITERS
REBECCA BRINKLEY
E. BUDDY DAMM
BRIAN DENGEL
D. SCOTT MACKENZIE
ADRIAN NOWOISKY
STEPHEN RADZEVICH
KARSTEN STAHL
THOMAS TOBIE
CHRISTIAN WEBER

Media solutions

PUBLISHED BY MEDIA SOLUTIONS, INC.
P. O. Box 1987 • Pelham, AL 35124
(800) 366-2185 • (205) 380-1580 fax
David C. Cooper PRESIDENT
Teresa Cooper OPERATIONS

Fig. 5.2 Facsimile of page 6 the August 2020 issue of the *Gear Solutions* magazine, where the marks of the removed article [3] are still available

Fig. 5.3 Facsimile of the article: [3]

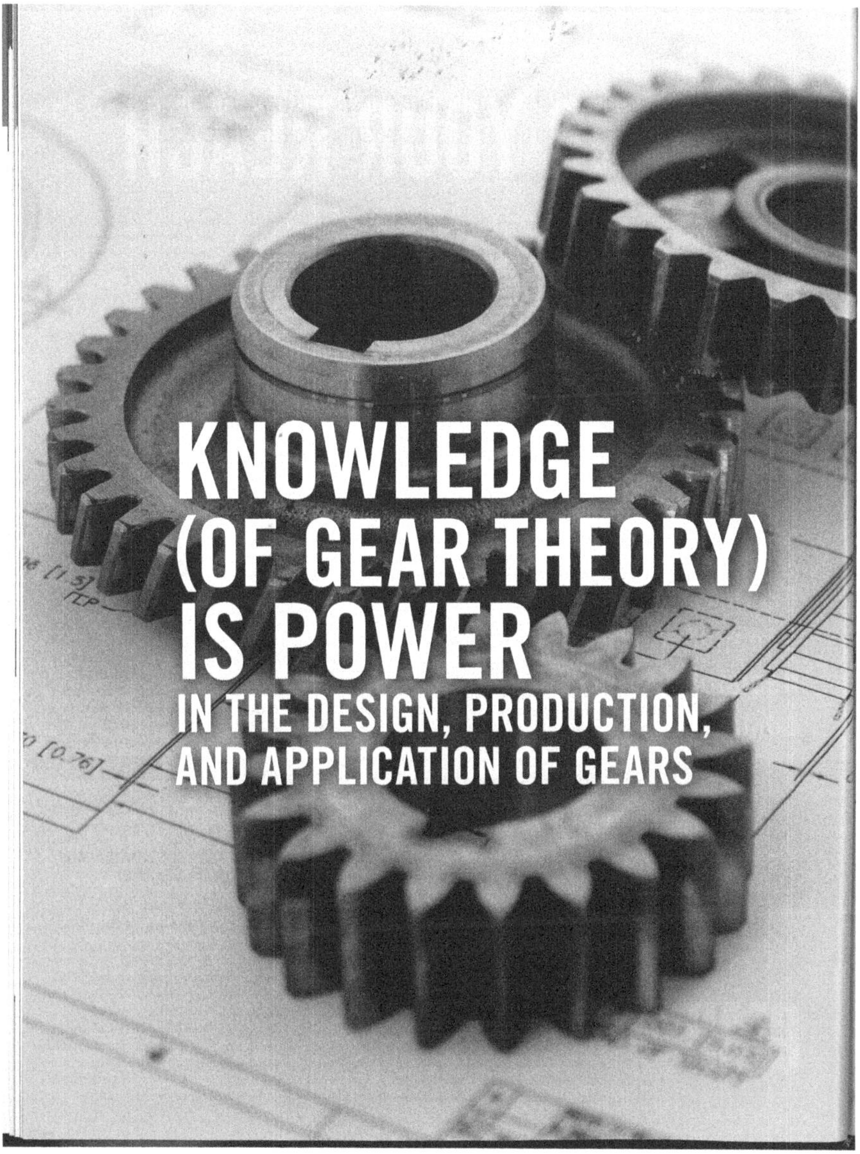

Fig. 5.3 (continued)

In this article, a few of the most critical accomplishments in the scientific theory of gearing will be discussed.

By STEPHEN P. RADZEVICH

The now-famous equation, "knowledge is power" ("scientia potestas est"), was coined by Francis Bacon (January 22, 1561-April 9, 1626). Formulated as early as 1597, this famous equation was proven to be correct for centuries. Knowledge in any and all areas of human activity gives one a strong advantage over those less knowledgeable ones. Gearing is no exception. Knowledgeable and skilled gear engineers always have a greater chance for success doing business, rather than less knowledgeable and less skilled ones. This article is aimed at the promotion of knowledge (namely, the promotion of the "scientific theory of gearing") to gear practitioners, and, in this way, to equip them with a means to achieving success running a business in the gear industry.

INTRODUCTION

The necessity of scientific foundations of gearing for design and manufacturing has long been realized by gear practitioners. Gears that operate on parallel axes of rotation of a gear and a mating pinion are the first to attract attention of mathematicians and gear theoreticians. It was Leonhard Euler who, in the mid-18th century, proved that an involute tooth profile fits the best gears that operate on parallel axes of rotation of a gear and a mating pinion.

Having started his own career in gearing in 1975, the author had many opportunities to get familiar with state-of-the-art innovations in both gear design and gear production. Shortly after graduating from college, the author was faced with: (a) numerous specific problems in gearing, and (b) the absence of an adequate means to solve these problems. The decades of research in the field of gearing theory by the author has resulted in a 2018 monograph titled "Theory of Gearing: Kinematics, Geometry, and Synthesis" [10]. It definitely makes sense to the gear community to get familiar with the capabilities that the proposed theory of gearing provides gear practitioners. In this article, a few of the most critical accomplishments in the scientific theory of gearing will be briefly discussed. The author concludes many in the gear community may be unfamiliar with the key accomplishments in the scientific theory of gearing. A recently published (2019) *Gear Solutions* article [15] (and not this article only) is taken as a representative example that proves this conclusion. Unfortunately, there are plenty of publications of this sort in current journals and magazines for gear engineers and scientists. Moreover, the article [15] (as well as other similar publications) may mislead gear practitioners, offering wrong approaches in the field of gear design and gear manufacturing. All the ambiguities in the practice of gear design and gear production should be eliminated. An old Chinese proverb, "The beginning of wisdom is to call things by their right names," is a good place to start.

MOTIVATION

This article was necessitated by the recently published article: Stadtfeld, H.J., "Why are Today's Hypoids the Perfect Crossed-Axes Gear Pairs?" *Gear Solutions* magazine, May 2019, pp. 42-50 [15]. These comments pertain to the key accomplishments in the scientific theory of gearing. The comments could be confusing to inexperienced gear community members, especially those who attempt to study the gear theory on their own and apply the theory in their practical work.

Here and below in this text, the article [15], is used solely as a representative example of an improper understanding of the scientific theory of gearing by a majority of gear practitioners, including those who may call themselves "gear theoreticians." An additional reason to reference the article [15] is that it claims to be written on behalf of established gear theory and manufacturing (see page 43 in [15]): "This write-up is meant to be a response on behalf of established gear theory and manufacturing.") No reference is provided in [15] as to whether this "established gear theory" is available to the reader in the public domain.

Recommendation to the readers of the current article: Except for section 1, the rest of the sections in this article intentionally reflect the article [15]. This is to help the reader better understand what these two articles are about, their differences, and why the approach outlined in the current article is advantageous over the approach recommended in the article [15]. When reading the current article, it is recommended to have the article [15] for comparison.

1 THE INTRODUCTORY SECTION: NOVIKOV GEARING (CONFORMAL GEARING)

The evolution of gear art is heavily associated with Ernest Wildhaber. No doubt, Wildhaber is a gifted inventor and a famous gear engineer. Despite that, Wildhaber has never been recognized as a gear scientist, as is loosely claimed in [15] (page 43). Because of that, Wildhaber cannot be granted with the title of "the father of modern gear theory" (see page 43 in [15]). Both these statements are incorrect: Wildhaber was NOT a gear scientist at all, and, therefore, it is incorrect to refer to him as the father of modern gear theory. Wildhaber is credited with NO key accomplishments in the scientific theory of gearing (see [10] for details). A difference between the terms "a

August 2020 39

Fig. 5.3 (continued)

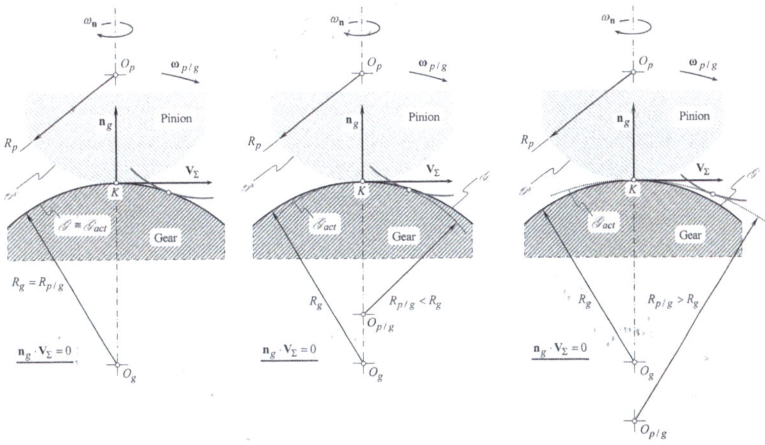

Figure 1: Examples of (a) fulfillment, and violation: (b) interference, and (c) separation, of a gear, $\mathscr{G}$, and its mating pinion, $\mathscr{P}$, tooth flanks: the "Shishkov equation of contact, $n_g \bullet V_\Sigma = 0$" is fulfilled in all three cases (a) through (c), while the law of conjugacy is violated in the cases (b) and (v).

gifted inventor and a famous gear engineer" and "a gear scientist" must be noted.

The earlier performed comparison of "Novikov gearing" [2], and Wildhaber "Helical gearing" [1] reveals these two gear systems are completely different, and, moreover, these gear systems are not compatible with one another [10, 7, 9]. On top of that, the term "Wildhaber-Novikov gearing" is incorrect by nature, as these two gear systems (namely, Wildhaber "Helical gearing" [1] and "Novikov gearing" [2]) cannot be combined into a common gear system (see [10, 7, 9], and others, for details). It is a huge mistake to adopt the term "Wildhaber-Novikov gearing."

"Novikov gearing" can be viewed as a reduced case of a parallel-axes involute gear pair, while Wildhaber "Helical gearing" cannot ([3, 6] and numerous other sources).

Today's gear professionals should know the difference between "Novikov gearing" [2] and Wildhaber "Helical gearing" [1].

2 NEWS ABOUT HYPOID GEARS?

The article [15] proves today's hypoid gearsets are approximate by nature, and, thus, they are not geometrically-accurate gearsets. No alterations can be even anticipated.

Further, no attempt is taken in the article [15] to "question the creditability of the scientists and gear engineers who worked on the theory and its improvement over approximately 100 years." No gear theory was developed during these 100 years, and no key scientific accomplishments are attained at this period of time. "Trial-and-error method" is the main – and perhaps only – tool used for "over approximately 100 years." But this is not a theory of gearing. Instead, this is a collection of more-or-less reasonable solutions to separate gear problems. Combined together, these solutions do not form a gear theory.

3 THE THREE FUNDAMENTAL LAWS OF GEARING

Leonhard Euler can be credited for originating geometrically-accurate gears (even though the term "geometrically-accurate gears" was introduced much later [10]). It is proven [10] that, in order to be referred to as "geometrically-accurate," the gears have to fulfill three fundamental laws of gearing. The fulfillment of three fundamental laws of gearing is necessary and sufficient to refer to a gearset as a "geometrically-accurate gearset." To the best possible extent, all three fundamental laws of gearing are discussed in [10]. These laws of gearing are briefly outlined here:

The first fundamental law of gearing: "At every point of contact of tooth flanks of a gear and a mating pinion, the vector of their instantaneous relative motion has to be orthogonal to the common perpendicular at every instant of time."

In this form, the first fundamental law of gearing is valid for gearing of all designs that operate on parallel axes, intersecting axes, and crossing axes of rotation of a gear and its mating pinion.

Numerous interpretations of the first fundamental law of gearing are known. The most practical and extensively adopted way of interpretation of the first fundamental law of gearing is to analytically describe this law of gearing in the form of the "Shishkov equation of contact, $n_g \bullet V_\Sigma = 0$." The "Shishkov equation of contact, $n_g \bullet V_\Sigma = 0$" is known to practically every gear engineer. Professor V.A. Shishkov was the first to express (~1948) the first fundamental law of gearing in the form of the dot product of a unit vector of a common perpendicular, n_g, by the vector of the resultant linear velocity, V_Σ, of the tooth flanks in relation to one another. Both the vectors, n_g and V_Σ, are calculated at the contact point of a gear and its mating pinion tooth flanks.

The second fundamental law of gearing is often referred to as the "conjugate action law." When the "Shishkov equation of contact,

40 gearsolutions.com

Fig. 5.3 (continued)

Figure 2: Rotationally-positive external parallel-axes gear pair (Pat. No. 163857, USSR, A Helical Gearing./B.V. Shitikov, N.A. Bayazitov, Int. Cl. F06h, Filed: February 25, 1963, Published: July 22, 1964.).

$\mathbf{n}_g \bullet \mathbf{V}_\Sigma = 0$" is fulfilled, the "conjugate action law" can either be fulfilled, or it can be violated. In Figure 1, an example of interaction of local patches of the tooth flanks, $\mathcal{G}$ and $\mathcal{P}$, is schematically illustrated. The tooth flanks, $\mathcal{G}$ and $\mathcal{P}$, make contact at point, K. The radii of curvature of the interacting tooth flanks at contact point, K, equal to R_g and R_p, correspondingly (see Figure 1a). The centers of curvature of the tooth profiles, $\mathcal{G}$ and $\mathcal{P}$, are denoted by O_g and O_p, correspondingly. In the instantaneous motion of the pinion, $\mathcal{P}$, in relation to the gear, $\mathcal{G}$, the pinion performs an instantaneous rotation, $\omega_{p/g}$, about the point O_g. The radius of curvature of the generated actual gear tooth flank, $\mathcal{G}_{act}$, equals to $R_{p/g} = R_g$. In this scenario, the second fundamental law of gear is fulfilled, and the actual tooth flank, $\mathcal{G}_{act}$, is identical to the desirable gear tooth flank, $\mathcal{G}$, as shown in Figure 1a.

Note: In Figure 1a, the rotation, ω_n, of the pinion, $\mathcal{P}$, in relation to the gear, $\mathcal{G}$, about the contact perpendicular, $\mathbf{n}_g$, is not prohibited by the second fundamental law of gearing.

If the instantaneous rotation is performed either about the center $O_{p/g}$ (when $R_{p/g} < R_g$, see Figure 1b), or about the center $O_{p/g}$ (when $R_{p/g} > R_g$, see Figure 1c), the second fundamental law of gear is violated, and the actual tooth flank, $\mathcal{G}_{act}$, differs from the desirable gear tooth flank, $\mathcal{G}$.

Reminder: The first fundamental law of gearing (as well as the "Shishkov equation of contact, $\mathbf{n}_g \bullet \mathbf{V}_\Sigma = 0$") is fulfilled in all three cases shown in Figure 1, while the second fundamental law of gearing is fulfilled only in the first case illustrated in Figure 1a. The schematic in Figure 1 is helpful for understanding the difference between the first and second fundamental laws of gearing and prevents the making of unsubstantiated conclusions in this regard.

Important: When the second fundamental law of gearing is fulfilled, the first fundamental law of gearing is always fulfilled, and not vice versa.

With that said, it makes sense to go back to the article [15] and compare the results illustrated in Figure 1 with a loosely made statement (see page 44 in [15]): "The second fundamental law of gearing, ... , is a redundant relationship to the first gearing law, and it is limited to cylindrical gears with parallel axes and straight bevel gears without hypoid offset. In this case, it adds nothing to the first gearing law; conjugacy is already given by the relationships required in the first gearing law." It is correct to question whether a gear practitioner has a chance for success in the design and production of today's sophisticated gear systems if he or she strictly follows the afore mentioned quote (see page 44 in [15]).

In the most general case of gearing (namely, in the case of "crossed-axes gearing", or "C_a – *gearing*", for simplicity) the second fundamental law of gearing is formulated as:

The second fundamental law of gearing: "In order to smoothly transmit a uniform rotary motion from a driving shaft to a driven shaft by means of gear teeth, perpendiculars to the tooth flanks of the interacting teeth at all points of their contact must intersect the axis of instantaneous rotation in the gear pair."

In a reduced case of gears that operate on parallel axes of rotation of a gear and its mating pinion (namely, in the case of "parallel-axes gearing", or "P_a – *gearing*", for simplicity) this fundamental law of gearing is formulated as:

The second fundamental law of (parallel-axes) gearing: "In parallel-axes gearing, in order to smoothly transmit a uniform rotary motion from a driving shaft to a driven shaft by means of gear teeth, perpendiculars to the tooth flanks of the teeth at all points of their contact must pass through a stationary point located on the line of centers, namely, the pitch point P; the pitch point subdivides the center-distance reciprocal to the angular velocities of the gear and the pinion." (Robert Willis, 1841 [16])

These days, the second fundamental law of gearing (in a case of parallel-axes gearing) is commonly referred to as "Camus-Euler-Savari fundamental law of gearing" (or " fundamental law of gearing," for simplicity).

The second fundamental law of gearing is also often referred to as "conjugate action law" – and this is also correct.

The design of "Alpha Worm Gearing" (New Venture Gear, Syracuse, NY; US Pat. No. 6,148,683, Nov. 21, 2000) is a good example of the violation of the second fundamental law of gearing. Violation of the second fundamental law of gearing is the main reason for this design fail, and it has caused major financial damage to the gear industry.

The third fundamental law of gearing relates to the distribution of the gear teeth over the periphery of a gear and a mating pinion. In the most general case of gearing (in the case of crossed-axes gearing), the third fundamental law of gearing is formulated as:

The third fundamental law of gearing: "In order to smoothly transmit a uniform rotary motion from a driving shaft to a driven shaft by means of gear teeth, at every instant of time the angular base pitch of a gear, and that of a mating pinion, both have to be equal to the operating angular base pitch in the gear pair."

In a reduced case of "parallel-axes gearing," this fundamental law of gearing is formulated as:

The third fundamental law of gearing (in parallel-axes gearing): "In parallel-axes gearing, in order to smoothly transmit a uniform rotary motion from a driving shaft to a driven shaft by means of gear teeth, at every instant of time base pitch of a gear and that of a mating pinion, both must be equal to the operating base pitch in the gear pair."

Fig. 5.3 (continued)

All the geometrically-accurate gears (that is, all the precision gears – with no exception) meet all three fundamental laws of gearing – this is a must.

4 PERFECT CONJUGACY IN STRAIGHT BEVEL GEARSETS

The title of the section "Perfect Conjugacy in Straight Bevel Gearsets" in the article [15] may be confusing to readers, as "conjugacy" by nature cannot be "perfect" or "imperfect." Gearsets of all kinds are either conjugate, or they are not conjugate. The term "perfect conjugacy" could be considered analogous to saying "a bit pregnant."

It is loosely claimed (see page 44 in [15]) that: "The plane of action cannot be extended beyond its tangential contacting line with the base element as shown incorrectly .... The plane of action exists where tooth engagement is possible." Gearsets that feature the plane of action that extends far beyond its tangential contacting line with the base element have been known for decades. As an example, a parallel-axes gear set of this particular kind is depicted in Figure 2 [14]. Later on, this concept was evolved (~2008) by Prof. S.P. Radzevich to the cases of intersected-axes gearing, as well as to crossed-axes gearing [10].

In a set of articles [11], [12], and [13], the preliminary results of testing of low-tooth-count spiral bevel gears of a novel design are discussed. Without going into details of design and manufacture of spiral bevel gears of this design (proposed by Prof. S.P. Radzevich, ~2008, and designed using the earlier developed scientific theory of gearing [10]), it is right to stress the following:

1. As shown in Figure 3, the predicted contact patch is perfect (see Figure 7a on page 21 in [13]).
2. As shown in Figure 4, the predicted contact pattern is perfect (see Figure 9 on page 23 in [13]).
3. As shown in Figure 5, the contact pattern in the roll test is also perfect (see Figure 10 on page 23 in [13]), and it indicates excellent correlation with the predicted one.
4. No lapping was used to finish the gears.
5. No adjusting for the axial configuration of the gear and the mating pinion is required, and the gears are ready to run.

In today's production of precision gears, the use of the lapping process is inevitable. Those knowledgeable in the scientific theory of gearing can eliminate dirty and obsolete lapping processes from finishing tooth flanks of gears for precision gearsets.

Concerning the statements on page 45 in [15] that state: "The Octoid is the analog function of an involute, and it provides to bevel gears the same advantage as an involute provides to cylindrical gears. ... bevel gears have a trapezoidal generating profile. The straight rack ... becomes a ring ... as shown in Figure 6." It has been known for a while that the tooth profile of a crown rack for generating a geometrically-accurate bevel gear has to be shaped in the form of a curve illustrated in (see Figure 6) [10], and NOT in the form of a straight-sided crown rack. Only approximate bevel gears can be generated by the just-mentioned straight-sided crown rack "shown in Figure 6" [15]. No gearing with the so-called "octoidal" path of contact is possible at all: If one assumes the path of contact shaped in a form of an octoid curve, this inevitably entails violation of the second fundamental law of gearing – it is important to bear this in mind. The so-called "octoid gearing" is a mistake that often travels from one publication to another.

Concerning the statement on page 46 in [15] that states: "Conjugacy is the basis of all gearsets manufactured in high volume on dedicated manufacturing machines." All the bevel gears manufactured in high volume on dedicated manufacturing machines are generated by means of straight-sided crown rack (shown in Figure 6, page 45 in [15]), and, thus, all of them are approximate gears.

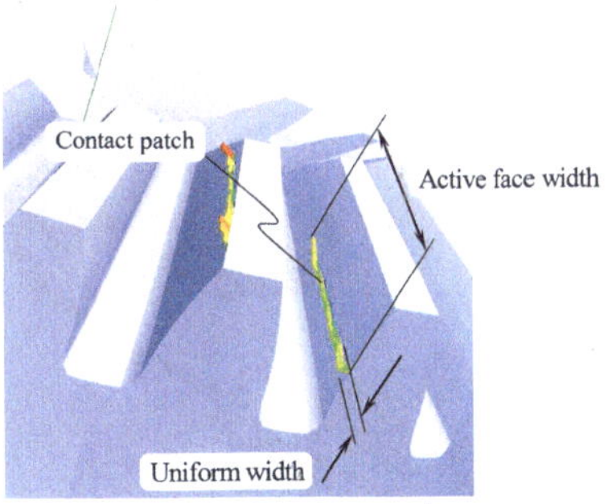

Figure 3: Predicted contact patch in geometrically-accurate intersected-axes spiral bevel gears.

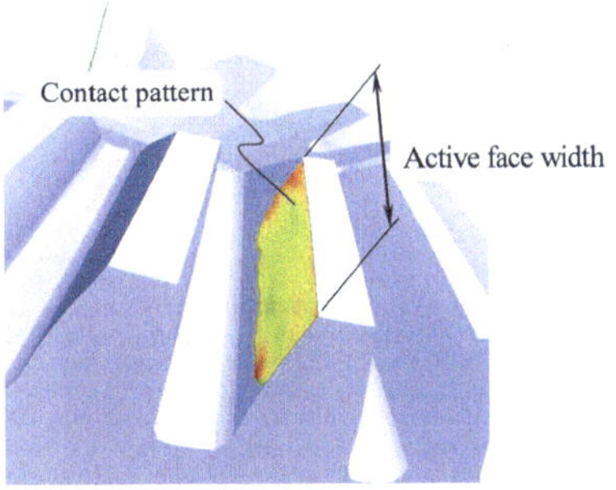

Figure 4: The predicted contact pattern in geometrically-accurate intersected-axes gearing.

Only approximate bevel gears can be generated by just mentioned straight-sided crown rack "shown in Figure 6" [15]. As the generated bevel gears are approximate, no conjugate bevel gear pairs can be composed of these gears.

Concerning the statement on page 46 in [15] that states: "Conjugate bevel gearsets cannot be used for a power transmission because manufacturing tolerances and ..." Today's gear industry is capable of manufacturing precision helical involute gears that are extensively used, for example, in the design of turbine reducers. The gear industry is NOT capable of manufacturing precision bevel gears, and this is the main reason for precision bevel gears NOT being extensively used in the industry. The bottom-line is: Precision bevel gears are NOT extensively used in the industry solely because the gear industry is not capable of producing (for a reasonable cost) these kinds of gears.

Concerning the statement on page 46 in [15] that states: "The right amount of crowning makes a gearset quiet and gives it the required load carrying capacity". How can the "right amount" be determined for a particular bevel gear application? The desirable amount of crowning can be determined by means of a "trial and error method" - this is the only tool developed by the "established

42 gearsolutions.com

Fig. 5.3 (continued)

Figure 5: Contact pattern in geometrically-accurate spiral bevel gears in roll test.

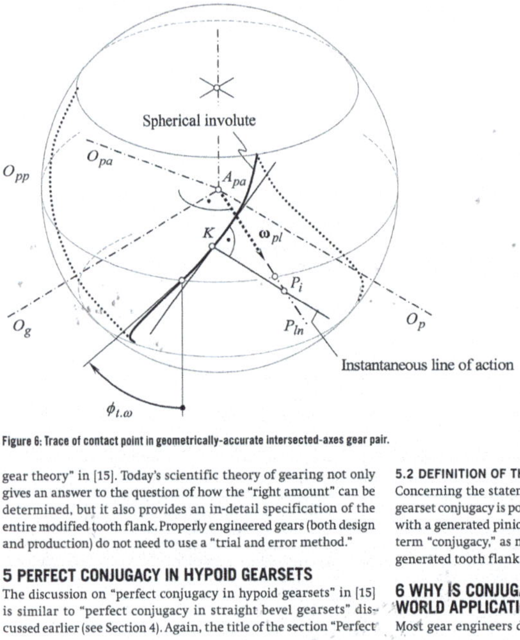

Figure 6: Trace of contact point in geometrically-accurate intersected-axes gear pair.

gear theory" in [15]. Today's scientific theory of gearing not only gives an answer to the question of how the "right amount" can be determined, but it also provides an in-detail specification of the entire modified tooth flank. Properly engineered gears (both design and production) do not need to use a "trial and error method."

5 PERFECT CONJUGACY IN HYPOID GEARSETS

The discussion on "perfect conjugacy in hypoid gearsets" in [15] is similar to "perfect conjugacy in straight bevel gearsets" discussed earlier (see Section 4). Again, the title of the section "Perfect Conjugacy in Straight Bevel Gearsets" in the article [15] may be confusing to readers as "conjugacy," by nature, can neither be "perfect"' nor "imperfect." Gearsets of all kinds are either conjugate, or they are not conjugate.

Concerning the statement on page 46 in [15] that states: "... the pitch elements of crossed axes hypoid gears are drawn as cones." It fails to describe one of the key features of the modern scientific theory of gearing [10]: In order to design gears for hypoid gearsets, no pitch elements are required at all. The term: "... the pitch elements are hyperboloids" is obsolete and is no longer used in the modern scientific theory of gearing [10].

Concerning the statement on page 47 in [15] that states: "No plane of action exists between two hyperbolic base elements." It is correct, as it proves there is no place for hyperboloids in the kinematics and the geometry of gears with crossing axes of rotation of a gear and its mating pinion.

Concerning the statement on page 47 in [15] that says: "The correct surface of action is curved and wrapped, as shown in Figure 10." Surface of action in hypoid gearing (as well as in gearing of all other kinds) is a plane through the axis of instantaneous rotation, or it is a plane through the common perpendiculars constructed at all points of the line of contact between the interacting tooth flanks of a gear and its mating pinion. The normal forces of interaction in gearing are commonly described by means of vectors. A vector is a quantity possessing both magnitude and direction. Vectors are described by means of straight-line segments with a specified direction. This leaves no room for "curved surface of action."

5.1 CONJUGACY BETWEEN MESHING FLANKS

Concerning the statement on page 47 in [15] that says: "The term conjugate is used in mathematics for two or more surfaces that contact each other along a line. Since the 1970s, the term conjugate has also been employed in gear technology literature to define..." Line contact between two surfaces (namely, between tooth flanks of two mating gears) is not sufficient (and is not always necessary) to refer to the pair of surfaces as to "conjugate surfaces." In addition, the second fundamental law of gearing has to be fulfilled for conjugate surfaces, which is a must. Moreover, it has been discovered over the last 10 years that two surfaces (two tooth flanks) can be conjugate to one another even when they are in point contact.

5.2 DEFINITION OF THE CONJUGATE GEAR PAIR

Concerning the statement on page 47 in [15] that says: "... hypoid gearset conjugacy is possible with a non-generated gear that meshes with a generated pinion." This reveals a poor understanding of the term "conjugacy," as no conjugate action is possible between non-generated tooth flank and generated tooth flank.

6 WHY IS CONJUGACY NOT DESIRABLE FOR REAL WORLD APPLICATIONS?

Most gear engineers don't doubt whether the conjugacy of inter-

August 2020 43

Fig. 5.3 (continued)

acting tooth flanks is desirable. Conjugacy is always desirable – especially in high-power-density transmissions and in high RPM-gearsets. Unfortunately, with no lapping process in finishing the tooth flanks, the gear industry is not capable of producing precision gears for real-world gearsets. For over a decade, a concept of crossed-axes gearsets that are insensitive to the mating gears axes displacements (both, the linear displacements, as well as the angular displacements) were introduced to the public [10]. Gears of this system are commonly referred to as "S_{pr} – *gearing*." The "S_{pr} – *gearing*" is a kind of point contact conjugate gearing capable of absorbing all the linear and angular displacements of a gear and its mating pinion axes of rotation, as long as the actual values of these displacements are within the prespecified tolerances for the accuracy of the gearset. At the same time, the "S_{pr} – *gearing*" features the highest possible power density being transmitting by the gearset.

7 TRANSMISSION OPERATIONS

Only time will tell if it is practical "... to place the hypoid gearset between motor and transmission ...?" (see page 49 in [15]). Looking to the future, one can conclude that the vehicle powertrain is not the only potential application of gears with crossing axes of rotation of a gear and a mating pinion.

8 IS LAPPING AN ATTEMPT TO MAKE HYPOID GEARS CONJUGATE?

Eventually, the lapping process in the production of gears for bevel and hypoid gearsets will be eliminated [6]. The gear-grinding process is a No. 1 candidate for the future of gear-finishing operations. The sooner the key accomplishments in the scientific theory of gearing reaches gear practitioners, the sooner gear lapping will be replaced by gear grinding. This negates the statement on page 49 in [15] that says: "This reveals a misperception about the reason for lapping."

An in-depth familiarity with the key accomplishments in the scientific theory of gearing will make it possible to eliminate the gear-lapping process and replace it with the gear-grinding process. This is a reliable evolution of the gear design practice and the gear-finishing processes in particular.

9 SUMMARY

It is important to note that over the last 100 years, the gear industry has failed to:

- Eliminate lapping process.
- Avoid the necessity of adjustment of bevel/hypoid gears in pairs (that makes the gears not-self-replaceable).
- Eliminate the necessity of pairing of gears that operate on intersected, as well as on crossing axes of rotation of a gear and its mating pinion.
- Solve the problem of excessive gear noise excitation and vibration generation.

CONCLUSION

In conclusion, it is likely the gear industry could end up wasting funds and time following the direction of evolution outlined in the article [15].

In the article [15] there is no evidence of:

- Whether the third fundamental law of gearing is understood by the author.
- The importance of the concept of the "operating base pitch" in a gear pair is realized.
- The concept of a gear, of a mating pinion base pitch in intersected-axes gearing, and in crossed-axes gearing, along with the concept of the "operating base pitch" in intersected-axes, and in crossed-axes gearsets as realized by the author.

Not all the inconsistences in [15] are addressed here – only some of the larger issues. However, a more detailed analysis (if necessary) can be undertaken.

REFERENCES

[1] Pat. USA, No. 1,601,750, Helical Gearing, /E. Wildhaber, Patented: October 5, 1926, Filed: November 2, 1923.

[2] Pat. USSR, No. 109,113, Gear Pairs and Cam Mechanisms Having Point System of Meshing. /M.L. Novikov, National Classification 47h, 6; Filed: April 19, 1956, published in Bull. of Inventions No.10, 1957.

[3] Radzevich, S.P., "An Examination of High-Conformal Gearing", Gear Solutions, February, 2018, pages 31-39.

[4] Radzevich, S.P., "Conjugate Action Law in Intersected-Axes Gear Pairs and in Crossed-Axes Gear Pairs", Gear Solutions magazine, June 2020, pages 42-48.

[5] Radzevich, S.P., "Design Features of Perfect Gears for Crossed-Axes Gear Pairs", Gear Solutions, February, 2019, pp. 36-43.

[6] Radzevich, S.P., Gear Cutting Tools: Science and Engineering, 2nd Edition, CRC Press, Boca Raton, FL, 2017, 606 pages.

[7] Radzevich, S.P., Dudley's Handbook of Practical Gear Design and Manufacture, Third Edition, CRC Press, Boca Raton, FL, 2016, 629 pages.

[8] Radzevich, S.P., High-Conformal Gearing: Kinematics and Geometry, 2nd edition, Elsevier, Amsterdam, 2020, 506 pages.

[9] Radzevich, S.P., "On the Inconsistency of the Term "Wildhaber-Novikov Gearing": A New Look at the Concept of "Novikov Gearing"", Appendix A, pp. 487-501, in: Radzevich, S.P., (Editor), Advances in Gear Design and Manufacture, CRC Press, Boca Raton, Florida, 2019, 549 pages.

[10] Radzevich, S.P., Theory of Gearing: Kinematics, Geometry, and Synthesis, 2nd edition, revised and expanded, CRC Press, Boca Raton, FL, 2018, 934 pages.

[11] Radzevich, S.P., et al, "Preliminary Results of Testing of Low-Tooth-Count Bevel Gears of a Novel Design. Part 1", Gear Solutions, October 2014, pp. 25-26.

[12] Radzevich, S.P., et al, "Preliminary Results of Testing of Low-Tooth-Count Bevel Gears of a Novel Design. Part 2", Gear Solutions, December 2014, pp. 20-21.

[13] Radzevich, S.P., et al, "Preliminary Results of Testing of Low-Tooth-Count Bevel Gears of a Novel Design. Part 3", Gear Solutions, January 2015, pp. 20-23.

[14] Shitikov, B.V., Bayazitov, N.A., A Helical Gearing, Pat. No. 163857, USSR, Int. Cl. F06h, Filed: February 25, 1963, Published: July 22, 1964. [see also: Bayazitov, N.A., Helical Gears with a New Type of Gearing, Ph.D. Thesis, Kazan', Kazan' Technological & Chemical Institute, 1964.].

[15] Stadtfeld, H.J., "Why are Today's Hypoids the Perfect Crossed-Axes Gear Pairs?", Gear Solutions magazine, May 2019, pp. 42-50.

[16] Willis, R., Principles of Mechanisms, Designed for the Use of Students in the Universities and for Engineering Students Generally, London, John W. Parker, West Stand, Cambridge: J. & J.J. Deighton, 1841, 446 pages.

ABOUT THE AUTHOR Stephen P. Radzevich, Dr.(Eng.)Sci, can be reached at radzevich@usa.com or at (586) 292-7209.

Fig. 5.3 (continued)

References

1. Radzevich, S. P. (2022). *Theory of gearing: Kinematics, geometry, and synthesis* (3rd ed., 1192 pages). CRC Press.
2. Stadtfeld, H. J. (2019, May). Why are today's hypoids the perfect crossed-axes gear pairs? *Gear Solutions*, pp. 42–50.
3. Radzevich, S. P. (2020, August). Knowledge (of gear theory) is power in the design, production, and application of gears. *Gear Solutions*, pp. 38–44.

Chapter 6
Improving the Accuracy of Generation of Face Screw Surfaces of Teeth of Hobs

Oleksandr A. Okhrimenko, Michael Storchak, Yurii M. Danylchenko, and Olha V. Kholiavik

6.1 Introduction

Currently, gear transmissions are an integral part of many modern devices, installations, and machines and largely determine their design and performance characteristics.

With the development of science and technology, the requirements for them increase accordingly.

With increasing requirements for the performance characteristics of gears transmissions, the question arises of increasing the accuracy of gears and reducing the costs of their manufacture.

Analysis of gear designs shows that about 65% of all gear transmissions used are spur gears. Among the existing methods for cutting cylindrical gears, the most widespread is the process of cutting with worm milling cutters.

Research and practice in the use of hob gear cutters have proven that the accuracy of the profile of hob cutters depends on the geometry of their cutting part. In turn, its performance depends on the geometry of the cutting part. Thus, the tasks of increasing the accuracy and performance of hob cutters are interconnected. To solve these problems, first of all, it is necessary to analyze the possible errors that arise in the process of cutting gears and the reasons for their occurrence, the role of methods for designing worm milling cutters on their accuracy and performance.

O. A. Okhrimenko · Y. M. Danylchenko · O. V. Kholiavik (✉)
National Technical University of Ukraine "Igor Sikorsky Kyiv Polytechnic Institute", Institute of Mechanical Engineering, Kyiv, Ukraine

M. Storchak
Institute for Machine Tools, Universityof Stuttgart, Stuttgart, Germany

S. P. Radzevich (ed.), *Gear Accuracy*, Mechanical Engineering Series,
https://doi.org/10.1007/978-3-031-74798-4_6

When transforming the generating surface (ITS) of a cutter into a real tool, select the surface of the cutter tooth that treatable milling and the one that not treatable milling, specifying the initial position of these surfaces by the geometry parameters of the cutting part in the tool coordinate system.

However, due to the peculiarities of the shaping of these surfaces during the manufacture of the cutting part of the hob cutter, the cutting edge thus obtained has the form of a spatial line and does not lie on the surface of the initial tool surface (ITS) of the cutter, which was taken as the basis for the design [5, 6, 9, 12].

This leads to the introduction of additional errors into the wheel profile by such a cutter. The issue of the influence of the shape of the cutting edge of the cutter obtained in this way on the parameters of the gear surface has not yet been considered due to the lack of work in this direction.

The influence of the shape of the cutting edge on the surface of the wheel is considered through the parameters of the manufacturing accuracy of these surfaces themselves, forming the cutting edge and normalized using the corresponding deviations that are specified by GOST 9324-80 "Finishing cutters for cutting evolvent gear wheels."

The influence of the shape of the cutting edge on the surface of the wheel is considered through the parameters of the manufacturing accuracy of these surfaces themselves, forming the cutting edge and normalized by the corresponding deviations that are specified by GOST 9324-80.

An important role is plade by the geometry of the rake surface of the tooth, normalized by the corresponding standard, and therefore it is first necessary to consider the features of the shaping of the surfaces of hob cutters that treatable remilling—the front surface [5, 7, 12].

6.2 Features of the Technology for Forming the Front Linear Helical Surfaces of the Teeth of Hobs

In hob cutters, the front surface can be re-sharpened. The front surface of the cutter is formed by milling chip grooves with disk cutters, and after heat treatment this surface is ground (sharpened). It is the sharpening operations that have a significant impact on the accuracy and quality of the hobs [5, 12].

As a rule, hobs are ground along the front surface of the teeth with the end or conical side of the grinding wheel. The first sharpening scheme is used for sharpening cutters with straight flutes and a flat front surface of the teeth, the other scheme is also used for sharpening cutters with helical chip flutes.

The use of the conical side of the grinding wheel as a working surface when sharpening multi-pass and precision single-pass leads to a relatively large curvature of the generatrix of the helical surface (which is unacceptable). One of the common methods for avoiding distortion of the tool profile when sharpening it is special shaped dressing of the profile of grinding wheels [1, 7], implemented using special dressing devices, the design of which uses special or universal copiers.

There are two methods for dressing a grinding wheel: copying and kinematic dressing. When using copying devices, it is necessary to determine the profile curve of the grinding wheel, which forms a theoretically accurate rake surface. But usually this surface is very complex, so it is approximated by an arc of a circle, or using several conjugate circles.

The work [9] presents dependencies by which the circle profile curve can be calculated when shaping the front surface of a hob cutter.

Since the surface of the grinding wheel must not intersect the nominal helical surface, the wheel must have a convex profile with a theoretical arrow of convexity f_u at the design point and a radius of curvature R_u of the axial profile. Curvature f_u values are calculated depending on the settings of the machine and the cutter being sharpened:

$$f_u = \frac{h^2 p^2 (r_u - h\cos\alpha_u)}{2\left(p_b^2 + r_1^2\right)^2 \cdot \left(\sin\alpha - h\cos\frac{\alpha_u}{R_u}\right)} \tag{6.1}$$

where

p_b helical parameter of front surface;

α_u angle of the circle profile at the point of contact on the initial cylinder of radius r_1;

h distance from the starting cylinder to the design point of the generatrix; $h > 0$ for $r_i > r_1$ and $h < 0$ for $r_i < r_1$—radius of the circle at the point of contact with the initial helix of the surface being ground.

In first approximation:

$$r_{u1} = r_{um} - \left(r_1 - r_f + 0,5\text{ m}\right)\cos\alpha_u \tag{6.2}$$

where

r_{um} maximum radius of the working surface of the circle;

r_f radius of the cylinder of the cutter teeth recesses;

m module of the generating worm, value *0.5* m—the required deepening of the working surface of the circle relative to the cylinder of the cutter recesses.

The deviation f_{vi} of the generatrix of the front surface from the radial straight line is determined by the difference between the required value f_u and its actual implementation when profiling a circle. When approximating the axial profile of a circle with a circular arc of radius R_u or a curve close to it:

$$f_{vi} = \left(\frac{h^2}{2R_u}\right) - f_u. \tag{6.3}$$

When using a conical grinding wheel, dependence (6.3) takes the form:

$$f_{vi} = \frac{h^2 p_b^2 (r_{u1} - h\cos\alpha_u)}{2\sin\alpha_u \left(p_b^2 + r_1^2\right)^2}. \quad (6.4)$$

The sign "–" before the fraction shows that when sharpening with a conical wheel, the deflection is always directed towards the body of the cutter tooth.

The optimal radius of curvature of the grinding wheel is determined by the formula:

$$R_u = \frac{\left(p_b^2 + r^2\right)^2 \sin\alpha_u}{r_{u1} p_b^2}. \quad (6.5)$$

In this case, the deviation of such a generatrix from the radial straight line is determined by the formula:

$$f_{vi} = \frac{h^2 (r_{u1} - R_u \sin\alpha_u)}{2\, r_{u1} R_u (R_u \mathrm{tg}\alpha_u - h)}. \quad (6.6)$$

When using sharpening machines with a copier in the form of a steel strip, separate profiling of the working surface of the abrasive tool is possible, i.e. the circle profile curve consists of two smoothly conjugate circles with radii R_{ua} i R_{uf} Fig. 6.1a. These radii are determined according to the dependencies:

$$\begin{aligned} R_{ua} &= R_u\left(1 + \frac{h_{\max}\cos\alpha_u}{r_{u1}}\right) - h_{\max}\mathrm{ctg}\alpha_u, \\ R_{uf} &= R_u\left(1 + \frac{h_{\max}\cos\alpha_u}{r_{u1}}\right) + h_{\max}\mathrm{ctg}\alpha_u. \end{aligned} \quad (6.7)$$

Fig. 6.1 Editing a circle with help two arcs

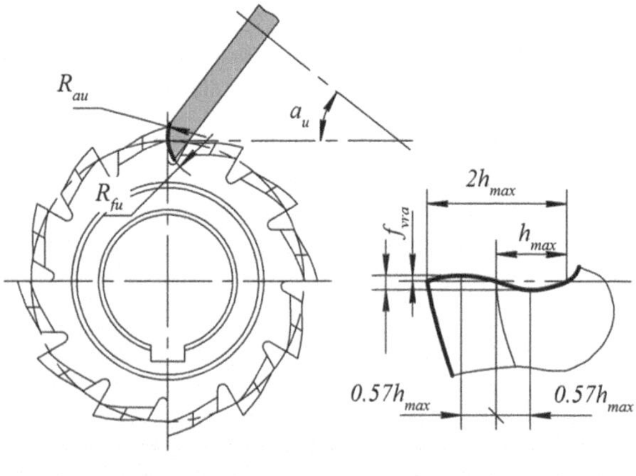

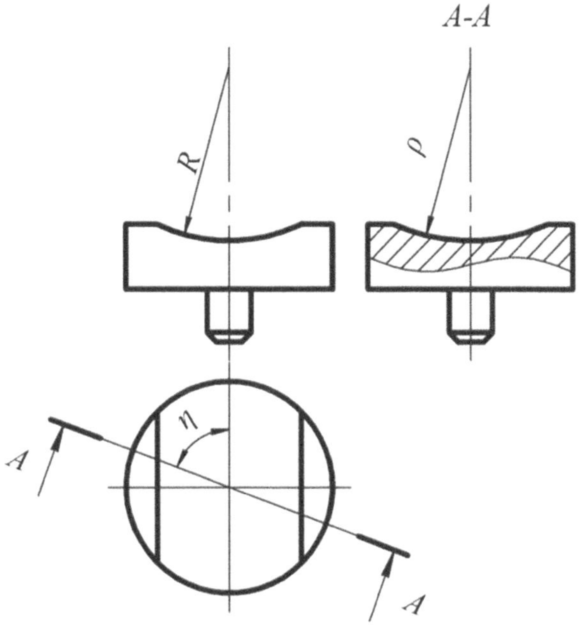

Fig. 6.2 Device with versatile heel

In this case, the deviation of such a generatrix from the radial straight line is determined by the formula:

$$f_{vi} = \frac{4\,h_{\max}^3 (r_{u1} - R_u \sin \alpha_u)}{27\,R_u^2 \mathrm{tg}\alpha_u r_{u1}}. \quad (6.8)$$

The cutter will have a convex profile (see Fig. 6.1).

The design of modern sharpening machines provides for the dressing of the grinding wheel by the copying method using universal copiers.

One of these copiers is shown in Fig. 6.2 [9]. This device is a heel with a concave cylindrical working surface r_k. The device ensures that the specified radius of curvature of the profile is obtained using a copier that sets the rotation at an angle φ_k.

Another device for editing a circle on a copier is shown in Fig. 6.3. The device has a flexible steel plate, which is given a given deflection using stops. The probe of the diamond holder slides along the copier, and a curved surface is formed on the wheel during editing [9].

Editing using the kinematic bending method when sharpening the front surfaces. The essence of the method is that the tool that edits reproduces the generatrix of the front surface of the hob cutter and, during the profiling process, performs a screw movement identical to the movement of the front surface of the teeth relative to the grinding wheel.

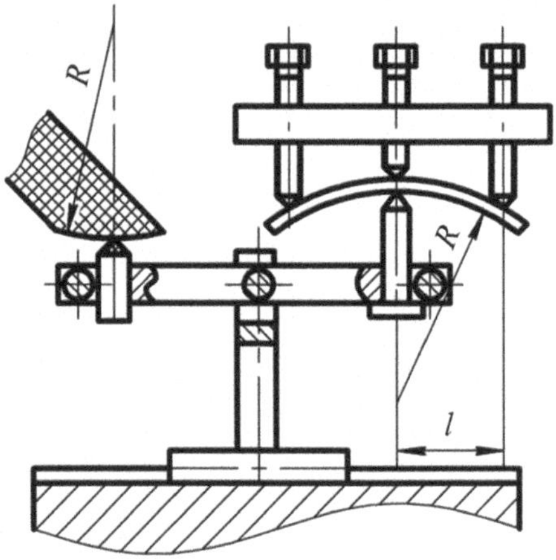

Fig. 6.3 Device with flexible strip

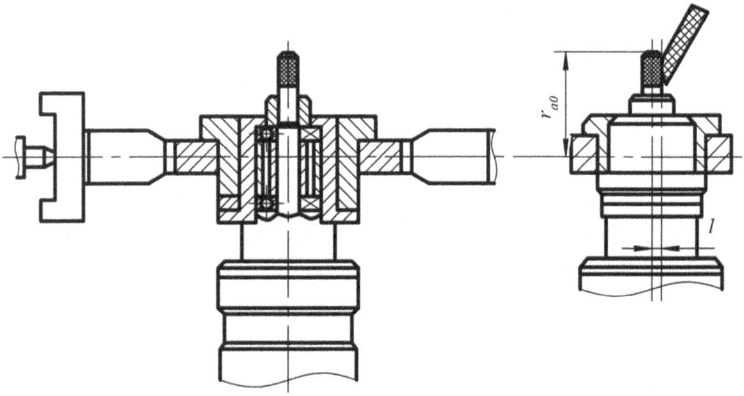

Fig. 6.4 Device for kinematic straightening

An example of a device is shown in Fig. 6.4 [9].

When profiling a circle using the kinematic bypass method using a steel knurling tool as a dressing tool, it is performed in the following order: first, the circle is corrected onto a cone to reduce the load on the knurl at the profiling site, then performed preliminary and final profiling.

When grinding helical front surfaces of hob cutters, especially multi-start ones and relatively small diameter ones, in addition to the errors described above, additional errors may occur due to cutting or under-profiling of the active section when the surface of the wheel and the cutter contacts outside the profiling zone.

The phenomenon of pruning the profile of the generatrix occurs in those areas where $\omega_f > \omega_u$ if the event that the section of the helical surface with a perpendicular plane to the generatrix intersects the surface of the grinding wheel, i.e. The surface of the grinding wheel intersects the surface of the cutter. To avoid the pruning phenomenon, it is necessary to limit the diameter of the wheel, as well as the angle α_u, and if this is not possible, then increase the installation angle of the grinding wheel axis ω_u, or switch to the kinematic dressing method.

The phenomenon of under-profiling of the generatrix of the front surface is manifested in the fact that in the circle of the cylinder of the depressions, contact is possible between the circle and the ground surface on the outer edge of the circle instead of its profile part. In order for the formative front surface to be profiled to the full height of the cutter tooth, it is necessary to insert the grinding wheel into the cavity so that the contact point between the wheel and the cutter is located on the working surface of the abrasive tool. Also, under-profiling can be completely eliminated if you set the axis of the grinding wheel to an angle ω_u equal to the angle ω_f of the inclination of the front surface on the cylinder of dimples.

The process of sharpening the front surface is a complex technological process, so errors in deviation of the machine settings from the calculated ones are inevitable. Deviations of these parameters should be limited by tolerance: $\alpha_u^{+\delta au}$, $r_{u1}^{+\delta ru1}$, $R_u^{+\delta Ru}$, where $+\delta au$, $+\delta ru1$, and $+\delta Ru$—are tolerance fields of the corresponding parameters.

All these questions are also relevant for manufacturers of special equipment used for sharpening worm cutters [13–15] and they implement the algorithms described above to improve the accuracy of the obtained surface of the cutter tooth.

The proposed methods do not define the limits of application of one or another straightening method and its impact on profile errors during sharpening. Also, the problem of determining the errors of the cutter surface profile during sharpening is solved in a simplified manner, namely, they lead it to a solution in the plane, which introduces errors into the calculation and does not consider their influence on the gear profile, this is due to the fact that this problem was considered separately from the problem of determining the rear surface and formation of the cutting edge of the cutter. Therefore, these questions need clarification. To determine the parameters of the cutting edge, it is necessary to consider the state of the issue of determining the shaping of the non-grinding surfaces of hob cutters.

6.3 Improving the Accuracy of Standard Cutters

In technological processes where hob cutters are used, new cutters are not always made to produce a new type of gear, or new cutters are ordered, but, if necessary, they try to use hob cutters that are already available, since this tool manufacture is expensive.

Therefore, the question arises whether, by choosing the setting parameters of the wheel while maintaining the rectilinear shape of the grinding wheel profile when shaping the front helical surface, it is possible to reduce the deviation from the straightness of the front surface of the cutter and increase accuracy, reduce the cost of manufacturing and operating such a tool. To do this, it is necessary to consider the shaping of the front surface of the tooth when sharpening a cutter and analyze the setting parameters of the grinding wheel when sharpening a cutter. The design diagram for shaping is shown in Fig. 6.5 [1, 2, 7, 8, 10, 11].

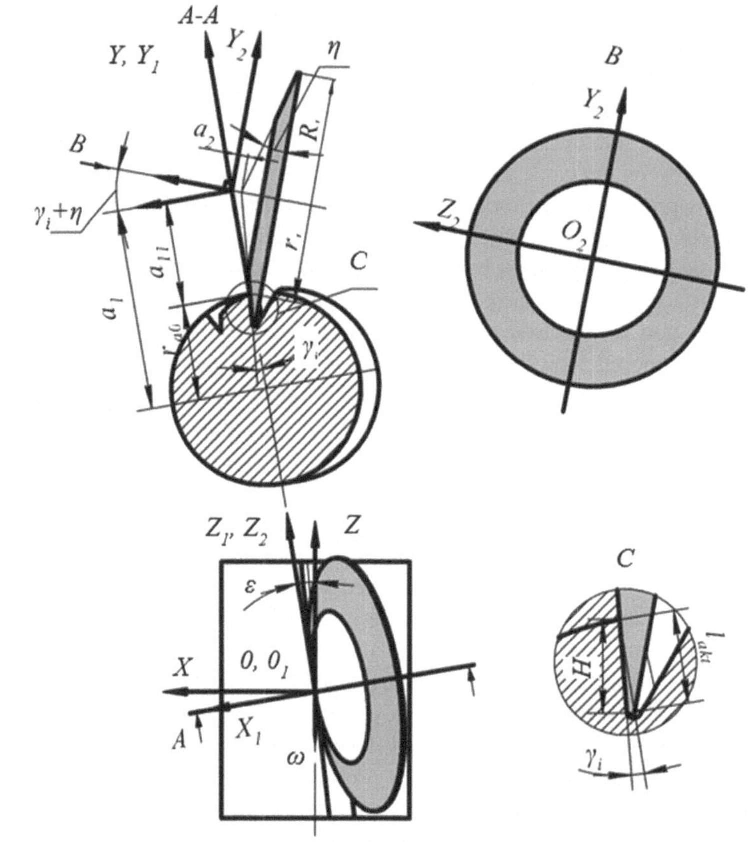

Fig. 6.5 Installing the grinding wheel when forming the front surface of the cutter tooth

6.3.1 Deviation from the Linearity of the Front Helical Surface of the Cutter Tooth

With the selected coordinate system, the constant setting parameters are (see Fig. 6.5):

l_{akt} height along the front surface of the part of the tooth that is sharpened;
r_{a0} cutter tooth tips radius;
H projection l_{akt} per axis Y, Y_1, tooth height;
η grinding wheel profile angle;
a_1 center-to-center distance between systems $S(X, Y, Z)$ and $S_1(X_1, Y_1, Z_1)$;
a_2 displacement of the beginning of the conical surface of a circle relative to the coordinate center in the system $S_2(X_2, Y_2, Z_2)$;

R_{kr} radius (diameter) of the grinding wheel in the projection of its contact with the lowest point of the cutter tooth that is being sharpened, in height;
r_{kr} radius of the grinding wheel in the projection of its contact with the highest point of the cutter tooth that is being sharpened, in height;
p_1 helical parameter of chip flute.

Parameters to be optimized:

ε grinding wheel rotation angle;
γ_i front instrumental angle of cutter (installation angle).

The contact equation is as follows:

$$
\begin{gathered}
l_{akt} = \frac{H}{\cos\gamma_i}; \qquad a_{11} = \frac{R_{kr} - \dfrac{H\sin\eta}{\cos\gamma_i}}{\cos(\eta+\gamma_i)};\\
r_{kr} = R_{kr} - H\frac{\sin\eta}{\cos\gamma_i};\\
a_2 = \frac{(R_{kr}\cos\gamma_i - H\cos\eta)(\mathrm{tg}(\eta+\gamma_i) - \mathrm{tg}\eta)}{\cos\gamma_i};\\
a_1 = r_a + a_{11};\\
k_1 + k_2 + k_3 + k_4 = 0;\\
k_1 = p_1\cos(\eta+\gamma_i) + l\cos t\sec^2\eta\sin(\eta+\gamma_i);\\
k_2 = -(a_1\cos t + (-a_2\cos t + p_1\sin t\sin(\eta+\gamma_i))\mathrm{tg}\eta);\\
k_3 = a_1\cos(\eta+\gamma_i) + l\sec^2\eta\sin t;\\
k_4 = (p_1\cos t + \sin t\,(a_2 - a_1\sin(\eta+\gamma_i)))\mathrm{tg}\eta.
\end{gathered}
\tag{6.9}
$$

where

l,t independent variable parameters that describe the conical surface of the grinding wheel, and are responsible for the radius and angular location of the point of the conical surface that is being considered, respectively.

By changing the parameter l within the selected limits, substituting into the contact Eq. (6.9) finding the parameter value t.

Substituting the found values into the equation of the conical surface, we find the coordinates of the property—the line of contact of the conical side of the circle with the front helical surface of the cutter tooth.

The coordinates of the end section points are calculated according to the following dependencies:

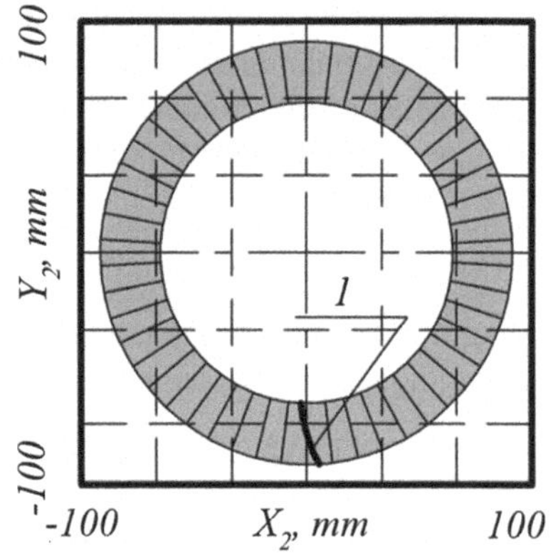

Fig. 6.6 Contact line of the sharpening wheel with the front helical surface of the cutter tooth: *1* contact line

$$\rho_{\text{char}} = \begin{bmatrix} -l\text{tg}\eta - a_2 \\ l\sin t \\ l\cos t \end{bmatrix};$$

$$\Theta_{\text{char}} = -\frac{z_{\text{char}}\sin\varepsilon - (x_{\text{char}}\cos(\eta+\gamma_i) - y_{\text{char}}\cos(\eta+\gamma_i))\sin\varepsilon}{p_1}; \tag{6.10}$$

$$x_{oc} = (z_{\text{char}}\sin\varepsilon + (x_{\text{char}}\cos(\eta+\gamma_i) - y_{\text{char}}\sin(\eta+\gamma_i))\cos\varepsilon)\cos\Theta_{\text{char}}$$
$$+(x_{\text{char}}\sin(\eta+\gamma_i) + y_{\text{char}}\cos(\eta+\gamma_i) + a\sin\Theta_{\text{char}});$$

$$y_{oc} = -(z_{\text{char}}\sin\varepsilon + (x_{\text{char}}\cos(\eta+\gamma_i) - y_{\text{char}}\sin(\eta+\gamma_i))\cos\varepsilon)\sin\Theta_{\text{char}}$$
$$+(x_{\text{char}}\sin(\eta+\gamma_i) + y_{\text{char}}\cos(\eta+\gamma_i) + a)\cos\Theta_{\text{char}}.$$

where

x_{oc}, y_{oc} coordinates of the end cross-section points of the cutter tooth.

For this purpose, the issue of finding the front surface of the cutter tooth when shaping it with a conical sharpening wheel was resolved. In Fig. 6.6 shows an example of calculating the contact line for the following cutter and wheel parameters (m = 10, d_{a0} = 150 mm—the diameter of the vertices of the cutter teeth, D_{kr} = 200 mm—diameter of the sharpening wheel, $\gamma_i = 0°$, ε = 4,58°—corresponds to lift angle the helix on the dividing cylinder, η = 20°, $H = h$ + 0,5 m, where h—cutter tooth height), as seen, it is a spatial line located on the conical surface of a circle.

Having received the coordinates of the points of the helical surface of the cutter, described by the contact equation, it is necessary to find the parameters of the resulting helical surface, namely the coordinates of the points of the end cross-section of such a helical surface; the greatest deviation of these points from the straight line passing through the center of the cutter axis will characterize the deviation of the profile from straightness.f_γ.

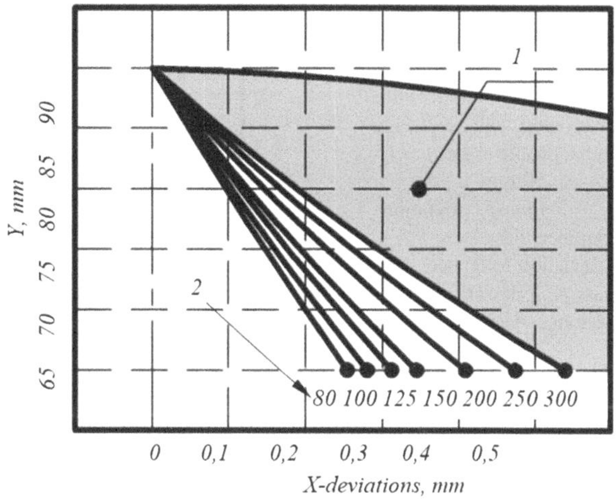

Fig. 6.7 Cutter front surface profile: $m = 10$ mm, $d_{a0} = 180$ mm, $\eta = 15°$: *1* cutter tooth, *2* diameter sharpening wheel

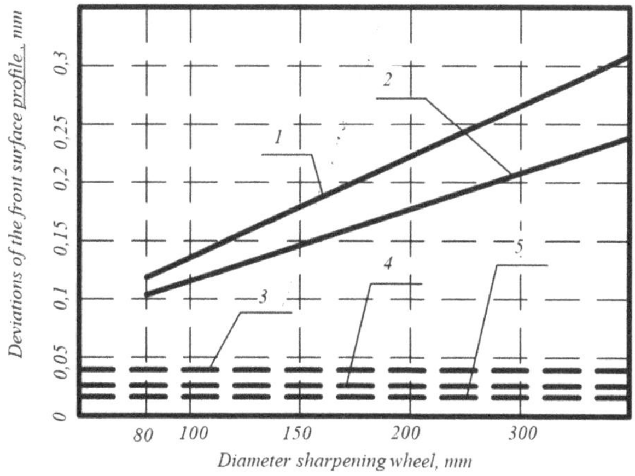

Fig. 6.8 Maximum deviations of the front surface profile for a cutter with a 3 mm module, when sharpening with a wheel with a profile angle $\eta = 15°$: *1* $r_{a0} = 40$ mm, *2* $r_{a0} = 56$ mm, *3* deviation tolerance class B, *4* deviation tolerance class A, *5* deviation tolerance class AA

Based on the given dependencies, calculations were made of the deviation of the front surface profile (see Figs. 6.7, 6.8, 6.9 and 6.10) for cutters of different modules, different diameters, sharpened with circles with different diameters and profile angles; the circle during sharpening was set for the front angle $\gamma = 0°$, the reversal angle of the sharpening circle corresponded to the angle of inclination of the cutter helix on the dividing cylinder, $H = h + 0{,}5$ m, the deviation was indicated on the length of the working section of the cutter tooth.

In accordance to the results of the calculations, cutters sharpened with circles, the surface of which is conical, have a deviation directed into the body of the cutter tooth, in addition, the larger the diameter sharpening wheel, the greater the deviation of the front surface profile and all milling cutters are sharpened with circles without curvilinear dressing at $\gamma_i = 0°$ in 2,5…3 times exceed the straightness tolerance according to GOST 9324-80 (see Figs. 6.7, 6.8, 6.9 and 6.10), therefore additional methods are needed that can reduce this deviation.

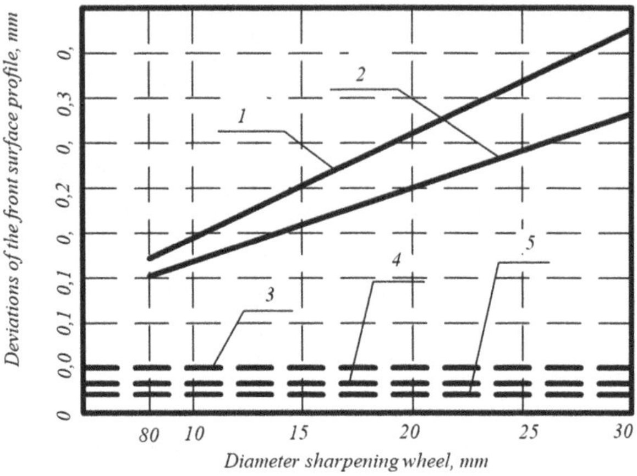

Fig. 6.9 Maximum deviations of the front surface profile for a cutter with a 5 mm module, when sharpening with a wheel with a profile angle $\eta = 15°$: *1* $r_{a0} = 50$ mm, *2* $r_{a0} = 70$ mm, *3* deviation tolerance class B, *4* deviation tolerance class A, *5* deviation tolerance class AA

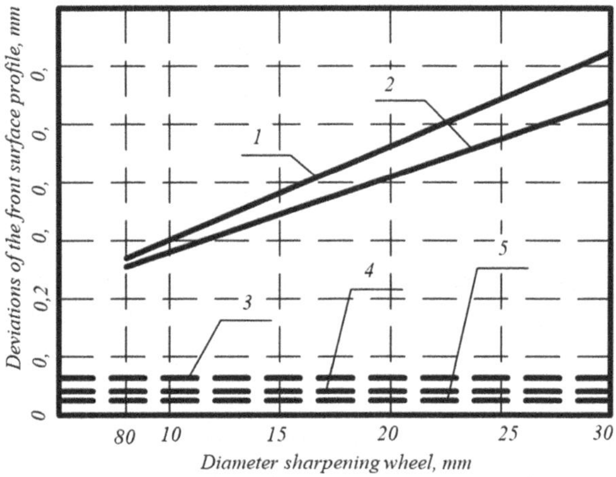

Fig. 6.10 Maximum deviations of the front surface profile for a cutter with a 10 mm module, when sharpening with a wheel with a profile angle $\eta = 15°$: *1* $r_{a0} = 75$ mm, *2* $r_{a0} = 90$ mm, *3* deviation tolerance class B, *4* deviation tolerance class A, *5* deviation tolerance class AA

The effect of changing the rotation angle on the profile of the front surface of the cutter tooth is shown in Fig. 6.11, angle ε varied within the limits corresponding to the angle of inclination of the helix of the chip flute on the diameter of the tops of the teeth to the angle of inclination of the helix by the diameter of the cavity.

As a result of changing the rotation angle, when it decreases, the deviation of the front surface profile decreases, but at the same time the depth of the groove will increase, which will not be formed (Fig. 6.12), however, as calculations show, this value does not exceed the amount by which the wheel is deepened when sharpening the front surface and at least it is *0,5* m, which is added when the sharpening wheel is deepened to the working height of the cutter tooth.

Also, the deviation of the front surface profile depends on the angle of the sharpening wheel profile; this angle is standard for sharpening disc wheels and takes certain values of 15°; 20°; 30° with which, according to the standards, they are manufactured for both el'bor borazon material and electrocorundum wheels.

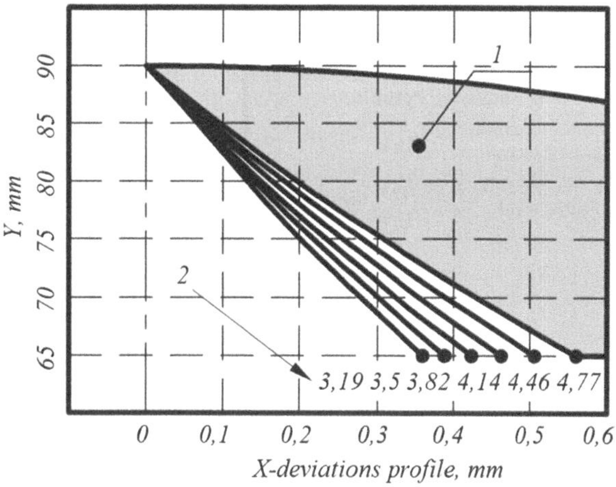

Fig. 6.11 Profile of the front surface of the cutter depending on the angle of rotation ε of the sharpeningwheel $m = 10$ mm, $d_{a0} = 180$ mm, $D_{kr} = 200$ mm, $\eta = 15°$, $\gamma = 0°$, the angle of inclination of the helix on the dividing cylinder is $b_f = 3{,}703°$: *1* cutter tooth, *2* ε

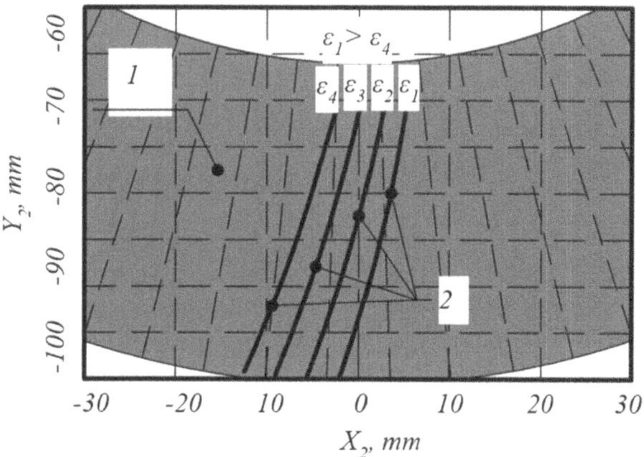

Fig. 6.12 The shape and location of the contact lines of the sharpening wheel and the helical groove depending on the reversal angle, $m = 10$ mm, $d_{a0} = 180$ mm, $D_{kr} = 200$ mm, $\eta = 15°$, $\gamma = 0°$, the angle of inclination of the helix on the dividing cylinder is $b_f = 3{,}703°$: *1* sharpening wheel, *2* contact lines

Wheels with a large value of η from 30° are not used since the angle of deployment of the chip flute of hob cutters does not exceed 35°. Fig. 6.13 shows the change in the profile of the front surface depending on the tilt angle profile of the sharpening wheel.

According to the calculation results, when the profile angle changes towards increasing values, the profile deviation of the front surface increases and it is directed into the body of the cutter tooth (see Fig. 6.13), the front surface of the cutter tooth is convex.

To reduce the deviation, it is proposed to change the installation angle of the sharpening wheel γ, for example (see Fig. 6.14) shows a change in the profile of the front surface and the deviation of the front surface for a cutter $m = 10$ mm, for the diameter of the sharpening wheel $D_{kr} = 200$ mm.

As a result of changing the installation angle γ (see Fig. 6.14) it is shown that for a cutter $m = 10$ mm, $d_{a0} = 180$ mm, circle diameter $D_{kr} = 200$ mm, $\eta = 15°$ it is

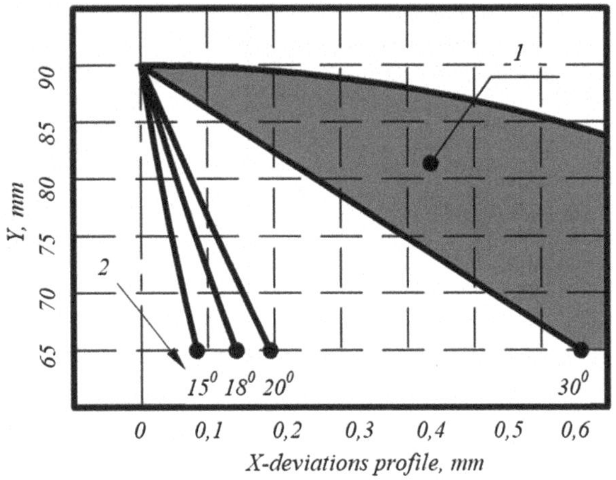

Fig. 6.13 Profile of the front surface of the cutter depending on the angle of the sharpening wheel profile η, $m = 10$ mm, $d_{a0} = 180$ mm, $D_{kr} = 200$ mm, $\gamma_i = 0$: *1* cutter tooth, *2* η

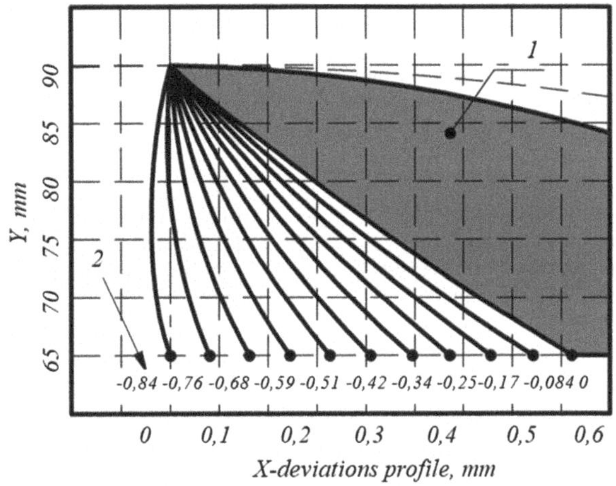

Fig. 6.14 Cutter front surface profile, $m = 10$ mm, $d_{a0} = 180$ mm, $D_{kr} = 200$ mm, $\eta = 15°$, depending on the angle of installation of the circle γ_i: *1* cutter tooth, *2* γ_i

possible to reduce the deviations of profile's of front surface from 0,41 mm to 0,019 mm, which less than the deviation according to GOST 9324-80, which is for class AA 0,025 mm. In practice, this is realized by rotating the cutter on the mandrel when sharpening by an amount at which the tip of the cutter tooth will be at a distance from the zero position by an amount $h_{\text{offset}} = r_a/\sin\gamma$ (see Fig. 6.15).

In this case, the profile of the front surface will be convex and extend out from the body of the tooth, in this case, the profile of the front surface will be convex and extend out from the body of the tooth. The tolerance for installing the circle should be directed into the body of the part, based on the calculation results, the angle of installation of the circle should be within $\gamma_i[-0.820 : -0.860]$ (see Fig. 6.16). That the deviation of the tooth profile does not exceed the tolerance, both in the tooth body and in the opposite direction, then the displacement value will lie within the limits $h[1,28 \text{ mm} : 1,35 \text{ mm}]$.

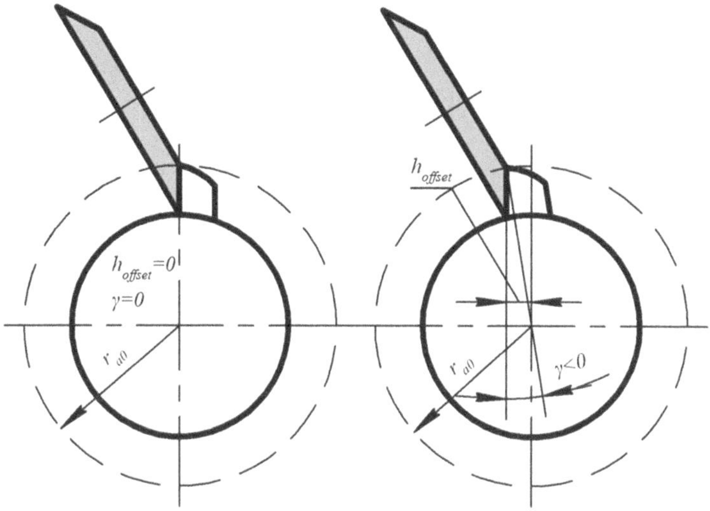

Fig. 6.15 Installing the cutter when sharpening

Fig. 6.16 The profile of the front surface of the cutter is within tolerance, $m = 10$ mm, $d_{a0} = 180$ mm, $D_{kr} = 200$ mm, $\eta = 15°$, depending on the angle of installation of the circle γ_i: *1* cutter tooth, *2* γ_i, *3* tolerance

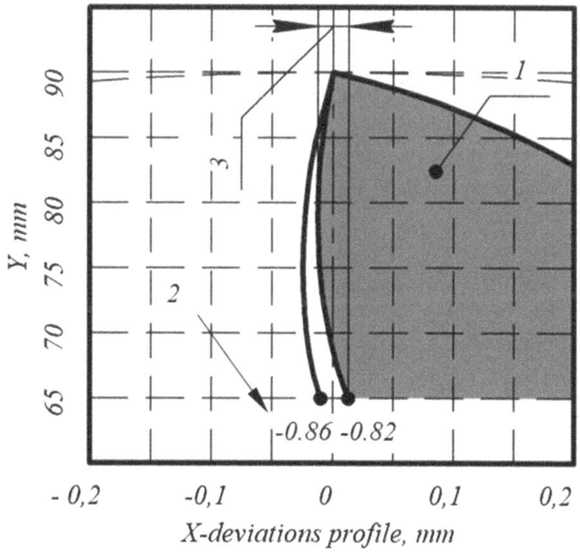

The tolerance is 0,07 mm, technological tolerance by offset, given that the sharpening operation is critical in the manufacture of the tool, can be assigned as half the tolerance from the calculated value $h_{tech.} = 1,28^{+0,035}$ mm, these values can be technologically realized when sharpening such milling cutter.

When switching to less precise cutters class *A*, *B*, *C* the calculated installation angle tolerance can be increased, since the accuracy of the front surface profile for them below.

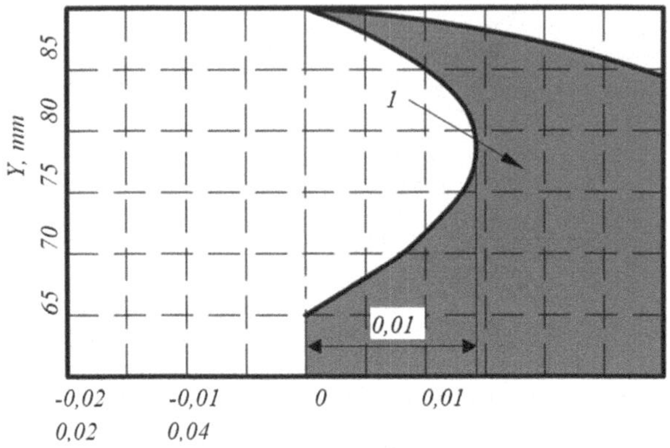

Fig. 6.17 Cutter front surface profile, $m = 10$ mm, $d_{a0} = 180$ mm, $D_{kr} = 200$ mm, $\eta = 15°$, $\varepsilon = 0°$, $\gamma_i = -0{,}772°$: *1* cutter tooth

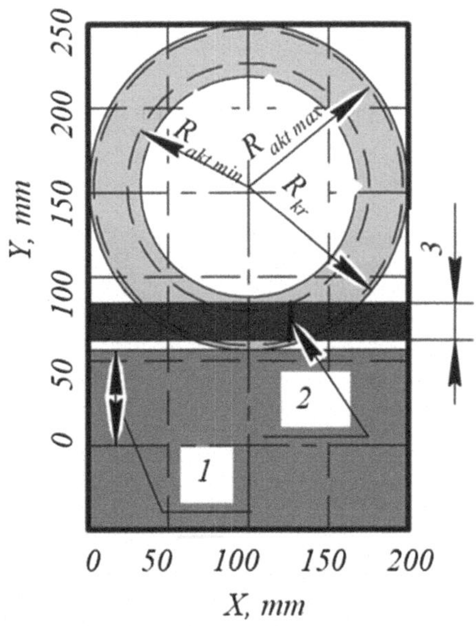

Fig. 6.18 The location of the contact line of the sharpening wheel and the helical groove when $\varepsilon = 0$, $m = 10$ mm, $d_{a0} = 180$ mm, $D_{kr} = 200$ mm, $\eta = 15°$, $\gamma_i = -0{,}772°$: *1* r_{f0}, *2* contact line, *3* *H*

One of the options for sharpening milling cutters can be considered when the angle of reversal of the sharpening wheel is $\varepsilon = 0$, the wheel is positioned as when sharpening cutters with straight grooves. According to the calculation results, it was found that by changing the angle of installation of the circle γ it is possible to achieve a result in which the deviation of the front surface profile (see Fig. 6.17) will be less in value than for the case when the angle of reversal of the circle corresponded to the angle of elevation of the helix on the dividing cylinder.

The minimum achieved deviation of the front surface profile from straightness for the case under consideration was $f_\gamma = 0{,}014$ mm (see Fig. 6.18), which is less than for the setting parameters obtained when $\varepsilon \neq 0$, $f_\gamma = 0{,}019$ mm, in addition, when $\varepsilon = 0$ the profile of the front surface of the tooth is concave, and not convex as for an angle $\varepsilon \neq 0$.

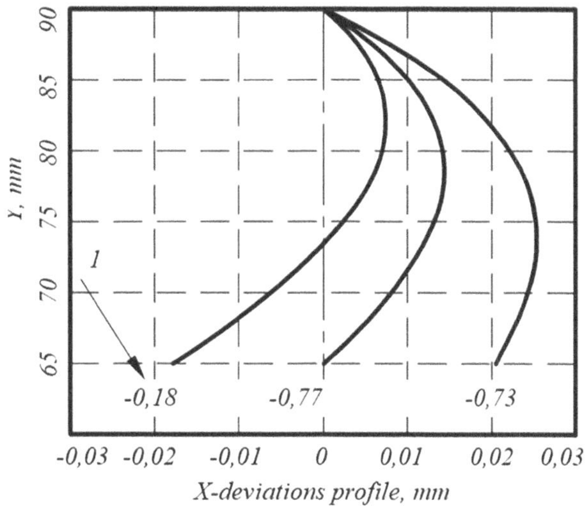

Fig. 6.19 The profile of the front surface of the cutter is within the calculated tolerance, depending on the angle of installation of the wheel γ_i: $m = 10$ mm, $d_{a0} = 180$ mm, $D_{kr} = 200$ mm, $\eta = 15°$, $\varepsilon = 0$: *1* γ_i

However, in such a case, it is necessary to check the possibility of completely processing the front surface of the tooth along its entire active length with the existing diameter of the sharpening wheel and the deepening its into the helical groove during sharpening.

According to the results of calculations with the value of the sharpening wheel deepening $H = h + 0.5$ m, for the case shown in Fig. 6.19 with the radius of the circle $R_{kr} = 100$ mm, the radius of the circle that forms the lowest point of the front surface of the tooth, located at a distance h =2, 5 m from the top of the tooth is $R_{aktmax} = 97$, 1 mm this is less than the radius of the circle, which means it is quite possible machine the groove to the full working height of the tooth.

Tolerance for the installation of the wheel for cutters of class AA at $\varepsilon = 0$ according to the calculation results for ($m = 10$ mm, $d_{a0} = 180$ mm, $D_{kr} = 200$ mm, $\eta = 15°$) the angle of installation of the wheel should have a value within $\gamma_i[-0.730 : -0.810]$ (see Fig. 6.20), this is twice as much as in the case indicated in Fig. 6.13, then the displacement value will be within h_{offset} [1,15 mm: 1.27 mm], the tolerance will be 0,12 mm, and the working size will be equal to $h_{offset} = 1,15^{+0,06}$.

For sharpening schemes in which $\varepsilon = 0$, it is advisable to use wheels with smaller values of the profile angle η and check whether there is any undercutting of the back of the tooth of cutter when sharpening by non-working part of the wheel by milling cutter, and also take into account when choosing the design parameters of cutters—the angle of deployment of the chip flute of the cutter.

It has been proven that due to the installation parameters of the sharpening wheel, it is possible to reduce the deviation of the front surface from straightness by a conical wheel without the use of curved dressing.

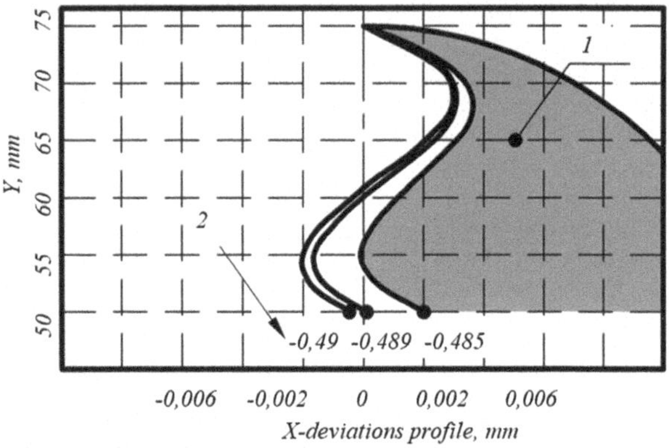

Fig. 6.20 Cutter front surface profile, $m = 10$ mm, $d_{a0} = 150$ mm, $D_{kr} = 200$ mm, $\eta = 15°$, $\gamma_i = 0°$, the angle of inclination of the helix on the dividing cylinder is $b_f = 4, 5886°$, $\varepsilon = 1, 28°$

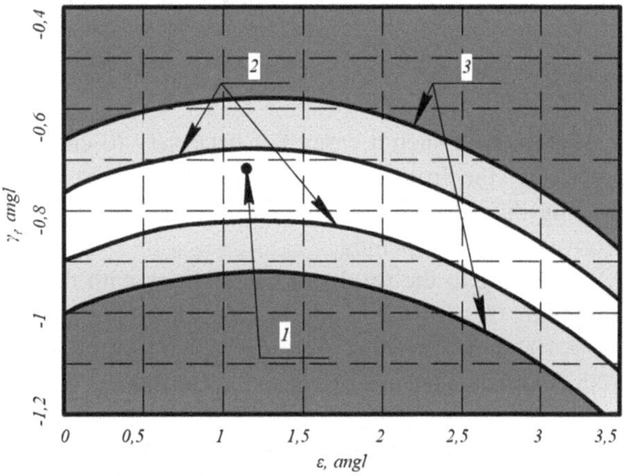

Fig. 6.21 Permissible limits for changing installation parameters of sharpening wheel $m = 3$ mm, $d_{a0} = 80$ mm, $D_{kr} = 80$ mm, $\eta = 15°$, $\gamma = 0°$, the angle of inclination of the helix on the dividing cylinder is $b_f = 2, 3715°$: *1* minimum deviation 0,00006 mm, *2* deviation tolerance 0,01 mm, *3* deviation tolerance 0,025 mm

6.3.2 Calculation of Optimal Installation Parameters of Grinding Wheels When Sharpening Milling Cutters

In Figs. 6.20, 6.21, 6.22, and 6.23 it is shown that by changing the installation parameters of the wheel, we obtain different profiles of the front surface, which are characterized by a maximum deviation from straightness f_γ [7, 8]. Depending on the angle of rotation of the grinding wheel ε, we obtain a different profile of the front surface of the tooth from the convex one Fig. 6.20 for $\varepsilon = b_f$—the angle of elevation of the helix on the dividing cylinder, to the concave for $\boldsymbol{\varepsilon} = 0^0$ Fig. 6.21, however, the maximum deviation f_γ will be within the tolerance.

Therefore, the question arises whether it is possible to calculate the setting parameters for sharpening using a wheel so that the front surface profile has zero curvature $f_\gamma = 0$. To do this, it was necessary to solve the optimization problem—for what values of the parameters ε and γ—the front mounting angle of the hob cutter—the deviation f_γ will be the smallest. Usin the problem of finding such setting

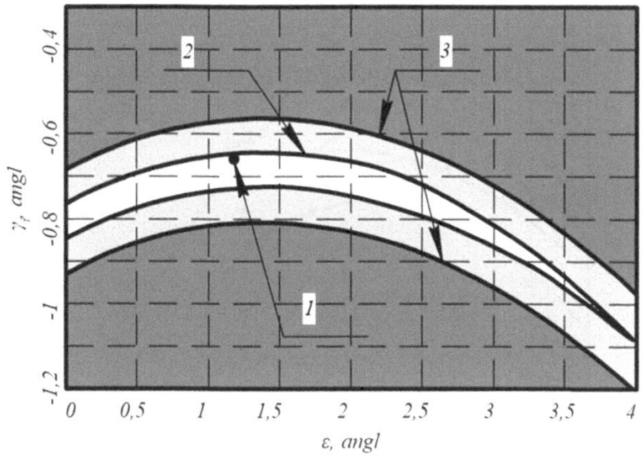

Fig. 6.22 Permissible limits for changing installation parameters of sharpening wheel, $m = 5$ mm, $d_{a0} = 100$ mm, $D_{kr} = 125$ mm, $\eta = 15°$, $\gamma = 0°$, the angle of inclination of the helix on the dividing cylinder is $b_f = 3, 2788°$: *1* minimum deviation 0,0004 mm, *2* deviation tolerance 0,01 mm, *3* deviation tolerance 0,03 mm

parameters for which the value of f_γ would be minimal was solved [1, 3, 7, 8]. As calculations have shown, there is a value of the setting parameters ε and γ for which the value of f_γ is less than Fig. 6.22.

Calculations have shown that it is impossible to obtain a profile of the front surface of a tooth with zero curvature when it is shaped by the conical surface of a sharpening wheel; however, through a rational choice of the installation parameters of the wheel during sharpening, these deviations can be significantly reduced, and the profile of the front surface of the tooth will consist of (see Fig. 6.20) from both the convex area at the base of the tooth and the bulging area at the top of the tooth of the cutter. The angle of reversal of the wheel for the case shown in Fig. 6.23 $\varepsilon = 1, 28°$ for the angle of the rise of the helix line of the cutter on the dividing cylinder $b_f = 4, 59°$, and according to the values they will not go beyond the tolerance limits according to the standard for monitoring the parameter f_γ. For the given calculation example, the largest error value for the parameters ($\varepsilon = 1, 28°$, $\gamma_i = -0, 4850$) was $\Delta = 0, 0037$ mm which is significantly less than the tolerance for manufacturing for such class A cutters, which is 0,04 mm.

Calculations were made of the permissible limits for changing the setting parameters of the wheel when sharpening hobs for various modules (3 mm, 5 mm, 10 mm) of class A (see Figs. 6.21, 6.22 and 6.23).

Permissible limits for the set-up parameters for the tolerance for straightness of the front surface, as well as the limits of the technological tolerance for manufacturing are shown, this is due to the fact that in technology by manufacturing it is customary consider tolerance for manufacturing, tolerance by the installation of the tool to reduce by three times from the nominal, thereby providing a margin for other errors from the components of the technological system.

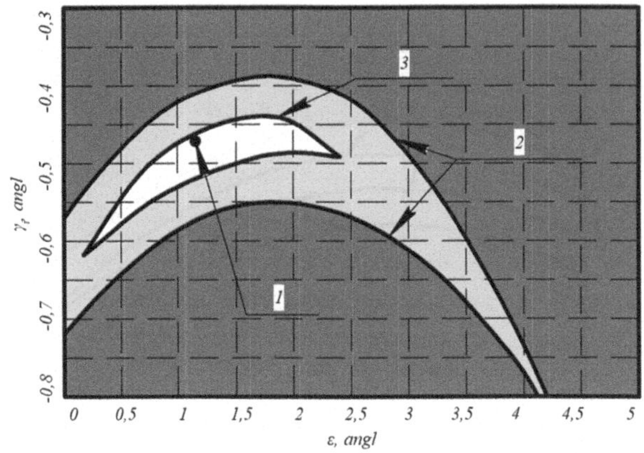

Fig. 6.23 Permissible limits for changing the installation parameters of the sharpening wheel when sharpening of hobs, $m = 10$ mm, $d_{a0} = 150$ mm, $D_{kr} = 200$ mm, $\eta = 15°$, $\gamma = 0°$, the angle of inclination of the helix on the dividing cylinder is $b_f = 4,5886°$: *1* minimum deviation 0,004 mm, *2* deviation tolerance 0,015 mm, *3* deviation tolerance 0,04 mm

Calculations show (see Figs. 6.21, 6.22, and 6.23) that the tolerance limits are curvilinear, therefore for each case of sharpening they must be calculated depending on the primary setting parameters of the wheel ε, or γ_i.

More severe influence on the variation of deviation from straightmess is caused by a change in the angle of installation of the wheel $\gamma_{\kappa p}$ than a change in the rotation angle of the wheel ε, than a change in the rotation angle of the wheel.

6.4 Sharpening Milling Cutters by Conical Wheels Without Curved Dressing

An experimental test of the proposed method of sharpening cutters was carried out. When sharpening hobs, the design parameters are indicated in accordence to the specification $m = 6$, $d_{a0} = 112$ mm—diameter of the tips of the cutter teeth, $\gamma = 0°$, $\varepsilon = 3\,°\,32^{'}$.

Analysis of the calculations showed that the conical shaping surface of hob cutters has a deviation directed into the body of the cutter tooth. In addition, the larger the diameter of the sharpening wheel, the greater the deviation of the front surface profile and all milling cutters are sharpened with circles without curved dressing at $\gamma = 0^0$ y 2,5...3 times higher than the straightness tolerance according to GOST 9324-80 for which this value is $f_\gamma = 0,04$ mm (see Fig. 6.24), therefore additional methods are needed to reduce this deviation.

Calculations for the option of sharpening a milling cutter $m = 6$, $d_{a0} = 112$ mm, $\varepsilon = 3°32'$, with a sharpening wheel with a diameter of 200 mm and a profile angle of

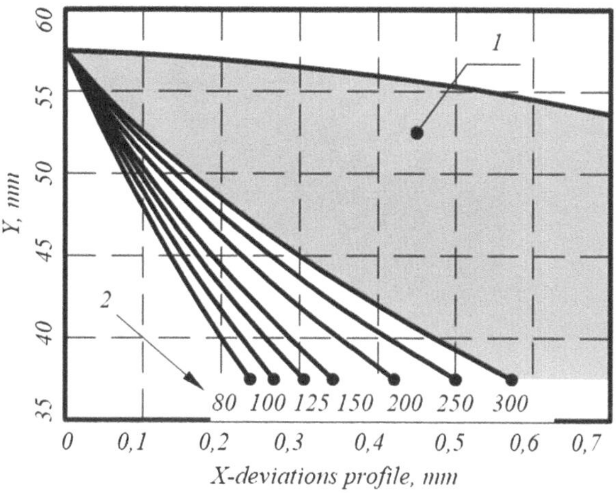

Fig. 6.24 Profile of the front surface of the cutter depending on the diameter of the sharpening wheel, $m = 6$ mm, $d_{a0} = 112$ mm, $\gamma = 0°$, $\varepsilon = 3° 32'$, $\eta = 15°$: *1* cutter tooth, *2* diameter of the sharpening wheel

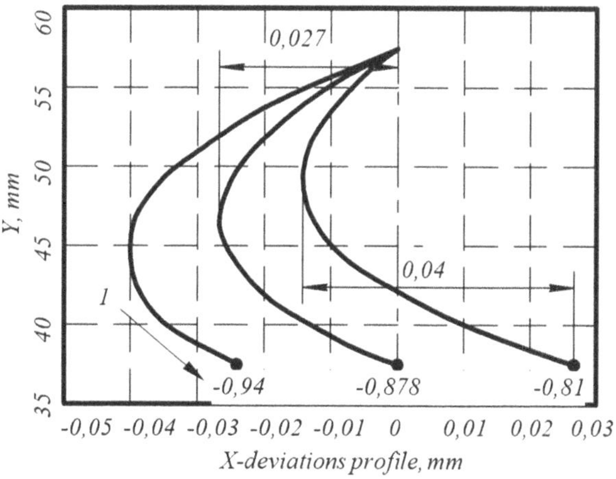

Fig. 6.25 The profile of the front surface of the cutter is within the calculated tolerance, $m = 6$, $d_{a0} = 112$ mm, $\gamma = 0°$, $\varepsilon = 3° 32'$, $D_{kr} = 200$ mm, $\eta = 15°$: *1* γ_i

15° according to the technological process of the customer's enterprise showed (see Fig. 6.25) that it is possible to achieve the required deviation accuracy due to changing the sharpening wheel installation angle γ_i [−0.910; −0.810].

When sharpening, the diagram for setting the wheel relative to the cutter is shown in Fig. 6.26.

The calculation of the displacement of the cutter tip when sharpening relative to the vertical plane is: $h = 0.89^{+0,\ 14.}$

Considering that the value Δ is calculated, and the system is also affected by other factors from other errors, the calculated tolerance Δ = 0, 14 must be reduced by half: $\Delta_{cal} = 0, 07$ mm, the working size that is installed on the machine when sharpening will be $h = 0.89^{+0.07}$.

To check the obtained calculations, hob cutters were sharpened using the developed method.

Fig. 6.26 Sharpening the milling cutter

Sharpening was carried out on a sharpening machine model 3Б662

The cutter and wheel were installed on the machine according to the calculated values (see Fig. 6.26).

Used when sharpening the cutter: wheel 3П 200 ×10 × 38A 10-П C2 7 K5 35 m/s A1 class GOST 2424-83.

Machine setup parameters:

- circle speed—16 m/s;
- longitudinal feed—4 m/min;
- angle of reversal of the wheel—3° 32′;
- pitch of helical flutes of the cutter—4917.1 mm;
- displacement of the cutter tooth tip from the vertical axial position—$0{,}89^{+0{,}07}$;
- the amount of deepening of the wheel into the workpiece—17 mm;
- allowance for one pass—0,015 mm;
- lubricating and cooling process medium—liquid Ukrinol-2

Measuring tool: time type indicator: division value 0,002 mm.

After sharpening milling cutters, the error of the front surface profile was measured; for the cutter of the proposed design it was $f_\gamma = 0,03$ mm, for the standard cutter $f_\gamma = 0{,}035$ mm, which is less than the established tolerance value according to GOST 9324-80 0,04 mm.

Thus, the proposed method of sharpening hobs provides the necessary accuracy of the cutters in terms of the accuracy of the front surface and it is more technologically advanced since there is no need to edit the wheel along a curved surface. The wheel is straightened only along a straight generatrix of the conical surface, which greatly simplifies the manufacture and subsequent operation of such a tool.

However, given the modern development of computer technology and the widespread use of CNC machines in the manufacture of tools, the problem of determining the tool profile for geometrically accurate processing of a ruled helical surface has not been fully solved.

6.5 Geometrically Precise Shaping of the Front Helical Surface of Hob Cutters During Sharpening

To obtain an accurate front surface, correction of the profile of the forming tool is used. An exact solution to this problem is based on solving the problem of shaping the front helical surface of a cutter tooth with a disk tool [1, 8].

We will solve the problem of precise shaping of a ruled helical surface taking into account the installation parameters of the disk tool and the spatial nature of shaping. To determine the profile of a disk tool, consider the following coordinate systems (see Fig. 6.27), which are shown in the initial state,

where

$S(X, Y, Z)$	system associated to the milling cutter;
$S_1(X_1, Y_1, Z_1)$	auxiliary coordinate system;
$S_2(X_2, Y_2, Z_2)$	system associated with disk tool.

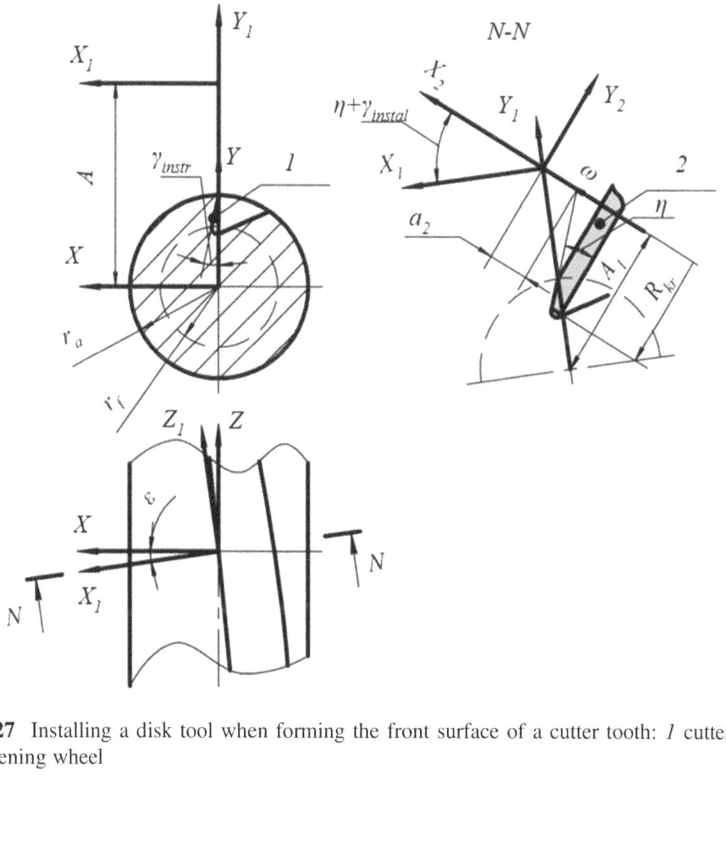

Fig. 6.27 Installing a disk tool when forming the front surface of a cutter tooth: *1* cutter tooth, *2* sharpening wheel

With the selected coordinate systems, the installation technological parameters will be:

l_{akt} height along the front surface of the part of the tooth being sharpened;
r_a cutter tooth tips radius;
r_f radius of the cavity point of the cutter teeth;
H projection l_{akt} on axis Y, Y_1, tooth height;
η profile angle of conical grinding wheel;
ε grinding wheel reversal angle;
A center-to-center distance between systems $S(X, Y, Z)$ and $S_1(X_1, Y_1, Z_1)$;
A_2 the smallest distance from the axis X_2 to the center of the coordinate system $S(X, Y, Z)$;
a_2 displacement of the beginning of the conical surface of a circle relative to the coordinate center in the system $S_2(X_2, Y_2, Z_2)$;
γ_i front instrumental angle of the cutter;
γ_{instr} angle of installation of the circle relative to the front surface of the cutter tooth;
R_{kr} the radius of the sharpening wheel in the projection of its contact with the lowest point of the cutter tooth in height, which is sharpened;
r_{kr} radius of the sharpening wheel in the projection of its contact with the highest point of the cutter tooth that is being sharpened in height.

The connection between coordinate systems is determined by dependencies $S(X, Y, Z)$ and $S_2(X_2, Y_2, Z_2)$:

$$\begin{aligned}
X_2 &= Z\sin\varepsilon + (X\cos(\eta + \gamma_{\text{install}}) - Y\sin(\eta + \gamma))\cos\varepsilon;\\
Y_2 &= A + X\sin(\eta + \gamma_{\text{install}}) + Y\cos(\eta + \gamma_{\text{install}});\\
Z_2 &= Z\cos\ \varepsilon + (X\cos(\eta + \gamma_{\text{install}}) - Y\sin(\eta + \gamma))\sin\varepsilon.
\end{aligned} \tag{6.11}$$

The installation parameters by making are determined by the following equations:

$$\begin{aligned}
&tg\gamma_{\text{install}} = \frac{tg\gamma_i}{\cos\varepsilon}; \qquad a_1 = \frac{R_{kr} - \dfrac{(r_a - r_j)\sin\eta}{\cos\gamma_{\text{install}}}}{\cos(\eta + \gamma_{\text{install}})};\\
&r_{kr} = R_{kr} - \frac{(r_a - r_j)\sin\eta}{\cos\ \gamma_{\text{install}}}; \qquad a = r_a + a_1;\\
&a_2 = \frac{(R_{kr}\cos\gamma_{\text{install}} - (r_a - r_f)\cos\eta\ )(tg(\eta + \gamma_{\text{install}}) - tg\eta)}{\cos\gamma_{\text{install}}}.
\end{aligned} \tag{6.12}$$

Based on the results of work [11], the equation of the ruled helical front surface of a cutter tooth will be written in the form:

$$X_p = (u\sin\gamma_{\text{instr}} - r_a \sin\gamma_{\text{instr}} \cos\gamma_{\text{instr}}) \cos t + \left(u\cos\gamma_{\text{instr}} + r_a \sin^2\gamma_{\text{instr}}\right) \sin t,$$
$$X_p = (u\sin\gamma_{\text{instr}} - r_a \sin\gamma_{\text{instr}} \cos\gamma_{\text{instr}}) \sin t + \left(u\cos\gamma_{\text{instr}} + r_a \sin^2\gamma_{\text{instr}}\right) \cos t,$$
$$Z_p = p_1\, t. \tag{6.13}$$

where

t independent parameter of the helical surface that determines the position of the rectilinear generatrix in space;
u independent parameter of the helical surface, determining the position of the point on the rectilinear generatrix of the helical surface;
p_1 spiral parameter of chip separating groove.

For cutters with a front zero instrumental angle $\gamma_{\text{instr}} = 0$ the equation of the ruled helical front surface of the cutter tooth will be written:

$$\begin{cases} X_p = u \sin t \\ Y_p = u \cos t \\ Z_p = p_1 t \end{cases} \tag{6.14}$$

Let us define the initial tool surface I of the disk tool as the envelope of the helical surface during its rotation around a given axis $X_{2.}$ To do this, we write the contact equation in kinematic form.

Equation of the normal vector to the ruled helical front surface of a cutter tooth:

- with an arbitrary rake angle in the system $S(X, Y, Z)$ will be:

$$\vec{N} = \cos(t + \gamma_{\text{instr}})\,\vec{i} - \sin(t + \gamma_{\text{instr}})\,\vec{j} - \frac{u}{p_1}\vec{k}. \tag{6.15}$$

- with zero instrumental front angle:

$$\vec{N} = \cos t\,\vec{i} - \sin t\,\vec{j} - \frac{u}{p_1}\vec{k}. \tag{6.16}$$

The speed of an arbitrary point on a helical surface during its rotation around the X_2 axis in the system $S(X, Y, Z)$ in general form:

$$\vec{V} = \vec{\omega} \times \vec{R} + \overrightarrow{A_1} \times \vec{\omega} \tag{6.17}$$

where

ω unit vector of angular velocity of rotation around the X_2 axis in the coordinate system $S(X, Y, Z)$;
R radius vector of a random point of the helical surface;

A_1 radius vector of the minimum distance from the center of coordinates of the system $S(X, Y, Z)$ to the direction of action of the vector ω.

The parameters included in Eq. (6.17) can be determined from the following mathematical dependencies:

$$\begin{aligned}\vec{\omega} &= \cos(\eta + \gamma_{\text{install}}) \cos\varepsilon\ \vec{i}\ + \sin(\eta + \gamma_{\text{install}}) - \cos(\eta + \gamma_{\text{install}}) \sin\varepsilon\ \vec{k}\,,\\ \overrightarrow{A_1} &= -A \sin\left(\eta + \gamma_{\text{уст}}\right) \cos\left(\eta + \gamma_{\text{уст}}\gamma_{\text{install}}\right) \cos\varepsilon\ \vec{i}\ + A\cos^2\left(\eta + \gamma_{\text{install}}\right)\\ &\quad \vec{j}\ - A\sin\left(\eta + \gamma_{\text{install}}\right) \cos\left(\eta + \gamma_{\text{install}}\right) \sin\varepsilon\ \vec{k}\,.\end{aligned} \tag{6.18}$$

In expanded form, equation [4–6] will be written:

- linear helical front face of cutter tooth with arbitrary front angle:

$$\begin{aligned}\vec{V} &= \left(v_x, v_y, v_z\right),\\ v_x &= \cos\left(\eta + \gamma_{\text{instr}}\right) \times \sin\varepsilon\Big(-A + r_a \sin\gamma_{\text{install}} \sin\left(\gamma_{\text{install}} + t\right)\\ &\quad + u\cos\left(\gamma_{\text{install}} + t\right)\Big) + p_1 t \sin\left(\eta + \gamma_{\text{instr}}\right),\\ v_y &= \cos\left(\eta + \gamma_{\text{instr}}\right)\Big(p_1 t\cos\varepsilon + \sin\varepsilon\Big(u\sin\left(\gamma_{\text{install}} + \text{t}\right)\\ &\quad - r_a \sin\gamma_{\text{install}} \cos\left(\gamma_{\text{install}} + t\right)\Big)\Big),\\ v_z &= \cos\left(\eta + \gamma_{\text{instr}}\right) \times \cos\varepsilon\Big(-A + r_a \sin\gamma_{\text{install}} \sin\left(\gamma_{\text{install}} + t\right)\\ &\quad + u\cos\left(\gamma_{\text{install}} + t\right)\Big) + p_1 t \sin\left(\eta + \gamma_{\text{instr}}\right).\end{aligned} \tag{6.19}$$

- linear helical rake face of the cutter tooth with zero front angle:

$$\begin{aligned}\vec{V} &= \cos\eta \sin\varepsilon\ (u\cos t - A)\\ &\quad + p_1 t \sin\eta\ \vec{i}\\ &\quad - \cos\eta\ (p_1 t\cos\varepsilon + u\sin t - r_a \sin t \sin\varepsilon)\ \vec{j}\\ &\quad + \cos\eta\cos\varepsilon\ (u\cos t - A) - u\ \sin\ \eta \sin t\ \vec{k}\,.\end{aligned} \tag{6.20}$$

Equation of the contact will be written:

- ruled helical front surface of the cutter tooth with an arbitrary front angle:

$$\vec{N} \bullet \vec{V} = 0 = \frac{1}{p_1}\left(\cos\left(\eta + \gamma_{\text{instr}}\right) \right.$$
$$\times \left(\cos\varepsilon \left(u \left(A - u\cos\left(\gamma_{\text{install}} + t\right)\right) + \sin \left(p_1 t \cos\varepsilon + u \sin t - r_a \sin t \sin\varepsilon \right) \right.\right.$$
$$\times \left(p_1^2 t - r_a u \sin\gamma_{\text{install}} \right)\Big) + \sin\left(\eta + \gamma_{\text{instr}}\right)$$
$$\left.\times \left(\cos\left(\gamma_{\text{install}} + t\right)\left(p_1^2 t - r_a u \sin\gamma_{\text{install}}\right) + u^2 \sin\left(\gamma_{\text{install}} + t\right)\right)\right). \tag{6.21}$$

– ruled helical rake face of the cutter tooth with zero front angle:

$$\vec{N} \bullet \vec{V} = 0 = \frac{1}{p_1}\Big(\cos\eta \cos\varepsilon \left(u(A - u\cos t) + {p_1}^2 t \sin t \right)$$
$$+ p_1 \cos\eta \sin\varepsilon \left(u - a\cos t \right) + \sin\eta \left({p_1}^2 t \cos t + u^2 \sin t \right)\Big). \tag{6.22}$$

We will find the solution to the contact equation numerically, specifying a change in the parameter u within the limits that describes the working sector of the surface and find the value of the parameter t. Substituting the found values of the parameters u, t into Eqs. (6.13) and (6.14), we obtain the value points of the characteristic—the contact of the original instrumental surface with the helical surface.

Substituting the coordinates of the characteristic into the equation of connection between the coordinate systems S and S_2, we obtain the characteristic in the system associated with the original instrumental surface. The coordinates of the points of the axial profile of the initial instrumental surface of the disk tool in the system $S_2(X_2, Y_2, Z_2)$ are determined by the dependence:

$$R_{\text{instr}} = \sqrt{Y_{2XXA}^2 + Z_{2XXA}^2}; \quad X_{\text{instr}} = X_{2XXA}. \tag{6.23}$$

The final operation of manufacturing a milling cutter is sharpening it along the front surface with a conical circle, therefore it is necessary to calculate the profile curve of the circle (see Fig. 6.28) along its rectilinear generatrix using the following dependencies:

$$\begin{cases} X_{\text{пр}} = -X_{\text{instr}} \sin\eta + R_{\text{instr}} \cos\eta \\ Y_{\text{пр}} = X_{\text{instr}} \cos\eta + R_{\text{instr}} \sin\eta \end{cases} \tag{6.24}$$

For example, the circle profile curve is calculated Fig. 6.29. for sharpening a helical chip-separating groove on the surface of a hob cutter $m = 10$ mm, $d_a = 180$ mm with step parameters 7532 mm, $r_a = 90$ mm, $r_f = 65$ mm, $R_{kr} = 100$ mm, $\varepsilon = 3, 7°$, $\gamma_{\text{instr}} = 0$ at different angles of the circle profile $\eta = 15°, 20°, 25°, 30°$.

Fig. 6.28 Scheme for straightening the grinding wheel profile: *1* working area, *2* profile curve is calculated

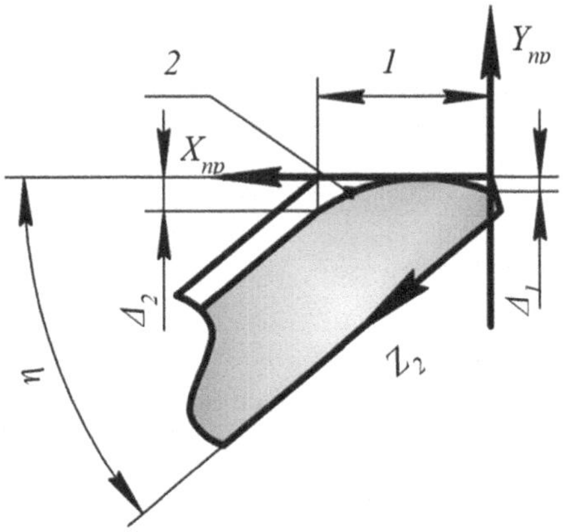

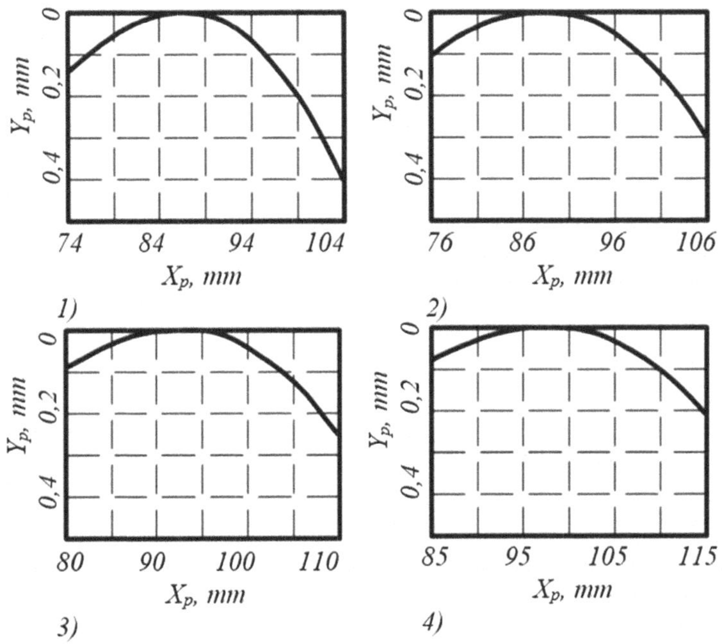

Fig. 6.29 Profile of a wheel when sharpening a screw surface: Step 7532 mm, $r_a = 90$ mm, $r_f = 65$ mm, $R_{kr} = 100$ mm, $\varepsilon = 3.7°$, $\gamma_{\text{instr}} = 0$; Circle profile angle η: *1* 150, *2* 200, *3* 250; *4* 300

According to the calculation results, with a decrease in the profile angle of the conical grinding wheel, the amount of wheel dressing increases, and according to the value $\Delta_2 < \Delta_1$.

The influence of the grinding wheel dimensions on the shape of the sharpening wheel profile curve was calculated (Fig. 6.30). According to the calculation results, with increasing diameter of the circle, the amount of editing increases. According to the calculation results, with increasing diameter of the circle, the amount of editing increases.

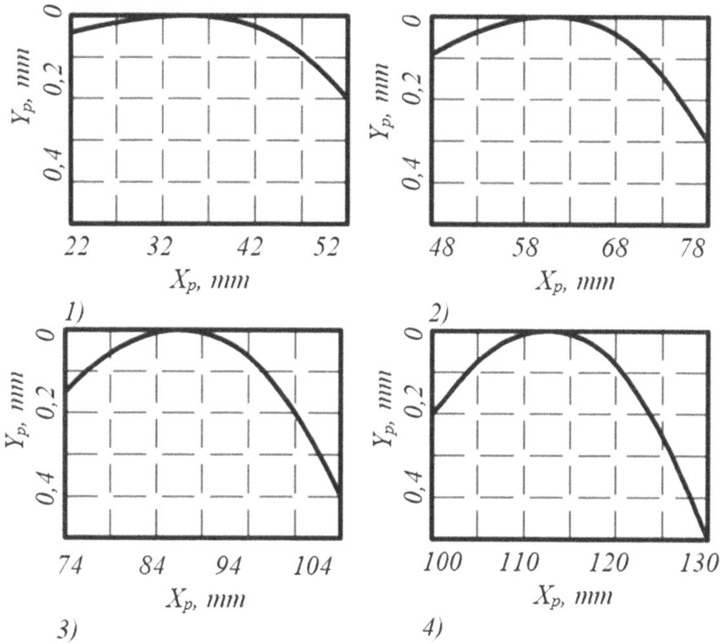

Fig. 6.30 Wheel profile curve when sharpening a helical surface with parameters: step 7532 mm, $r_a = 90$ mm, $r_f = 65$ mm, $\eta = 150$, $\varepsilon = 3.7°$, $\gamma_{\text{instr}} = 0$; radius of sharpening wheel R_{kr}: *1* 50 mm; *2* 75 mm, *3* 100 mm, *4* 125 mm

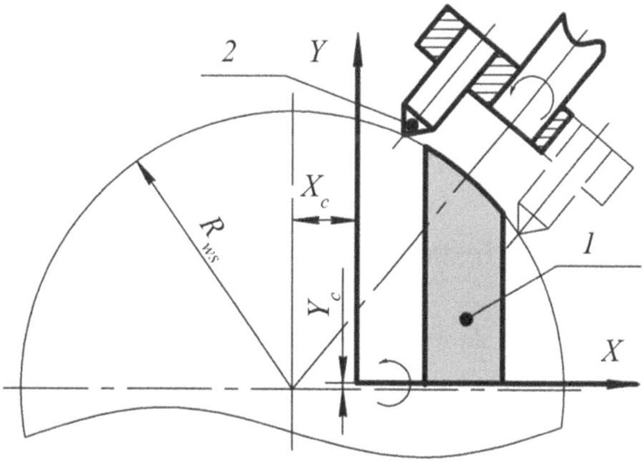

Fig. 6.31 Scheme for dressing a grinding wheel on a spherical surface: *1* cutting tool, *2* sharpening wheel

For sharpening machines that do not have a CNC dressing mechanism in their design, it is practical to calculate the parameters for forming the required wheel profile curve using a spherical wheel dressing scheme Fig. 6.31. For this, it is necessary to calculate the parameters for replacing the curved profile of a circle using Eq. (6.25).

A feature of the replacement is that the profile of the replacement wheel must pass through a point that has a minimum deviation of the value of the editing value Δ, equal to zero at the point $[X_k\ Y_k]$ and this circle has a tangent at this point at the angle of the profile of the rectilinear generatrix of the circle, which for the selected coordinate system (Fig. 6.31) $\psi = \pi/2 + \eta$.

Assuming that the circle profile is obtained in the form of a set of points $[X_i\ Y_i]$, де $i = 1...N$, N—is the number of design points, then we will find the parameters of the replacement circle using method indefinite multipliers the Lagrange of by solving the following system of equations:

$$
\begin{gathered}
F(A,B,C,\lambda_1,\lambda_2)=\sum_{i=1}^{N}\left[\left(X_i^2+Y_i^2-AX_i-BY_i+C\right)^2+\lambda_1\left(X_i^2+Y_i^2-AX_i-BY_i+C\right)\right.\\
\left.+\lambda_2(2X_i+2Y_i tg\psi-A-Btg\psi)\right],\frac{\partial F(A,B,C,\lambda_1,\lambda_2)}{\partial A}=0\\
\frac{\partial F(A,B,C,\lambda_1,\lambda_2)}{\partial B}=0,\frac{\partial F(A,B,C,\lambda_1,\lambda_2)}{\partial C}=0,\\
\frac{\partial F(A,B,C,\lambda_1,\lambda_2)}{\partial \lambda_1}=0,\frac{\partial F(A,B,C,\lambda_1,\lambda_2)}{\partial \lambda_2}=0,\\
\frac{\partial F(A,B,C,\lambda_1,\lambda_2)}{\partial C}=0,\\
X_c=A/2, Y_c=B/2, R_{ws}=\sqrt{X_c^2+Y_c^2-C}.
\end{gathered}
\tag{6.25}
$$

where

- X_i, Y_i coordinates of the points of the curvilinear profile of the sharpening wheel,
- N number of points of the curvilinear profile of the sharpening wheel;
- X_c, Y_c coordinates of the center of the replacement circle;
- $tg\psi$ the angle of inclination of the tangent to the circle at one of the end points,
- R_{ws} radius wheel of substitution.

6.6 Conclusions

The example shows the calculation of the circle profile curve for sharpening a screw surface with step parameters of 7532 mm, $r_a = 90$ mm, $r_f = 65$ mm, $R_{kr} = 100$ mm, $\varepsilon = 3.7°$, $\gamma_{instr} = 0$, $\eta = 15°$ for the diagram in Fig. 6.31, with the number substitute points $N = 100$, the parameters of the substitute circle were calculated and the following values were obtained: $X_c = -3686, 6$ mm, $Y_c = -898{,}6$ mm, $R_{sub} = 3793, 6$ mm. The calculated values are technologically implemented on a sharpening machine and this method can be used when calculating dressing parameters using the specified method.

Based on the results of the work, the deviation of the profile of the front surface of the cutter from straightness with helical grooves was established, depending on the installation parameters of the wheels conical shape.

Rational values of the installation parameters of sharpening wheels conical shape have been determined, which make it possible to reduce the deviation of the tooth profile of the front helical surface of the cutter from straightness.

Dependencies have been established to determine the curvilinear profile of a disk sharpening tool for processing theoretically accurate ruled helical surfaces of chip-separating grooves of hob cutters with an arbitrary front angle.

Equations are determined for replacing such a curved profile with an arc of a circle when replacing the surface of a circle with a curvilinear profile to a spherical one.

References

1. Achmad, A., & Wu, Y.-R. (2022). Analytical design of disk-type milling cutter with multiple inserts and CNC milling simulation of screw rotors considering grinding stock. *Mechanism and Machine Theory, 170*, 104724. https://doi.org/10.1016/j.mechmachtheory.2022.104724. ISSN 0094-114X.
2. Dayuan, W., Yan, P., Pei, J., Yingtao, S., Zhou, H., Yi, R., & Guilong, H. (2022). Quality estimation method for gear hobbing based on attention and adversarial transfer learning, 110383. *Measurement, 188*. https://doi.org/10.1016/j.measurement.2021.110383. ISSN 0263-2241.
3. Guba, N., Hüsemann, T., & Karpuschewski, B. (2020). Influence of gear hobbing feed marks on the resulting gear quality after discontinuous profile grinding. *CIRP Journal of Manufacturing Science and Technology, 31*, 314–321. https://doi.org/10.1016/j.cirpj.2020.06.005. ISSN 1755-5817.
4. Radzevich, S. P. (2014). *Generation of surfaces: Kinematic geometry of surface machining* (1st ed.). CRC Press. https://doi.org/10.1201/b16398
5. Radzevich, S. P. (2010). *Gear cutting tools: Fundamentals of design and computation* (1st ed.). CRC Press. https://doi.org/10.1201/9781439819685
6. Radzevich, S. P. (2022). *Theory of gearing: Kinematics, geometry, and synthesis* (3rd ed.). CRC Press. https://doi.org/10.1201/9781003311744
7. Ravska, N. S. (2009). Sharpening of worm gear cutters/N. S. Ravska, O. A. Okhrimenko, and Y. R. Dubyk. *Bulletin of ZSTU, 1*(29), 42–48. https://library.ztu.edu.ua/e-copies/Zbirnyk/Process_obrobky_6/160.pdf
8. Ravska, N. S. (2012). Features of forming the screw groove spiral drills with a disk tool straight form/Ravska, N. S., Okhrimenko, O. A. Plivak. *Visnyk NTUU KPI Mechanical Engineering, 63*, C254–C265. https://ela.kpi.ua/server/api/core/bitstreams/5e067850-2e05-49c4-9547-0e621 70056a1/content/
9. Sandler, A. I. (2021). *Theory and practice of worm gear production of general type* (2nd ed., p. 348c). Infra-Engineering. ISBN 978-5-9729-0534-8.
10. Shilong Wang, Guolong Li, Chi Ma, Chapter 3 – Method to reduce the principal errors in high-performance gear machining, Shilong Wang, Guolong Li, Chi Ma, , Elsevier, 2024, 87–141. doi:https://doi.org/10.1016/B978-0-443-16056-1.00003-7. ISBN 9780443160561.
11. Shuying, Y. A. N. G., & Weifang, C. H. E. N. (2021). Modeling and experiment of grinding wheel axial profiles based on gear hobs. *Chinese Journal of Aeronautics, 34*(6), 141–150. https://doi.org/10.1016/j.cja.2020.06.030. ISSN 1000-9361.
12. Rodin, P. R. (1982) Technology of gear cutting tools manufacturing (P. R. Rodin, V. I. Klimov, & S. B. Yakubson, Eds.) (208p.). Tekhnika
13. https://multigrind.com/en/industries-and-applications/industries/tool-manufacturing
14. https://doimak.es/portfolio/af-series/#1677674545161-6e15ac5e-0c87
15. https://multigrind.com/en/multigrind

Chapter 7
Analysis of Gear Pairs Composed of Eccentrically Mounted Circular Gears (EM Gears)

Boris M. Klebanov and Yuriy V. Shekhtman

7.1 Introduction

A variable gear ratio in a toothed gear pair is achieved by the use of noncircular gears [1]. To manufacture such gears, the tooth cutting machine must be equipped with special precision devices that provide the required law of motion of the work piece relative to the cutter (a hob or a shaper). The machine designers, wanting to avoid such a difficulty, pay attention to the possibility of using, instead, conventional involute gears with the axis of rotation displaced relative to their geometric center (see Fig. 7.1). In this case, all technological problems associated with the cutting of noncircular gears are eliminated. However, this method has its drawbacks: first, special measures must be taken to prevent the disengagement of the gears, and second, in order to maintain the constant distance between geometric centers of the gears, the distance between their axes of rotation must change. That means that the axis of rotation of at least one gear should be floating. In a batch production, the use of noncircular gears is preferable. But in a single-piece production, it can be appropriate to achieve the required nonuniform rotation using a pair of round involute gears mounted eccentrically on their shafts—EM gears.

Dr. Yuriy V. Shekhtman passed away before the publication of this book. His untimely passing is deeply felt by all who knew him, and his contributions to the field will be profoundly missed.

B. M. Klebanov (✉)
Dallas, USA

Y. V. Shekhtman
New York, USA

S. P. Radzevich (ed.), *Gear Accuracy*, Mechanical Engineering Series,
https://doi.org/10.1007/978-3-031-74798-4_7

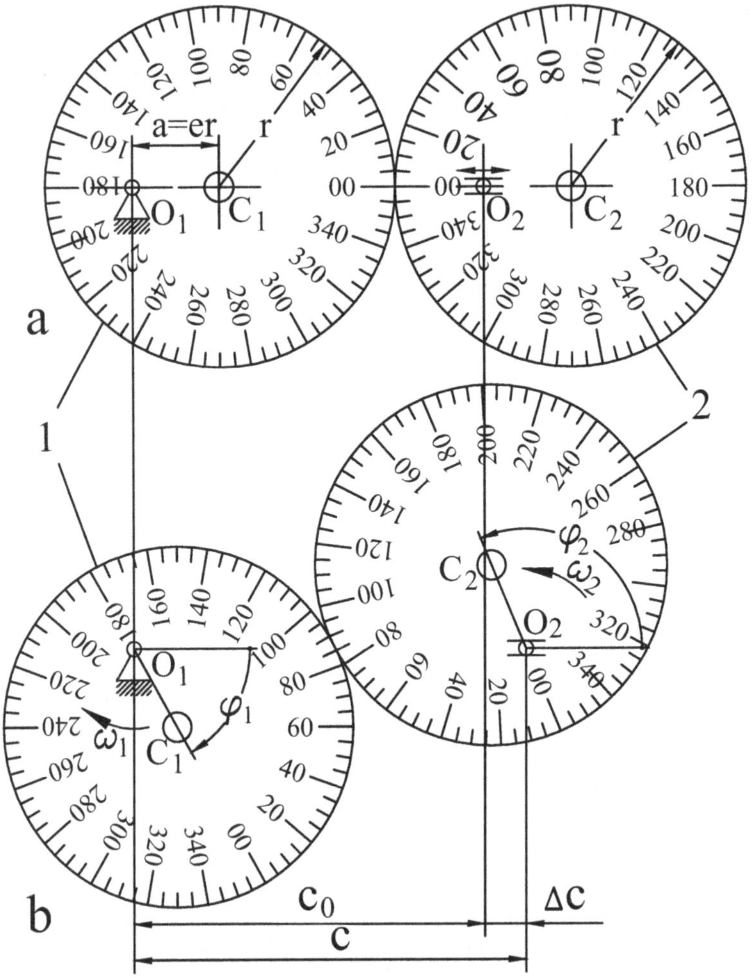

Fig. 7.1 EM gear pair in the initial position (**a**) and rotated (**b**)

7.2 Kinematics and Dynamics of the EM Gear Pairs

In Fig. 7.1a, the gears are represented by their operating pitch circles 1 and 2, which should always touch each other and roll over each other without slipping. For better visualization, the periphery of each circle is provided with a scale marked in angular degrees. Owing to these scales, one can see that, in any angular position of the gears, the indicators of the angle at the point of touch of circles 1 and 2 coincide, and this indicates the absence of slipping between the circles.

Since the numbers of teeth are equal ($z_1 = z_2$), radius r of each gear equals half of the distance between their geometric centers c_0:

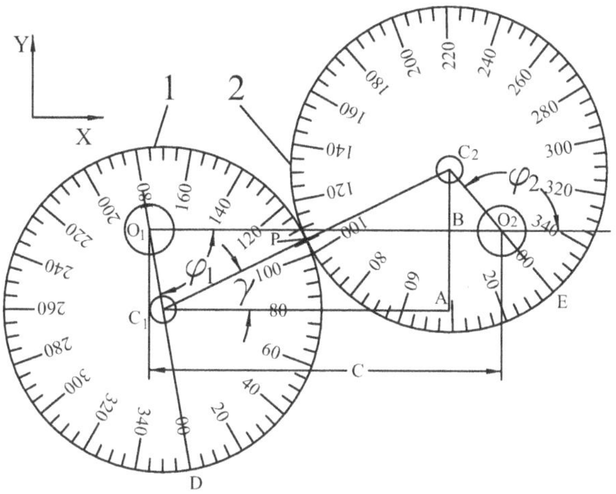

Fig. 7.2 To the determination of function $\varphi_2 = f(\varphi_1)$. (Hereinafter, the letter f denotes some functions of the arguments designated in brackets. The functions in each case can be different)

$$r = 0.5\, c_0.$$

The axes of rotation O_1 and O_2 of both gears are shifted relative to their geometric centers by the same value $a = er$, the dimensionless factor e is called eccentricity. The location of axis O_1 is fixed, while axis O_2 is floating, it has the possibility to move along the straight line that connects centers O_1 and O_2, i.e., in the horizontal direction in Fig. 7.1. The distance between axes O_1 and O_2 at $\varphi_1 = \varphi_2 = 0$ is $c_0 = 2r$.

Figure 7.1a presents the initial position. In this position, $\varphi_1 = 0$ and $\varphi_2 = 0$. Figure 7.1b shows that when gears 1 and 2 are rotated by angles φ_1 and φ_2 from the starting position, the distance c between the centers of rotation O_1 and O_2 changes by the value of Δc.

Note The rectilinear trajectory of motion of the floating center O_2 is not a must. For example, if the O_2 axis is fixed in a lever that can swing about its hinge, then point O_2 will move along the arc of a circle.

To find the ratio between angles φ_1 and φ_2, we need to make additional geometrical constructions, see Fig. 7.2. Here: $C_1O_1 = C_2O_2 = er$.

In $\Delta C_1C_2A : C_1C_2 = 2r$; $AC_2 = er\,(\sin\varphi_1 + \sin\varphi_2)$;

$$\sin\gamma = \frac{AC_2}{C_1C_2} = \frac{e}{2}(\sin\varphi_1 + \sin\varphi_2). \tag{7.1}$$

It is obvious from the drawing that:

$$\angle DC_1P = \varphi_1 + \gamma;\ \angle EC_2P = \varphi_2 - \gamma;$$

From the condition of the absence of relative slip between circles 1 and 2, $\angle DC_1P = \angle EC_2P$, and, hence,

$$\gamma = \frac{\varphi_2 - \varphi_1}{2}. \tag{7.2}$$

From Eqs. (7.1) and (7.2) we obtain the relation between angles φ_1 and φ_2:

$$\sin\frac{\varphi_2 - \varphi_1}{2} = \frac{e}{2}(\sin\varphi_1 + \sin\varphi_2).$$

The diagrams of $\varphi_2 = f(e, \varphi_1)$, $(\varphi_2 - \varphi_1) = f(e, \varphi_1)$, and $\gamma = f(e, \varphi_1)$ obtained using a computer program are presented in Figs. 7.3, 7.4, and 7.5, respectively.

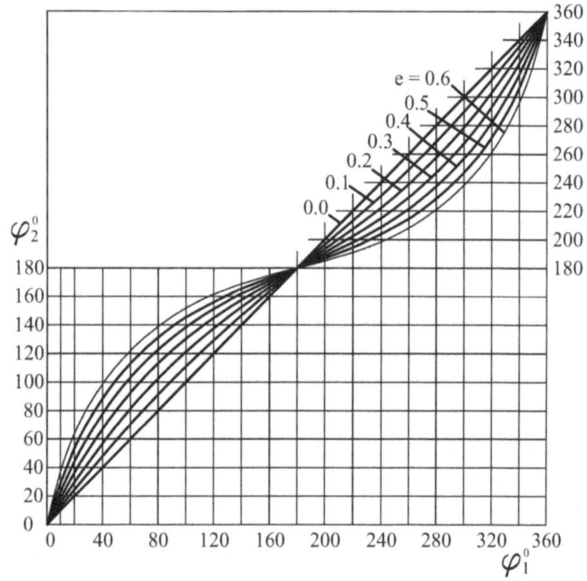

Fig. 7.3 Diagrams of functions $\varphi_2 = f(e, \varphi_1)$

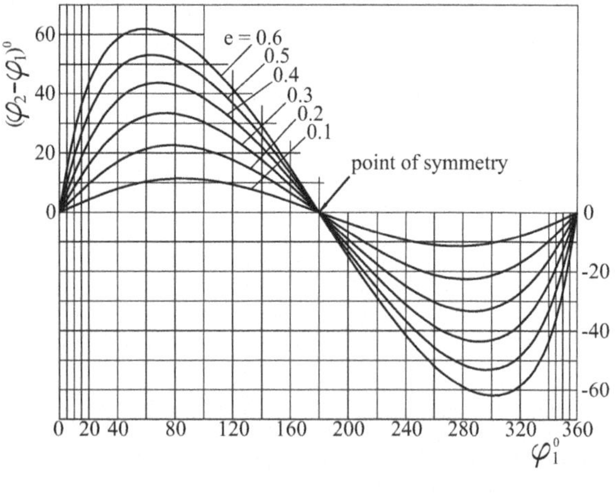

Fig. 7.4 Diagrams of functions $(\varphi_2 - \varphi_1) = f(e, \varphi_1)$

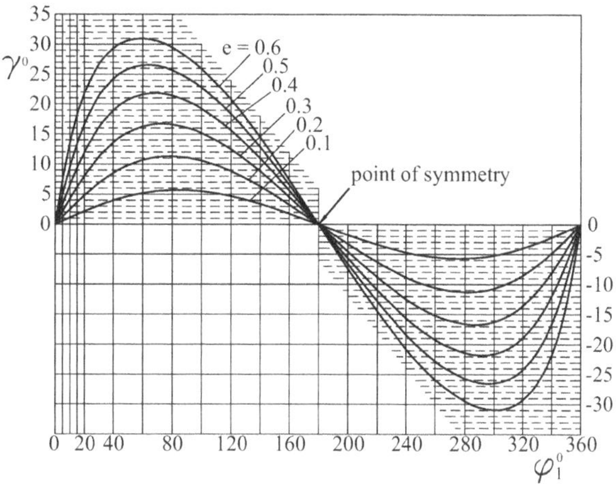

Fig. 7.5 Diagrams of functions $\gamma = f(e, \varphi_1)$

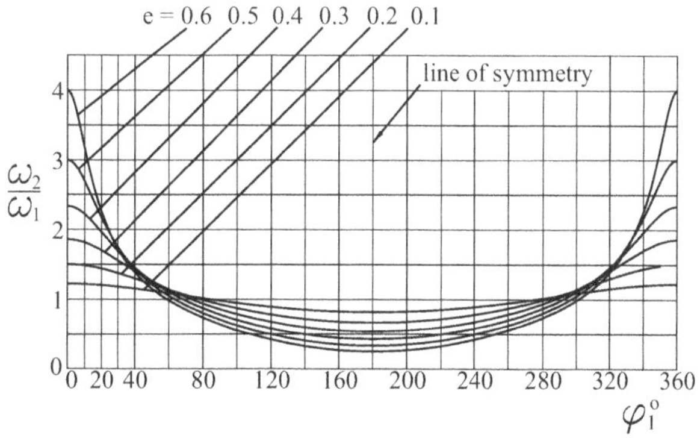

Fig. 7.6 Diagrams of functions $\omega_2/\omega_1 = f(e, \varphi_1)$

The functions of the angular speed ω_2 and angular acceleration ε_2 at $\omega_1 =$ constant were obtained by double differentiation of the rotation angle function $\varphi_2 = f(e, \varphi_1)$. The diagrams of functions $\omega_2/\omega_1 = f(e, \varphi_1)$ and $\varepsilon_2/\omega_1^2 = f(e, \varphi_1)$ are presented in Figs. 7.6 and 7.7, respectively.

The ratio of the largest angular speed of the driven gear to the smallest one ($\omega_{2\max}/\omega_{2\min}$) at $\omega_1 =$ const is diagrammed in Fig. 7.8 as a function of eccentricity e.

The distance c between the rotation centers O_1 and O_2, as can be seen from Fig. 7.2, equals:

$$c = er\,(\cos\varphi_1 - \cos\varphi_2) + 2r\,\cos\gamma.$$
$$\frac{c}{c_0} = \frac{c}{2\,r} = \frac{e}{2}(\cos\varphi_1 - \cos\varphi_2) + \cos\gamma. \tag{7.3}$$

Function $c/(2r) = f(e, \varphi_1)$ is diagrammed in Fig. 7.9. The linear speed v and acceleration $\dot{v}$ of point O_2 are obtained by double differentiation of this function.

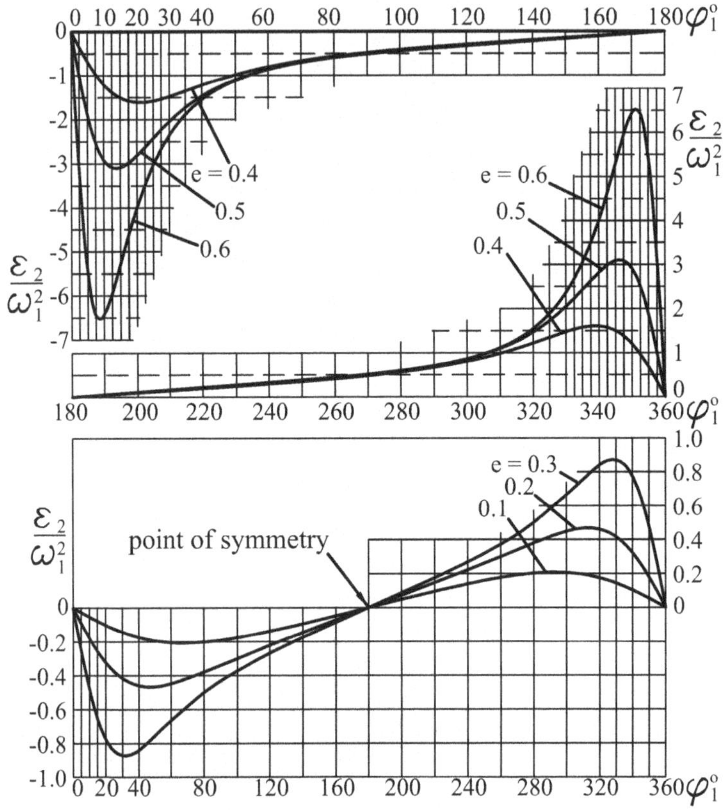

Fig. 7.7 Diagrams of functions $\varepsilon_2/\omega_1^2 = f(e, \varphi_1)$. (The dimensionality of ω_1 is rad/s)

Fig. 7.8 Diagram of function $\omega_{2\max}/\omega_{2\min} = f(e)$

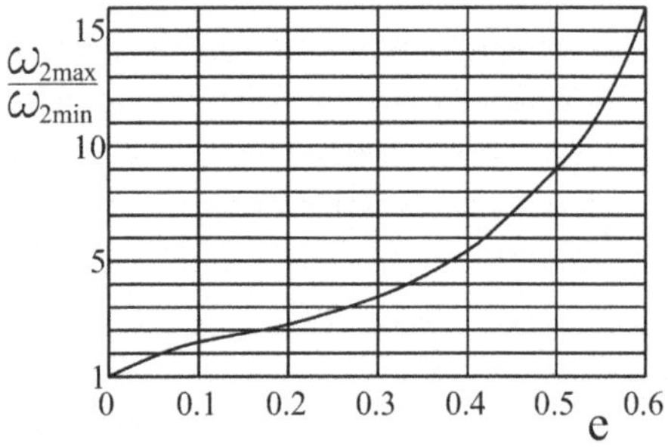

Parameter $v/(r\omega_1) = f(e, \varphi_1)$ is diagrammed in Fig. 7.10, and parameter $\dot{v}/\left(r\,\omega_1^2\right)$—in Fig. 7.11.

The efficiency of the gearing depends on the angle of pressure; we denote it ψ. Pressure angle is the angle between the direction of the force applied by a driving body to the driven body and the direction of motion of the point of a driven body where the driving force is applied. In the best case, when these directions coincide

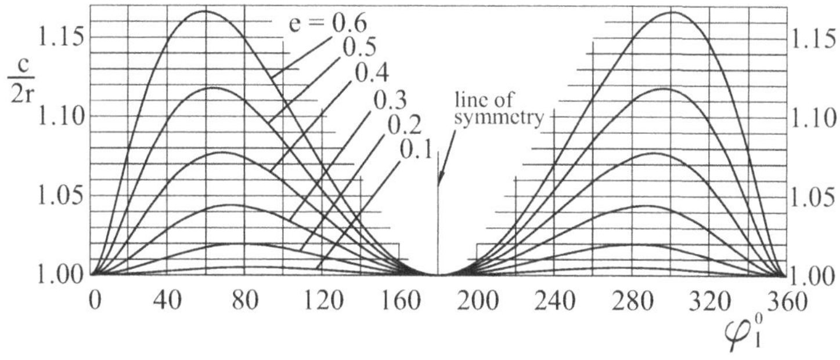

Fig. 7.9 Diagrams of functions $c/(2r) = f(e, \varphi_1)$

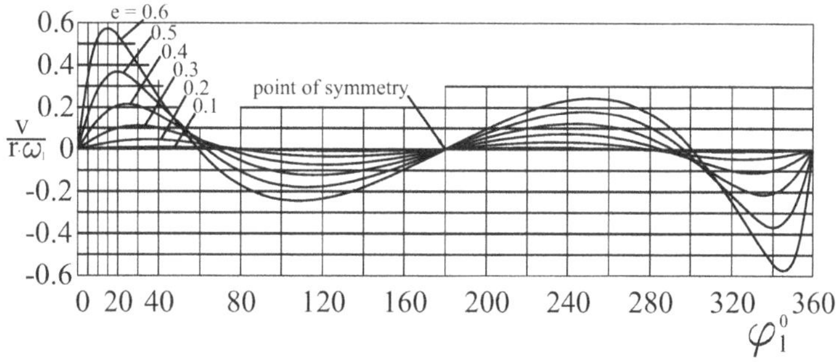

Fig. 7.10 Diagrams of functions $v/(r\omega_1) = f(e, \varphi_1)$. (The dimensionalities are: r—mm ω_1—rad/s)

(that is, the pressure angle $\psi = 0$), the efficiency of mechanism is maximal. With the growing pressure angle, the efficiency of the mechanism is reduced, thus, it is important not to exceed the recommended maximum of $\psi = 50°$ [1].

Finding the pressure angle requires construction of additional lines, see Fig. 7.12. Here line n is the line of action that is normal to the working profiles of the engaged teeth. Thus, the vector of the driving force applied to the teeth of gear 2 is directed along the line n.

Angle β between line n and center line C_1C_2 equals $90° - \alpha_t$, where α_t is the transverse pressure angle of the gearing. In this article, we set $\alpha_t = 20°$, so $\beta = 70°$.

Vector V_S that shows the direction of motion of the driven tooth at point P is the sum of two vectors: vector V_C of the circumferential speed that is perpendicular to radius O_2P, and vector V_{O_2} of the linear speed of point O_2. Thus, the angle between vector V_S and the line of action n is the sought-for pressure angle ψ.

The diagrams of function $\psi = f(e, \varphi_1)$ are presented in Fig. 7.13. From the diagrams we can see that the e value should not usually exceed 0.5. At $e = 0.6$, angle ψ exceeds 50° in the interval from $\varphi_1 = 55°$ to $\varphi_1 = 140°$ and increases to a maximum value of 57°. If in this interval the load is small or reversed, the possibility of using gears with $e = 0.6$ shall be examined.

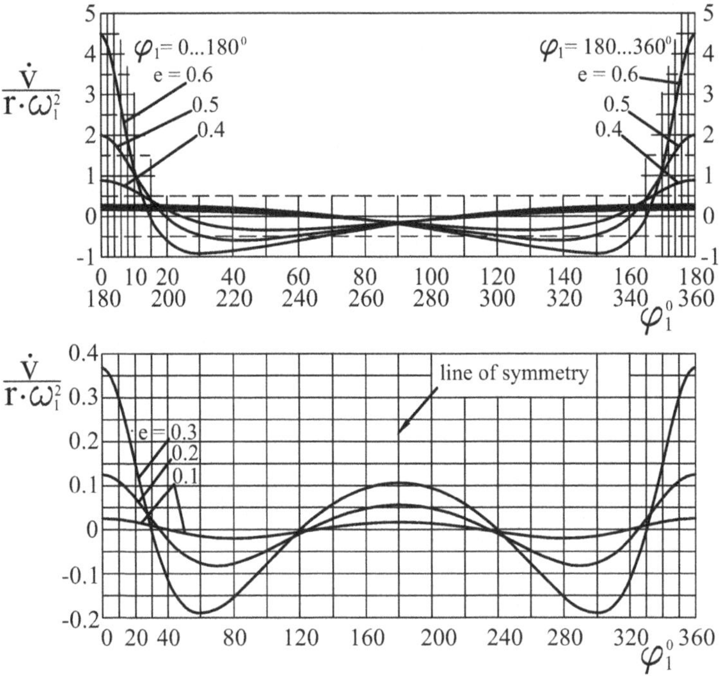

Fig. 7.11 Diagrams of functions $\dot{v}/\left(r\omega_1^2\right) = f(e, \varphi_1)$. (The dimensionalities are: r—mm, ω_1—rad/s)

Fig. 7.12 Determination of pressure angle ψ

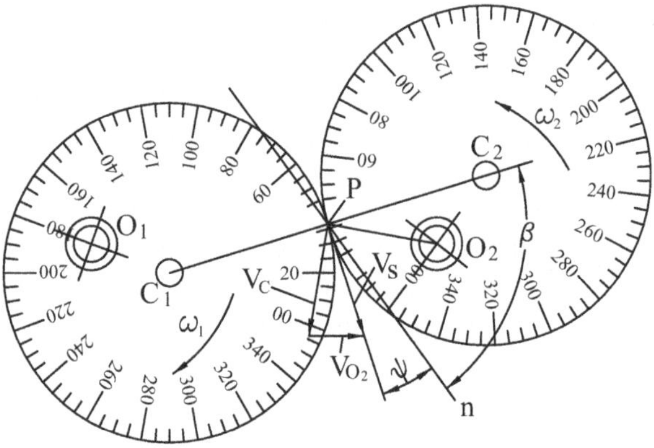

7.3 Requirements for the Mechanism with EM Gears

The mechanism with the EM gear pair must satisfy two specific requirements:

1) The distance between the geometric centers of the gears must be kept constant: $C_1C_2 = 2r = \text{const}$, see Fig. 7.2.

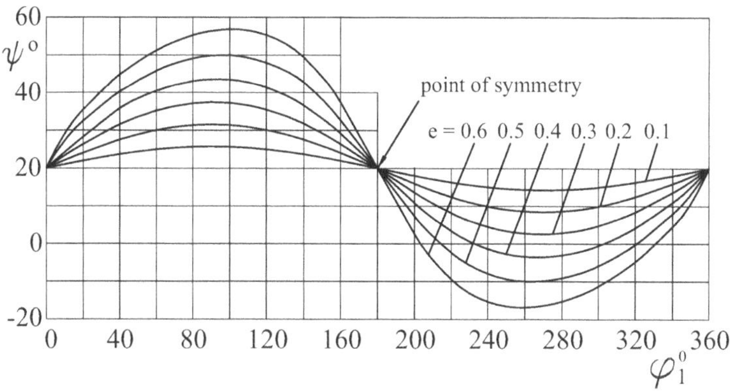

Fig. 7.13 Diagrams of function $\psi = f(e, \varphi_1)$

2) The axis of rotation of one of the gears (in this case—of the driven gear) must be allowed to move along a prescribed path; while moving, the axes of rotation of the gears should remain parallel to avoid misalignment of the teeth.

One of the options for such a mechanism is presented schematically in Fig. 7.14. Here 1 and 2 are the driving and the driven gears, 3—hinged link that connects the geometrical centers C_1 and C_2 of the gears and thus fulfills the first requirement. Fixed housing 4 contains the shaft of gear 1 with its bearings. Floating housing 5, which contains the shaft of gear 2 with its bearings, is made as a lever that can swing around its axis O_3. Since the angle of the lever oscillation is small, the deviation of the trajectory of point O_2 from the linear path can be negligible. As an option, housing 5 can be made in the form of a carriage that slides along a straight guide. It is important to note that the guide should provide a strictly progressive motion of the carriage, i.e., such a motion in which there is no turn of the carriage that causes misalignment of the axes of rotation of the gears.

The position of point O_2 on its route is determined uniquely by the geared linkage mechanism, which includes gears 1 and 2 and link 3.

7.4 Forces in the EM Gear Pairs

The totality of forces and torques acting on the gears and their shafts is shown in Fig. 7.15. Here 1 and 2 are the driving and the driven gears, and 3 is the link that connects the geometrical centers of the gears. Line n is the line of action, like in Fig. 7.12. Figure 7.15 shows two angular positions of the gears: at $\varphi_1 = 20\,°$ (a) and $\varphi_1 = 260\,°$ (b). It is made to explain the differences in calculations related to the first half-turn ($\varphi_1 = 0...180°$) and the second half-turn ($\varphi_1 = 180\,°...360°$), see below.

Some of the forces and torques presented in Fig. 7.15 are known from the Requirements Specification (RS) for the design of the drive and some data are

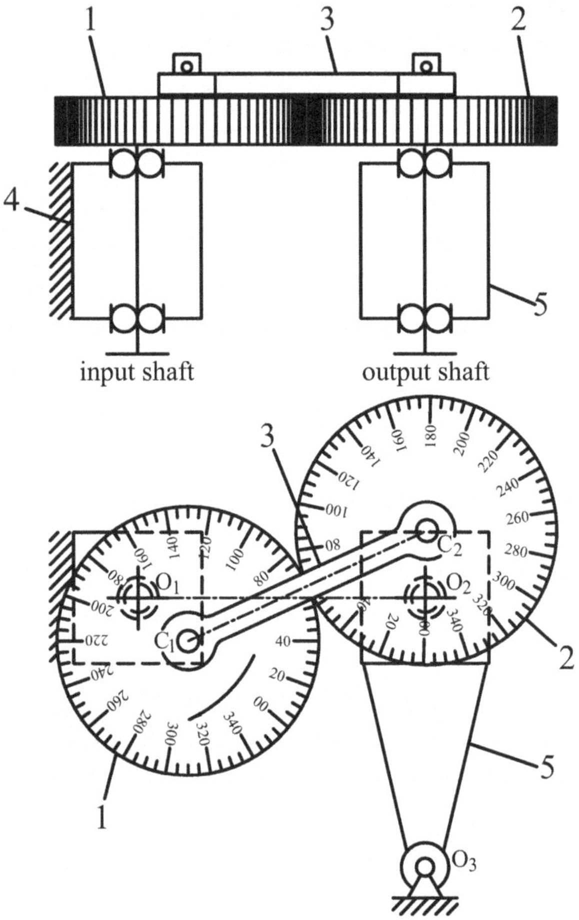

Fig. 7.14 Layout of a gear drive with EM involute gears

obtained during preliminary calculations; however, the definition of all forces and torques requires the consideration of the equations of dynamic equilibrium in accordance with the *d'Alembert's principle*. Three equations shall be written for each gear:

- the sum of all torques equals zero;
- the sum of projections of all forces on the X axis equals zero;
- the sum of projections of all forces on the Y axis equals zero.

To write the equations of equilibrium, we need to know the directions of the forces and torques. The following rules for signs are adopted here:

- torques which act in the direction of rotation are given sign "plus"; torques acting in the opposite direction are given sign "minus";

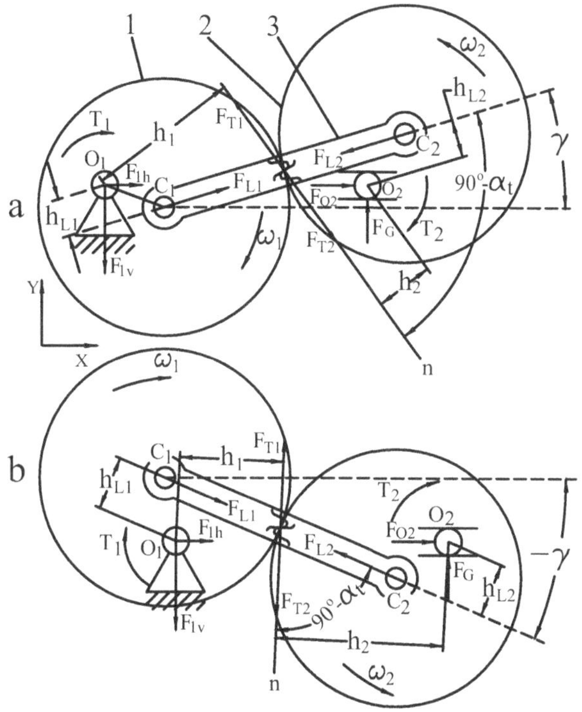

Fig. 7.15 Forces and torques acting on the gears and their shafts

- link forces F_{L1} and F_{L2} that are aimed at bringing the gears closer to each other (as shown in Fig. 7.15) are given sign "plus";
- about the signs of other forces see explanations below.

We shall start writing the equations of dynamic equilibrium from gear 2 where the output torque magnitude is given and the full dynamic torque can be determined approximately during the preliminary calculations. Here are the torques and forces applied to gear 2:

- T_2= torque applied to the body of gear 2; it is the sum of the resistance torque of the driven mechanism (T_{res}) and the torque associated with the inertia of unevenly rotating masses (T_{in}); thus, $T_2 = T_{res} + T_{in}$. Since this torque is directed against rotation, its sign in the equation of equilibrium should be negative.
- F_{T2}= tooth force of gear 2; this force is driving, so it must be directed in such a way as to create a torque $F_{T2}h_2$ in the direction of rotation. The sign of this torque is positive.
- h_2= arm of force F_{T2} relative to point O_2, see Fig. 7.15. It can be taken from diagrams given in Fig. 7.16 or calculated by the formula:

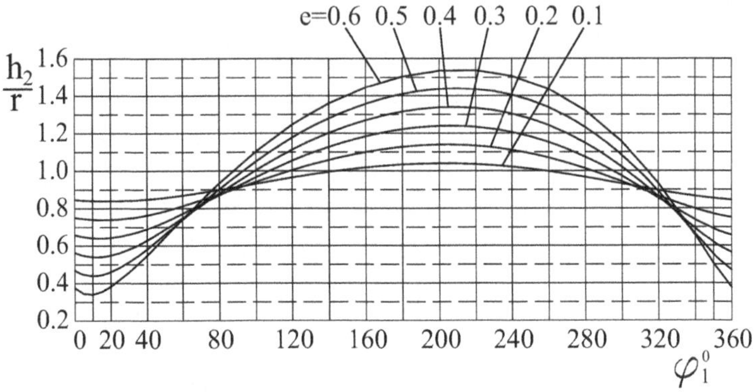

Fig. 7.16 Diagrams of $h_2/r = f(e, \varphi_1)$

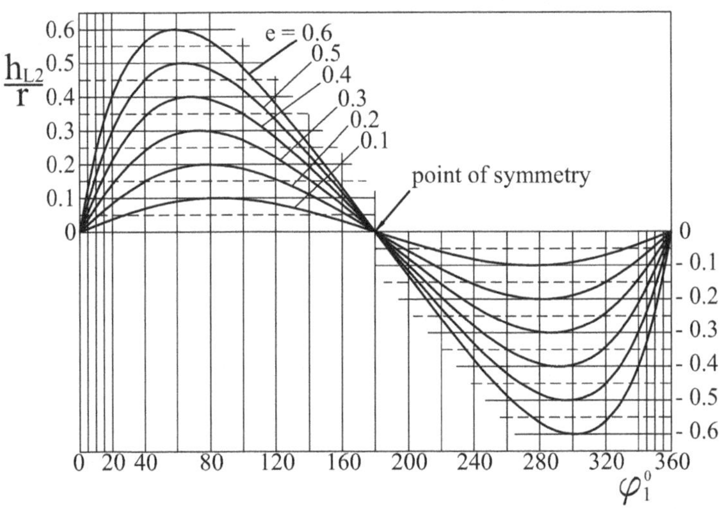

Fig. 7.17 Diagrams of $h_{L2}/r = f(e, \varphi_1)$

$$\frac{h_2}{r} = \left[\cos\gamma - e\ \cos\varphi_2 + \frac{e\ \sin\varphi_1 - \sin\gamma}{\tan(90° - \alpha_t - \gamma)}\right] \sin(90° - \alpha_t - \gamma). \qquad (7.4)$$

(***Note*****:** in all our calculations it is accepted $\alpha_t = 20°$.)

- F_{L2}= link force. Force F_{L2} creates torque $F_{L2}h_{L2}$ that acts in the direction of rotation during the first half-turn (Fig. 7.15a), and the sign of this torque is positive. During the second half-turn (Fig. 7.15b) torque $F_{L2}h_{L2}$ is directed against rotation, and its sign shall be negative.
- h_{L2}= arm of force F_{L2} relative to point O_2. It can be taken from diagrams in Fig. 7.17 or calculated by the formula:

$$\frac{h_{L2}}{r} = e \ \sin(\varphi_1 + \gamma); \tag{7.5}$$

As we see, h_{L2} changes its sign from positive to negative when switching from the first half-turn to the second half-turn. This automatically gives the correct sign to the moment $F_{L2}h_{L2}$ in the equation of equilibrium.

- F_{O2}= force that is applied to axle O_2. It causes the acceleration of the gear 2 assembly (with its shaft, bearings, housing etc.) in its oscillatory linear motion. $F_{O_2} = m\dot{v}$, where m= mass of the oscillating assembly, $\dot{v}$= acceleration that can be taken from diagrams given in Fig. 7.11. The direction of this force depends on the sign of acceleration that is clear from Fig. 7.11. The positive acceleration $\dot{v}$ corresponds to the direction of force F_{O_2} in the positive direction of the X axis, and vice versa.
- F_G= guide supporting force in point O_2. This force is directed upwards, because it is created mainly by the vertical component of force F_{T2} directed downwards.

Here is the set of three Equations of dynamic equilibrium for gear 2 with three unknown quantities: F_{T2}, F_{L2}, and F_G:

$$F_{T2}h_2 + F_{L2}h_{L2} - T_2 = 0; \tag{7.6}$$

$$F_{T2}\cos(90° - \alpha_t - \gamma) - F_{L2}\cos\gamma + F_{O_2} = 0; \tag{7.7}$$

$$F_{T2}\sin(90° - \alpha_t - \gamma) + F_{L2}\sin\gamma - F_G = 0. \tag{7.8}$$

Note
Pay attention: within the second half-turn, angle γ is negative, see Figs. 7.5 and 7.15a, b.

After torque T_2 is determined, the dimensions of gears (their radius r and face width), mass moment of inertia of gear 2 and the mass of its assembly are obtained in the preliminary design (see Sect. 7.8.1). Then the unknown quantities F_{T2}, F_{L2}, and F_G can be determined from Eqs. 7.6, 7.7, and 7.8, see example in Sect. 7.8.

Gear 1 is loaded with the following forces and torques, see Fig. 7.15:

- T_1= torque applied to the shaft of gear 1; it is the driving torque. Since this torque acts in the direction of rotation, its sign in the equation of equilibrium should be positive.
- F_{T1}= tooth force; this force is equal in magnitude and opposite in direction to force F_{T2}. It creates torque $F_{T1}h_1$ in the direction opposite to rotation; the sign of this torque is negative.
- h_1= arm of force F_{T1} relative to point O_1. It can be taken from diagrams given in Fig. 7.18 or calculated by formula:

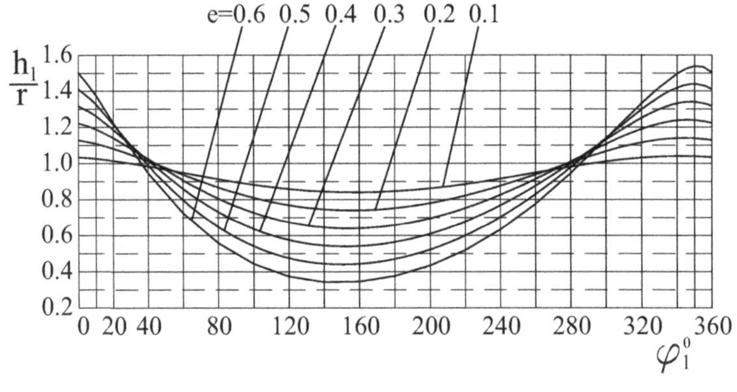

Fig. 7.18 Diagrams of $h_1/r = f(e, \varphi_1)$

$$\frac{h_1}{r} = \left[\frac{e\ \sin(\varphi_1 + \gamma)}{\sin\gamma} - (e\ \sin\varphi_1 - \sin\gamma)\left(\frac{1}{\tan\gamma} + \frac{1}{\tan(90^\circ - \alpha_t - \gamma)}\right)\right] \sin(90^\circ - \alpha_t - \gamma); \tag{7.9}$$

F_{L1}= link force; this force is equal in magnitude and opposite in direction to force F_{L2}. Force F_{L1} creates torque $F_{L1}h_{L1}$ that is directed against rotation during the first half-turn (Fig. 7.15a), and the sign of this torque is negative. During the second half-turn (Fig. 7.15b) torque $F_{L1}h_{L1}$ acts in the direction of rotation, and its sign shall be positive.

- h_{L1}= arm of force F_{L1} relative to point O_1; $h_{L1} = h_{L2}$ with its signs.
- F_{1h} and F_{1v} are the horizontal and the vertical components of the support in point O_1. Their directions are understandable, but if the designer chooses an incorrect direction, the equations of dynamic equilibrium will give the negative magnitude of these forces.

And here is the set of three equations of dynamic equilibrium for gear 1 from which three unknown quantities, T_1, F_{1h}, and F_{1v}, can be determined:

$$T_1 - F_{T1}h_1 - F_{L1}h_{L1} = 0; \tag{7.10}$$

$$F_{T1}\cos(90^\circ - \alpha_t - \gamma) - F_{L1}\cos\gamma - F_{1h} = 0; \tag{7.11}$$

$$F_{T1}\, sin\left(90^0 - \alpha_t - \gamma\right) + F_{L1} sin\gamma - F_{1v} = 0; \tag{7.12}$$

Note

Among the forces applied to gears 1 and 2, there should also be considered the gravity forces of these gears applied at their mass centers, as well as the torques of the gravity forces relative to the axes of rotation of the gears. However, because the gravity forces of the gears are negligible in comparison with the forces applied to their teeth, it was decided to exclude them from consideration.

7.5 Mass Moments of Inertia of Rotating Bodies

Mass moment of inertia of a body is determined by the formula

$$J = \int r^2 dm, \tag{7.13}$$

where each element of the body mass (dm) is multiplied by a square of its distance to the axis (r^2). The integral spreads to the entire body volume. For a solid cylindrical body with diameter d and width b the mass moment of inertia relative to its central axis is expressed by formula:

$$J_{\text{cyl}} = \frac{\gamma}{g}\frac{\pi d^4}{32} b\ \text{N} \cdot \text{mm} \cdot \text{s}^2, \tag{7.14}$$

where

γ = specific weight of material (for steels, $\gamma = 7.7 \cdot 10^{-5}$ N/mm^3);
g = acceleration of gravity ($g = 9.81 \cdot 10^3$ mm/s^2);
d and b are in mm.

Mass moment of a solid cylinder with diameter d and width b relative to the axis that is parallel to the central axis and displaced from the center by er (similar to the eccentrically mounted gear) equals:

$$J_{\text{gear}} = \frac{\gamma}{g}\frac{\pi d^4 b}{16}\left(e^2 + 0.5\right) \text{N} \cdot \text{mm} \cdot \text{s}^2. \tag{7.15}$$

Equations 7.14 and 7.15 are true for solid cylinders. Mass moments of inertia of drive elements with more complicated shape shall be calculated using the basic Eq. 7.13.

7.6 Tooth Force F_{T2} and Torque T_2 Applied to the Driven Shaft

Tooth force F_{T2} depends mainly on the torque T_2 applied to the driven gear. In addition, force F_{T2} is influenced by forces F_{L2} and F_{O_2}. Since the magnitudes of these forces are unknown, and their estimated influence is relatively small, we neglect them for preliminary calculations keeping in mind to take them into account with the final calculation. However, to be precise in our definitions, we denote F_{Tt} the tooth force associated only with the torque T_2.

Force F_{Tt} is found as follows:

$$F_{Tt} = \frac{T_2}{h_2}. \tag{7.16}$$

Torque T_2 on the driven shaft consists of two components (see above):

$$T_2 = T_{\text{res}} + T_{\text{in}}.$$

Torque T_{res} is given in the Requirement Specification (RS). Torque T_{in} is given by

$$T_{\text{in}} = J_{\text{eq.}} \varepsilon_2. \tag{7.17}$$

Here:

- $J_{\text{eq.}}$ = equivalent mass moment of inertia of all rotating parts that are in kinematic connection with the driven gear 2 and follow the changes in its speed of rotation;
- ε_2 = angular acceleration of the driven gear; it is obtained by multiplying the dimensionless value ε_2/ω_1^2 taken from Fig. 7.7 by ω_1^2 that is given in the RS or chosen by the designer.

Note Depending on the sign of ε_2, torque T_2 can be increased or reduced by torque $T_{\text{in.}}$. At higher rotational speed, torque T_2 may become negative in some angular positions. In this case, the teeth work with the reverse profiles. This mode of operation should be avoided, since it is associated with impact loads in the gear meshing. But if it is unavoidable, then it must be taken into account that when the teeth work with the reverse profiles, the geometric characteristics of the gear pair, such as h_1, h_2, and others, change.

Different elements of the transmission and the driven equipment may rotate with different speeds. Consider, for example, the scheme presented in Fig. 7.19. Here 1 and 2 are the EM gears, and there is one more rotating member i, which is connected with the shaft of gear 2 through gear drive 4 with the gear ratio $u_i = \omega_i/\omega_2 =$ const. To take into account the influence of member i on the torque T_{in}, let us use the principle of virtual work and consider only the inertia torques. According to the condition of dynamic equilibrium, the work of inertia torques on an elementary angular movement should be zero. For the scheme shown in Fig. 7.19, this condition is expressed by the following equation (for our purpose, only the inertial masses of gear 2 and member i are taken into account, the others are neglected):

$$T_{\text{in}} \delta\varphi_2 - J_2 \varepsilon_2 \delta\varphi_2 - J_i \varepsilon_i \delta\varphi_i = 0. \tag{7.18}$$

Here

$\delta\varphi_2$ = elementary angular displacement of gear 2;
$\delta\varphi_i$ = elementary angular displacement of member i;
ε_2 = angular acceleration of gear 2;
ε_i = angular acceleration of member i;

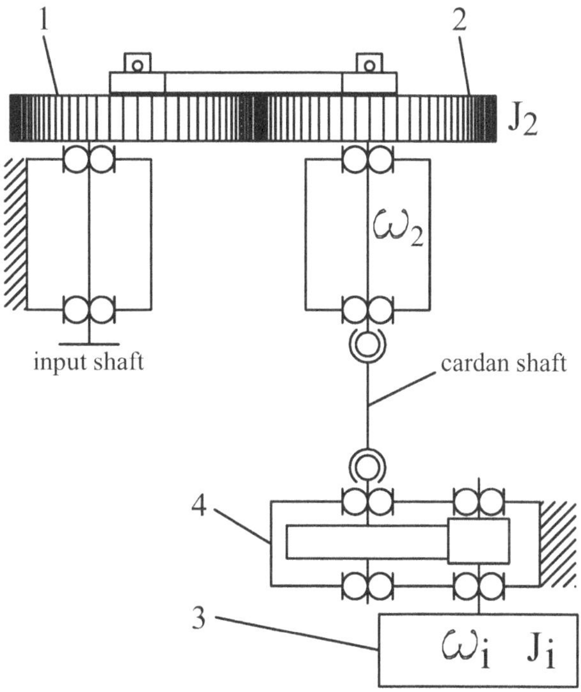

Fig. 7.19 Drive elements rotating with different angular speed

J_2 and J_i= mass moments of inertia of gear 2 and member i, respectively;
T_{in}= outside torque, which balances the inertial torques created by unevenly rotating masses of gear 2 and member i.

If the shafts are rigid, it is obvious that $\delta\varphi_i = \delta\varphi_2 \cdot u_i$, and it is easy to prove that $\varepsilon_i = \varepsilon_2 \cdot u_i$. Substituting these relations into Eq. 7.18 and cancelling $\delta\varphi_2$, we obtain the following equation for the scheme presented in Fig. 7.19:

$$T_{in} = \varepsilon_2\left(J_2 + J_i u_i^2\right).$$

Based on this equation, for the general case, the sum of all partial inertial torques created by the unevenly rotating members of the gear drive and the driven equipment can be expressed as follows:

$$T_{in} = \varepsilon_2 \sum_{i=1}^{k} J_i \cdot u_i^2, \tag{7.19}$$

where

J_i = mass moment of inertia of rotating member i driven by gear 2, including gear 2 itself;

$u_i = \omega_i/\omega_2 =$ angular speed ratio between member i and gear 2;
$\varepsilon_2 = f(\varphi_1)=$ angular acceleration of the driven gear 2, see Fig. 7.7.

Considering Eqs. 7.17 and 7.19, the expression for the equivalent mass moment of inertia is given by:

$$J_{\text{eq.}} = \sum_{i=1}^{k} J_i \cdot u_i^2.$$

It is desirable to make the kinematic chains as short and rigid as possible. In the case of relatively large rotating masses connected by long, flexible shafts, the calculation of torsional vibrations should be made using computer programs to determine the critical frequencies.

7.7 Approximate Formulas for Determining the Sizes of Gears Limited by the Resistance to Pitting

In this section, which is related to the calculations for strength, we need to indicate the dimensions that will be used:

- lengths in millimeters (mm);
- forces in Newtons (N);
- torques in N·mm;
- stresses in N/mm².

Contact stress factor K, which is directly proportional to the square of the surface stress in the contact of teeth (*Hertzian stress*), is mostly used for estimative calculation of gears for strength limited by their resistance to pitting. This factor is determined for the involute gears as follows [2]:

$$K = \frac{2\,T_{1p}(u+1)}{b_g\, d_1^2\, u}\ \frac{\text{N}}{\text{mm}^2},$$

where:

$T_{1p} =$ torque applied to the pinion shaft (N·mm);
$u = z_2/z_1 =$ gear ratio (in the case of EM gears, $u = 1$);
$d_1 =$ reference diameter of the pinion (mm);
$b_g =$ length of contact line, for straight teeth it equals the gear face width (mm).

Denoting $b_g/d_1 = \psi$, we obtain for $u = 1$:

$$K = \frac{4\,T_{1p}}{\psi\, d_1^3}\ \frac{N}{\text{mm}^2}, \tag{7.20}$$

Notes

1. Torque T_{1p} in Eq. 7.20 is understood as the moment of tooth force F_{Tt} relative to the geometric center of the gear and should be determined as follows:

$$T_{1p} = F_{Tt} r \ \cos \alpha_t. \tag{7.21}$$

Here

F_{Tt} = tooth force determined from Eq. 7.16;
r = radius of the operating pitch circle of gear 2;
α_t = pressure angle of the gearing (here it was accepted equal to 20°).
$d_1 = 2r$.

2. After the full force F_T is determined, it shall be used for the final calculation for strength.

The peculiarity of determining the required size of any gear rotating unevenly is that the torque transmitted by the meshing teeth also depends on the inertial mass of the gear itself, the size of which has yet to be determined. Therefore, an iterative process is required. It consists of at least three iterations:

at first, the dimensions of gears are determined without taking into account the inertia of the driven gear for the position in which the value of $(T_{res} + T_{in})/h_2$ is maximum;
then, the required size of the gears is determined in the position of the greatest positive acceleration, taking into account the additional torque caused by the inertia of the driven gear as well;
and finally, the strength of the gears shall be checked using tooth force F_T obtained from Eqs. 7.6 and 7.7.

Iteration process can be convergent or divergent. Main factor that contributes to the rapid convergence of iterations is the decrease of rotational speed and the reduction of the inertial masses of all moving elements of the mechanism, including the decrease in the sizes of gears by using stronger materials for their teeth.

7.8 Example of Design of an EM Gear Drive

It is required to design an EM gear pair, which has the maximum angular speed of the driven shaft seven times higher than its minimum angular speed. The service load on the driven shaft T_{res}= 100 Nm, rotational speed n_1= 180 rpm (ω_1= 18.85 s^{-1}). The straight teeth with core hardness 280...320 HB are induction hardened to 50...52 HRC, so the admissible contact stress factor $K_{adm.}$= 3.5 N/mm^2 [2] Equivalent mass moment of inertia of the parts connected to the driven gear, such as the shaft, coupling, and others is given: J_M= 5 N·mm·s^2.

7.8.1 Gear Design

From Fig. 7.8 we find that for $\omega_{2max}/\omega_{2min} = 7$, $e = 0.44$. Analyzing the diagrams of Figs. 7.7 and 7.16, we see that at $\varphi_1 \approx 345°$ there is a combination of maximum acceleration and relatively small value of shoulder h_2. This combination promises the maximum relation T_2/h_2, thus, we can choose this angular position for the first approximation. From Fig. 7.7 we find that in this position $\varepsilon_2/\omega_1^2 \approx 2.2$; $\varepsilon_2 = 2.2 \cdot 18.85^2 = 781.7\ \text{s}^{-2}$.

$$T_2 = T_{res} + J_M \varepsilon_2 = 100 \cdot 10^3 + 5 \cdot 781.7 = 103906\ \text{Nmm}.$$

From Fig. 7.16: at $\varphi_1 = 345\ °$, $h_2 \approx 0.61\ r$. Hence,

$$F_{Tt} = \frac{T_2}{h_2} = \frac{103906}{0.61\ r} = \frac{170.3 \cdot 10^3}{r}\ \text{N}. \tag{7.22}$$

From Eq. 7.21:

$$T_{1p} = F_{Tt} r\ \cos \alpha_t = 170.3 \cdot 10^3 \cos 20\ ° = 160 \cdot 10^3\ \text{Nmm}.$$

To find the size $d_1 = 2\ r$ from Eq. 7.20, we have to preset the ψ value. As the gears are overhang mounted, their face width should be reduced to decrease the unevenness of load distribution along the tooth length. Let us set $\psi = 0.2$. Now, from Eq. 7.20, we obtain:

$$d_1 = \sqrt[3]{\frac{4\ T_{1p}}{K\ \psi}} = \sqrt[3]{\frac{4 \cdot 160 \cdot 10^3}{3.5 \cdot 0.2}} = 97\ \text{mm}.$$

The face width $b_g = 0.2 \cdot d_1 = 19.4$ mm. These dimensions are minimal.

For the second approximation, we choose the same angular position, however, with the additional inertia of the driven gear. The mass moment of inertia of the gear with the dimensions obtained as the first approximation equals (see Eq. 7.15):

$$J_{\text{gear}} = \frac{7.7 \cdot 10^{-5}}{9,81 \cdot 10^3} \frac{\pi \cdot 97^4 \cdot 19.4}{16} \left(0.44^2 + 0.5\right) = 1.84\ \text{N} \cdot \text{mm} \cdot \text{s}^2.$$

The total moment of inertia J_T equals:

$$J_T = J_{\text{gear}} + J_M = 1.84 + 5 = 6.84\ \text{Nmm} \cdot \text{s}^2.$$

Additional torque applied to the driven gear due to angular acceleration:
$T_{\text{in}} = \varepsilon_2\ J_T = 781.7 \cdot 6.84 = 5347$ Nmm.
Thus, $T_2 = T_{\text{res}} + T_{\text{in}} = 105{,}347$ N mm. From Eq. 7.16:

$$F_{Tt} = \frac{105,347}{0.61\ r} = \frac{172.7 \cdot 10^3}{r}\ \text{N}.$$

$$T_{1p} = 172.7 \cdot 10^3 \cos 20° = 162.3 \cdot 10^3\ \text{Nmm}.$$

$$d_1 = \sqrt[3]{\frac{4 \cdot 162.3 \cdot 10^3}{3.5 \cdot 0.2}} = 97.5\ \text{mm}.$$

For the third approximation, let's check the position with minimum value of h_2: at $\varphi_1 = 10°$, $h_2/r \approx 0.5$. Here $\varepsilon_2/\omega_1^2 \approx -2$; $\varepsilon_2 \approx -2 \cdot 18.85^2 = -710.6\ \text{s}^{-2}$, and $T_{\text{in}} = \varepsilon_2 \cdot J_T = -710.6 \cdot 6.84 = -4860$ Nmm.

$$T_2 = 100 \cdot 10^3 - 4860 = 95.2 \cdot 10^3\ \text{Nmm};$$

$$F_{Tt} = \frac{95.2 \cdot 10^3}{0.5\ r} = \frac{190.4 \cdot 10^3}{r}\ \text{N}.$$

$$T_{1p} = 190.4 \cdot 10^3 \cos 20° = 178.9 \cdot 10^3\ \text{Nmm}.$$

$$d_1 = \sqrt[3]{\frac{4 \cdot 178.9 \cdot 10^3}{3.5 \cdot 0.2}} = 100.7\ \text{mm}.$$

$$b_g = 100.7 \cdot 0.2 = 20.2\ \text{mm}.$$

The third iteration gives the final dimensions of the gears: $r = 0.5d_1 = 50$ mm; $b_g = 20$ mm.

The gear module can be chosen using the approximate formula [2]:

$$m \geq \frac{10\ T_{1p}}{d_1 b_g \sigma_b} = \frac{10 \cdot 178.9 \cdot 10^3}{100 \cdot 20 \cdot 230} \geq 3.89\ \text{mm}.$$

(Here $\sigma_b = 230\ \text{N/mm}^2$ is the bending stress admissible for induction hardened teeth in an approximate calculation [2].) The number of teeth may be odd or even. In the first case, the center of rotation can be displaced symmetrically to a tooth or to a tooth space, but identically on both gears. In the second case, when the numbers of teeth are even, the displacement of the rotation center must be on one gear symmetrical to a tooth, and on the other gear symmetrical to a tooth space. Let us choose here the following dimensions of the gears: $m = 4$ mm, $z = 25$ with no profile shift.

Thus, finally: $m = 4$ mm; $z_1 = z_2 = 25$; $b_g = 20$ mm; $r = 50$ mm.

7.8.2 Determination of Forces Using Equations Given in Sect. 7.4

The geometry of the EM gear drive changes as it rotates. Therefore, the forces are different in each angular position of the drive even if the output torque is constant. The designer is to choose for calculation the positions where, in his opinion, the forces may be maximum, or he can make a computer program to obtain the diagrams of forces for the complete cycle (360°). Here we choose as an example the position in which the tooth force is maximum. It is characterized by rotation angle $\varphi_1 = 10\,°$.

Now we have to estimate the known members of Eqs. 7.6, 7.7, and 7.8. Here are two factors that are independent on the angular position: the total moment of inertia of rotating masses J_T and the mass of the oscillating assembly of gear 2.

$$J_T = J_M + J_{\text{gear}};$$

$$J_{\text{gear}} = \frac{7.7 \cdot 10^{-5}}{9,81 \cdot 10^3} \frac{\pi \cdot 100^4 \cdot 20}{16} \left(0.44^2 + 0.5\right) = 2.14 \text{ N mm s}^2;$$

$$J_T = 5 + 2.14 = 7.14 \text{ N mm s}^2.$$

To know the mass of the driven gear assembly, we need to design this assembly and calculate its weight. For our purpose, let's instead make some quick estimation. The weight of the gear equals about 1.3 kg, half of the universal joint shaft for this torque weighs 1 kg... Let's suppose that the entire assembly, together with the shaft, bearings, housing, coupling, and half of the link, weighs, say, 7 kg = 68.64 N. Hence, the mass of the assembly m equals:

$$m = \frac{68,64}{9.81 \cdot 10^3} = 7 \cdot 10^{-3} \frac{\text{N} \cdot \text{s}^2}{\text{mm}}.$$

The values of known members of Eqs. 7.6, 7.7 and 7.8 that depend on rotation angle φ_1 are as follows (at $\varphi_1 = 10°$):

- from Fig. 7.16: $h_2/r \approx 0.5$, hence, $h_2 = 0.5 \cdot 50 = 25$ mm;
- from Fig. 7.17: $h_{L2}/r \approx 0.13$, hence, $h_{L2} = 0.13 \cdot 50 = 6.5$ mm;
- from Fig. 7.7: $\varepsilon_2/\omega_1^2 \approx -2$, hence, $\varepsilon_2 = -2 \cdot 18.85^2 = -710.6 \text{ s}^{-2}$;
- from Fig. 7.5: $\gamma \approx 8°$;
- from Fig. 7.11: $\dot{v}/\left(r\omega_1^2\right) \approx 0.8$,

hence, $\dot{v} = 0.8 \cdot 50 \cdot 18.85^2 = 1.42 \cdot 10^4 \text{ mm/s}^2$.

$$F_{O_2} = m \cdot \dot{v} = 7 \cdot 10^{-3} \cdot 1.42 \cdot 10^4 = 99.4 \text{ N}.$$

$$T_{\text{in}} = J_T \varepsilon_2 = -7.14 \cdot 710.6 = -5074 \text{ Nmm};$$

$$T_2 = T_{\text{res}} + T_{\text{in}} = 100 \cdot 10^3 - 5074 = 94,926 \text{ Nmm}.$$

Here are Eqs. 7.6 and 7.7 with all known and unknown members:

$$25\, F_{T2} + 6.5\, F_{L2} - 94,926 = 0$$

$$F_{T2} \cos\, 62° - F_{L2} \cos\, 8° + 99.4 = 0;$$

Solving these equations, we obtain: $F_{L2} = 1692$ N; $F_{T2} = 3357$ N.

Force F_G is obtained from Eq. 7.8 by substituting there the calculated values of F_{L2} and F_{T2}:

$$3357 \sin 62^\circ + 1692 \sin 8^\circ - F_G = 0.$$
$$F_G = 3200 \text{ N}.$$

The following data are required for Eqs. 7.10, 7.11 and 7.12 related to dynamic equilibrium of gear 1:

- from Fig. 7.18: $h_1/r \approx 1.3$, hence, $h_1 = 1.3 \cdot 50 = 65$ mm;
- from the calculation for gear 2: $h_{L1} = h_{L2} = 6.5$ mm;
- from Fig. 7.5: $\gamma \approx 8°$;
- forces: $F_{L1} = F_{L2} = 1692$ N; $F_{T1} = F_{T2} = 3357$ N;
- the directions of forces, except F_{1h} and F_{1v}, are known. For F_{1h} and F_{1v} the direction shall be chosen arbitrarily, and if the direction is chosen incorrectly, the calculated forces will have sign "minus".

Here are equations with the known data substituted:

Equation 7.10:

$$T_1 - 3357 \cdot 65 - 1692 \cdot 6.5 = 0;$$
$$T_1 = 229.2 \cdot 10^3 \text{ N mm};$$

Equation 7.11:

$$3357 \cos 62^\circ - 1692 \cos 8^\circ - F_1 h = 0;$$
$$F_{1h} = -99.5 \text{ N};$$

Equation 7.12:

$$3357 \sin 62^\circ + 1692 \sin 8^\circ - F_{1v} = 0;$$
$$F_{1v} = 3200 \text{ N}.$$

Let us do the same calculations for $\varphi_1 = 345°$:

- from Fig. 7.16: $h_2/r \approx 0.68$, hence, $h_2 = 0.68 \cdot 50 = 34$ mm;
- from Fig. 7.17: $h_{L2}/r \approx -0.25$, hence, $h_{L2} = -0.25 \cdot 50 = -12.5$ mm;
- from Fig. 7.7: $\varepsilon_2/\omega_1^2 \approx 2.1$, hence, $\varepsilon_2 = 2.1 \cdot 18.85^2 = 746.2 \text{ s}^{-2}$;
- from Fig. 7.5: $\gamma \approx -14°$;
- from Fig. 7.11: $\dot{v}/\left(r\omega_1^2\right) \approx 0$, hence, $\dot{v} = 0$, and $F_{O2} = 0$.

$$T_{\text{in}} = J_T \varepsilon_2 = 7.14 \cdot 746.2 = 5328 \text{ N mm};$$
$$T_2 = T_{\text{res}} + T_{\text{in}} = 100 \cdot 10^3 + 5328 = 105{,}328 \text{ N mm}.$$

Here are the Eqs. 7.6 and 7.7 with all known and unknown members:

$$34\,F_{T2} - 12.5\,F_{L2} - 105{,}328 = 0;$$
$$F_{T2}\cos\,84^\circ\; - F_{L2}\cos(-14^\circ) = 0;$$

Solving these equations, we obtain: $F_{L2} = 347.5\ N$; $F_{T2} = 3226$ N.

Force F_G is obtained from Eq. 7.8 by substituting there the calculated values of F_{L2} and F_{T2}:

$$3226\ \sin\,84^\circ + 347.5\ \sin(-14^\circ) - F_G = 0.$$
$$F_G = 3124\ \text{N}.$$

The following data are required for Eqs. 7.10, 7.11 and 7.12 related to dynamic equilibrium of gear 1:

- from Fig. 7.8: $h_1/r \approx 1.36$, hence, $h_1 = 1.36 \cdot 50 = 68$ mm;
- from the calculation for gear 2: $h_{L1} = h_{L2} = -12.5$ mm;
- from Fig. 7.5: $\gamma \approx -14°$;
- forces: $F_{L1} = F_{L2} = 347.5$ N; $F_{T1} = F_{T2} = 3226$ N;
- the directions of forces, except F_{1h} and F_{1v} are known. For F_{1h} and F_{1v} the direction is chosen arbitrarily.

Here are equations with the known data substituted:

Equation 7.10:

$$T_1 - 3226 \cdot 68 + 347.5 \cdot 12.5 = 0;$$
$$T_1 = 215 \cdot 10^3\ \text{N mm};$$

Equation 7.11:

$$3226\ \cos\,84^\circ\; - 347.5\ \cos(-14^\circ) - F_{1h} = 0;$$
$$F_{1h} = 0;$$

Equation 7.12:

$$3226\ \sin\,84^\circ + 347.5\ \sin(-14^\circ) - F_{1v} = 0;$$
$$F_{1v} = 3124\ \text{N}.$$

Note When calculating the strength of the shafts and bearings of gears 1 and 2, it should be taken into account that the overhanging forces F_T and F_L act not in the same plane.

7.8.3 Final Calculation of the Gears for Resistance to Pitting

From Sect. 7.8.2, the maximum tooth force $F_{T2} = 3357$ N. From Eqs. 7.21 and 7.20 we determine the real value of parameter K:

$$T_{1p} = 3357 \cdot 50 \cdot \cos\ 20^{\circ} = 158 \cdot 10^3\ \text{Nmm};$$
$$K_{\text{real}} = \frac{4 \cdot 158 \cdot 10^3}{0.2 \cdot 100^3} = 3.16\ \frac{\text{N}}{\text{mm}^2}.$$

This is less by 9% than the admissible maximum (3.5 N/mm^2). The face width can be slightly reduced (by about 2 mm), but we would live it unchanged.

7.8.4 Link Bearing Design

The maximum link force $F_L = 1692$ N. Let us select ball bearings for the link to provide service life of 20,000 h at 180 rpm.

$$L = \left(\frac{C}{F_L}\right)^3,$$

where

L= nominal rating life (in millions of revolutions),
C= dynamic load rating (kN),
F_L = radial load (kN).

$$L = 20{,}000 \cdot 60 \cdot 180 = 216 \cdot 10^6\ \text{rev}.$$
$$C = F_L L^{1/3} = 1692 \cdot 216^{1/3} = 10.15\ \text{kN}.$$

Keeping in mind that the minimum value of the link force $F_L = 347.5$ N,deep groove ball bearing with d= 12 mm, D= 37 mm, and B= 12 mm that has dynamic load rating C= 9.65 kN and static load rating C_0= 4.15 kN is suitable for these conditions. For more exact calculation, the diagram of $F_L = f(\varphi_1)$ should be built in order to determine the equivalent load and choose a smaller bearing.

7.8.5 Lever Length Choice

In the mechanism presented in Fig. 7.14, the unit of driven gear 2 is designed as part of lever 5. The longer the lever, the lesser the deviation of the trajectory of center O_2 from the straight line. From Fig. 7.9 we find that at $e = 0.44$, the maximum displacement of center O_2 equals approximately $0.092 \cdot 2r = 0.092 \cdot 2 \cdot 50 = 9.2$ mm. If, for example, the lever length (distance O_2O_3) equals 100 mm, the swing angle of the lever equals 5.27°, and the deviation of the trajectory of center O_2 from the straight line equals 0.1 mm only.

Now the drawing of the gear drive can be made, and the final calculations can be performed using the final dimensions and weights of all elements of the mechanism.

7.9 Conclusion

In this research, the formulas for calculating the kinematics of gear pairs composed of circular involute gears mounted eccentrically on their shafts are developed. The diagrams obtained using computer programs allow quick determination of kinematic and dynamic characteristics, as well as the geometric parameters required for calculations for strength. Formulas for gear design allow quick estimation of the dimensions of gears. The example of design of such a gear pair is presented for the case when the torque caused by the resistance of the driven mechanism is constant. If the torque changes with the angle of rotation, the number of angular positions, in which the calculation for strength is required, may increase—depending on the torque diagram.

The use of a computer program allows the rapid visualization of all characteristics of the gear drive, including factor K, as functions of angle φ_1 in small intervals, for example, one degree, at any torque diagram on the shaft of the driven gear.

References

1. Litvin, F. L., et al. (2009). *Noncircular gears: Design and generation.* Cambridge University Press.
2. Klebanov, B. M., & Groper, M. (2016). *Power mechanisms of rotational and cyclic motions.* CRC Press.

Chapter 8
Envelopes in Gearing: A Novel Development in Classical Differential Geometry

Stephen P. Radzevich

8.1 Preamble

The problem of determining envelope surfaces in gear design and gear-machining processes is not a new one—this problem has been known for decades. For a long while, this problem was intuitively realized by proficient gear experts, but at the same time, it could not even be correctly formulated. Where this problem originated from has yet to be determined. The author of this chapter still remembers that when he was a university student (~1975), one of his professors mentioned (among others) the consequences that not understanding this problem has on gear designers and manufacturers. This problem was not correctly realized by any of the lead gear experts at that time (~1975). Moreover, once the university's mechanical engineering students were informed about this strange problem, the problem itself was not correctly understood before 1975. The author failed in trying to identify the names of gear experts who deserve to be credited with formulating the problem at hand.

The problem of determining envelope surfaces in gear design and gear-machining processes was correctly formulated and then solved by the author in 2008 [3]. All three kinds of gearing—(a) parallel-axes gearing (P_a−*gearing*); (b) intersected-axes gearing (I_a−*gearing*); and (c) crossed-axes gearing (C_a−*gearing*)—are covered by the derived solution to the problem.

S. P. Radzevich (✉)
Machine Design Department, National Technical University of Ukraine "Igor Sikorsky Kyiv Polytechnic Institute", Sterling Heights, MI, USA

S. P. Radzevich (ed.), *Gear Accuracy*, Mechanical Engineering Series,
https://doi.org/10.1007/978-3-031-74798-4_8

8.2 Introduction

Consider, for example, interactions with the involute tooth profiles of a gear and a mating pinion in parallel-axes involute gearing.

8.2.1 Interactions with Tooth Flanks in Parallel-Axes Involute Gearing

A uniform rotary motion can be smoothly transmitted by means of gear teeth in parallel-axes involute gearing. A schematic of a parallel-axes gear pair associated with a corresponding equivalent pulley-and-belt transmission is shown in Fig. 8.1.

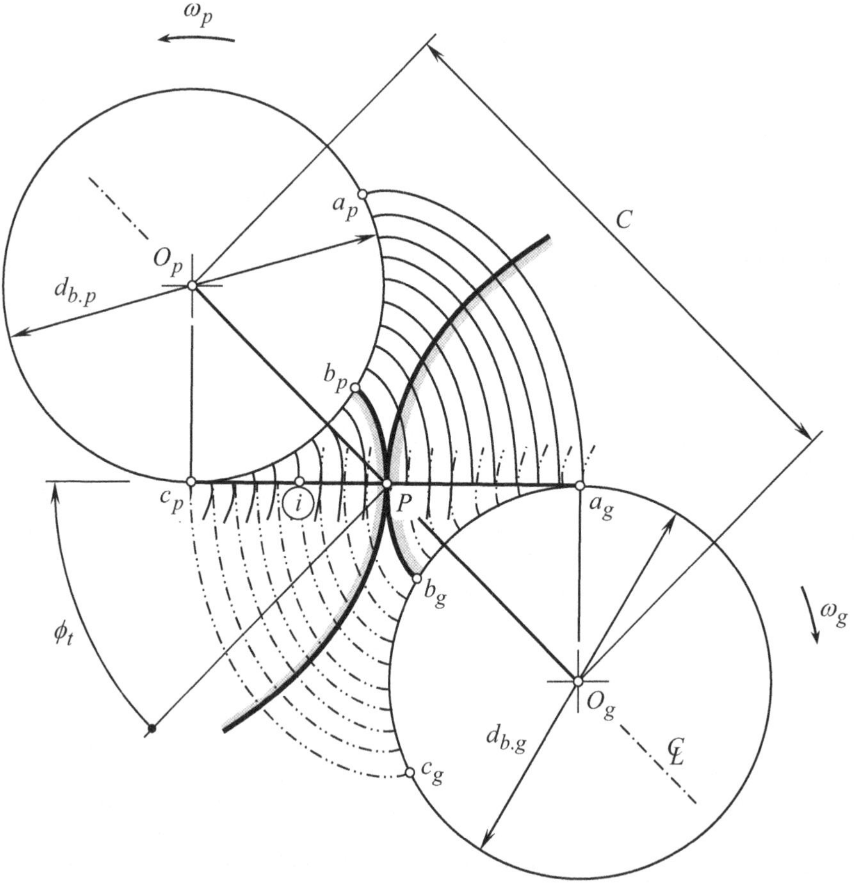

Fig. 8.1 Interaction of involute profiles in transverse section of parallel-axes gearing

The belt that is stretched between the two pulleys corresponds to the line of action, $a_g c_p$, in Fig. 8.1. A distance traveled by point i on the belt corresponds with the one described by the point of contact between the tooth flanks along the line of action. The angle formed perpendicular to the centerline, ℄, with the line of action is the transverse pressure angle, ϕ_t.

When the gear involute tooth profile, $\mathcal{G}$ (which starts at point b_g), is given, the pinion tooth profile, $\mathcal{P}$ (which starts at point b_p), can be generated as an envelope to a family of consecutive positions of the gear tooth profile, $\mathcal{G}$, in its motion in relation to a reference system associated with the pinion, $\mathcal{P}$, and vice versa (this is crucial for the upcoming analysis). Also, the gear tooth profile, $\mathcal{G}$, can be generated as an envelope to a family of consecutive positions of the pinion tooth profile, $\mathcal{P}$, in its motion relative to a reference system associated with the gear.

Generated in such a scenario, gear and pinion tooth flanks are commonly referred to as reversibly enveloping tooth flanks (or as *R_e–tooth flanks*: $\mathcal{G}$ and $\mathcal{P}$ for simplicity). Conjugate tooth flanks, $\mathcal{G}$ and $\mathcal{P}$, refer to the tooth flanks, $\mathcal{G}$ and $\mathcal{P}$, of such a geometry.

Important *Uniform motion transmission between two parallel axes is possible only by conjugating tooth profiles when the line of action at any and all configurations of the mating gears passes through a motionless point, known as the pitch point, P.*

In another example, consider the generation of the interacting tooth profiles in the gear-machining mesh in the hobbing of straight-sided splines.

8.2.2 Interactions with Tooth Flanks in Straight-Sided Spline–Spline Hob Mesh

One more example of an interaction with noninvolute tooth profiles can be found when designing a hob for machining straight-sided splines. The hob design is based on an auxiliary rack (or basic rack), the teeth of which are entwined in mesh with the splines of the spline shaft. The tooth profile of the auxiliary rack is commonly generated as an envelope to a family of consecutive positions of the straight-sided spline profile when the pitch circle of the spline rolls with no sliding over the pitch line associated with the rack.

The determination of the coordinates of the points of the tooth profile of an auxiliary rack conjugated to a spline shaft is commonly executed by using the method of common perpendiculars. An example of how this kind of problem can be solved in the case under consideration is illustrated in Fig. 8.2.

The spline profile is associated with a pitch circle of a radius, $r_{w.\ sp}$. The pitch line of the auxiliary rack to be determined is tangent to the pitch circle of the spline shaft. The point of tangency for the pitch line and for the pitch circle, $r_{w.\ sp}$, is the pitch point in the rolling motion of the spline shaft and the rack. The pitch point is designated as P.

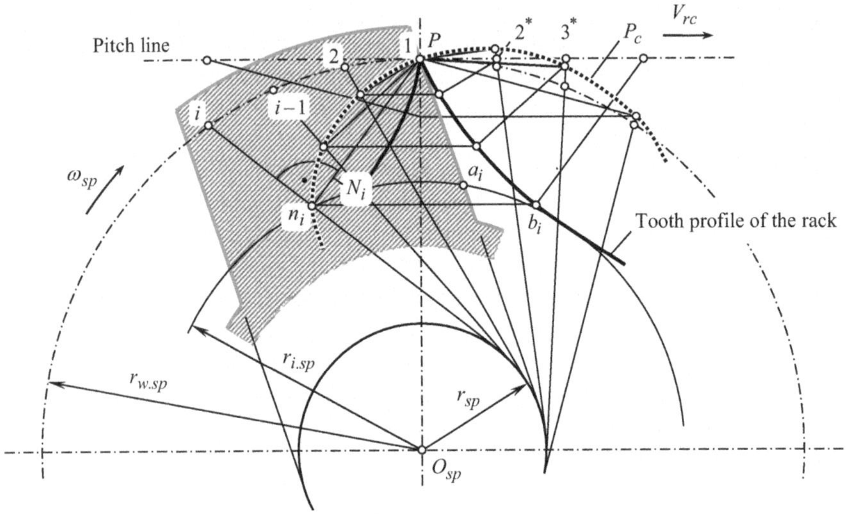

Fig. 8.2 Generation of a rack tooth profile as an envelope to a family of consecutive positions of the lateral profile of a spline in a straight-sided spline shaft. (*Source*: F. Reuleaux's *Lehrbuch der Kinematik*, Braunschweig, 1875)

The spline shaft is rotated on its axis of rotation, O_{sp}. The angular velocity of this rotation is designated as ω_{sp}. The auxiliary rack is associated with the pitch line. The rack travels straight and forward together with the pitch line. The linear velocity of the rack is designated as V_{rc}.

Let us assume that at the initial configuration of the pitch circle and the pitch line, the profile of the spline passes through the pitch point, P. This profile is at a distance, r_{sp}, from the axis of rotation, O_{sp}, of the spline shaft. Practically, the straight-line profile is tangent to a circle of the radius r_{sp}. The radius, r_{sp}, is equal to half of the spline thickness of the spline shaft.

When the spline shaft rotates, the lateral spline profile is rotated together with the spline shaft. Consequently, the spline shaft lateral profile passes through points 1, 2, ..., $(i-1)$, i. Point 1 coincides with the pitch point, P. The pitch line travels straight and forward. In this motion, the pitch point consequently occupies positions 1^*, 2^*, 3^*, ... The distance $1^* - 2^*$, $2^* - 3^*$, ... between consequent locations of the pitch point is equal to the lengths of the arcs $\smile$1–2, $\smile$2–3, ... of the pitch circle of the spline shaft. This is because the pitch line of the auxiliary rack is rolling with no sliding over the pitch circle of the spline shaft.

At every chosen location of the spline lateral profile, perpendiculars to the profile are constructed so as to pass through the pitch point, P. For example, a perpendicular, N_i, is normal to the spline profile at its i–th location. Point n_i is the point of tangency of the lateral profile of the spline and the rack tooth profile.

Constructed in this way, the plurality of points for various configurations of the lateral spline profile are located within the path of contact, P_c. In the section of the path of contact, P_c, by a transverse plane, it is traced by the yet-to-be-determined contact point in the rolling motion of the given spline shaft and the rack. The path of contact is determined in a stationary reference system. In a reference system associated with the spline shaft, all the points are situated within the lateral spline profile of the spline shaft.

When the spline shaft rotates, points of the lateral profile consequently cross the path of contact, P_c. At these instants of time, the points coincide with the corresponding points of the rack tooth profile. An arbitrary point, n_i, within the path of contact, P_c, corresponds to the point of contact in the i–th location of the lateral profile of the spline. If this point, n_i, returns to the initial position of the spline by means of rotation through the angle of the arc, $\overset{\frown}{Pi}$, then this point occupies the position of point a_i.

Similarly, in a reference system associated with the rack, contact points are located within the tooth profile of the rack, which has yet to be determined.

Let us assume that an arbitrary point, n_i, taken from the path of contact, P_c, is associated with the pitch line, P_{ln}. To determine the location of this point in the initial instance of time, the pitch line, together with point n_i travels a distance that is equal to the arc length, $\overset{\frown}{Pi}$, in a direction opposite to the direction of the straight motion of the rack tooth in its rolling motion. After this transition has been completed, point n_i occupies the position of point b_i. Point b_i is located on the tooth profile of the rack. All points of the rack tooth profile are constructed similarly to that of point b_i. By connecting the constructed points by using a smooth curve, the rack tooth profile can be determined.

The aforementioned approach for the determination of the tooth profile of an auxiliary rack (as the envelope to a family of a spline shaft profile) is commonly adopted. However, this method is inaccurate by nature—it can be used only for an approximate generation of the auxiliary rack tooth profile. The generated tooth profile of the auxiliary rack is assumed here to be able to generate the spline profile of the straight-sided spline shaft when a problem that is inverse to the original problem is considered. The latter is not correct. Because these envelope profiles do not conjugate to one another, no straight-sided spline profile can be generated in the inverse rolling of the auxiliary rack in relation to spline shafts. In practice, instead of a straight-sided spline profile, a curved profile of the splines is generated (see Fig. 8.3).

This discussion reveals that using the method of common perpendiculars returns a tooth profile of the auxiliary rack that is an envelope to a family of consecutive positions of the spline shaft profile but that is not a conjugate to it.

When a gear involute tooth profile, $\mathcal{G}$ (which starts at point b_g, within the gear-based circle of diameter $d_{b.\ g}$), is given, the pinion tooth profile, $\mathcal{P}$ (which starts at point b_p, within the pinion-based circle of diameter $d_{b.\ p}$), can be generated as an envelope to a family of consecutive positions of the gear tooth profile, $\mathcal{G}$, in its motion in relation to a reference system associated with the pinion, $\mathcal{P}$, and vice versa. Also, the gear tooth profile, $\mathcal{G}$, can be generated as an envelope to a family of consecutive positions of the pinion tooth profile, $\mathcal{P}$, in its motion in relation to a reference system associated with the gear. Generated in such a scenario, the straight-

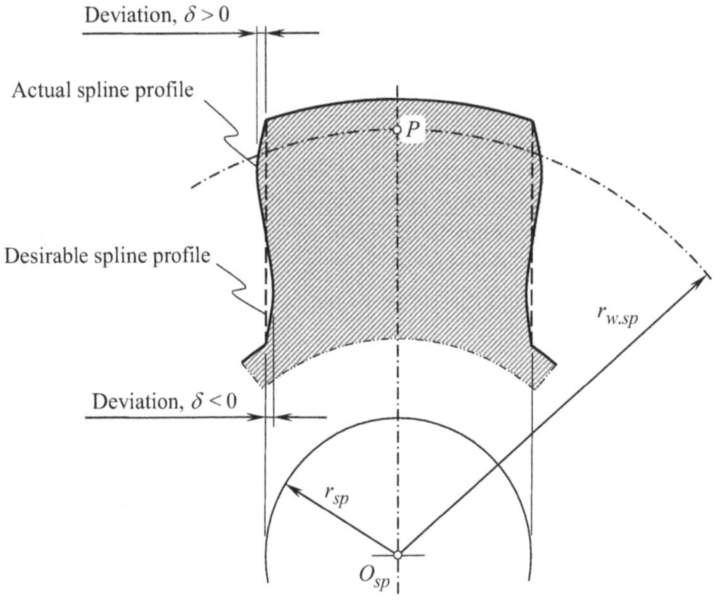

Fig. 8.3 Deviation, δ, of the actual lateral profile of a spline shaft from its desirable profile (resulting from qualitative, not quantitative, analysis)

sided spline and the spline hob are commonly referred to as not reversibly enveloping tooth flanks. Tooth flanks of such a geometry are not conjugates to one another.

8.3 Generation of Envelope Surfaces: State of the Art

Two approaches to generating envelope surfaces are currently used in science and engineering. One of the approaches was developed in the differential geometry of surfaces and another in the kinematic geometry of surfaces [5].

8.3.1 Generation of Envelope Surfaces in Differential Geometry

A gear tooth flank, $\mathcal{G}$, can be given by using an equation in vector form:

$$\mathbf{r}_g = \mathbf{r}_g(U_g, \ V_g) \tag{8.1}$$

where U_g and V_g are the Gaussian parameters of the gear tooth flank, $\mathcal{G}$.

If the gear tooth flank, $\mathbf{r}_g$, travels and θ is the parameter of this motion, then a family of consecutive positions of the tooth flank is generated. In a certain reference system, $X_gY_gZ_g$, this family of the tooth flank, $\mathbf{r}_g$, can be described by the following equation:

$$\mathbf{r}_g = \mathbf{r}_g(U_g,\ V_g,\ \theta) \tag{8.2}$$

In this reference system, $X_gY_gZ_g$, the envelope surface to the family of consecutive positions of the tooth flank, $\mathbf{r}_g$, is described by the following set of equations:

$$\begin{cases} \mathbf{r}_g = \mathbf{r}_g(U_g,\ V_g,\ \theta) \\ \dfrac{\partial \mathbf{r}_g(U_g,\ V_g,\ \theta)}{\partial \theta} = 0 \end{cases} \tag{8.3}$$

An equation of the pinion tooth flank, $\mathcal{P}$, can be determine after the second equation in the set of equations (8.3) has been solved with respect to the motion parameter, θ, after which the derived solution for θ is substituted into the first equation in the set of equations (8.3).

8.3.2 Generation of Envelope Surfaces in Kinematic Geometry

In the second half of the nineteenth century (~1875), Franz Reuleaux published [4] a well-known principle, according to which the relative motion of two surfaces in contact along a common perpendicular is zero.

The condition of contact requires that at the contact point, a gear and mating pinion tooth flanks that interact with one another have to travel at equal velocities along their common perpendicular; that is, the equality $\mathbf{V}_g \equiv \mathbf{V}_p$ must be valid (see Fig. 8.4). The relative velocity of the interacting tooth flanks equals zero in this case ($\mathbf{V}_{rel} = 0$). In a case where $\mathbf{V}_g < \mathbf{V}_p$, components 1 and 2 separate from each other (see Fig. 8.5), and the relative velocity is a positive value ($\mathbf{V}_{rel} > 0$). Inversely, in a case where $\mathbf{V}_g > \mathbf{V}_p$, components 1 and 2 interfere with each other (see Fig. 8.6), and

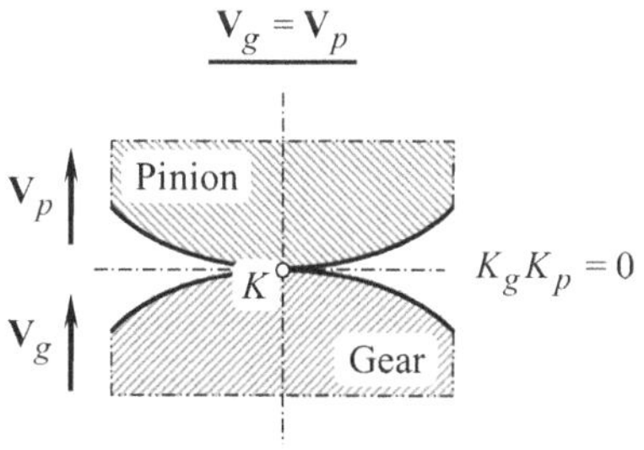

Fig. 8.4 Condition of contact of a gear and mating pinion tooth flanks: correct contact with the tooth flanks, $\mathbf{V}_g = \mathbf{V}_p$

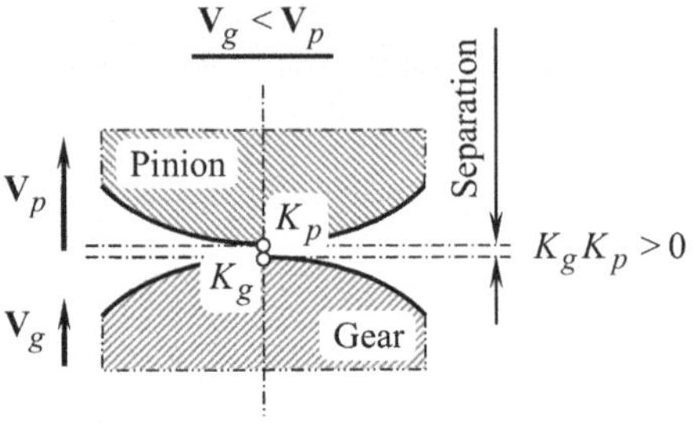

Fig. 8.5 Condition of contact of a gear and mating pinion tooth flanks: separation of the tooth flanks, $\mathbf{V}_g < \mathbf{V}_p$

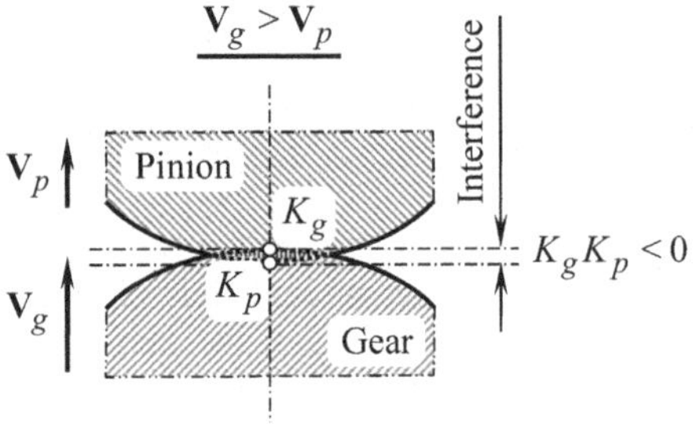

Fig. 8.6 Condition of contact of a gear and mating pinion tooth flanks: interference between the tooth flanks, $\mathbf{V}_g > \mathbf{V}_p$

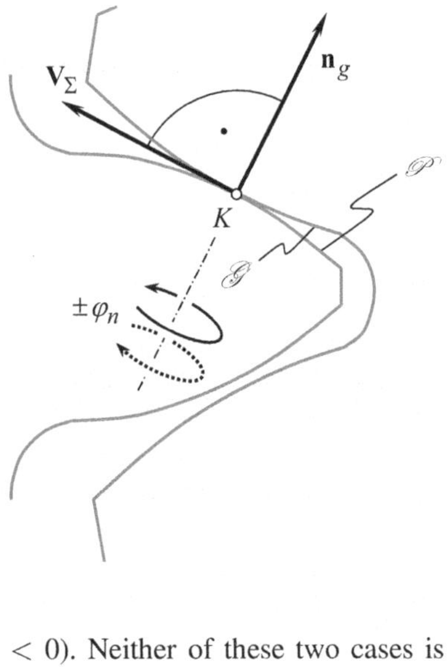

Fig. 8.7 Permissible instant relative motions in geometrically accurate gearing (general case of gearing)

the relative velocity is a negative value ($\mathbf{V}_{rel} < 0$). Neither of these two cases is permissible in perfect gearing.

The permissible instant relative motions of the tooth flanks, $\mathcal{G}$ and $\mathcal{P}$, in a gear pair are shown in Fig. 8.7. No relative motion of the tooth flanks, $\mathcal{G}$ and $\mathcal{P}$, along the common perpendicular, $\mathbf{n}_g$, is permissible, and the instant relative motion is permissible in any direction within the common tangent plane through contact point K. Here, a swivel relative motion, $\pm\varphi_n$, of the tooth flanks, $\mathcal{G}$ and $\mathcal{P}$, around the axis along the common perpendicular, $\mathbf{n}_g$, also meets the requirement specified by the

Shishkov equation of contact, $\mathbf{n}_g \cdot \mathbf{V}_\Sigma = 0$. However, not all the kinds of the swivel motion of the tooth flanks in Fig. 8.7 are permissible. For example, no swivel relative motion is permissible on an axis that either intersects with or crosses a straight line along the common perpendicular, $\mathbf{n}_g$, because the condition of contact, $\mathbf{n}_g \cdot \mathbf{V}_\Sigma = 0$, in these cases would be violated. The swivel motion, $\pm\varphi_n$, of the tooth flanks is not necessary to transmit a rotation from a driving shaft to a driven shaft. However, motion of this nature, $\boldsymbol{\omega}_\mathbf{n}$, is observed in spatial gearing, namely in C_a−gearing. Therefore, when the sliding of the interacting tooth flanks is discussed, for example, swivel motion must be incorporated into the analysis.

The condition of contact is important to the theory of gearing. Unfortunately, who should be credited with this important accomplishment remains unknown (perhaps because the condition of contact was discovered for a more general case of interaction between machine elements and not for the specific purposes of gears).

Later on (in the late 1940s), V.A. Shishkov proposed analytically expressing this principle in the form of an equation of contact, $\mathbf{n} \cdot \mathbf{v} = 0$. Here, $\mathbf{n}$ is the unit normal vector at the contact point, K, of the interacting tooth flanks, $\mathcal{G}$ and $\mathcal{P}$, and $\mathbf{v}$ is the unit vector of the speed of relative motion of the gear and the mating pinion tooth flanks, $\mathcal{G}$ and $\mathcal{P}$. This equation, $\mathbf{n} \cdot \mathbf{v} = 0$, is commonly referred to as the Shishkov equation of contact, $\mathbf{n} \cdot \mathbf{v} = 0$.

The accomplishments by Reuleaux and Shishkov are very helpful when determining a pinion tooth flank, $\mathcal{P}$, as an envelope to a family of consecutive positions of a traveling gear tooth flank, $\mathcal{G}$, or, vice versa, when determining a gear tooth flank, $\mathcal{G}$, as an envelope to a family of consecutive positions of the traveling pinion tooth flank, $\mathcal{P}$. Such a replacement of the conjugate tooth flanks, $\mathcal{G}_{cnj}$ and $\mathcal{P}_{cnj}$, with the enveloping tooth flanks, $\mathcal{G}_{env}$ and $\mathcal{P}_{env}$, is not always valid, because all conjugate tooth flanks are envelopes to one another, but not all envelope tooth flanks are conjugates to one another. As a consequence of such a replacement, the gear/pinion tooth flanks are now loosely determined as envelope surfaces and not as conjugate surfaces; the latter is not permissible for precision gearing, high-power-density gearing, or other kinds of sophisticated gearings.

The difference between **conjugate** gear tooth flanks and **envelope** gear tooth flanks must be understood as much as possible. Determining envelope tooth flanks (either of a gear tooth flank, $\mathcal{G}$, or of a pinion tooth flank, $\mathcal{P}$) with the Shishkov equation of contact, $\mathbf{n} \cdot \mathbf{v} = 0$ works under the following conditions:

- The unit normal vector, $\mathbf{n}$, can be determined with no differentiation of equation for the family of the gear tooth flank, $\mathbf{r}_g = \mathbf{r}_g(U_g,\ V_g,\ \theta)$, with respect to the Gaussian parameters, U_g and V_g (or the same with no differentiation of equation for the family of the pinion tooth flank, $\mathbf{r}_p = \mathbf{r}_p(U_p,\ V_p,\ \theta)$, with respect to the Gaussian parameters, U_p and V_p)—which is possible for the surfaces of simple geometries
- The unit vector of the speed of relative motion, $\mathbf{v}$, of the gear and the mating pinion tooth flanks, $\mathcal{G}$ and $\mathcal{P}$, can be determined with no differentiation of either equation $\mathbf{r}_g = \mathbf{r}_g(U_g,\ V_g,\ \theta)$ or equation $\mathbf{r}_p = \mathbf{r}_p(U_p,\ V_p,\ \theta)$ with respect to the

motion parameter, θ—which is possible when the kinematics of the relative motion of the tooth flanks, $\mathcal{G}$ and $\mathcal{P}$, is simple.

In a more general case of interaction between the functional surfaces of machine elements, using the Shishkov equation of contact, $\mathbf{n} \cdot \mathbf{v} = 0$, is less effective or even useless.

8.4 Generation of Envelope Surfaces Under Constraints on Relative Motion

Rolling relative motion, swivel relative motion, and combined rolling/swivel relative motion are considered here as the principal constraints on the relative motion of the components when gears operate and in gear-machining processes.

8.4.1 Generation of Envelopes in Rolling Relative Motion

The functional dependence of elementary motions is the key feature of the rolling relative motion in gearing. This is due to rolling without the sliding of the axodes of the interacting gears, which occurs when a gear pair is operating or when gear teeth are machined.

The elementary motions of the pinion tooth flank, $\mathcal{P}$, relative to the stationary gear tooth flank, $\mathcal{G}$, must be synchronized with one another to allow the common perpendicular to permanently pass through a point on the axis of instant rotation, P_{ln}—as illustrated in Fig. 8.8. Here, in Fig. 8.8, the section of the tooth flanks, $\mathcal{G}$ and $\mathcal{P}$, along a normal plane through the vector of linear velocity, $\mathbf{V}_\Sigma$, is shown.

The equation

$$|\,\mathbf{V}_\Sigma\,| = |\,\boldsymbol{\omega}_{p/g}\,| \cdot R_g \tag{8.4}$$

is composed on the basis of the analysis of the instantaneous kinematics of gearing [1–3, 5]. Here, $\boldsymbol{\omega}_{p/g}$ is the vector of the instant rotation of the pinion tooth flanks, $\mathcal{P}$, in relation to the motionless gear tooth flank, $\mathcal{G}$.

In Figs. 8.9, 8.10 and 8.11, several examples of interactions between the two local patches of tooth flanks, $\mathcal{G}$ and $\mathcal{P}$, are schematically illustrated. The Shishkov equation of contact, $\mathbf{n}_g \cdot \mathbf{V}_\Sigma = 0$, is fulfilled in all three cases, which are shown in Figs. 8.9, 8.10 and 8.11.

The tooth flanks, $\mathcal{G}$ and $\mathcal{P}$, make contact at point K. The radii of the curvature of the interacting tooth flanks at contact point K are equal to R_g and R_p, respectively (see Fig. 8.9). The centers of the curvature of the tooth profiles, $\mathcal{G}$ and $\mathcal{P}$, are denoted by o_g and o_p, respectively . In the instant motion of the pinion, $\mathcal{P}$, in relation to the gear, $\mathcal{G}$, the pinion performs an instant rotation, $\omega_{p/g}$, on point o_g. The radius of the

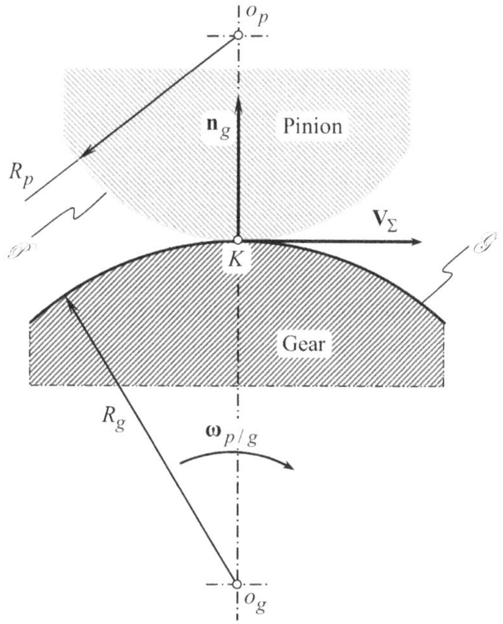

Fig. 8.8 Required synchronization of the elementary motions of the tooth flanks, $\mathcal{G}$ and $\mathcal{P}$, in rolling relative motion

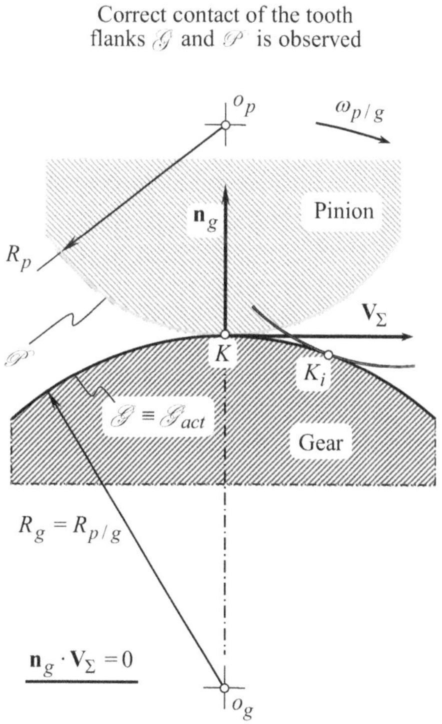

Fig. 8.9 The law of conjugacy is fulfilled

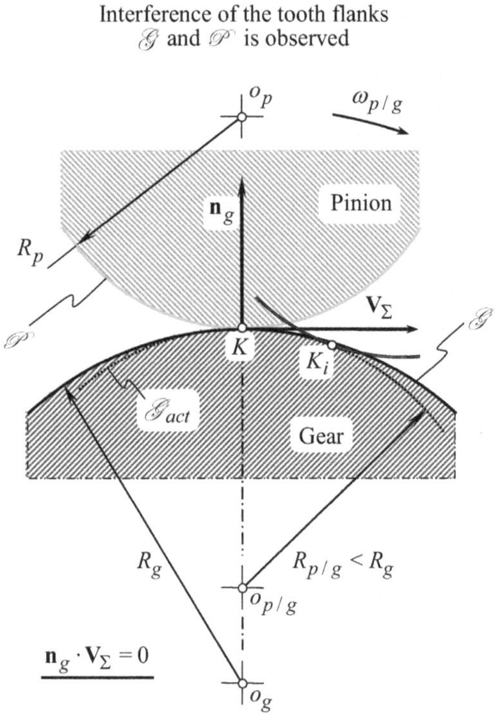

Fig. 8.10 The law of conjugacy is violated—interference between the tooth flanks is observed

curvature of the generated actual gear tooth flank, $\mathcal{G}_{act}$, is equal to $R_{p/g} \equiv -R_g$. In this scenario, the second fundamental law of gearing is fulfilled, and the actual tooth flank, $\mathcal{G}_{act}$, is identical to the desired gear tooth flank, $\mathcal{G}$, as shown in Fig. 8.9.

Note In Figs. 8.9, 8.10 and 8.11, a rotation, $\omega_{\mathbf{n}}$, of the pinion, $\mathcal{P}$, in relation to the gear, $\mathcal{G}$, on the contact perpendicular, $\mathbf{n}_g$, is not prohibited by the second fundamental law of gearing.

If the instantaneous rotation were performed either on the center $O_{p/g}$ when $R_{p/g} < R_g$(see Fig. 8.10) or on the center $o_{p/g}$ when $R_{p/g} > R_g$ (see Fig. 8.11), the second fundamental law of gearing would be violated and the actual tooth flank, $\mathcal{G}_{act}$, would differ from the desired gear tooth flank, $\mathcal{G}$.

Reminder The first fundamental law of gearing (as well as the Shishkov equation of contact, $\mathbf{n}_g \cdot \mathbf{V}_\Sigma = 0$) is fulfilled in all three cases, which are shown in Figs. 8.9, 8.10 and 8.11, whereas the second fundamental law of gearing is fulfilled only in the first case, which is illustrated in Fig. 8.11.

This discussion results in the following analytical description of the conjugate tooth flanks, $\mathcal{G}$ and $\mathcal{P}$: When two surfaces roll in relation to one another, then the envelope to a family of consecutive positions of the traveling surface is described by a set of three equations, as follows:

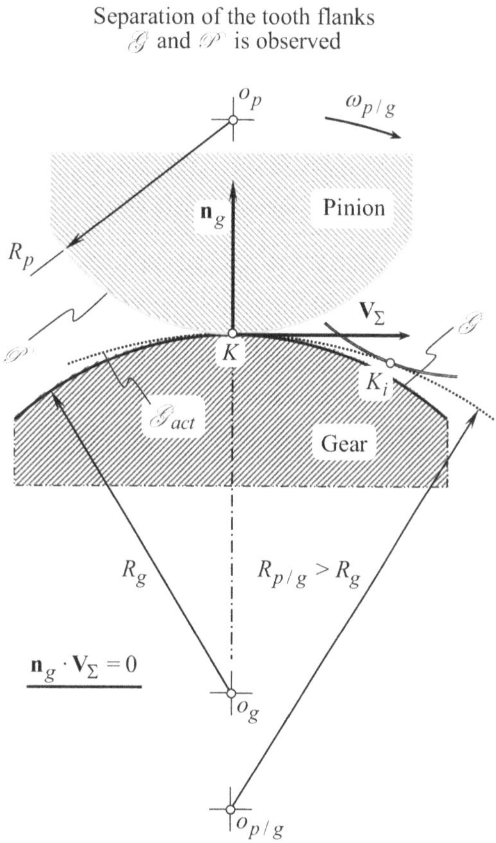

Fig. 8.11 The law of conjugacy is violated—separation of the tooth flanks is observed

$$\begin{cases} \mathbf{r}_g = \mathbf{r}_g \left(U_g,\ \ V_g,\ \ \theta \right) \\ \dfrac{\partial\, \mathbf{r}_g}{\partial \theta} \left(U_g,\ \ V_g,\ \ \theta \right) = 0 \\ R_{envelope} = C - R_{traveling} \end{cases} \tag{8.5}$$

where

$\mathbf{r}_g\,(U_g,\ \ V_g,\ \ \theta)$	– the position vector of a point of the gear tooth flank, $\mathcal{G}$
$U_g,\ V_g$	– the Gaussian parameters of the gear tooth flank, $\mathcal{G}$
$R_{envelope}$	– the radius of the curvature of the traveling gear tooth flank (algebraic value of the radius of curvature)
$R_{traveling}$	– the radius of curvature of the enveloping gear tooth flank (algebraic value of the radius of the curvature)
C	– the center distance

One can play with the enveloping parameter, θ. Only those "envelopes" that fulfill the requirement

$$R_{envelope} = C - R_{traveling} \tag{8.6}$$

are "real envelopes." R_e*−surfaces* [5] under any kind of relative motion of the surfaces (the condition specified in Eq. 8.5) must be fulfilled at all points of the desired line of contact, *LC*, of the tooth flanks, $\mathcal{G}$ and $\mathcal{P}$. The rest of the surfaces represent envelopes for the nonrolling relative motions. This analysis is performed in the plane through the axes of rotation, o_g and o_p. An equivalent analysis can be performed in the plane of action, *PA*—that is, the plane through the axes of instant rotation, o_g and o_p (the plane tangent to the base cylinders).

The schematic shown in Figs. 8.9, 8.10 and 8.11 is helpful for understanding the difference between the first and the second fundamental laws of gearing and prevents the drawing of incorrect conclusions in this context.

8.4.2 Generation of Envelopes in Swivel Relative Motion

The analysis in this subsection is similar to that on rolling relative motion because it can also be performed on the swivel relative motion of the tooth flanks, $\mathcal{G}$ and $\mathcal{P}$. Actually, the swivel motion, $\pm\varphi_n$, itself does not need to transmit a steady rotation from the driving shaft to a driven shaft.[1] However, motion of this nature, $\boldsymbol{\omega}_\mathbf{n}$, is observed, for example, in crossed-axes gearing, namely in C_a−gearing as a consequence of the complex relative motion of gears (see Fig. 8.12) [1–3, 5]. C_a*−gearing* is the most general kind of gearing with a constant angular velocity ratio.

For a specified gear pair, the vector of swivel relative motion, $\boldsymbol{\omega}_\mathbf{n}$, can be expressed in terms of the rotation, $\boldsymbol{\omega}_g$, of the gear; the rotation, $\boldsymbol{\omega}_p$, of the mating pinion; and the design parameters of the gear pair.

Crucially, although the equation of contact, $\mathbf{n}_g \cdot \mathbf{V}_\Sigma = 0$, was proposed by V.A. Shishkov in the mid-twentieth century, the nature of the condition of contact had already been properly understood by scholars in the time of Leonardo da Vinci [3] and even earlier.

Fulfillment of the first fundamental law of gearing is necessary but insufficient for smoothly transmitting a uniform rotary motion from a driving shaft to a driven shaft.

[1] In the mid-1980s, S.P. Radzevich introduced the permissible relative rotation, $\pm\varphi_n$, of a contact perpendicular, $\mathbf{n}_g$ [1–3, 5]. A permissible relative rotation, $\pm\varphi_n$, can be either intentionally introduced (e.g., into the kinematics of NC machining sculptured part surfaces, etc.), or it can be introduced into the kinematics of a real interaction with part surfaces (e.g., in C_a*−gearing*). In the second case, a rotation, $\pm\varphi_n$ (or $\boldsymbol{\omega}_n$), is just a component of the resultant relative motion of two tooth flanks, $\mathcal{G}$ and $\mathcal{P}$, that interact with one another. A vector of the resultant relative motion can be dissolved into a few components, and the rotation, $\pm\varphi_n$ (or $\boldsymbol{\omega}_n$) is one of these components.

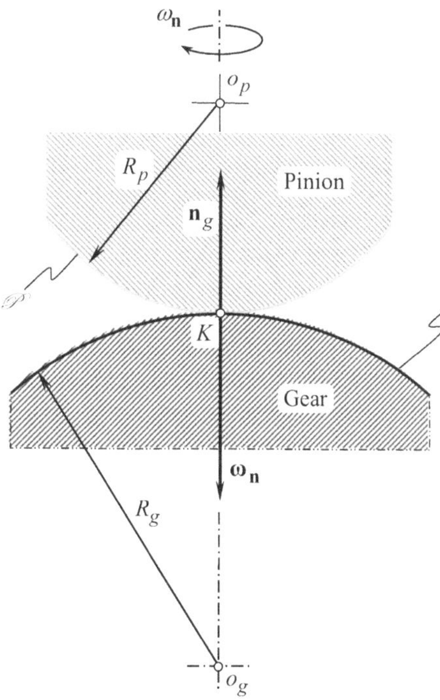

Fig. 8.12 Swivel relative motion of tooth flanks, $\mathcal{G}$ and $\mathcal{P}$, in crossed-axes gearing

Geometrically accurate gears of all designs must meet the first fundamental law of gearing.

Tooth flanks, $\mathcal{G}$ and $\mathcal{P}$, of a gear and a mating pinion that meet the requirements imposed by the first fundamental law of gearing are envelopes to one another. Envelope surfaces of this geometry are referred to as the envelope surfaces of the first kind[2] [5].

An illustrative example of the indicatrix of conformity $Cnf_R(\mathcal{G}/\mathcal{P})$ at the point of contact of a gear and mating pinion tooth flanks, $\mathcal{G}$ and $\mathcal{P}$, is shown in Fig. 8.13. This example corresponds to the cases featuring the contact of a hyperbolic local patch of the gear tooth flank, $\mathcal{G}$, and a convex elliptic-like local patch of the pinion tooth flank, $\mathcal{P}$. In this example (see Fig. 8.13), the corresponding curvature indicatrices $Crv(\mathcal{G})$ and $Crv(\mathcal{P})$ at the point of contact of the interacting tooth flanks, $\mathcal{G}$ and $\mathcal{P}$, are depicted as well. The imaginary (phantom) branches of Dupin's indicatrix

[2] Differentiating between envelope surfaces of the first kind and those of the second kind is necessary to generate an envelope surface to a family of surfaces, where the enveloping motion of a traveling surface must be independent and can be expressed solely in terms of the geometry of the surface to be generated. In the case of rolling/swivel motion, an enveloping motion is not independent. Therefore, the envelope surface can be generated not for any surface but rather only for a unique surface that is a conjugate to the envelope surface to be generated. Conjugate pairs of surfaces are always envelopes to one another, but not vice versa.

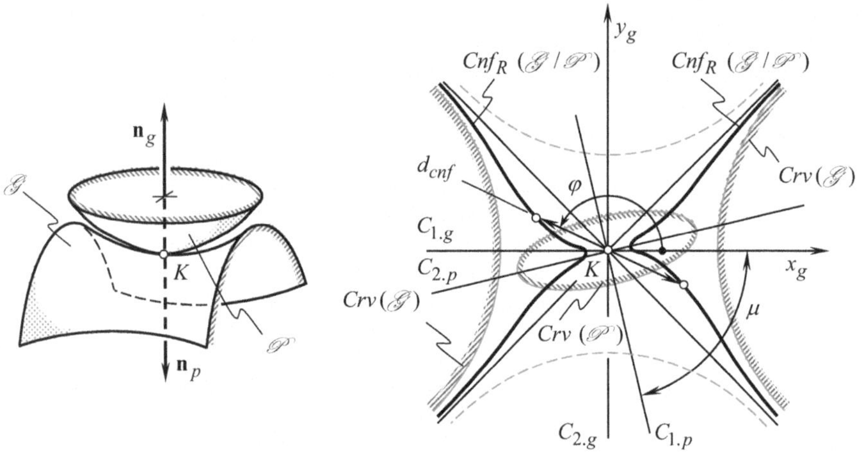

Fig. 8.13 An example of indicatrix of conformity, $Cnf_R(\mathcal{G}/\mathcal{P})$, at the point of contact of smooth regular tooth flanks $\mathcal{G}$ and $\mathcal{P}$ in the first order of tangency

$Dup(\mathcal{G})$ for the hyperbolic local patch of the part surface $\mathcal{G}$ are shown in dashed lines.

This discussion results in the following analytical description of the permissible swivel motion of tooth flanks, $\mathcal{G}$ and $\mathcal{P}$:

$$\begin{cases} \mathbf{r}_g = \mathbf{r}_g\left(U_g,\ V_g,\ \theta\right) \\ \dfrac{\partial \mathbf{r}_g}{\partial \theta}\left(U_g,\ V_g,\ \theta\right) = 0 \\ d_{cnf}^{\min} = 0 \end{cases} \tag{8.7}$$

where the minimum diameter of the indicatrix of conformity, $Cnf_R(\mathcal{G}/\mathcal{P})$, is designated as $d_{cnf}^{\min}$.

The current value of the diameter, d_{cnf}, of the indicatrix of conformity, $Cnf_R(\mathcal{G}/\mathcal{P})$, depends on the geometry of the interacting tooth flanks, $\mathcal{G}$ and $\mathcal{P}$, at the point of their contact, K, and on the actual value of the angle, μ, of the tooth flanks, $\mathcal{G}$ and $\mathcal{P}$, in the local relative orientation of the same contact point, K. The condition, $d_{cnf}^{\min} = 0$, is fulfilled in the linear and locally linear kinds of contact of the tooth flanks, $\mathcal{G}$ and $\mathcal{P}$ (in locally linear contact, two curves feature a common tangent and equal curvatures of an opposite sign).

This problem of generating envelopes in swivel relative motion has not yet been thoroughly investigated.

8.4.3 Generation of Envelopes in Combined Rolling/Swivel Relative Motion

When a combined rolling/sliding relative motion is used to generate envelopes, the rolling relative motion and the swivel relative motion of the gear tooth flanks need to be incorporated into the analysis. This problem has not yet been thoroughly investigated.

8.5 Conclusion

This chapter dealt with envelope surfaces used in gear design and gear-machining processes, specifically those utilizing the gear-generating principle either when rotations of two gears are in mesh or when rotations of a gear and a gear-cutting tool are in sync. The problem of determining envelope surfaces in gear design and gear-machining processes was correctly formulated and then solved by the author in 2008 [3]. The core of the proposed method of determining envelope surfaces in gear design and gear-machining process was laid out in the chapter. The surfaces considered in this chapter are commonly referred to as reversibly enveloping surfaces or as R_e*–surfaces*, for simplicity [5]. All three kinds of gearing—(a) parallel-axes gearing (P_a*–gearing*); (b) intersected-axes gearing (I_a*–gearing*); and (c) crossed-axes gearing (C_a*–gearing*)—were covered by the derived solution to the problem.

References

1. Radzevich, S. P. (1990). *Differential-geometric method of surface generation* (300p.), Doctoral Dissertation, Dneprodzerzhinsk, Dneprodzerzhinsk Industrial Institute.
2. Radzevich, S. P. (2014). *Generation of surfaces: Kinematic geometry of surface machining* (738 pages). CRC Press.
3. Radzevich, S. P. (2022). *Theory of gearing: Kinematics, geometry, and synthesis* (3rd ed., revised and expanded, 1208 pages). CRC Press.
4. Reuleaux, F. (1963). *The kinematics of machinery: Outlines of a theory of machines*, Translated and Edited by Alexander B. W. Kennedy, and with a New Introduction by Eugene S., Ferguson, Dover Publishers, Inc. (622 pages). [Translated from: Reuleaux, F., *Theoretische Kinematik: Grundziige einer Theori des Maschinenwesens*, 1875].
5. Radzevich, S. P. (2019). *Geometry of surfaces: A practical guide for mechanical engineers* (2nd ed.). Springer. © Springer Nature Switzerland AG (2020), XXVI, 329 pages, 182 illustrations.

Bibliography

Radzevich, S. P. (Ed.). (2019). *Advances in Gear Design and Manufacture* (549 pages). CRC Press.

Radzevich, S. P. (2018, February). An examination of high-conformal gearing. *Gear Solutions* (pp. 31–39).

Radzevich, S. P. (2019, February). Design features of perfect gears for crossed-axes gear pairs. *Gear Solutions Magazine* (pp. 36–43).

Radzevich, S. P. (Ed.). (2021). *Dudley's handbook of practical gear design and manufacture* (4th ed.). CRC Press. 1170 pages, 718 B/W Illustrations.

Radzevich, S. P. (2020). *High-conformal gearing: Kinematics and geometry* (2nd ed., 506 pages). Elsevier.

Radzevich, S. P. (2020, August). Knowledge (of gear theory) is power in the design, production, and application of gears. *Gear Solutions Magazine* (pp. 38–44). [A file in .pdf format with this article can be requested from the author (at no charge)]. A reprint of the article can be found out in: Radzevich, S.P., "Poor Understanding of the Scientific Theory of Gearing by the Majority of Gear Scientists and Engineers", Chapter 9, pages 193–214 in: Radzevich, S.P., *Novikov/Conformal Gearing: Scientific Theory and Practice* (526 pages), Springer, 2022.

Radzevich, S. P. (2022, November 15). *Novikov/conformal gearing: Scientific theory and practice*. Springer. 33+493 (526) pages.

Radzevich, S. P. (Ed.). (2021, June 25). *Recent advances in gearing: Scientific theory and applications* (1st ed., 2022 edition, 569 pages). Springer.

Radzevich, S. P. (2021, February). The commonalities and differences between helical "low-tooth-count gears" and "multiple-start worms". *Gear Solutions Magazine* (pp. 34-39).

Radzevich, S. P., & Storchak, M. G. (Eds.). (2022). *Advances in gear theory and gear cutting tool design* (500 pages). Springer.

Appendices

Appendix A: Elements of Vector Calculus

The *vector*, the key to all the theory of gearing, is a triple real number (in most computer languages these are usually called *floating point numbers*) and is noted in a **bold** typeface, that is, **A** or **a**.

Care must be taken to differentiate between two types of vectors:

- ***Position vector*:** A position vector runs from the origin of coordinate (0, 0, 0) to a point (X, Y, Z) and its length gives the distance of the point from the origin. Its components are given by (X, Y, Z). The essential concept to understand about a position vector is that it is anchored to specific coordinates (points in space). The set of points that are used to describe the shape of all part surfaces can be thought of as position vectors.
- ***Direction vector*:** A direction vector differs from a position vector in that it is ***not*** anchored to specific coordinates. Frequently, direction vectors are used in a form where they have unit length; in this case they are said to be *normalized*. The most common application of a direction vector in the theory of part surface generation is to specify the orientation of a surface or ray direction. For this, we use a direction vector at right angles (*normal*) and pointing away from the part surface. Such *normal* vectors are also the key in many calculations in the theory of gearing.

Vector calculus is a powerful tool for solving many of the geometrical and kinematical problems that pertain to the design and generation of gears and gearsets. In this book, ***vectors*** *are understood as quantities that have magnitude and direction and obey the law of addition.*

S. P. Radzevich (ed.), *Gear Accuracy*, Mechanical Engineering Series,
https://doi.org/10.1007/978-3-031-74798-4

A.1 Fundamental Properties of Vectors

The distance-and-direction interpretation suggests a powerful way to visualize a vector, and that is as a directed line segment or arrow. The length of the arrow (at some predetermined scale) represents magnitude of the vector, and the orientation of the segment and placement of the arrowhead (at one end of the segment or the other) represents its direction.

Vectors possess certain properties, the set of which is commonly interpreted as the set of fundamental properties of vectors.

A.1.1 Addition

Given two vectors **a** and **b**, their sum (**a** + **b**) is graphically defined by joining the tail of **b** to the head of **a**. Then, the line from the tail of **a** to the head of **b** is the sum **c** = (**a** + **b**).

A.1.2 Equality

Two vectors are equal when they have the same magnitude and direction. Position of the vectors is unimportant for equality.

A.1.3 Negation

The vector −**a** has the same magnitude as **a** but opposite direction.

A.1.4 Subtraction

From the properties of *addition* and *negation*, the following $\mathbf{a} - \mathbf{b} = \mathbf{a} + (-\mathbf{b})$ can be defined.

A.1.5 Scalar Multiplication

The vector $k\mathbf{a}$ has the same direction as **a**, with a magnitude k times that of **a**. k is called a scalar as it changes the scale of the vector **a**.

A.2 Mathematical Operations over Vectors

The following rules and mathematical operations can be determined from the above-listed fundamental properties of vectors.

A.2.1 Components of Vectors

Let's assume that a set of three vectors **a**, **b**, **c** and two scalars k and t are given. Then, vector addition and scalar multiplication have the following properties:

$$\mathbf{a}+\mathbf{b}=\mathbf{b}+\mathbf{a} \tag{A.1}$$

$$\mathbf{a}+(\mathbf{b}+\mathbf{c})=(\mathbf{a}+\mathbf{b})+\mathbf{c} \tag{A.2}$$

$$k(t\mathbf{a})=kt\mathbf{a} \tag{A.3}$$

$$(k+t)\mathbf{a}=k\mathbf{a}+t\mathbf{a} \tag{A.4}$$

$$k(\mathbf{a}+\mathbf{b})=k\mathbf{a}+k\mathbf{b} \tag{A.5}$$

Magnitude a of a vector **a** is:

$$a=|\mathbf{a}|=\sqrt{a_x^2+a_y^2+a_z^2} \tag{A.6}$$

where a_x, a_y, and a_z are the scalar components of **a**.

A unit vector $\overline{\mathbf{a}}$ in the direction of a vector **a** is:

$$\overline{\mathbf{a}}=\frac{\mathbf{a}}{|\mathbf{a}|}=\frac{\mathbf{a}}{a} \tag{A.7}$$

The components $\overline{a}_x$, $\overline{a}_y$, and $\overline{a}_z$ of a unit vector $\overline{\mathbf{a}}$ are also the direction cosines of the vector $\overline{\mathbf{a}}$:

$$\cos\alpha=\overline{a}_x \tag{A.8}$$

$$\cos\beta=\overline{a}_y \tag{A.9}$$

$$\cos\gamma=\overline{a}_z \tag{A.10}$$

It is common practice to denote the components $\overline{a}_x$, $\overline{a}_y$, and $\overline{a}_z$ by l, m and n accordingly.

A.2.2 Scalar Product (or Dot Product) of Two Vectors

The formula:

$$\mathbf{a} \cdot \mathbf{b} = a_x b_x + a_y b_y + a_z b_z = |\mathbf{a}||\mathbf{b}| \cos\angle(\mathbf{a}, \mathbf{b}) \tag{A.11}$$

is commonly used for calculation of scalar product of two vectors **a** and **b**.

Equation (A.11) can also be represented in the form:

$$\mathbf{a} \cdot \mathbf{b} = [\mathbf{a}]^T \cdot [\mathbf{b}] = \begin{bmatrix} a_x & a_y & a_z \end{bmatrix} \cdot \begin{bmatrix} b_x \\ b_y \\ b_z \end{bmatrix} \tag{A.12}$$

Angle $\angle(\mathbf{a}, \mathbf{b})$ between two vectors **a** and **b** is calculated from:

$$\angle(\mathbf{a}, \mathbf{b}) = \cos^{-1}\left(\frac{\mathbf{a} \cdot \mathbf{b}}{|\mathbf{a}||\mathbf{b}|}\right) \tag{A.13}$$

Scalar product of two vectors **a** and **b** features the following properties:

$$\mathbf{a} \cdot \mathbf{a} = |\mathbf{a}|^2 \tag{A.14}$$

$$\mathbf{a} \cdot \mathbf{b} = \mathbf{b} \cdot \mathbf{a} \tag{A.15}$$

$$\mathbf{a} \cdot (\mathbf{b} + \mathbf{c}) = \mathbf{b} \cdot \mathbf{a} + \mathbf{b} \cdot \mathbf{c} \tag{A.16}$$

$$(k\mathbf{a}) \cdot \mathbf{b} = \mathbf{a} \cdot (k\mathbf{b}) = k(\mathbf{a} \cdot \mathbf{b}) \tag{A.17}$$

If **a** is perpendicular to **b**, then:

$$\mathbf{a} \cdot \mathbf{b} = 0 \tag{A.18}$$

A.2.3 Vector Product (or Cross Product) of Two Vectors

Vector product of two vectors can be calculated from the formula:

$$\mathbf{a} \times \mathbf{b} = \left(a_y b_z - a_z b_y\right)\mathbf{i} + (a_z b_x - a_x b_z)\mathbf{j} + \left(a_x b_y - a_y b_x\right)\mathbf{k} \tag{A.19}$$

Here, in Eq. (A.19), **i**, **j**, and **k** are unit vectors in the *X*, *Y*, and *Z* directions of the reference system *XYZ*, in which the vectors **a** and **b** are specified.

Vector product posseses the following property: in case $\mathbf{a} \times \mathbf{b} = \mathbf{c}$, then the vector $\mathbf{c}$ is perpendicular to a plane through the vectors $\mathbf{a}$ and $\mathbf{b}$.

Vector product of two vectors $\mathbf{a}$ and $\mathbf{b}$ features the following properties:

$$\mathbf{a} \times \mathbf{b} = \begin{vmatrix} \mathbf{i} & \mathbf{j} & \mathbf{k} \\ a_x & a_y & a_z \\ b_x & b_y & b_z \end{vmatrix} \tag{A.20}$$

$$\mathbf{a} \times \mathbf{b} = |\mathbf{a}||\mathbf{b}|\,\mathbf{n} \sin\angle(\mathbf{a}, \mathbf{b}) \tag{A.21}$$

where unit normal vector to the plane through the vectors $\mathbf{a}$ and $\mathbf{b}$ is denoted by $\mathbf{n}$:

$$|\mathbf{a} \times \mathbf{b}| = |\mathbf{a}||\mathbf{b}| \sin\angle(\mathbf{a}, \mathbf{b}) \tag{A.22}$$

Coordinates of the vector product $\mathbf{a} \times \mathbf{b}$ can also be expressed in the form:

$$|\mathbf{a} \times \mathbf{b}| = \begin{bmatrix} 0 & -a_z & a_y \\ a_z & 0 & -a_x \\ -a_y & a_x & 0 \end{bmatrix} \cdot \begin{bmatrix} b_x \\ b_y \\ b_z \end{bmatrix} = \begin{bmatrix} -a_z b_y + a_y b_z \\ -a_x b_z + a_z b_x \\ -a_y b_x + a_x b_y \end{bmatrix} \tag{A.23}$$

$$\mathbf{a} \times \mathbf{b} = -\mathbf{b} \times \mathbf{a} \tag{A.24}$$

$$\mathbf{a} \times (\mathbf{b} + \mathbf{c}) = \mathbf{a} \times \mathbf{b} + \mathbf{a} \times \mathbf{c} \tag{A.25}$$

$$(k\mathbf{a}) \times \mathbf{b} = \mathbf{a} \times (k\mathbf{b}) = k(\mathbf{a} \times \mathbf{b}) \tag{A.26}$$

$$\mathbf{i} \times \mathbf{j} = \mathbf{k}, \mathbf{j} \times \mathbf{k} = \mathbf{i}, \mathbf{k} \times \mathbf{i} = \mathbf{j} \tag{A.27}$$

If $\mathbf{a}$ is parallel to $\mathbf{b}$, then:

$$\mathbf{a} \times \mathbf{b} = 0 \tag{A.28}$$

A.2.4 Triple Scalar Product of Three Vectors

The product $(\mathbf{a} \times \mathbf{b}) \cdot \mathbf{c}$ is commonly referred to as *triple scalar product* of three vectors $\mathbf{a}$, $\mathbf{b}$, and $\mathbf{c}$.

Triple scalar product of three vectors $\mathbf{a}$, $\mathbf{b}$, and $\mathbf{c}$ features the following properties:

$$(\mathbf{a} \times \mathbf{b}) \cdot \mathbf{c} = (\mathbf{b} \times \mathbf{c}) \cdot \mathbf{a} = (\mathbf{c} \times \mathbf{a}) \cdot \mathbf{b} \tag{A.29}$$

$$(\mathbf{b} \times \mathbf{c}) \cdot \mathbf{a} = \mathbf{a} \cdot (\mathbf{b} \times \mathbf{c}) \tag{A.30}$$

$$(\mathbf{a} \times \mathbf{b}) \cdot \mathbf{c} = \mathbf{a} \cdot (\mathbf{b} \times \mathbf{c}) \tag{A.31}$$

$$\mathbf{a} \cdot (\mathbf{b} \times \mathbf{c}) = \begin{vmatrix} a_x & a_y & a_z \\ b_x & b_y & b_z \\ c_x & c_y & c_z \end{vmatrix} \tag{A.32}$$

A.2.5 Triple Vector Product of Three Vectors

The product $(\mathbf{a} \times \mathbf{b}) \times \mathbf{c}$ is commonly referred to as *triple vector product* of three vectors **a**, **b**, and **c**.

The product $(\mathbf{a} \times \mathbf{b}) \times \mathbf{c}$ can be evaluated by two vector products. However, it also can be evaluated in a more simply way by use of the identity:

$$(\mathbf{a} \times \mathbf{b}) \times \mathbf{c} = (\mathbf{a} \cdot \mathbf{c})\mathbf{b} - (\mathbf{b} \cdot \mathbf{c})\mathbf{a} \tag{A.33}$$

It should be mentioned here that in general, the triple vector products $(\mathbf{a} \times \mathbf{b}) \times \mathbf{c}$ and $\mathbf{a} \times (\mathbf{b} \times \mathbf{c})$ are not equal:

$$(\mathbf{a} \times \mathbf{b}) \times \mathbf{c} \neq \mathbf{a} \times (\mathbf{b} \times \mathbf{c}) \tag{A.34}$$

Analytical interpretation of many problems and results in the field of gear science are getting much simpler when vector calculus is used.

A.2.6 Lagrange Equation for Vectors

For the purposes of calculation of mixed product of vectors, **a** and **b**, an equation:

$$(\mathbf{a} \times \mathbf{b}) \cdot (\mathbf{a} \times \mathbf{b}) = (\mathbf{a} \cdot \mathbf{a})(\mathbf{b} \cdot \mathbf{b}) - (\mathbf{a} \cdot \mathbf{b})^2 \tag{A.35}$$

is used.

Equation (A.35) is due to *Lagrange*[1].

[1] *Joseph-Louis* ***Lagrange*** (January 25, 1736 – April 10, 1813), a famous French mathematician, astronomer and mechanician.

A.3 On the Similarity and Difference Between Vectors and Matrices

A vector, **a**, is commonly represented in the form:

$$\mathbf{a} = \mathbf{i}a + \mathbf{j}b + \mathbf{k}c \tag{A.36}$$

That same vector, **a**, allows for matrix representation in one of the following forms:

$$\mathbf{a} = \begin{bmatrix} a \\ b \\ c \end{bmatrix} \tag{A.37}$$

$$\mathbf{a} = \begin{bmatrix} a \\ b \\ c \\ 1 \end{bmatrix} \tag{A.38}$$

$$\mathbf{a} = \begin{bmatrix} 1 & 0 & 0 & a \\ 0 & 1 & 0 & b \\ 0 & 0 & 1 & c \\ 0 & 0 & 0 & 1 \end{bmatrix} \tag{A.39}$$

Operation over vectors in matrix representation [see the form Eq. (A.39)] is preferred, as multiple coordinate system transformations are often required to be executed.

Vectors obey the commutative law, that is, the equalities, $\mathbf{a} \times \mathbf{b} = \mathbf{b} \times \mathbf{a}$, $\mathbf{a} \cdot \mathbf{b} = \mathbf{b} \cdot \mathbf{a}$, are valid for vectors. This is not always applicable to matrices.

Two kinds of products are valid for vectors, that is, dot product ($\mathbf{a} \cdot \mathbf{b}$), and cross product ($\mathbf{a} \times \mathbf{b}$) of vectors, that is not valid with respect to matrices.

Appendix B: Elements of Differential Geometry of Surfaces

Discussion in this book is focused primarily on the elements of theory of gearing.

The gear and the pinion tooth flanks and their motion in space in relation to one another are analytically described in a reference system. An orthogonal *Cartesian*[2]

[2] *René* ***Descartes*** (March 31, 1596 – February 11, 1650), (Latinized form: *Renatus* ***Cartesius***;), a French mathematician, philosopher, and writer.

reference system is a major kind of reference system that is commonly used for this purpose. Mutually perpendicular coordinate axes of a *Cartesian coordinate system* are conventionally labeled as X, Y, and Z.

In a *Cartesian reference system*, the axes can be oriented in either a left or right-handed sense. A right-handed *Cartesian reference system* is preferred, and all algorithms and formulae used in this book assume a right-handed convention.

A coordinate system provides a numerical frame of reference for the three-dimensional space in which the theory is developed. Two coordinate systems are particularly useful to us: the ubiquitous *Cartesian* (*XYZ*) *rectilinear system* and the spherical polar (r, θ, φ) or angular system. *Cartesian coordinate systems* are the most commonly used, but angular coordinates are often helpful as well.

B.1 Specification of a Gear Tooth Flank

A gear tooth flank can be uniquely determined by two independent variables. Therefore, we give a gear tooth flank $\mathcal{G}$ (see Fig. B.1), in most cases, by expressing

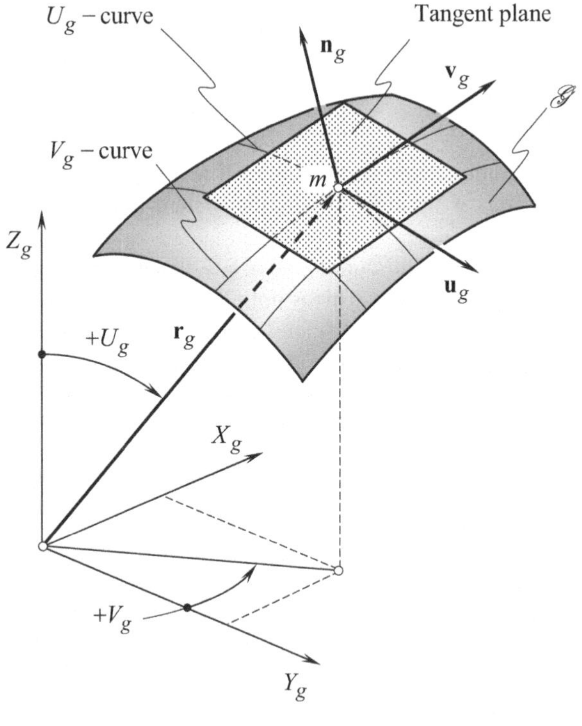

Fig. B.1 Principal parameters of local topology at point of gear tooth flank, $\mathcal{G}$

its rectangular coordinated X_g, Y_g, and Z_g, as functions of two *Gaussian*[3] *coordinates* U_g and V_g in a certain closed interval[4]:

$$\mathcal{G} \quad \Rightarrow \quad \mathbf{r}_g = \mathbf{r}_g(U_g,\ V_g) = \begin{bmatrix} X_g(U_g,\ V_g) \\ Y_g(U_g,\ V_g) \\ Z_g(U_g,\ V_g) \\ 1 \end{bmatrix} \tag{B.1}$$

$U_{1.\,g} \leq U_g \leq U_{2.\,g}$; $V_{1.\,g} \leq V_g \leq V_{2.\,g}$where:

$\mathbf{r}_g$ – is the position vector of point of the gear tooth flank, $\mathcal{G}$
U_g and V_g – are the curvilinear coordinates (*Gaussian coordinates*) of the gear tooth flank, $\mathcal{G}$
X_g, Y_g, Z_g – are the *Cartesian* coordinates of point of the gear tooth flank, $\mathcal{G}$
$U_{1.\,g}$, $U_{2.\,g}$ – are the boundary values of the closed interval of the U_g– parameter
$V_{1.\,g}$, $V_{2.\,g}$ – are the boundary values of the closed interval of the V_g– parameter

The parameters U_g and V_g must enter into Eq. (B.1) independently, which means that the matrix:

$$\mathbf{M} = \begin{bmatrix} \dfrac{\partial X_g}{\partial U_g} & \dfrac{\partial Y_g}{\partial U_g} & \dfrac{\partial Z_g}{\partial U_g} \\ \dfrac{\partial X_g}{\partial V_g} & \dfrac{\partial Y_g}{\partial V_g} & \dfrac{\partial Z_g}{\partial V_g} \end{bmatrix} \tag{B.2}$$

has a rank 2.

Positions, where the rank is 1 or 0 are singular points; when the rank at all points is 1, then Eq. (B.1) represents a curve.

Other methods of surface's specification are known as well. Specification of a gear tooth flank by

- an equation in explicit form
- an equation in implicit form, and
- a set of parametric equations

are among the most frequently used in practice methods of surfaces specification.

It is assumed here and below that any given kind of a gear tooth flank, $\mathcal{G}$, specification can be converted either into the vector form or into the matrix form of its specification as it is following from Eq. (B.1).

[3]*Johan Carl Friedrich* ***Gauss*** (April 30, 1777 – February 23, 1855) – a famous German mathematician and physical scientist.

[4]All the equations that are valid for the gear tooth flank, $\mathcal{G}$, are also valid for the pinion tooth flank, $\mathcal{P}$.

B.2 Tangent Vectors and Tangent Plane; Unit Normal Vector

The following notation is proven to be convenient in the consideration below.

The first derivatives of $\mathbf{r}_g$ with respect to *Gaussian coordinates* U_g and V_g are designated as:

$$\frac{\partial \mathbf{r}_g}{\partial U_g} = \mathbf{U}_g \tag{B.3}$$

$$\frac{\partial \mathbf{r}_P}{\partial V_P} = \mathbf{V}_P \tag{B.4}$$

and for the unit tangent vectors:

$$\mathbf{u}_g = \frac{\mathbf{U}_g}{|\, \mathbf{U}_g \,|} \tag{B.5}$$

$$\mathbf{v}_g = \frac{\mathbf{V}_g}{|\, \mathbf{V}_g \,|} \tag{B.6}$$

correspondingly[5].

Direction of the tangent line to the U_g−coordinate line through a given point m on the gear tooth flank, $\mathcal{G}$, is specified by the unit tangent vector $\mathbf{u}_g$ (as well as by the tangent vector $\mathbf{U}_g$). Similarly, direction of the tangent line to the V_g−coordinate line through that same point m on a gear tooth flank $\mathcal{G}$ is specified by the unit tangent vector $\mathbf{v}_g$ (as well as by the tangent vector $\mathbf{V}_g$).

Significance of the unit tangent vectors $\mathbf{u}_g$ and $\mathbf{v}_g$ becomes evident from the following considerations.

First, unit tangent vectors $\mathbf{u}_g$ and $\mathbf{v}_g$ yield an equation of the tangent plane to a gear tooth flank $\mathcal{G}$ at a specified point m:

$$\text{Tangent plane} \quad \Rightarrow \quad \begin{bmatrix} \left[\mathbf{r}_{t.p} - \mathbf{r}_g^m\right] \\ \mathbf{u}_g \\ \mathbf{v}_g \\ 1 \end{bmatrix} = 0 \tag{B.7}$$

where:

$\mathbf{r}_{t.\,p}$ – is the position vector of point of the tangent plane to a gear tooth flank $\mathcal{G}$ at a specified point m

$\mathbf{r}_g^m$ – is the position vector of point m on a gear tooth flank $\mathcal{G}$

[5] It is right point to underline here that the unit tangent vectors $\mathbf{u}_P$ and $\mathbf{v}_P$ are dimensionless values as it is following from Eqs. (B.5) and (B.6).

Second, tangent vectors yield an equation of the perpendicular $\mathbf{N}_g$, and of the unit normal vector $\mathbf{n}_g$ to a gear tooth flank $\mathcal{G}$ at a given point m:

$$\mathbf{N}_g = \mathbf{U}_g \times \mathbf{V}_g \tag{B.8}$$

and

$$\mathbf{n}_g = \frac{\mathbf{N}_g}{|\mathbf{N}_g|} = \frac{\mathbf{U}_g \times \mathbf{V}_g}{|\mathbf{U}_g \times \mathbf{V}_g|} = \mathbf{u}_g \times \mathbf{v}_g \tag{B.9}$$

When the order of the multipliers in Eqs. (B.8) and (B.9) is chosen properly, then the unit normal vector $\mathbf{n}_g$ (as well as the normal vector $\mathbf{N}_g$) is pointed outward of the bodily side of the surface $\mathcal{G}$.

B.3 Local Frame

Two unit tangent vectors $\mathbf{u}_g$ and $\mathbf{v}_g$ along with the unit normal vector $\mathbf{n}_g$ comprise a local frame $\mathbf{u}_g$, $\mathbf{v}_g$, $\mathbf{n}_g$ having the origin at current point m on a gear tooth flank $\mathcal{G}$. Unit tangent vector $\mathbf{u}_g$ is perpendicular to the unit normal vector $\mathbf{n}_g$ (that is, $\mathbf{u}_g \perp \mathbf{n}_g$), as well as unit tangent vector $\mathbf{v}_g$ is also perpendicular to the unit normal vector $\mathbf{n}_g$ (that is, $\mathbf{v}_g \perp \mathbf{n}_g$). Speaking generally, the unit tangent vectors $\mathbf{u}_g$ and $\mathbf{v}_g$ are not perpendicular to one another, they form a certain angle ω_g. In order to construct an orthogonal local frame, either the unit tangent vector $\mathbf{u}_g$ in the local frame ($\mathbf{u}_g$, $\mathbf{v}_g$, $\mathbf{n}_g$) must be substituted with a unit tangent vector $\mathbf{u}_g^*$ or the unit tangent vector $\mathbf{v}_g$ in that same local frame ($\mathbf{u}_g$, $\mathbf{v}_g$, $\mathbf{n}_g$) must be substituted with a unit tangent vector $\mathbf{v}_g^*$. For the calculation of the newly introduced unit tangent vectors $\mathbf{u}_g^*$ and $\mathbf{v}_g^*$ the following equations can be used:

$$\mathbf{u}_g^* = \mathbf{u}_g \times \mathbf{n}_g \tag{B.10}$$

$$\mathbf{v}_g^* = \mathbf{v}_g \times \mathbf{n}_g \tag{B.11}$$

It is convenient to choose that order of the multipliers in Eqs. (B.10) and (B.11), which preserves the orientation (the hand) of the original local frame ($\mathbf{u}_g$, $\mathbf{v}_g$, $\mathbf{n}_g$), namely, if the original local frame ($\mathbf{u}_g$, $\mathbf{v}_g$, $\mathbf{n}_g$) is right-hand-oriented, then the newly constructed local frame [either the local frame $\left(\mathbf{u}_g^*, \ \mathbf{v}_g, \ \mathbf{n}_g\right)$ or the local frame $\left(\mathbf{u}_g^*, \ \mathbf{v}_g, \ \mathbf{n}_g\right)$] should also be a right-hand-oriented local frame, and vice versa.

It should be pointed out here that another possibility to construct an orthogonal local frame is also available. Local frame of this kind is commonly referred to as *Darboux*[6] *frame*, and is briefly considered below in this section of the book.

Unit tangent vectors $\mathbf{u}_g$ and $\mathbf{v}_g$ to a surface $\mathcal{G}$ at point m are of critical importance when solving practical problems in the field of gearing. This statement is proven by numerous examples shown below.

B.4 Fundamental Forms of Surface

Consider two other important issues concerning the gear tooth flank geometry – both relate to intrinsic geometry in differential vicinity of surface point m.

B.4.1 First Fundamental Form of a Surface

The first issue is the so-called the *first fundamental form,* $\Phi_{1.\ g}$ of a gear tooth flank $\mathcal{G}$. The metric properties of a gear tooth flank $\mathcal{G}$ are described by the first fundamental form, $\Phi_{1.\ g}$, of the surface. Usually, the first fundamental form, $\Phi_{1.\ g}$, is represented as the quadratic form:

$$\Phi_{1.g} \quad \Rightarrow \quad ds_g^2 = E_g dU_g^2 + 2F_g dU_g dV_g + G_g dV_g^2 \tag{B.12}$$

Here, in Eq. (B.12) is designated:

s_g – is the linear element on a gear tooth flank $\mathcal{G}$ (s_g is equal to the length of a segment of a certain curve on a gear tooth flank $\mathcal{G}$)

E_g, F_g, G_g – are the fundamental magnitudes of the first order at a surface point

Equation (B.12) for the first fundamental form, $\Phi_{1.\ g}$, is known from many advanced sources. In the theory of gearing, another form of analytical representation of the first fundamental form, $\Phi_{1.\ g}$, is proven to be useful:

$$\Phi_{1.g} \quad \Rightarrow \quad ds_g^2 = [dU_g \quad dV_g \quad 0 \quad 0] \cdot \begin{bmatrix} E_g & F_g & 0 & 0 \\ F_g & G_g & 0 & 0 \\ 0 & 0 & 1 & 0 \\ 0 & 0 & 0 & 1 \end{bmatrix} \cdot \begin{bmatrix} dU_g \\ dV_g \\ 0 \\ 0 \end{bmatrix} \tag{B.13}$$

This kind of analytical representation of the first fundamental form $\Phi_{1.\ P}$ is proposed by Prof. *S.P. Radzevich* (~2008).

[6] *Jean Gaston* ***Darboux*** (August 14, 1842 – February 23, 1917), a French mathematician.

The practical advantage of Eq. (B.13) is that it can easily be incorporated into computer programs when multiple coordinate system transformations are used. The last is vital for the theory of gearing.

Fundamental magnitudes of the first order E_g, F_g, G_g, are calculated from the set of the following equations:

$$E_g = \mathbf{U}_g \cdot \mathbf{U}_g \tag{B.14}$$

$$F_g = \mathbf{U}_g \cdot \mathbf{V}_g \tag{B.15}$$

$$G_g = \mathbf{V}_g \cdot \mathbf{V}_g \tag{B.16}$$

Equations (B.14) through (B.16) can be represented in an expanded form:

$$E_g = \frac{\partial \mathbf{r}_g}{\partial U_g} \cdot \frac{\partial \mathbf{r}_g}{\partial U_g} = \frac{\partial X_g}{\partial U_g} \cdot \frac{\partial X_g}{\partial U_g} + \frac{\partial Y_g}{\partial U_g} \cdot \frac{\partial Y_g}{\partial U_g} + \frac{\partial Z_g}{\partial U_g} \cdot \frac{\partial Z_g}{\partial U_g} \tag{B.17}$$

$$F_g = \frac{\partial \mathbf{r}_g}{\partial U_g} \cdot \frac{\partial \mathbf{r}_g}{\partial V_g} = \frac{\partial X_g}{\partial U_g} \cdot \frac{\partial X_g}{\partial V_g} + \frac{\partial Y_g}{\partial U_g} \cdot \frac{\partial Y_g}{\partial V_g} + \frac{\partial Z_g}{\partial U_g} \cdot \frac{\partial Z_g}{\partial V_g} \tag{B.18}$$

$$G_g = \frac{\partial \mathbf{r}_g}{\partial V_g} \cdot \frac{\partial \mathbf{r}_g}{\partial V_g} = \frac{\partial X_g}{\partial V_g} \cdot \frac{\partial X_g}{\partial V_g} + \frac{\partial Y_g}{\partial V_g} \cdot \frac{\partial Y_g}{\partial V_g} + \frac{\partial Z_g}{\partial V_g} \cdot \frac{\partial Z_g}{\partial V_g} \tag{B.19}$$

Fundamental magnitudes of the first order, E_g, F_g, G_g, are functions of the U_g-, and V_g-coordinates of point of a gear tooth flank $\mathcal{G}$. In general form, these relationships can be represented in the form:

$$E_g = E_g(U_g, \ V_g) \tag{B.20}$$

$$F_g = F_g(U_g, \ V_g) \tag{B.21}$$

$$G_g = G_g(U_g, \ V_g) \tag{B.22}$$

It is important to point out here that fundamental magnitudes E_g and G_g are always positive (that is, $E_g > 0$, $G_g > 0$), and the fundamental magnitude F_g can be equal zero ($F_g \geq 0$). This results in that the first fundamental form, $\Phi_{1.\ g}$ at a point of a gear tooth flank $\mathcal{G}$, is always positively defined ($\Phi_{1.\ g} \geq 0$), and it cannot be of a negative value.

By use of the first fundamental form, $\Phi_{1.\ g}$, the following major parameters of geometry of a gear tooth flank $\mathcal{G}$ can be calculated:

(a) Length of a curve-line segment on a gear tooth flank $\mathcal{G}$
(b) Area of a portion of a gear tooth flank $\mathcal{G}$ that is bounded by a closed curve on the surface, and
(c) Angle between any two directions on a gear tooth flank $\mathcal{G}$.

Length, s_g, of a curve-line segment:

$$U_g = U_g(t) \tag{B.23}$$

$$V_g = V_g(t) \tag{B.24}$$

on a gear tooth flank, $\mathcal{G}$, is given by the equation:

$$s_g = \int_{t_0}^{t} \sqrt{E_g \left(\frac{dU_g}{dt}\right)^2 + 2F_g \frac{dU_g}{dt}\frac{dV_g}{dt} + G_g \left(\frac{dV_g}{dt}\right)^2}\, dt \tag{B.25}$$

$$t_0 \leq t \leq t_1$$

For calculation of square, $\mathcal{S}_g$, of a gear tooth flank $\mathcal{G}$ patch Σ, which is bounded by a closed curve on the surface $\mathcal{G}$, the following equation can be used:

$$\mathcal{S}_g = \iint_{\Sigma} \sqrt{E_g G_g - F_g^2}\, dU_g\, dV_g \tag{B.26}$$

Ultimately, value of the angle, ω_g, between two given directions through a certain point m on a gear tooth flank $\mathcal{G}$ can be calculated from one of the equations below:

$$\cos \omega_g = \frac{F_g}{\sqrt{E_g G_g}} \tag{B.27}$$

$$\sin \omega_g = \frac{H_g}{\sqrt{E_g G_g}} \tag{B.28}$$

$$\tan \omega_g = \frac{H_g}{F_g} \tag{B.29}$$

For the calculation of the discriminant, H_g, of the first fundamental form, $\Phi_{1.\ g}$, the following equation can be used:

$$H_g = \sqrt{E_g G_g - F_g^2} \tag{B.30}$$

It is assumed here that the discriminant, H_g, is always nonnegative, that is, $H_g = +\sqrt{E_g G_g - F_g^2}$.

The first fundamental form, $\Phi_{1.\ g}$, represents the length of a curve-line segment, and thus it is always of a nonnegative value, that is, the inequality $\Phi_{1.\ g} \geq 0$ is always valid.

The first fundamental form, $\Phi_{1.\ g}$, remains the same when the surface is banding. This is another important feature of the first fundamental form $\Phi_{1.\ g}$.

B.4.2 Second Fundamental Form of a Surface

The *second fundamental form*, $\Phi_{2.\ g}$ of a gear tooth flank $\mathcal{G}$ is another of the two above-mentioned issues. Curvature at point of a smooth regular surface, $\mathcal{G}$, is described by the second fundamental form $\Phi_{2.\ g}$.

Consider point K on a smooth regular part surface $\mathcal{G}$ (see Fig. B.2). Location of the point K is specified by two coordinates U_g and V_g. A line through the point K is entirely located within the surface $\mathcal{G}$. A nearby point m is located within the line through the point K. Location of the point m is specified by the coordinates $U_g + dU_g$ and $V_g + dV_g$ as it is infinitesimally close to the point K. The closest distance of approach of the point m to the tangent plane through the point K is expressed by the second fundamental form $\Phi_{2.\ g}$. Torsion of the curve Km is ignored. Therefore, the distance "a" is assumed equal to zero ($a = 0$).

The curvature at point of a smooth, regular part surface, $\mathcal{G}$, is described by the second fundamental form, $\Phi_{2.\ g}$. Usually, it is represented as the quadratic form (see Fig. B.2):

$$\Phi_{2.g} \quad \Rightarrow \quad -d\mathbf{r}_g \cdot d\mathbf{n}_g = L_g\, dU_g^2 + 2M_g\, dU_g\, dV_g + N_g\, dV_g^2 \tag{B.31}$$

Equation (B.31) is known from many advanced sources.

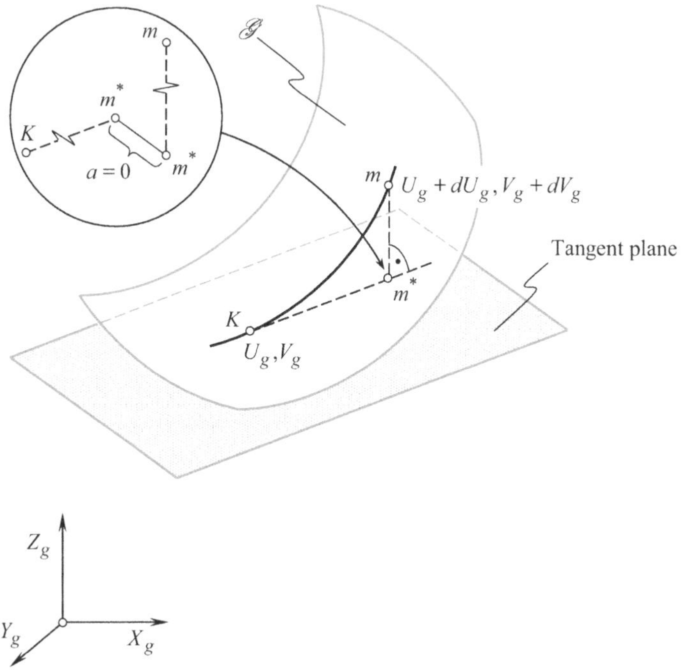

Fig. B.2 On the definition of the second fundamental form, $\Phi_{2.\ g}$, at point of smooth gear tooth flank, $\mathcal{G}$

In the theory of gearing, another analytical representation of the second fundamental form, $\Phi_{2.\ g}$, is proven to be useful:

$$\Phi_{2.g} \quad \Rightarrow \quad \begin{bmatrix} dU_g & dV_g & 0 & 0 \end{bmatrix} \cdot \begin{bmatrix} L_g & M_g & 0 & 0 \\ M_g & N_g & 0 & 0 \\ 0 & 0 & 1 & 0 \\ 0 & 0 & 0 & 1 \end{bmatrix} \cdot \begin{bmatrix} dU_g \\ dV_g \\ 0 \\ 0 \end{bmatrix} \tag{B.32}$$

This analytical representation of the second fundamental form, $\Phi_{2.\ P}$, is proposed by Prof. *S.P. Radzevich* (~2008).

Similar to Eq. (B.13), the practical advantage of Eq. (B.32) is that it can easily be incorporated into computer codes when multiple coordinate system transformations are used. The last is vital for the theory of gearing.

In Eq. (B.32), the parameters L_g, M_g, N_g designate fundamental magnitudes of the second order.

By definition, fundamental magnitudes of the second order are equal:

$$L_g = -\mathbf{U}_g \cdot \frac{\partial \mathbf{n}_g}{\partial U_g} \qquad = \mathbf{n}_g \cdot \frac{\partial \mathbf{U}_g}{\partial U_g} \tag{B.33}$$

$$M_g = \quad -\frac{1}{2}\left(\mathbf{U}_g \cdot \frac{\partial \mathbf{n}_g}{\partial V_g} + \mathbf{V}_g \cdot \frac{\partial \mathbf{n}_g}{\partial U_g}\right) \quad = \mathbf{n}_g \cdot \frac{\partial \mathbf{U}_g}{\partial V_g} = \mathbf{n}_g \cdot \frac{\partial \mathbf{V}_g}{\partial U_g} \tag{B.34}$$

$$N_g = -\mathbf{V}_g \cdot \frac{\partial \mathbf{n}_g}{\partial V_g} \qquad = \mathbf{n}_g \cdot \frac{\partial \mathbf{V}_g}{\partial V_g} \tag{B.35}$$

For the calculation of the fundamental magnitudes of the second order of a smooth regular gear tooth flank $\mathcal{G}$ the following equations are: used:

$$L_g = \frac{\frac{\partial \mathbf{U}_g}{\partial U_g} \times \mathbf{U}_g \cdot \mathbf{V}_g}{\sqrt{E_g G_g - F_g^2}} \tag{B.36}$$

$$M_g = \frac{\frac{\partial \mathbf{U}_g}{\partial V_g} \times \mathbf{U}_g \cdot \mathbf{V}_g}{\sqrt{E_g G_g - F_g^2}} = \frac{\frac{\partial \mathbf{V}_g}{\partial U_g} \times \mathbf{U}_g \cdot \mathbf{V}_g}{\sqrt{E_g G_g - F_g^2}} \tag{B.37}$$

$$N_g = \frac{\frac{\partial \mathbf{V}_g}{\partial V_g} \times \mathbf{U}_g \cdot \mathbf{V}_g}{\sqrt{E_g G_g - F_g^2}} \tag{B.38}$$

Equations (B.36) through (B.38) can be represented in expanded form:

$$L_g = \frac{\begin{vmatrix} \frac{\partial^2 X_g}{\partial U_g^2} & \frac{\partial^2 Y_g}{\partial U_g^2} & \frac{\partial^2 Z_g}{\partial U_g^2} \\ \frac{\partial X_g}{\partial U_g} & \frac{\partial Y_g}{\partial U_g} & \frac{\partial Z_g}{\partial U_g} \\ \frac{\partial X_g}{\partial V_g} & \frac{\partial Y_g}{\partial V_g} & \frac{\partial Z_g}{\partial V_g} \end{vmatrix}}{\sqrt{E_g G_g - F_g^2}} \tag{B.39}$$

$$M_g = \frac{\begin{vmatrix} \frac{\partial^2 X_g}{\partial U_g \partial V_g} & \frac{\partial^2 Y_g}{\partial U_g \partial V_g} & \frac{\partial^2 Z_g}{\partial U_g \partial V_g} \\ \frac{\partial X_g}{\partial U_g} & \frac{\partial Y_g}{\partial U_g} & \frac{\partial Z_g}{\partial U_g} \\ \frac{\partial X_g}{\partial V_g} & \frac{\partial Y_g}{\partial V_g} & \frac{\partial Z_g}{\partial V_g} \end{vmatrix}}{\sqrt{E_g G_g - F_g^2}} \tag{B.40}$$

$$N_P = \frac{\begin{vmatrix} \frac{\partial^2 X_g}{\partial V_g^2} & \frac{\partial^2 Y_g}{\partial V_g^2} & \frac{\partial^2 Z_g}{\partial V_g^2} \\ \frac{\partial X_g}{\partial U_g} & \frac{\partial Y_g}{\partial U_g} & \frac{\partial Z_g}{\partial U_g} \\ \frac{\partial X_g}{\partial V_g} & \frac{\partial Y_g}{\partial V_g} & \frac{\partial Z_g}{\partial V_g} \end{vmatrix}}{\sqrt{E_g G_g - F_g^2}} \tag{B.41}$$

Fundamental magnitudes of the second order, L_g, M_g, N_g, are also functions of the U_g-, and V_g-coordinates of point of a gear tooth flank $\mathcal{G}$. In general form, these relationships can be represented in the form:

$$L_g = L_g(U_g, \ V_g) \tag{B.42}$$

$$M_g = M_g(U_g, \ V_g) \tag{B.43}$$

$$N_g = N_g(U_g, \ V_g) \tag{B.44}$$

The discriminant, T_g, of the second fundamental form, $\Phi_{2.\ g}$, can be calculated from the following equation:

$$T_g = \sqrt{L_g N_g - M_g^2} \tag{B.45}$$

We now come to the theorem, which is essential justification for considering the differential geometry of surfaces in connection with the six fundamental magnitudes. It has been proven (1867) first by *Bonnet*[7], and may be enunciated as follows:

Theorem B.1 *When six fundamental magnitudes* E_g, F_g, G_g *and* L_g, M_g, N_g *are given, and when they fulfill the Gauss characteristic equation, and the two Mainardi*[8]*-Codazzi*[9] *relations, the fundamental magnitudes determine a gear tooth flank,* $\mathcal{G}$, *uniquely say as to its position and orientation in space.*

This theorem is commonly referred to as the *main theorem in the theory of surface*, or simply as *Bonnet theorem.* According to the main theorem of the theory of surfaces, that have the first and second fundamental forms identical must be either congruent or symmetrical to one another.

By use of six fundamental magnitudes, all the parameters of local geometry of a given part surface can be calculated.

B.5 Principal Directions at Point of Gear Tooth Flank

The direction of vectors, $\mathbf{T}_{1.\ g}$ and $\mathbf{T}_{2.\ g}$, at a point on a gear tooth flank $\mathcal{G}$, can be specified in terms of the ratio dU_g/dV_g. For the vectors of the first, $\mathbf{T}_{1.\ g}$, and for the second, $\mathbf{T}_{2.\ g}$, principal directions at point m of a smooth, regular surface $\mathcal{G}$, the corresponding values of the ratio dU_g/dV_g are calculated as roots of the quadratic equation:

$$\begin{vmatrix} E_g\,dU_g + F_g\,dV_g & F_g\,dU_g + G_g\,dV_g \\ L_g\,dU_g + M_g\,dV_g & M_g\,dU_g + N_g\,dV_g \end{vmatrix} = 0 \tag{B.46}$$

The first principal plane section, $C_{1.\ g}$, is perpendicular to a gear tooth flank $\mathcal{G}$ at current surface point m, and passes through the vector of the first principal direction $\mathbf{T}_{1.\ g}$. The second principal plane section, $C_{2.\ g}$, is orthogonal to a gear tooth flank $\mathcal{G}$ at current surface point m, and passes through the vector of the second principal direction $\mathbf{T}_{2.\ g}$.

The principal directions $\mathbf{T}_{1.\ g}$ and $\mathbf{T}_{2.\ g}$ can be identified at any and all points of a smooth, regular gear tooth flank $\mathcal{G}$ except of umbilic points, and in flatten points of the surface. At umbilic points of a surface, as well as at flatten points, principal directions cannot be identified.

[7] *Pierre Ossian* ***Bonnet*** (December 22, 1819 – June 22, 1892), a French mathematician.

[8] *Gaspare* ***Mainardi*** (June 27, 1800 – March 9, 1879) – an Italian mathematician.

[9] *Delfino* ***Codazzi*** (March 7, 1824 – July 21, 1873) – an Italian mathematician.

In the theory of gearing, it is often preferred to use not the vectors $\mathbf{T}_{1.\,g}$ and $\mathbf{T}_{2.\,g}$ of the principal directions, but, instead, to use the unit vectors $\mathbf{t}_{1.\,g}$ and $\mathbf{t}_{2.\,g}$ of the principal directions. The unit tangent vectors $\mathbf{t}_{1.\,g}$ and $\mathbf{t}_{2.\,g}$ are calculated from the equations:

$$\mathbf{t}_{1.g} = \frac{\mathbf{T}_{1.g}}{|\mathbf{T}_{1.g}|} \tag{B.47}$$

$$\mathbf{t}_{2.g} = \frac{\mathbf{T}_{2.g}}{|\mathbf{T}_{2.g}|} \tag{B.48}$$

correspondingly.

Unit tangent vectors $\mathbf{t}_{1.\,g}$ and $\mathbf{t}_{2.\,g}$ of principal directions at point m on a gear tooth flank $\mathcal{G}$ along with unit normal vector $\mathbf{n}_g$ at that same point m comprise an orthogonal local frame ($\mathbf{t}_{1.\,g}$, $\mathbf{t}_{2.\,g}$, $\mathbf{n}_g$). All three unit vectors, namely, the vectors: $\mathbf{t}_{1.\,g}$, $\mathbf{t}_{2.\,g}$, and $\mathbf{n}_g$, are mutually perpendicular to one another. The local frame ($\mathbf{t}_{1.\,g}$, $\mathbf{t}_{2.\,g}$, $\mathbf{n}_g$) is commonly referred to as *Darboux frame*.

B.6 Curvatures at Point of Gear Tooth Flank

The first, $R_{1.\,g}$, and the second, $R_{2.\,g}$, principal radii of curvature at point of a gear tooth flank $\mathcal{G}$ are measured within the first and in the second principal plane sections, $C_{1.\,g}$ and $C_{2.\,g}$, accordingly. For the calculation of values of the principal radii of curvature, the following equation is commonly used:

$$R_g^2 - \frac{E_g N_g - 2F_g M_g + G_g L_g}{T_g} R_g + \frac{H_g}{T_g} = 0 \tag{B.49}$$

Remember that algebraic values of the radii of principal curvature, $R_{1.\,g}$ and $R_{2.\,g}$, relate to one another as $R_{2.\,g} > R_{1.\,g}$. In particular cases, at umbilic points on a gear tooth flank $\mathcal{G}$, no principal curvatures can be identified as all normal curvatures of the tooth surface, $\mathcal{G}$, at an umbilic point are equal to one another.

Another two important parameters of local topology at point of a gear tooth flank $\mathcal{G}$ are: Mean curvature, $\mathcal{M}_g$, and Intrinsic curvature (*Gaussian curvature* or *full curvature*), $\mathcal{G}_g$.

For the calculation of the curvatures $\mathcal{M}_g$ and $\mathcal{G}_g$, the following equations are: commonly used:

$$\mathcal{M}_g = \frac{k_{1.g} + k_{2.g}}{2} = \frac{E_g N_g - 2F_g M_g + G_g L_g}{2 \cdot \left(E_g G_g - F_g^2\right)} \tag{B.50}$$

$$\mathcal{G}_g = k_{1.g} \cdot k_{2.g} = \frac{L_g N_g - M_g^2}{E_g G_g - F_g^2} \tag{B.51}$$

The expressions for the mean curvature $\mathcal{M}_g$, and for the *Gaussian curvature,* $\mathcal{G}_g$, at point of tooth flank, $\mathcal{G}$:

$$\mathcal{M}_g = \frac{k_{1.g} + k_{2.g}}{2} \tag{B.52}$$

$$\mathcal{G}_g = k_{1.g} \cdot k_{2.g} \tag{B.53}$$

considered together yield a quadratic equation with respect to principal curvatures $k_{1.\ g}$ and $k_{2.\ g}$:

$$k_g^2 - 2\mathcal{M}_g k_g + \mathcal{G}_g = 0 \tag{B.54}$$

The following formulae

$$k_{1.g} = \mathcal{M}_g + \sqrt{\mathcal{M}_g^2 - \mathcal{G}_g} \tag{B.55}$$

$$k_{2.g} = \mathcal{M}_g - \sqrt{\mathcal{M}_g^2 - \mathcal{G}_g} \tag{B.56}$$

are the solutions to Eq. (B.54).

Here, in Eqs. (B.55) and (B.56), the first principal curvature of a gear tooth flank $\mathcal{G}$ at a current point m is designated as $k_{1.\ g}$, and $k_{2.\ g}$ designates the second principal curvature of a gear tooth flank $\mathcal{G}$ at that same point m.

The principal curvatures $k_{1.\ g}$, and $k_{2.\ g}$ are the reciprocals to the corresponding principal radii of curvature $R_{1.\ g}$, and $R_{2.\ g}$:

$$k_{1.g} = \frac{1}{R_{1.g}} \tag{B.57}$$

$$k_{2.g} = \frac{1}{R_{2.g}} \tag{B.58}$$

The first principal curvature, $k_{1.\ g}$, is always larger than the second principal curvature, $k_{2.\ g}$, of a gear tooth flank $\mathcal{G}$ at current point m, i.e., the inequality:

$$k_{1.g} > k_{2.g} \tag{B.59}$$

is always valid.

B.7 Brioschi *Formula for* Gauss *Curvature*

Gauss curvature at surface, $\mathcal{G}$, point m can be expressed in terms of just three fundamental magnitudes E_g, F_g, G_g, of the first order, and their derivatives of the first, and of the second order, that is, with no application of three fundamental magnitudes L_g, M_g, N_g, of the second order. For this purpose, *Brioschi*[10] formula for *Gauss* curvature can be used:

$$\mathcal{G}_g = \frac{\begin{vmatrix} -\frac{1}{2}\frac{\partial^2 E_g}{\partial V_g^2} + \frac{\partial^2 F_g}{\partial U_g \partial V_g} - \frac{1}{2}\frac{\partial^2 G_g}{\partial U_g \partial V_g} & \frac{1}{2}\frac{\partial E_g}{\partial U_g} & \frac{\partial F_g}{\partial U_g} - \frac{1}{2}\frac{\partial E_g}{\partial V_g} \\ \frac{\partial F_g}{\partial V_g} - \frac{1}{2}\frac{\partial G_g}{\partial U_g} & E_g & F_g \\ \frac{1}{2}\frac{\partial G_g}{\partial V_g} & F_g & G_g \end{vmatrix} - \begin{vmatrix} 0 & \frac{1}{2}\frac{\partial E_g}{\partial V_g} & \frac{1}{2}\frac{\partial G_g}{\partial U_g} \\ \frac{1}{2}\frac{\partial E_g}{\partial V_g} & E_g & F_g \\ \frac{1}{2}\frac{\partial G_g}{\partial U_g} & F_g & G_g \end{vmatrix}}{\left(E_g G_g - F_g^2\right)} \tag{B.60}$$

For an orthogonally parameterized surface (when the equality $F_g = 0$ is valid), *Gaussian* curvature is:

$$\mathcal{G}_g = -\frac{1}{2\sqrt{E_g G_g}}\left(\frac{\partial}{\partial U_g}\frac{\frac{\partial G_g}{\partial U_g}}{\sqrt{E_g G_g}} + \frac{\partial}{\partial V_g}\frac{\frac{\partial E_g}{\partial V_g}}{\sqrt{E_g G_g}}\right) \tag{B.61}$$

The *Brioschi* formula gives *Gaussian* curvature solely in terms of the first fundamental form, as illustrated by Eq. (B.60) and Eq. (B.61).

Other formats of representation of the *Brioschi* formula for *Gauss* curvature are also known.

B.8 Natural Form of Surface Representation

This brief consideration of major elements of part surface geometry makes possible the introduction of two definitions that are of critical importance for further discussion.

As it is already mentioned earlier in this section of the book, it is proven by *Bonnet* that a unique surface is specified by the first and the second fundamental forms if the *Gauss' characteristic equation* and the *Mainardi-Codazzi relations of compatibility* are satisfied, and those two surfaces that have identical first and second

[10] *Francesco* ***Brioschi*** (22 December, 1824 – 13 December, 1897) was an Italian mathematician.

fundamental forms are congruent[11]. Six fundamental magnitudes determine a surface uniquely, except as to position and orientation in space.

Specification of a surface in terms of the first and the second fundamental forms is usually called the *natural kind of surfaces representation*. In general form, this kind of part surfaces representation can be expressed by a set of two equations:

$$\left.\begin{array}{r}\text{Natural form}\\ \text{of surface } \mathcal{G} \text{ representation}\end{array}\right| \Rightarrow \mathcal{G} = \mathcal{G}(\Phi_{1.g}, \Phi_{2.g}) \begin{cases} \Phi_{1.g} = \Phi_{1.g}(E_g, F_g, G_g) \\ \Phi_{2.g} = \Phi_{2.g}(E_g, F_g, G_g, L_g, M_g, N_g) \end{cases} \tag{B.62}$$

Equation (B.62) can be derived from Eq. (B.1). A given gear tooth flank $\mathcal{G}$ can be expressed in both forms, namely, either by Eq. (B.19) or by Eq. (B.1).

B.9 A Few More Useful Equations

Many calculations of the parameters of geometry can be significantly simplified by use of the first and of the second fundamental forms of a smooth, regular part surface $\mathcal{G}$.

1. For the calculation of the actual value of radius, R_g, of normal curvature within a normal plane section through given point m on a gear tooth flank $\mathcal{G}$ and at a given direction, the following equation can be used:

$$R_g = \frac{\Phi_{1.g}}{\Phi_{2.g}} \tag{B.63}$$

2. *Euler formula* for the calculation of normal curvature, $k_{\theta.\ g}$, at point m in a direction that is specified by the angle, θ, can be represented as follows:

$$k_{\theta.g} = k_{1.g} \cos^2\theta + k_{2.g} \sin^2\theta \tag{B.64}$$

Here, in Eq. (B.64), θ is the angle that the normal plane section, C_g, makes with the first principal plane section, $C_{1.\ g}$. In other words, $\theta = \angle(\mathbf{t}_g, \mathbf{t}_{1.\ g})$; here $\mathbf{t}_g$ designates the unit tangent vector within the normal plane section C_g.

[11]Two surfaces with the identical first and second fundamental forms might also by symmetrical. Refer to the literature – Koenderink, J.J., *Solid Shape*, The MIT Press, Cambridge, MA, 1990, 699 pages. – on differential geometry of surfaces for details about this specific issue.

Equation (B.64) also is a perfect illustration of significant simplification of the calculations when fundamental magnitudes, E_g, F_g, G_g, of the first and L_g, M_g, N_g of the second order are used.

In order to get a profound understanding of differential geometry of surfaces, the interested reader may wish to go to advanced monographs in the field. Systematic discussion of the topic is available from many sources. The author would like to turn the reader's attention to the books by *doCarmo*, *Eisenhart*, *Stuik*, and others.

B.10 Illustrative Example

Consider an example of how analytical representation of a surface in a *Cartesian reference system* can be converted into the natural representation of that same surface.

A *Cartesian coordinate system* $X_gY_gZ_g$ is associated with a gear tooth flank, $\mathcal{G}$, as it is schematically shown in Fig. B.3.

Position vector of point, $\mathbf{r}_g$, of the gear tooth flank $\mathcal{G}$ can be represented as summa of three vectors:

$$\mathbf{r}_g = \mathbf{A} + \mathbf{B} + \mathbf{C} \tag{B.65}$$

Each of the vectors **A**, **B**, and **C** can be expressed in terms of projections onto the axes of the reference system$X_gY_gZ_g$. Then, Eq. (B.65) casts into the equation:

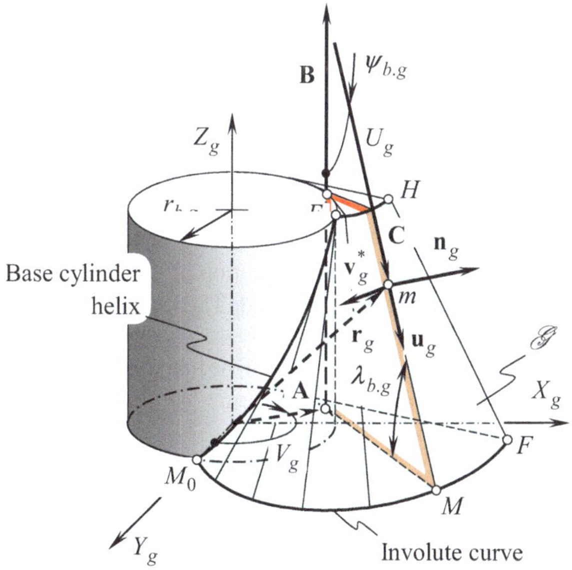

Fig. B.3 Derivation of the natural form of representation of gear tooth flank, *M*

$$\mathbf{r}_g(U_g, V_g) = \begin{bmatrix} r_{b.g} \cos V_g + U_g \cos \tau_{b.g} \sin V_g \\ r_{b.g} \sin V_g - U_g \sin \tau_{b.g} \sin V_g \\ r_{b.g} \tan \tau_{b.g} - U_g \sin \tau_{b.g} \\ 1 \end{bmatrix} \tag{B.66}$$

This yields the calculation of two tangent vectors $\mathbf{U}_g(U_g, V_g)$ and $\mathbf{V}_g(U_g, V_g)$, which are correspondingly equal:

$$\mathbf{U}_g(U_g, V_g) = \left[\cos\ \tau_{b.g} \sin V_g - \cos \tau_{b.g} \cos V_g - \sin \tau_{b.g} 0\right] \tag{B.67}$$

$$\mathbf{V}_g(U_g, V_g) = \begin{bmatrix} -r_{b.g} \sin V_g + U_g \cos \tau_{b.g} \cos V_g \\ r_{b.g} \cos V_g + U_g \cos \tau_{b.g} \sin V_g \\ r_{b.g} \tan \tau_{b.g} \\ 0 \end{bmatrix} \tag{B.68}$$

Substituting the derived vectors $\mathbf{U}_g$ and $\mathbf{V}_g$ into Eq. (B.14), one can come up with formulae for the calculation of the fundamental magnitudes of the first order:

$$E_g = 1 \tag{B.69}$$

$$F_g = -\frac{r_{b.g}}{\cos \tau_{b.g}} \tag{B.70}$$

$$G_g = \frac{U_g^2 \cos^4 \tau_{b.g} + r_{b.g}^2}{\cos^2 \tau_{b.g}} \tag{B.71}$$

These expressions can be substituted directly to Eq. (B.12) for the first fundamental form $\Phi_{1.\ g}$ of the gear tooth flank, $\mathcal{G}$:

$$\Phi_{1.g} \quad \Rightarrow \quad dU_g^2 - 2\frac{r_{b.g}}{\cos \tau_{b.g}} dU_g dV_g + \frac{U_g^2 \cos^4 \tau_{b.g} + r_{b.g}^2}{\cos^2 \tau_{b.g}} dV_g^2 \tag{B.72}$$

The derived expressions for the fundamental magnitudes E_g, F_g, and G_g [see Eqs. (B.69) through (B.71)] can also be substituted to Eq. (B.13). In this way a corresponding matrix representation of the first fundamental form $\Phi_{1.\ g}$ of the gear tooth flank, $\mathcal{G}$, can be calculated. The interested reader may wish to complete these formulae on his or her own.

The discriminant, H_g, of the first fundamental form of the gear tooth flank, $\mathcal{G}$, can be calculated from the expression:

$$H_g = U_g \cos \tau_{b.g} \tag{B.73}$$

In order to derive an equation for the second fundamental form, $\Phi_{2.\ g}$, at point of the gear tooth flank, $\mathcal{G}$, the second derivatives of the position vector of a point,

$\mathbf{r}_g(U_g, V_g)$, with respect to U_g- and V_g-parameters are necessary. The above-derived equations for the tangent vectors $\mathbf{U}_g$ and $\mathbf{V}_g$ [see Eqs. (B.67) and (B.68)] make it possible the following expressions for the derivatives under consideration:

$$\frac{\partial \mathbf{U}_g}{\partial U_P} = \begin{bmatrix} 0 \\ 0 \\ 0 \\ 1 \end{bmatrix} \tag{B.74}$$

$$\frac{\partial \mathbf{U}_g}{\partial V_g} \equiv \frac{\partial \mathbf{V}_g}{\partial U_g} = \left[\cos\tau_{b.g}\cos V_g \cos\tau_{b.g}\sin V_g 0 1\right] \tag{B.75}$$

$$\frac{\partial \mathbf{V}_g}{\partial V_g} = \begin{bmatrix} -r_{b.g}\cos V_g - U_g\cos\tau_{b.g}\sin V_g \\ -r_{b.g}\sin V_g + U_g\cos\tau_{b.g}\cos V_g \\ 0 \\ 1 \end{bmatrix} \tag{B.76}$$

Further, substitute these expressions [see Eqs. (B.74) through (B.76)] into Eqs. (B.36) through (B.38). After the necessary formulae transformations are complete, then Eqs. (B.36) through (B.38) cast into the set of formulae for the calculation of the fundamental magnitudes of the second order of the gear tooth flank, $\mathcal{G}$. This set of formulae is as follows:

$$L_g = 0 \tag{B.77}$$

$$M_g = 0 \tag{B.78}$$

$$N_g = -U_g \sin\tau_{b.g}\cos\tau_{b.g} \tag{B.79}$$

Further, after substituting Eqs. (B.77) through (B.79) into Eq. (B.31), an equation for the calculation of the second fundamental form of the gear tooth flank, $\mathcal{G}$, can be represented in the form:

$$\Phi_{2.g} \quad \Rightarrow \quad -d\mathbf{r}_g \cdot d\mathbf{N}_g = -U_g \sin\tau_{b.g}\cos\tau_{b.g} dV_g^2 \tag{B.80}$$

Similar to Eq. (B.72), the derived expressions for the fundamental magnitudes L_g, M_g, and N_g of the second order can be substituted into Eq. (B.32) for the second fundamental form $\Phi_{2.\ g}$. In this way a corresponding matrix representation of the second fundamental form, $\Phi_{2.\ g}$, of the surface $\mathcal{G}$ can be derived. The interested reader may wish to complete this formulae transformation on his or her own.

For the calculation of the discriminant, T_g, of the second fundamental form, $\Phi_{2.\ g}$, of the gear tooth flank, $\mathcal{G}$, the following expression is used:

Table B.1 Fundamental magnitudes of the first and the second order at point of involute gear tooth flank, $\mathcal{G}$

$E_g = 1$	$L_g = 0$
$F_g = -\frac{r_{b.g}}{\cos \tau_{b.g}}$	$M_g = 0$
$G_g = \frac{U_g^2 \cos^4 \tau_{b.g} + r_{b.g}^2}{\cos^2 \tau_{b.g}}$	$N_g = -U_g \sin \tau_{b.\ g} \cos \tau_{b.\ g}$

$$T_g = U_g \sin \tau_{b.g} \cos \tau_{b.g} \tag{B.81}$$

The natural representation of the gear tooth flank, $\mathcal{G}$, can be expressed in terms of the derived set of six equations for the calculation of the fundamental magnitudes of the first E_g, F_g, G_g and of the second L_g, M_g, and N_g (see Table B.1).

All major elements of local geometry of the gear tooth flank, $\mathcal{G}$, can be calculated based on the fundamental magnitudes, E_g, F_g, G_g, of the first, $\Phi_{1.\ P}$, and L_g, M_g, N_g, of the second $\Phi_{2.\ g}$, fundamental forms. Location and orientation of the gear tooth flank, $\mathcal{G}$, are the two parameters that remain indefinite.

Once a part surface is represented in natural form, that is, it is expressed in terms of six fundamental magnitudes of the first and of the second order, then further calculation of parameters of a gear tooth flank, $\mathcal{G}$, becomes significantly easier.

Appendix C: Change of Surface Parameters

When designing gears, it is often necessary to treat two or more surfaces simultaneously. For example, contact geometry at common point of two tooth flanks, $\mathcal{G}$ and $\mathcal{P}$, can be described analytically if initially the tooth flanks are parameterized properly.

When two surfaces $\mathbf{r}_i$ and $\mathbf{r}_j$ are necessarily been treated simultaneously, then it is required that they are not only represented in a common reference system, but the U_i- and V_i-parameters of one of the surfaces $\mathbf{r}_i = \mathbf{r}_i(U_i,\ V_i)$ have to be synchronized with the corresponding U_j- and V_j-parameters of another surface $\mathbf{r}_j = \mathbf{r}_j(U_j,\ V_j)$. The procedure of changing of surface parameters is used for this purpose. Use of the procedure allows representation of one of the surfaces, for example of the surface $\mathbf{r}_j = \mathbf{r}_j(U_j,\ V_j)$ in the terms of the U_i- and V_i-parameters, say as $\mathbf{r}_j = \mathbf{r}_j(U_i,\ V_i)$.

If the parameterization of a surface is transformed by the equations $U^* = U^*(U,\ V)$ and $V^* = V^*(U,\ V)$, we obtain the new derivatives:

$$\frac{\partial \mathbf{r}}{\partial U^*} = \frac{\partial \mathbf{r}}{\partial U} \cdot \frac{\partial U}{\partial U^*} + \frac{\partial \mathbf{r}}{\partial V} \cdot \frac{\partial V}{\partial U^*} \tag{C.1}$$

$$\frac{\partial \mathbf{r}}{\partial V^*} = \frac{\partial \mathbf{r}}{\partial U} \cdot \frac{\partial U}{\partial V^*} + \frac{\partial \mathbf{r}}{\partial V} \cdot \frac{\partial V}{\partial V^*} \tag{C.2}$$

so that:

$$\mathbf{A}^* = \left[\frac{\partial \mathbf{r}}{\partial U^*} \middle| \frac{\partial \mathbf{r}}{\partial V^*} \right] = \mathbf{A} \cdot \mathbf{J} \tag{C.3}$$

where:

$$\mathbf{J} = \begin{bmatrix} \frac{\partial U}{\partial U^*} & \frac{\partial U}{\partial V^*} \\ \frac{\partial V}{\partial U^*} & \frac{\partial V}{\partial V^*} \end{bmatrix} \tag{C.4}$$

is called the *Jacobian matrix* of the transformation.

It can be shown that the new fundamental matrix $\mathbf{G}^*$ is given by:

$$\mathbf{G}^* = \mathbf{A}^{*T}\mathbf{A}^* = \mathbf{J}^T\mathbf{A}^T\mathbf{A}\mathbf{J} = \mathbf{J}^T\mathbf{G}\mathbf{J} \tag{C.5}$$

From this equation we see by the properties of determinants that $|\mathbf{G}^*| = |\mathbf{J}|^2 |\mathbf{G}|$. Using this result, and Eqs. (C.1) and (C.2), we can show that the unit surface normal $\mathbf{n}$ is invariant under the transformation, as we could expect.

The transformation of the second fundamental matrix can similarly be shown to be given by:

$$\mathbf{D}^* = \mathbf{J}^T\mathbf{D}\mathbf{J} \tag{C.6}$$

by differentiating Eqs. (C.1) and (C.2) and using the invariance of $\mathbf{n}$. From Eqs. (C.5) and (C.6), it can be shown that the principal curvatures and directions are invariant under the transformation.

We conclude that the unit normal vector $\mathbf{n}$ and the principal directions and curvatures are independent of the parameters used, and are therefore geometric properties of the surface itself. They should be continuous if the surface is to be tangent and curvature continuous.

Appendix D: Applied Coordinate Systems and Linear Transformations

Consequent coordinate systems transformations can be easily described analytically by means of matrices. The use of matrices for the coordinate system transformation[12] can be traced back to the mid of 1940s[13] when Dr. *S.S. Mozhayev*[14] began describing of coordinate system transformations by means of matrices.

Below, coordinate systems transformation is briefly discussed from the standpoint of its implementation in the theory of gearing.

D.1 Coordinate System Transformation

Homogeneous coordinates utilize a mathematical trick to embed three-dimensional coordinates and transformations into a four-dimensional matrix format. As a result, inversions or combinations of linear transformations are simplified to inversions or multiplication of the corresponding matrices.

D.1.1 Homogeneous Coordinate Vectors

Instead of representing each point $\mathbf{r}(x,\ y,\ z)$ in three-dimensional space with a single three-dimensional vector,

$$\mathbf{r} = \begin{bmatrix} x \\ y \\ z \end{bmatrix} \tag{D.1}$$

homogeneous coordinates allow each point $\mathbf{r}(x,\ y,\ z)$ to be represented by any of an infinite number of four-dimensional vectors:

[12] Matrices were introduced into mathematics by *A. Cayley* in 1857. They provide a compact and flexible notation particularly useful in dealing with linear transformations, and they present an organized method for the solution of systems of linear differential equations.

[13] Application of matrices for the purposes of analytical representation of coordinate system transformation should be credited to Dr. *S.S. Mozhayev* [Mozhayev, S.S., *General Theory of Cutting Tools*, Doctoral Thesis, Leningrad, Leningrad Polytechnic Institute, 1951, 295 pages]. Dr. *S.S. Mozhayev* began using matrices for this purpose in the mid of 1940s. Later on, matrix approach for coordinate system transformation has been used by *Denavit* & *Hartenberg*, as well as by many other researchers.

[14] ***S.S. Mozhayev*** – is a soviet scientist known mainly for his accomplishments in the theory of cutting tool design.

$$\mathbf{r} = \begin{bmatrix} T \cdot x \\ T \cdot y \\ T \cdot z \\ T \end{bmatrix} \tag{D.2}$$

The three-dimensional vector corresponding to any four-dimensional vector can be calculated by dividing the first three elements by the fourth, and a four-dimensional vector corresponding to any three-dimensional vector can be created by simply adding a fourth element and setting it equal to one.

D.1.2 Homogeneous Coordinate Transformation Matrices of the Dimension 4 × 4

Homogeneous coordinate transformation matrices operate on four-dimensional homogeneous vector representations of traditional three-dimensional coordinate locations. Any three-dimensional linear transformation (translation, rotation, and so forth) can be represented by a 4 × 4 homogeneous coordinate transformation matrix. In fact, because of the redundant representation of three-space in a homogeneous coordinate system, an infinite number of different 4 × 4 homogeneous coordinate transformation matrices are available to perform any given linear transformation. This redundancy can be eliminated to provide a unique representation by dividing all elements of a 4 × 4 homogeneous transformation matrix by the last element (which will become equal to one). This means that a 4 × 4 homogeneous transformation matrix can incorporate as many as 15 independent parameters. The generic format representation of a homogeneous transformation equation for mapping the three-dimensional coordinate $(x_1, \ y_1, \ z_1)$ to the three-dimensional coordinate $(x_2, \ y_2, \ z_2)$ is:

$$\begin{bmatrix} T^* \cdot x_2 \\ T^* \cdot y_2 \\ T^* \cdot z_2 \\ T^* \end{bmatrix} = \begin{bmatrix} T^* \cdot a & T^* \cdot b & T^* \cdot c & T^* \cdot d \\ T^* \cdot e & T^* \cdot f & T^* \cdot g & T^* \cdot h \\ T^* \cdot i & T^* \cdot j & T^* \cdot k & T^* \cdot m \\ T^* \cdot n & T^* \cdot p & T^* \cdot q & T^* \end{bmatrix} \cdot \begin{bmatrix} T \cdot x_2 \\ T \cdot y_2 \\ T \cdot z_2 \\ T \end{bmatrix} \tag{D.3}$$

If any two matrices or vectors of this equation are: known, the third matrix (or vector) can be calculated, and then the redundant T element in the solution can be eliminated by dividing all elements of the matrix by the last element.

Various transformation models can be used to constraint the form of the matrix to transformations with fewer degrees of freedom.

D.1.3 Translations

Translation of a coordinate system is one of the major linear transformations used in the theory of gearing. Translations of the coordinate system $X_2Y_2Z_2$ along axes of the coordinate system $X_1Y_1Z_1$ are depicted in Fig. D.1. Translations can be analytically described by the homogeneous transformation matrix of dimension 4 × 4.

For an analytical description of translation along coordinate axes, the operators of translation $\mathbf{Tr}\,(a_x,\ X)$, $\mathbf{Tr}\,(a_y,\ Y)$, and $\mathbf{Tr}\,(a_z,\ Z)$ are used. These operators yield matrix representation in the form:

$$\mathbf{Tr}\,(a_x, X) = \begin{bmatrix} 1 & 0 & 0 & a_x \\ 0 & 1 & 0 & 0 \\ 0 & 0 & 1 & 0 \\ 0 & 0 & 0 & 1 \end{bmatrix} \tag{D.4}$$

$$\mathbf{Tr}\,\left(a_y, Y\right) = \begin{bmatrix} 1 & 0 & 0 & 0 \\ 0 & 1 & 0 & a_y \\ 0 & 0 & 1 & 0 \\ 0 & 0 & 0 & 1 \end{bmatrix} \tag{D.5}$$

$$\mathbf{Tr}\,(a_z, Z) = \begin{bmatrix} 1 & 0 & 0 & 0 \\ 0 & 1 & 0 & 0 \\ 0 & 0 & 1 & a_z \\ 0 & 0 & 0 & 1 \end{bmatrix} \tag{D.6}$$

Here, in Eq. (D.4) through Eq. (D.6), the parameters a_x, a_y, and a_z are signed values that denote the distance of translation along the corresponding axis.

Consider two coordinate systems, $X_1Y_1Z_1$ and $X_2Y_2Z_2$, displaced along the X_1- axis at a distance a_x as schematically depicted in Fig. D.1a. Point m in the reference system $X_2Y_2Z_2$ is given by the position vector $\mathbf{r}_2(m)$. In the coordinate system,

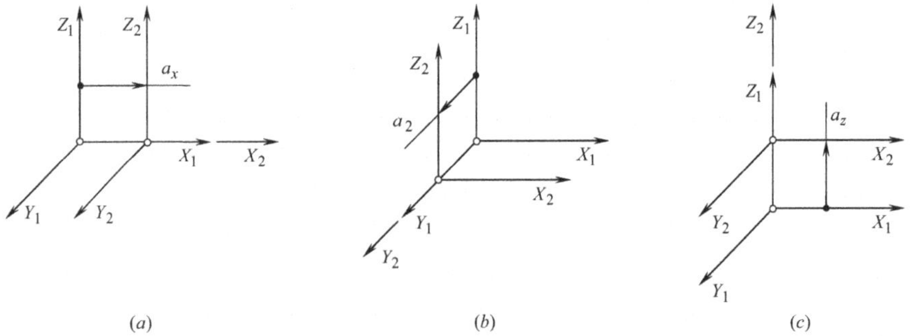

Fig. D.1 Analytical description of the operators of translations $\mathbf{Tr}\,(a_x,\ X)$, $\mathbf{Tr}\,(a_y,\ Y)$, $\mathbf{Tr}\,(a_z,\ Z)$ along the coordinate axes of a *Cartesian reference system XYZ*

$X_1Y_1Z_1$, that same point m can be specified by the position vector $\mathbf{r}_1(m)$. Then the position vector $\mathbf{r}_1(m)$ can be expressed in terms of the position vector $\mathbf{r}_2(m)$ by the equation:

$$\mathbf{r}_1(m) = \mathbf{Tr}(a_x, X) \cdot \mathbf{r}_2(m) \tag{D.7}$$

Equations similar to Eq. (D.7) are valid for the operators $\mathbf{Tr}(a_y, \ Y)$, and $\mathbf{Tr}(a_z, \ Z)$ of the coordinate system transformation. The latter is schematically illustrated in Fig. D.1b, c.

Use of the operators of translation $\mathbf{Tr}(a_x, \ X)$, $\mathbf{Tr}(a_y, \ Y)$, and $\mathbf{Tr}(a_z, \ Z)$ makes possible the introduction of an operator $\mathbf{Tr}(a, \ \mathbf{A})$ of a combined transformation. Suppose that point, p, on a rigid body goes through a translation describing a straight line from a point p_1 to a point p_2 with a change of coordinates of $(a_x, \ a_y, \ a_z)$. This motion of point, p, is analytically described with a resultant translation operator $\mathbf{Tr}(a, \ \mathbf{A})$:

$$\mathbf{Tr}(a, \ \mathbf{A}) = \begin{bmatrix} 1 & 0 & 0 & a_x \\ 0 & 1 & 0 & a_y \\ 0 & 0 & 1 & a_z \\ 0 & 0 & 0 & 1 \end{bmatrix} \tag{D.8}$$

The operator $\mathbf{Tr}(a, \ \mathbf{A})$ of the resultant coordinate system transformation can be interpreted as the operator of translation along an arbitrary axis having the vector $\mathbf{A}$ as the direct vector.

An analytical description of translation of the coordinate system $X_1Y_1Z_1$ in direction of an arbitrary vector $\mathbf{A}$ to the position of $X_2Y_2Z_2$ can be composed from Fig. D.2. The operator of translation $\mathbf{Tr}(a, \mathbf{A})$ of that particular kind can be

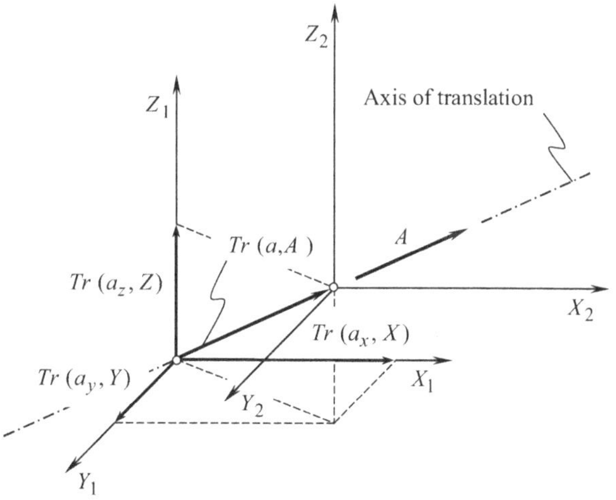

Fig. D.2 Analytical description of an operator, $\mathbf{Tr}(a, \mathbf{A})$, of translation along an arbitrary axis (vector $\mathbf{A}$ is the direction vector of the axis)

expressed in terms of the operators $\mathbf{Tr}(a_x, X)$, $\mathbf{Tr}(a_y, Y)$, and $\mathbf{Tr}(a_z, Z)$ of elementary translations:

$$\mathbf{Tr}(a, \mathbf{A}) = \mathbf{Tr}(a_z, Z) \cdot \mathbf{Tr}(a_y, Y) \cdot \mathbf{Tr}(a_x, X) \tag{D.9}$$

Evidently, the axis along the vector **A** is always the axis through the origins of both the reference systems $X_1Y_1Z_1$ and $X_2Y_2Z_2$.

Any and all coordinate system transformations that do not change the orientation of a geometrical object are referred to as *orientation-preserving transformation*, or *direct transformation*. Therefore, transformation of translation is an example of a direct transformation.

D.1.4 Rotation About a Coordinate Axis

Rotation of a coordinate system about a coordinate axis is another major linear transformation used in the theory of part surface generation. A rotation is specified by an axis of rotation and the angle of the rotation. It is a fairly simple trigonometric calculation to obtain a transformation matrix for a rotation about one of the coordinate axes.

Possible rotations of the coordinate system $X_2Y_2Z_2$ about the axis of the coordinate system $X_1Y_1Z_1$ are illustrated in Fig. D.3.

For analytical description of rotation about a coordinate axis, the operators of rotation $\mathbf{Rt}(\varphi_x, X_1)$, $\mathbf{Rt}(\varphi_y, Y_1)$, and $\mathbf{Rt}(\varphi_z, Z_1)$ are used. These operators of linear transformations yield representation in the form of homogeneous matrices:

$$\mathbf{Rt}(\varphi_x, X_1) = \begin{bmatrix} 1 & 0 & 0 & 0 \\ 0 & \cos & & \varphi_x \sin \varphi_x 00 - \sin \varphi_x \cos \varphi_x 00001 \end{bmatrix} \tag{D.10}$$

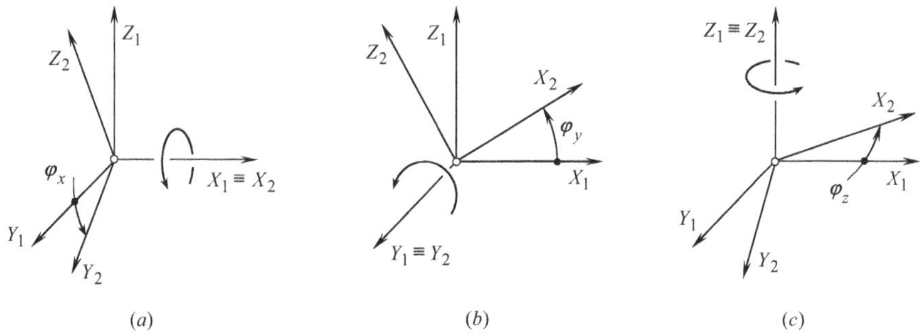

Fig. D.3 Analytical description of the operators of rotation $\mathbf{Rt}(\varphi_x, X)$, $\mathbf{Rt}(\varphi_y, Y)$, and $\mathbf{Rt}(\varphi_z, Z)$ about a coordinate axis of a reference system $X_1Y_1Z_1$

$$\mathbf{Rt}\left(\varphi_y, Y_1\right) = \left[\cos\ \varphi_y 0 \sin \varphi_y 00100 - \ \sin \varphi_y 0 \cos \varphi_y 00001\right] \tag{D.11}$$

$$\mathbf{Rt}\left(\varphi_z, Z_1\right) = \left[\cos\ \varphi_z \sin \varphi_z 00 - \ \sin \varphi_z \cos \varphi_z 0000100001\right] \tag{D.12}$$

Here φ_x, φ_y, and φ_z are signed values that denote the corresponding angles of rotations about a corresponding coordinate axis: φ_x is the angle of rotation around the X_1-axis (pitch) of the *Cartesian* coordinate system $X_1Y_1Z_1$; φ_y is the angle of rotation around the Y_1-axis (roll), and φ_z is the angle of rotation around the Z_1-axis (yaw) of that same *Cartesian* reference system $X_1Y_1Z_1$.

Rotation about a coordinate axis is illustrated in Fig. D.3.

Consider two coordinate systems $X_1Y_1Z_1$ and $X_2Y_2Z_2$, which are turned about X_1-axis through an angle φ_x as shown in Fig. D.3a. In the reference system $X_2Y_2Z_2$, point m is given by a position vector $\mathbf{r}_2(m)$. In the coordinate system $X_1Y_1Z_1$, that same point m can be specified by the position vector $\mathbf{r}_1(m)$. Then, the position vector $\mathbf{r}_1(m)$ can be expressed in terms of the position vector $\mathbf{r}_2(m)$ by the equation:

$$\mathbf{r}_1(m) = \mathbf{Rt}(\varphi_x, X) \cdot \mathbf{r}_2(m) \tag{D.13}$$

Equations those similar to that above Eq. (D.13) are also valid for other operators $\mathbf{Rt}(\varphi_y, Y)$ and $\mathbf{Rt}(\varphi_z, Z)$ of the coordinate system transformation. These elementary coordinate system transformations are schematically illustrated in Fig. D.3b, c accordingly.

D.1.5 Rotation About an Arbitrary Axis Through the Origin

When a rotation is to be performed around an arbitrary vector based at the origin, the transformation matrix must be assembled from a combination of rotations about the *Cartesian* coordinate.

Two different approaches for analytical description of a rotation about an arbitrary axis through the origin are discussed below.

D.1.5.1 Conventional Approach

Analytical description of rotation of the coordinate system $X_1Y_1Z_1$ about an arbitrary axis through the origin to the position of a reference system $X_2Y_2Z_2$ is illustrated in Fig. D.4. It is assumed here that the rotation is performed about the axis having a vector $\mathbf{A}_0$ as the direction vector. The operator $\mathbf{Rt}(\varphi_A,\ \mathbf{A}_0)$ of rotation of that kind can be expressed in terms of the operators $\mathbf{Rt}(\varphi_x, X)$, $\mathbf{Rt}(\varphi_y, Y)$, and $\mathbf{Rt}(\varphi_z, Z)$ of elementary rotations:

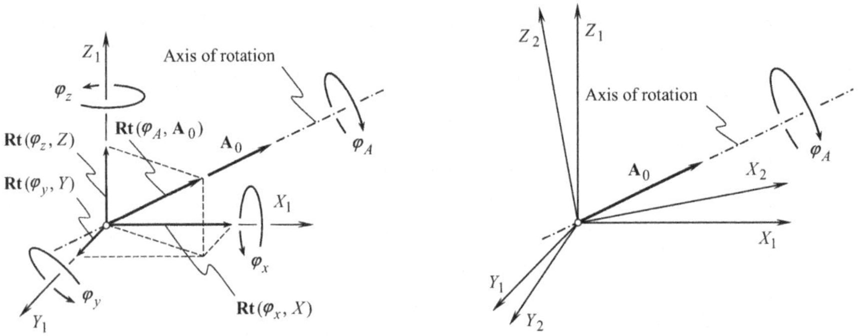

Fig. D.4 Analytical description of the operator $\mathbf{Rt}(\varphi_A, \mathbf{A})$ of rotation about an arbitrary axis through the origin of a *Cartesian coordinate system* $X_1Y_1Z_1$ (the vector $\mathbf{A}$ is the direction vector of the axis of rotation)

$$\mathbf{Rt}(\varphi_A,\ \mathbf{A}_0) = \mathbf{Rt}(\varphi_z, Z) \cdot \mathbf{Rt}(\varphi_y, Y) \cdot \mathbf{Rt}(\varphi_x, X) \tag{D.14}$$

Evidently, the axis of rotation (a straight line along the vector $\mathbf{A}_0$) is always an axis through the origin.

The operators of translation and of rotation also yield linear transformations of other kinds as well.

D.1.5.2 Euler Transformation

Euler transformation is a well-known kind of linear transformation used extensively in mechanical engineering. The linear transformation of this kind is analytically described by the operator $\mathbf{Eu}(\psi,\ \theta,\ \varphi)$ of *Euler*[15] *transformation.*

The operator $\mathbf{Eu}(\psi,\ \theta,\ \varphi)$ is expressed in terms of three *Euler angles* (or *Eulerian angles*) ψ, θ, and φ. Configuration of an orthogonal *Cartesian coordinate system* $X_1Y_1Z_1$ in relation to another orthogonal *Cartesian coordinate system* $X_2Y_2Z_2$ is defined by the *Euler angles* ψ, θ, and φ. These angles are shown in Fig. D.5.

The line of intersection of the coordinate plane X_1Y_1 of the first reference system by the coordinate plane X_2Y_2 of the second reference system is commonly referred to as *nodal line* (or *line of nodes*). In Fig. D.5 the line *OK* is the *nodal line*. It is assumed here and below that the line of nodes, *OK*, and the axes Z_1 and Z_2 form a frame of that same orientation as the reference systems $X_1Y_1Z_1$ and $X_2Y_2Z_2$ do.

The *Euler angle*, φ is referred to as the *angle of pure rotation.* This angle is measured between the X_1-axis and the *nodal line*, *OK*. The angle of pure rotation, φ,

---

[15] *Leonhard* ***Euler*** (April 15, 1707 – September 18, 1783), a famous Swiss mathematician and physicist who spent most of his life in Russia and Germany.

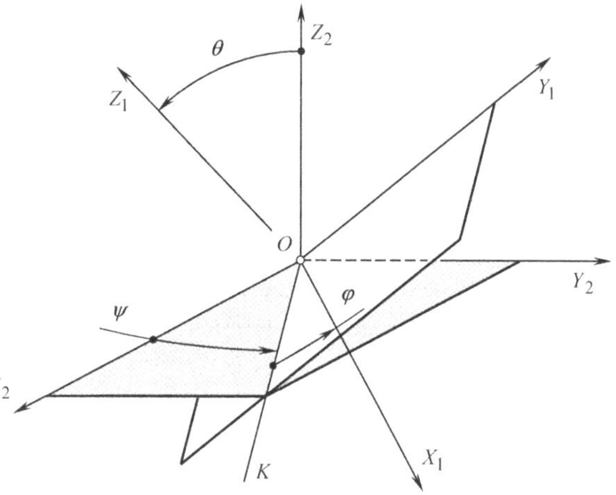

Fig. D.5 *Euler angles*

is measured within the coordinate plane X_1Y_1 in the direction of shortest rotation from the axis X_1 to the axis Y_1.

The *Euler angle,* θ is referred to as the *angle of nutation*. The angle of nutation, θ, is measured between the axes Z_1 and Z_2. Actual value of this angle never exceeds 180°.

The *Euler angle,* ψ is referred to as the *angle of precession*. The angle of precession, ψ, is measured in the coordinate plane X_2Y_2. This the angle between the line of nodes, OK, and the X_2−axis. Direction of the shortest rotation from the axis X_2 to the axis Y_2 is the direction in which the angle of precession is measured.

In case, when the angle of nutation is equal either $\theta = 0^\circ$ or $\theta = 180^\circ$, then the *Euler angles* are not defined.

The operator $\mathbf{Eu}\,(\psi,\ \theta,\ \varphi)$ of *Eulerian transformation* allows for the following matrix representation:

$$\mathbf{Eu}(\psi,\theta,\varphi) = \begin{bmatrix} -\sin\psi\cos\theta\sin\varphi + \cos\psi\cos\varphi & \cos\psi\cos\theta\sin\varphi + \sin\psi\cos\varphi & \sin\theta\sin\varphi & 0 \\ -\sin\psi\cos\theta\cos\varphi - \cos\psi\sin\varphi & \cos\psi\cos\theta\cos\varphi - \sin\psi\cos\varphi & \sin\theta\cos\varphi & 0 \\ \sin\theta\sin\varphi & -\cos\psi\cos\theta & \cos\theta & 0 \\ 0 & 0 & 0 & 1 \end{bmatrix} \tag{D.15}$$

It is important to stress here on difference between the operator $\mathbf{Eu}\,(\psi,\ \theta,\ \varphi)$ of *Eulerian transformation*, and between the operator $\mathbf{Rt}\,(\psi_A,\ \mathbf{A}_0)$ of rotation about an arbitrary axis through the origin.

The operator $\mathbf{Rt}\,(\psi_A,\ \mathbf{A})$ of rotation about an arbitrary axis through the origin can result in that same final orientation of the coordinate system $X_2Y_2Z_2$ in relation to the coordinate system $X_1Y_1Z_1$ as the operator $\mathbf{Eu}\,(\psi,\ \theta,\ \varphi)$ of *Euler transformation* does. However, the operators of linear transformations $\mathbf{Rt}\,(\psi_A,\ \mathbf{A}_0)$ and $\mathbf{Eu}\,(\psi,\ \theta,\ \varphi)$ are the operators of completely different nature. They can result in identical coordinate system transformation, but they are not equal to one another.

D.1.6 Rotation About an Arbitrary Axis Not Through the Origin

The transformation corresponding to rotation of a coordinate system through an angle φ around an arbitrary vector not through the origin cannot readily be written in a form similar to the rotation matrices about the coordinate axes.

The desired transformation matrix is obtained by combining a sequence of elementary translation and rotation matrices. (Once a single 4×4 matrix has been obtained representing the composite transformations, it can be used in the same way as any other transformation matrix).

Rotation of the coordinate system $X_1Y_1Z_1$ to a configuration, which the coordinate system $X_2Y_2Z_2$ possesses, can be performed about a corresponding axis that features an arbitrary configuration in space (see Fig. D.6). The vector **A** is the direction vector of the axis of the rotation. The axis of the rotation is not a line through the origin.

The operator of linear transformation of this particular kind $\mathbf{Rt}\,(\psi_A,\ \mathbf{A})$ can be expressed in terms of the operator $\mathbf{Tr}\,(a,\ \mathbf{A})$ of translation along, and of the operator $\mathbf{Rt}\,(\psi_A,\ \mathbf{A}_0)$ of rotation about an arbitrary axis through the origin:

$$\mathbf{Rt}\,(\varphi_A, \mathbf{A}) = \mathbf{Tr}\,(-b,\ \mathbf{B}^*) \cdot \mathbf{Rt}\,(\varphi_A, \mathbf{A}_0) \cdot \mathbf{Tr}\,(b,\ \mathbf{B}) \tag{D.16}$$

Here, in Eq. (D.16) is designated:

$\mathbf{Tr}\,(b,\ \mathbf{B})$ – is the operator of translation along the shortest distance of approach of the axis of rotation and origin of the coordinate system

$\mathbf{Tr}\,(-b,\ \mathbf{B}^*)$ – is the operator of translation in the direction opposite to the translation $\mathbf{Tr}\,(b,\ \mathbf{B})$ after the rotation $\mathbf{Rt}\,(\psi_A,\ \mathbf{A})$ is completed.

In order to determine the shortest distance of approach, B, of the axis of rotation (that is, the axis along the direction vector **B**) and origin of the coordinate system, consider the axis (**B**) through two given points $\mathbf{r}_{B.1}$ and $\mathbf{r}_{B.2}$.

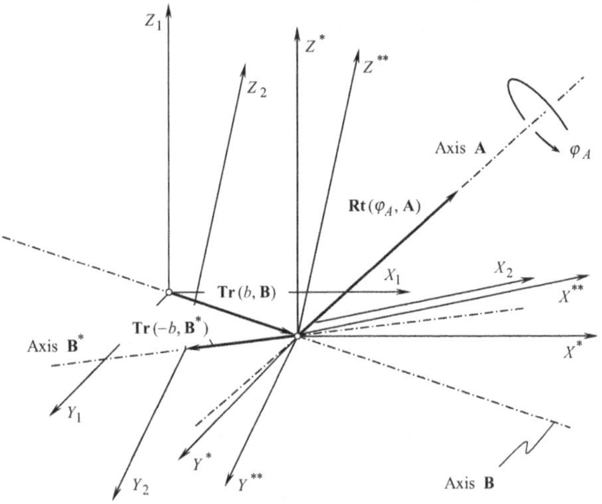

Fig. D.6 Analytical description of the operator, $\mathbf{Rt}\,(\varphi_A,\ \mathbf{A})$, of rotation about an arbitrary axis not through the origin (vector **A** is the direction vector of the axis of rotation)

The shortest distance between a certain point $\mathbf{r}_0$ and the straight line through the points $\mathbf{r}_{B.1}$ and $\mathbf{r}_{B.2}$ can be calculated from the following formula:

$$B = \frac{|\,(\mathbf{r}_2 - \mathbf{r}_1) \times (\mathbf{r}_1 - \mathbf{r}_0)\,|}{|\,\mathbf{r}_2 - \mathbf{r}_1\,|} \tag{D.17}$$

For the origin of the coordinate system, the equality $\mathbf{r}_0 = 0$ is observed. Then,

$$B = |\,\mathbf{r}_1\,| \cdot \sin\angle[\mathbf{r}_1,\ (\mathbf{r}_2 - \mathbf{r}_1)] \tag{D.18}$$

Matrix representation of the operators of translation $\mathbf{Tr}\,(a_x, X)$, $\mathbf{Tr}\,(a_y, Y)$, $\mathbf{Tr}\,(a_z, Z)$ along the coordinate axes, together with the operators of rotation $\mathbf{Rt}\,(\varphi_x, X)$, $\mathbf{Rt}\,(\varphi_y, Y)$, $\mathbf{Rt}\,(\varphi_z, Z)$ about the coordinate axes is convenient for implementation in the gear science. Moreover, use of the operators is the simplest possible way to analytically describe the linear transformations.

D.1.7 Resultant Coordinate System Transformation

The operators of translation $\mathbf{Tr}\,(a_x, X)$, $\mathbf{Tr}\,(a_y, Y)$, and $\mathbf{Tr}\,(a_z, Z)$ together with the operators of rotation $\mathbf{Rt}\,(\varphi_x, X)$, $\mathbf{Rt}\,(\varphi_y, Y)$, and $\mathbf{Rt}\,(\varphi_z, Z)$ are used for the purpose of composing the operator $\mathbf{Rs}\,(1 \mapsto 2)$ of the resultant coordinate system transformation. The transition from the initial *Cartesian reference system* $X_1Y_1Z_1$ to other *Cartesian reference system* $X_2Y_2Z_2$ is analytically described by the operator $\mathbf{Rs}\,(1 \mapsto 2)$ of the resultant coordinate system transformation.

For example, the expression:

$$\mathbf{Rs}\,(1 \mapsto 5) = \mathbf{Tr}\,(a_x, X) \cdot \mathbf{Rt}\,(\varphi_z, Z) \cdot \mathbf{Rt}\,(\varphi_x, X) \cdot \mathbf{Tr}\,(a_y, Y) \tag{D.19}$$

indicates, that the transition from the coordinate system $X_1Y_1Z_1$ to the coordinate system $X_5Y_5Z_5$ is executed in the following four steps (see Fig. D.7):

- translation, followed by
- rotation $\mathbf{Rt}\,(\varphi_x, X)$, followed by
- second rotation $\mathbf{Rt}\,(\varphi_z, Z)$, and finally followed by the translation $\mathbf{Tr}\,(a_x, X)$.

Ultimately, the equality:

$$\mathbf{r}_1(m) = \mathbf{Rs}\,(1 \mapsto 5) \cdot \mathbf{r}_5(m) \tag{D.20}$$

is valid.

When the operator $\mathbf{Rs}\,(1 \mapsto t)$ of the resultant coordinate system transformation is specified, then the transition in the opposite direction can be performed by means of the operator $\mathbf{Rs}\,(t \mapsto 1)$ of the inverse coordinate system transformation. The following equality can be easily proven:

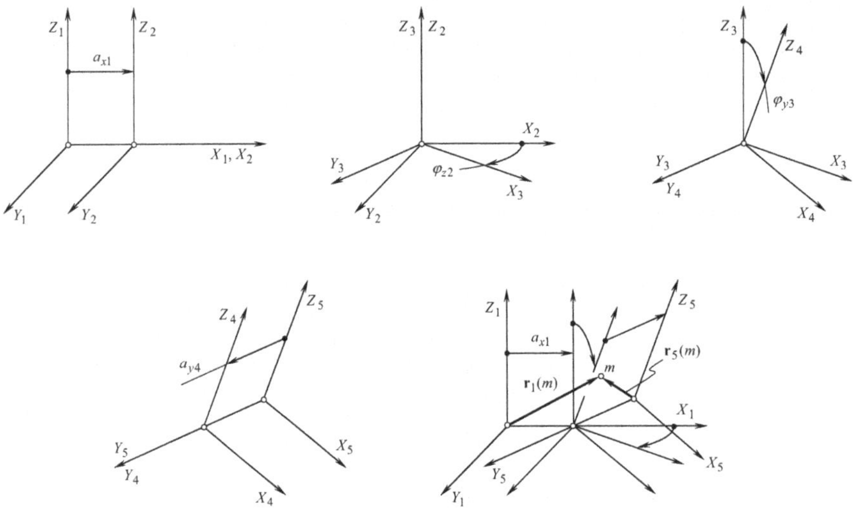

Fig. D.7 An example of the resultant coordinate system transformation, analytically expressed by the operator $\mathbf{Rs}\,(1 \mapsto 5)$

$$\mathbf{Rs}\,(t \mapsto 1) = \mathbf{Rs}^{-1}(1 \mapsto t) \tag{D.21}$$

In the above example illustrated in Fig. D.7, the operator $\mathbf{Rs}\,(5 \mapsto 1)$ of the inverse resultant coordinate system transformation can be expressed in terms of the operator $\mathbf{Rs}\,(1 \mapsto 5)$ of the direct resultant coordinate system transformation. Following Eq. (D.21), one can come up with the equation:

$$\mathbf{Rs}\,(5 \mapsto 1) = \mathbf{Rs}^{-1}(1 \mapsto 5) \tag{D.22}$$

It is easy to show that the operator $\mathbf{Rs}\,(1 \mapsto t)$ of the resultant coordinate system transformation allows for representation in the following form:

$$\mathbf{Rs}\,(1 \mapsto t) = \mathbf{Tr}\,(a,\ A) \cdot \mathbf{Eu}(\psi,\ \theta,\ \varphi) \tag{D.23}$$

The linear transformation $\mathbf{Rs}\,(1 \mapsto t)$ [see Eq. (D.23)] can also be expressed in terms of rotation about an axis $\mathbf{Rt}\,(\varphi_A,\ A)$, not through the origin [see Eq. (D.16)].

D.2 Complex Coordinate System Transformation

In particular cases of complex coordinate system transformations that are repeatedly used in practice, special purpose operators of coordinate system transformation can be composed of elementary operators of translation, and operators of rotation.

D.2.1 Linear Transformation Describing a Screw Motion About a Coordinate Axis

Operators for analytical description of screw motions about an axis of the *Cartesian coordinate system* are a particular case of the operators of the resultant coordinate system transformation.

By definition (see Fig. D.8), the operator $\mathbf{Sc}_x(\varphi_x,\ p_x)$ of a screw motion about X–axis of the *Cartesian* coordinate system XYZ is equal to:

$$\mathbf{Sc}_x(\varphi_x,\ p_x) = \mathbf{Rt}(\varphi_x, X) \cdot \mathbf{Tr}(a_x, X) \tag{D.24}$$

After substituting of the operator of translation $\mathbf{Tr}(a_x, X)$, and the operator of rotation $\mathbf{Rt}(\varphi_x, X)$ [see Eq. (D.10)], Eq. (D.24) casts into the expression:

$$\mathbf{Sc}_x(\varphi_x,\ p_x) = \begin{bmatrix} 1 & 0 & 0 & p_x \cdot \varphi_x \\ 0 & \cos\varphi_x & \sin\varphi_x & 0 \\ 0 & -\sin\varphi_x & \cos\varphi_x & 0 \\ 0 & 0 & 0 & 1 \end{bmatrix} \tag{D.25}$$

for the calculation of the operator of the screw motion $\mathbf{Sc}_x(\varphi_x,\ p_x)$ about X–axis.

The operators of screw motions $\mathbf{Sc}_y(\varphi_y,\ p_y)$ and $\mathbf{Sc}_z(\varphi_z,\ p_z)$ about Y– and Z–axis correspondingly are defined in the way similar to that the operator of the screw motion $\mathbf{Sc}_x(\varphi_x,\ p_x)$ is defined:

$$\mathbf{Sc}_y(\varphi_y,\ p_y) = \mathbf{Rt}(\varphi_y, Y) \cdot \mathbf{Tr}(a_y, Y) \tag{D.26}$$

$$\mathbf{Sc}_z(\varphi_z,\ p_z) = \mathbf{Rt}(\varphi_z, Z) \cdot \mathbf{Tr}(a_z, Z) \tag{D.27}$$

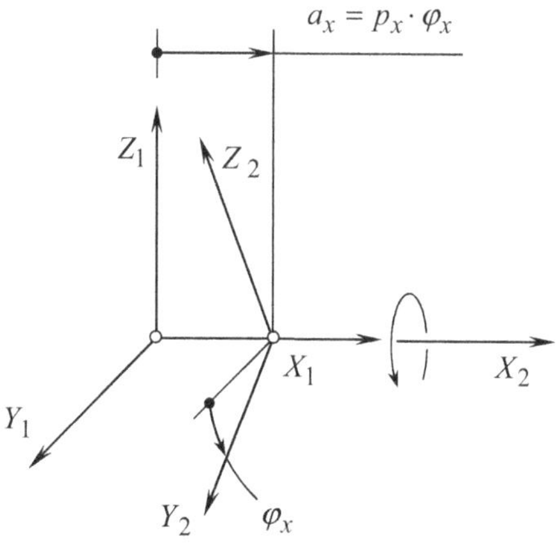

Fig. D.8 On analytical description of the operator of screw motion, $\mathbf{Sc}_x(\varphi_x,\ p_x)$

Using Eqs. (D.5) and (D.6) together with Eqs. (D.11) and (D.12), one can come up with the expressions:

$$\mathbf{Sc}_y(\varphi_y,\ p_y) = [\cos\ \varphi_y 0 -\ \sin\varphi_y 0010 p_y \cdot \varphi_y \sin\varphi_y 0 \cos\varphi_y 00001] \tag{D.28}$$

$$\mathbf{Sc}_z(\varphi_z,\ p_z) = [\cos\ \varphi_z \sin\varphi_z 00 -\ \sin\varphi_z \cos\varphi_z 00001 p_z \cdot \varphi_z 0001] \tag{D.29}$$

for the calculation of the operators of the screw motion $\mathbf{Sc}_y(\varphi_y,\ p_y)$ and $\mathbf{Sc}_z(\varphi_z,\ p_z)$ about $Y-$ and $Z-$axis.

Screw motions about a coordinate axis as well as screw surfaces are common in the theory of part surface generation. This makes it practical to use the operators of the screw motion $\mathbf{Sc}_x(\varphi_x,\ p_x)$, $\mathbf{Sc}_y(\varphi_y,\ p_y)$, and $\mathbf{Sc}_z(\varphi_z,\ p_z)$ in the theory of part surface generation.

In case of necessity, an operator of the screw motion about an arbitrary axis either through the origin of the coordinate system or not through the origin of the coordinate system can be derived following the method similar to that used for the derivation of the operators $\mathbf{Sc}_x(\varphi_x,\ p_x)$, $\mathbf{Sc}_y(\varphi_y,\ p_y)$ and $\mathbf{Sc}_z(\varphi_z,\ p_z)$.

D.2.2 Linear Transformation Describing Rolling Motion of a Coordinate System

One more practical combination of a rotation and of a translation is often used in the theory of gearing.

Consider a *Cartesian coordinate system* $X_1Y_1Z_1$ (see Fig. D.9). The coordinate system $X_1Y_1Z_1$ is traveling in the direction of the X_1-axis. Velocity of the translation is denoted by V. The coordinate system $X_1Y_1Z_1$ is rotated about its Y_1-axis simultaneously with the translation. Speed of the rotation is denoted as ω. Assume that the ratio V/ω is constant. Under such a scenario the resultant motion of the reference system $X_1Y_1Z_1$ to its arbitrary position $X_2Y_2Z_2$ allows interpretation in the form of rolling with no sliding of a cylinder of a radius R_w over the plane. The plane is parallel to the coordinate X_1Y_1-plane, and it is remote from it at the distance R_w. For the calculation of radius of the rolling cylinder the expression $R_w = V/\omega$ can be used.

Due to rolling of the cylinder of a radius, R_w, over the plane is performed with no sliding, a certain correspondence between the translation and the rotation of the coordinate system is established. When the coordinate system turns through a certain angle φ_y, then the translation of origin of the coordinate system along X_1-axis is equal to $a_x = \varphi_r \cdot R_w$.

Transition from the coordinate system $X_1Y_1Z_1$ to the coordinate system $X_2Y_2Z_2$ is analytically described by the operator of the resultant coordinate system transformation $\mathbf{Rs}(1 \mapsto 2)$. The operator $\mathbf{Rs}(1 \mapsto 2)$ is equal:

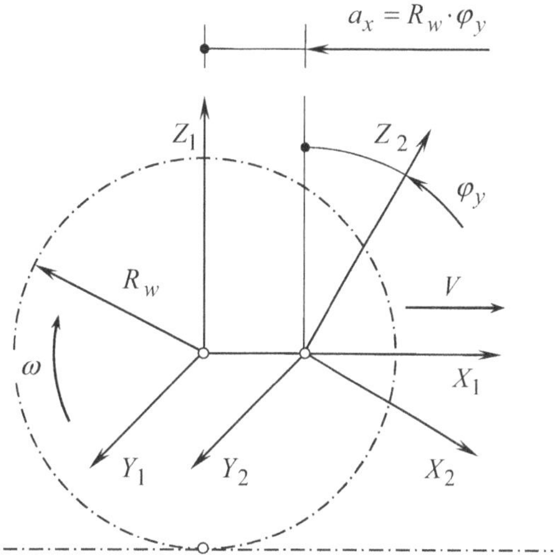

Fig. D.9 Illustration of transformation of rolling, $\mathbf{Rl}_x(\varphi_y,\ Y)$, of a coordinate system

$$\mathbf{Rs}\,(1 \mapsto 2) = \mathbf{Rt}\,(\varphi_y,\ Y_1) \cdot \mathbf{Tr}\,(a_x,\ X_1) \tag{D.30}$$

Here $\mathbf{Tr}\,(a_x,\ X_1)$ designates the operator of the translation along X_1-axis, and $\mathbf{Rt}$ $(\varphi_y,\ Y_1)$ is the operator of the rotation about Y_1-axis.

Operator of the resultant coordinate system transformation of the kind [see Eq. (D.30)] is referred to as the *operator of rolling motion over a plane.*

When the translation is performed along the X_1-axis, and the rotation is performed about the Y_1-axis, the operator of rolling is denoted as $\mathbf{Rl}_x(\varphi_y,\ Y)$. In this particular case, the equality $\mathbf{Rl}_x(\varphi_y,\ Y) = \mathbf{Rs}\,(1 \mapsto 2)$ [see Eq. (D.30)] is valid. Based on this equality, the operator of rolling over a plane $\mathbf{Rl}_x(\varphi_y,\ Y)$ can be calculated from the equation:

$$\mathbf{Rl}_x(\varphi_y,\ Y) = \left[\cos\,\varphi_y 0 - \sin\varphi_y a_x \cdot \cos\varphi_y 0100 \sin\varphi_y 0 \cos\varphi_y a_x \cdot \sin\varphi_y 0001\right] \tag{D.31}$$

While rotation remains about the Y_1-axis, the translation can be performed not along the X_1-axis, but along the Z_1-axis instead. For rolling of this kind the operator of rolling is equal:

$$\mathbf{Rl}_z(\varphi_y,\ Y) = \left[\cos\,\varphi_y 0 - \sin\varphi_y - a_z \cdot \sin\varphi_y 0100 \sin\varphi_y 0 \cos\varphi_y a_z \cdot \cos\varphi_y 0001\right] \tag{D.32}$$

For the cases when the rotation is performed about the X_1-axis, the corresponding operators of rolling are as follow:

$$\mathbf{Rl}_y(\varphi_x,\ X) = \begin{bmatrix} 1 & 0 & 0 & 0 \\ 0 & \cos \end{bmatrix} \varphi_x \sin \varphi_x a_y \cdot \cos \varphi_x 0 - \sin \varphi_x \cos \varphi_x - a_y \cdot \sin \varphi_x 0001 \tag{D.33}$$

for the case of rolling along the Y_1–axis, and

$$\mathbf{Rl}_z(\varphi_x,\ X) = \begin{bmatrix} 1 & 0 & 0 & 0 \\ 0 & \cos \end{bmatrix} \varphi_x \sin \varphi_x a_z \cdot \sin \varphi_x 0 - \sin \varphi_x \cos \varphi_x a_z \cdot \cos \varphi_x 0001 \tag{D.34}$$

for the case of rolling along the Z_1– axis.

Similar expressions can be derived for the case of rotation about the Z_1– axis:

$$\mathbf{Rl}_x(\varphi_z,\ Z) = [\cos\ \varphi_z \sin \varphi_z 0 a_x \cdot \cos \varphi_z - \sin \varphi_z \cos \varphi_z 0 a_x \cdot \sin \varphi_z 00100001] \tag{D.35}$$

$$\mathbf{Rl}_y(\varphi_z,\ Z) = [\cos\ \varphi_z \sin \varphi_z 0 a_y \cdot \sin \varphi_z - \sin \varphi_z \cos \varphi_z 0 a_y \cdot \cos \varphi_z 00100001] \tag{D.36}$$

Use of the operators of rolling Eq. (D.31) through Eq. (D.36) significantly simplifies analytical description of the coordinate system transformations.

D.2.3 Linear Transformation Describing Rolling of Two Coordinate Systems

In the theory of gearing, combinations of two rotations about parallel axes are of particular interest.

As an example, consider two *Cartesian coordinate systems*, $X_1Y_1Z_1$ and $X_2Y_2Z_2$, shown in Fig. D.10. The coordinate systems $X_1Y_1Z_1$ and $X_2Y_2Z_2$ are rotated about their axes, Z_1 and Z_2. The axes of the rotations are parallel to each other ($Z_1 \parallel Z_2$). The rotations ω_1 and ω_2 of the coordinate systems can be interpreted so that a circle of a certain radius R_1 that is associated with the coordinates system $X_1Y_1Z_1$, is rolling with no sliding over a circle of the corresponding radius R_2 that is associated with the coordinate system $X_2Y_2Z_2$. When the center-distance C is known, then radii, R_1 and R_2, of the circles (that is, of the *centrodes*) can be expressed in terms of the center-distance, C, and of the given rotations, ω_1 and ω_2. For the calculations, the following formulae:

Fig. D.10 On derivation of the operator of rolling, $\mathbf{Rr}_u(\varphi_1,\ Z_1)$, of two coordinate systems

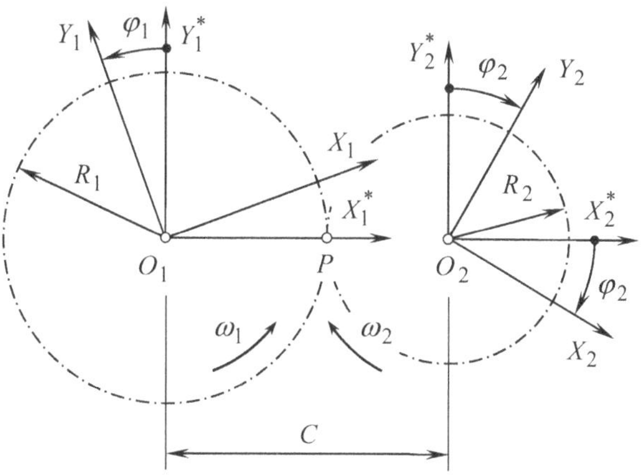

$$R_1 = C \cdot \frac{1}{1+u} \tag{D.37}$$

$$R_2 = C \cdot \frac{u}{1+u} \tag{D.38}$$

are used. Here, the ratio ω_1/ω_2 is denoted by u.

In the initial configuration, the X_1-, and X_2-axes align to each other. The Y_1- and Y_2-axes are parallel to each other. As shown in Fig. D.10, the initial configurations of the coordinate systems $X_1Y_1Z_1$ and $X_2Y_2Z_2$ are labeled as $X_1^*Y_1^*Z_1^*$, and $X_2^*Y_2^*Z_2^*$, correspondingly.

When the coordinate system $X_1Y_1Z_1$ turns through a certain angle φ_1, then the coordinate system $X_2Y_2Z_2$ turns through the corresponding angle φ_2. When the angle φ_1 is known then the corresponding angle φ_2 is equal to $\varphi_2 = \varphi_1/u$.

Transition from the coordinate system $X_2Y_2Z_2$ to the coordinate system $X_1Y_1Z_1$ can be analytically described by the operator of the resultant coordinate system transformation, $\mathbf{Rs}\,(1 \mapsto 2)$. In the case under consideration, the operator $\mathbf{Rs}$ $(1 \mapsto 2)$ can be expressed in terms of the operators of the elementary coordinate system transformations:

$$\mathbf{Rs}\,(1 \mapsto 2) = \mathbf{Rt}\,(\varphi_1,\ Z_1) \cdot \mathbf{Rt}\,(\varphi_1/u,\ Z_1) \cdot \mathbf{Tr}\,(-\,C,\ X_1) \tag{D.39}$$

Other equivalent combinations of the operators of elementary coordinate system transformations can result in that same operator $\mathbf{Rs}\,(1 \mapsto 2)$ of the resultant coordinate system transformation. The interested reader may wish to exercise deriving the equivalent expressions for the operator on his or her own $\mathbf{Rs}\,(1 \mapsto 2)$.

The operators of the resultant coordinate system transformations of the kind [see Eq. (D.39)] are referred to as the *operators of rolling motion over a cylinder*.

When rotations are performed around the Z_1-, and the Z_2-axis, the operator of rolling motion over a cylinder is designated as $\mathbf{Rr}_u(\varphi_1,\ Z_1)$. In this particular case, the equality $\mathbf{Rr}_u(\varphi_1,\ Z_1) = \mathbf{Rs}(1 \mapsto 2)$ [see Eq. (D.39)] is valid. Based on this equality, the operator of rolling $\mathbf{Rr}_u(\varphi_1,\ Z_1)$ over a cylinder can be calculated from the equation:

$$\mathbf{Rr}_u(\varphi_1,\ Z_1) = \left[\cos\left(\varphi_1 \cdot \frac{u+1}{u}\right) \sin\left(\varphi_1 \cdot \frac{u+1}{u}\right) 0 - C - \sin\left(\varphi_1 \cdot \frac{u+1}{u}\right) \cos\left(\varphi_1 \cdot \frac{u+1}{u}\right) 0000100001\right] \tag{D.40}$$

is derived.

For the inverse transformation, the inverse operator of rolling of two coordinate systems $\mathbf{Rr}_u(\varphi_2,\ Z_2)$ can be used. It is equal to $\mathbf{Rr}_u(\varphi_2,\ Z_2) = \mathbf{Rr}_u^{-1}(\varphi_1,\ Z_1)$. In terms of the operators of the elementary coordinate system transformations the operator $\mathbf{Rr}_u(\varphi_2,\ Z_2)$ can be expressed as:

$$\mathbf{Rr}_u(\varphi_2,\ Z_2) = \mathbf{Rt}(\varphi_1/u,\ Z_2) \cdot \mathbf{Rt}(\varphi_1,\ Z_2) \cdot \mathbf{Tr}(C,\ X_1) \tag{D.41}$$

Other equivalent combinations of the operators of elementary coordinate system transformations can result in that same operator $\mathbf{Rr}_u(\varphi_2,\ Z_2)$ of the resultant coordinate system transformation. The interested reader may wish to exercise on his or her own deriving the equivalent expressions for the operator $\mathbf{Rr}_u(\varphi_2,\ Z_2)$.

For the calculation of the operator of rolling of two coordinate systems $\mathbf{Rr}_u(\varphi_2,\ Z_2)$, the equation:

$$\mathbf{Rr}_u(\varphi_2,\ Z_2) = \left[\cos\left(\varphi_1 \cdot \frac{u+1}{u}\right) - \sin\left(\varphi_1 \cdot \frac{u+1}{u}\right) 0C \sin\left(\varphi_1 \cdot \frac{u+1}{u}\right) \cos\left(\varphi_1 \cdot \frac{u+1}{u}\right) 0000100001\right] \tag{D.42}$$

can be used.

Similar to that the expression [see Eq. (D.40)] is derived for the calculation of the operator of rolling $\mathbf{Rr}_u(\varphi_1,\ Z_1)$ around the Z_1-, and Z_2-axis, corresponding formulae can be derived for the calculation of the operators of rolling $\mathbf{Rr}_u(\varphi_1,\ X_1)$ and $\mathbf{Rr}_u(\varphi_1,\ Y_1)$ about parallel axes X_1 and X_2, as well as about parallel axes Y_1 and Y_2.

Use of the operators of rolling about two axes $\mathbf{Rr}_u(\varphi_1,\ X_1)$, $\mathbf{Rr}_u(\varphi_1,\ Y_1)$ and $\mathbf{Rr}_u(\varphi_1,\ Z_1)$ substantially simplifies analytical description of the coordinate system transformations.

D.2.4 Coupled Linear Transformation

It is right point to notice here that a translation, $\mathbf{Tr}(a_x,\ X)$, along the $X-$axis of a *Cartesian* reference system, XYZ, and a rotation, $\mathbf{Rt}(\varphi_x,\ X)$, about the axis X of that same coordinate system, XYZ, obey the commutative law, that is, these two coordinate system transformations can be performed in different order equally. This means that the operators, $\mathbf{Cp}_x(a_x,\ \varphi_x)$, of coupled linear transformations are invariant with respect to the order of the elementary transformations, namely, of the translation/

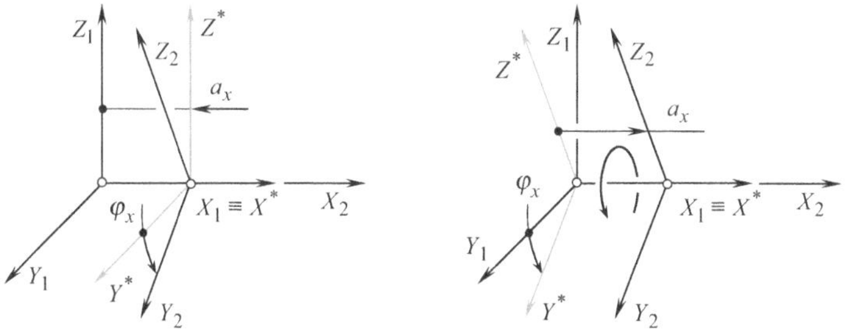

(a) (b)

Fig. D.11 On the equivalency of the linear transformations, $\mathbf{Rt}(\varphi_x,\ X) \cdot \mathbf{Tr}(a_x,\ X)$ and $\mathbf{Tr}(a_x,\ X) \cdot \mathbf{Rt}(\varphi_x,\ X)$, that comprise the operator, $\mathbf{Cp}_x(a_x,\ \varphi_x)$, of coupled linear transformation of a *Cartesian* reference system XYZ

rotation – this is important. It makes no difference whether the translation, $\mathbf{Tr}(a_x,\ X)$, is initially performed, which is followed by the rotation, $\mathbf{Rt}(\varphi_x,\ X)$, or the rotation, $\mathbf{Rt}(\varphi_x,\ X)$, is initially performed, which is followed by the translation, $\mathbf{Tr}(a_x,\ X)$. This is because the dot products $\mathbf{Rt}(\varphi_x,\ X) \cdot \mathbf{Tr}(a_x,\ X)$ and $\mathbf{Tr}(a_x,\ X) \cdot \mathbf{Rt}(\varphi_x,\ X)$ are identical to one another:

$$\mathbf{Rt}(\varphi_x,\ X) \cdot \mathbf{Tr}(a_x,\ X) \equiv \mathbf{Tr}(a_x,\ X) \cdot \mathbf{Rt}(\varphi_x,\ X) \tag{D.43}$$

This means that the translation from the coordinate system $X_1Y_1Z_1$ to the intermediate coordinate system $X^*Y^*Z^*$ followed by the rotation from the coordinate system $X^*Y^*Z^*$ to the finale coordinate system $X_2Y_2Z_2$ produces that same reference $X_2Y_2Z_2$ as in a case when the rotation from the coordinate system $X_1Y_1Z_1$ to the intermediate coordinate system $X^*Y^*Z^*$ followed by the translation from the coordinate system $X^*Y^*Z^*$ to the finale coordinate system $X_2Y_2Z_2$.

The validity of Eq. (D.43) is illustrated in Fig. D.11. The translation, $\mathbf{Tr}(a_x,\ X)$, that is followed by the rotation, $\mathbf{Rt}(\varphi_x,\ X)$, as shown in Fig. D.11a, is equivalent to the rotation, $\mathbf{Rt}(\varphi_x,\ X)$, that is followed by the translation, $\mathbf{Tr}(a_x,\ X)$ as shown in Fig. D.11b.

Therefore, the two linear transformations, $\mathbf{Tr}(a_x,\ X)$ and $\mathbf{Rt}(\varphi_x,\ X)$, can be coupled into a linear transformation:

$$\mathbf{Cp}_x(a_x,\ \varphi_x) = \mathbf{Rt}(\varphi_x,\ X) \cdot \mathbf{Tr}(a_x,\ X) \equiv \mathbf{Tr}(a_x,\ X) \cdot \mathbf{Rt}(\varphi_x,\ X) \tag{D.44}$$

The operator of linear transformation, $\mathbf{Cp}_x(a_x,\ \varphi_x)$, can be expressed in matrix form (see Fig. D.12a):

$$\mathbf{Cp}_x(a_x,\ \varphi_x) = \begin{bmatrix} 1 & 0 & 0 & a_x & \\ 0 & \cos & & & \varphi_x \sin \varphi_x 00 - \sin \varphi_x \cos \varphi_x 00001 \end{bmatrix} \tag{D.45}$$

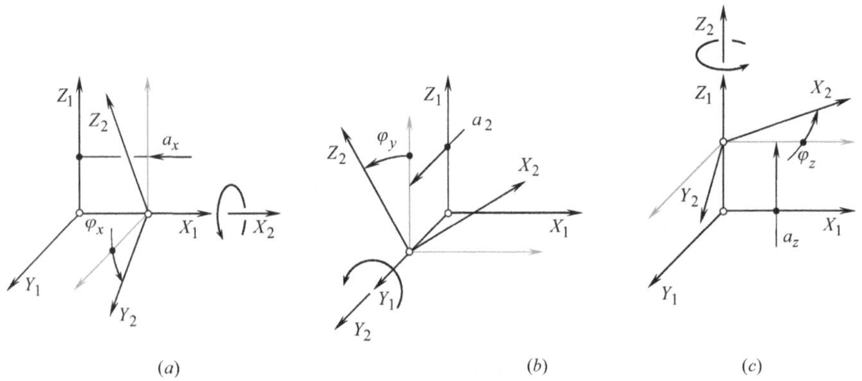

Fig. D.12 On the equivalency of linear transformations, $\mathbf{Rt}(\varphi_x, X) \cdot \mathbf{Tr}(a_x, X)$ and $\mathbf{Tr}(a_x, X) \cdot \mathbf{Rt}(\varphi_x, X)$, which comprise the operator, $\mathbf{Cp}_x(a_x, \varphi_x)$, of coupled linear transformation of a *Cartesian* reference system *XYZ*

This expression is composed based on Eq. (D.4) for the linear transformation $\mathbf{Tr}(a_x, X)$, and on Eq. (D.10) that describes the linear transformation $\mathbf{Rt}(\varphi_x, X)$.

Two reduced cases of operator of the linear transformation, $\mathbf{Cp}_x(a_x, \varphi_x)$, are distinguished.

First, it could happen that in a particular case the component, a_x, of the translation is zero, that is $a_x = 0$. Under such a scenario the operator of linear transformation, $\mathbf{Cp}_x(a_x, \varphi_x)$, reduces to the operator of rotation, $\mathbf{Rt}(\varphi_x, X)$, and the equality $\mathbf{Cp}_x(a_x, \varphi_x) = \mathbf{Rt}(\varphi_x, X)$ is observed in the case under consideration.

Second, it could happen that in a particular case the component, φ_x, of the rotation is zero, that is $\varphi_x = 0°$. Under such a scenario the operator of linear transformation, $\mathbf{Cp}_x(a_x, \varphi_x)$, reduces to the operator of translation, $\mathbf{Tr}(a_x, X)$, and the equality $\mathbf{Cp}_x(a_x, \varphi_x) = \mathbf{Tr}(a_x, X)$ is observed in the case under consideration.

The said is valid with respect to the translations and the rotations along and about the axes *Y* and *Z* of a *Cartesian reference system XYZ*. The corresponding coupled operators, $\mathbf{Cp}_y(a_y, \varphi_y)$ and $\mathbf{Cp}_z(a_z, \varphi_z)$, for linear transformations of these kinds can also be composed (see Fig. D.12b, c):

$$\mathbf{Cp}_y(a_y, \varphi_y) = [\cos\varphi_y 0 \sin\varphi_y 0010a_y - \sin\varphi_y 0 \cos\varphi_y 00001] \quad \text{(D.46)}$$

$$\mathbf{Cp}_z(a_z, \varphi_z) = [\cos\varphi_z \sin\varphi_z 00 - \sin\varphi_z \cos\varphi_z 00001a_z 0001] \quad \text{(D.47)}$$

In the operators of linear transformations, $\mathbf{Cp}_x(a_x, \varphi_x)$, $\mathbf{Cp}_y(a_y, \varphi_y)$, and $\mathbf{Cp}_z(a_z, \varphi_z)$, values of the translations a_x, a_y, and a_z, as well as values of the rotations φ_x, φ_y, and φ_z, are expressed in finite values (and not as continuous). The linear and angular displacements do not correlate to one another in time; thus, they do not comprise a screw. They are just a kind of couples of a translation along, and a rotation about a coordinate axis of a *Cartesian reference system*.

Introduction of the operators of linear transformation, $\mathbf{Cp}_x(a_x, \varphi_x)$, $\mathbf{Cp}_y(a_y, \varphi_y)$, and $\mathbf{Cp}_z(a_z, \varphi_z)$, makes the linear transformations easier as all the operators of the linear transformations become uniform.

The operators of linear transformation $\mathbf{Cp}_x(a_x, \varphi_x)$, $\mathbf{Cp}_y(a_y, \varphi_y)$, and $\mathbf{Cp}_z(a_z, \varphi_z)$ do not obey the commutative law. This is because rotation is not a vector by nature. Therefore, special care should be undertaken when treating rotations as vectors – when implementing coupled operators of linear transformations in particular.

The operators of coupled linear transformations $\mathbf{Cp}_x(a_x, \varphi_x)$, $\mathbf{Cp}_y(a_y, \varphi_y)$, and $\mathbf{Cp}_z(a_z, \varphi_z)$, [see Eq. (D.45) through Eq. (D.47)] can be used for the purpose of analytical description of a resultant coordinate system transformation. In this scenario, the operator, $\mathbf{Rs}(1 \mapsto t)$, of a resultant coordinate system transformation can be expressed in terms of all the operators $\mathbf{Cp}_x(a_x, \varphi_x)$, $\mathbf{Cp}_y(a_y, \varphi_y)$, and $\mathbf{Cp}_z(a_z, \varphi_z)$ by the following expression:

$$\mathbf{Rs}(1 \mapsto t) = \prod_{\substack{i=1 \\ j=x,y,z}}^{t-1} \mathbf{Cp}_j^i\left(a_j^i, \varphi_j^i\right) \tag{D.48}$$

In Eq. (D.48), only operators of coupled linear transformations are used.

D.2.5 An Example of Non-orthogonal Linear Transformation

Consider a non-orthogonal reference system $X_1Y_1Z_1$ having certain angle ω_1 between the axes X_1 and Y_1. Axis Z_1 is perpendicular to the coordinate plane X_1Y_1. Another reference system $X_2Y_2Z_2$ is identical to the first coordinate system $X_1Y_1Z_1$, and is turned about Z_1–axis through a certain angle φ. Transition from the reference system $X_1Y_1Z_1$ to the reference system $X_2Y_2Z_2$ can be analytically described by the operator of linear transformation:

$$\mathbf{Rt}_\omega(1 \to 2) = \left[\frac{\sin(\omega_1 + \varphi)}{\sin\omega_1} \frac{\sin\varphi}{\sin\omega_1} 00 - \frac{\sin\varphi}{\sin\omega_1} \frac{\sin(\omega_1 - \varphi)}{\sin\omega_1} 0000100001\right] \tag{D.49}$$

In order to distinguish the operator of rotation in the orthogonal linear transformation $\mathbf{Rt}(1 \to 2)$ from the similar operator of rotation in a non-orthogonal linear transformation $\mathbf{Rt}_\omega(1 \to 2)$, the subscript "$\omega$" is assigned to the latter.

When $\omega = 90°$, Eq. (D.49) casts into Eq. (D.12).

D.2.6 Conversion of Hand of Coordinate System

Application of matrix method of coordinate system transformation presumes that both of the reference systems "i" and "$(i \pm 1)$" are of the same hand. This means that it is assumed from the very beginning that both of them are either right-hand- or left-hand-oriented *Cartesian* coordinate systems. In the event the coordinate systems i and $(i \pm 1)$ are of opposite hand, say one of them is the right-hand-oriented coordinate system while another one is left-hand-oriented coordinate system, then one of the coordinate systems needs be converted into the oppositely oriented *Cartesian coordinate system.*

For the conversion of a left-hand-oriented *Cartesian coordinate system* into a right-hand-oriented coordinate system or vice versa, the operators of reflection are commonly used.

In order to change the direction of X_i axis of the initial coordinate system i to the opposite direction (in this case in the new coordinate system $(i \pm 1)$ the equalities $X_{i\pm1} = -X_i$, $Y_{i\pm1} \equiv Y_i$ and $Z_{i\pm1} \equiv Z_i$ are observed) the operator of reflection $\mathbf{Rf}_x(Y_i Z_i)$ is applied. The operator of reflection yields representation in matrix form:

$$\mathbf{Rf}_x(Y_i Z_i) = \begin{bmatrix} -1 & 0 & 0 & 0 \\ 0 & 1 & 0 & 0 \\ 0 & 0 & 1 & 0 \\ 0 & 0 & 0 & 1 \end{bmatrix} \tag{D.50}$$

Similarly, the use of the operators of reflections $\mathbf{Rf}_y(X_i Z_i)$ and $\mathbf{Rf}_z(X_i Y_i)$ change the directions of Y_i axis and Z_i axes onto opposite directions. The operators of reflections $\mathbf{Rf}_y(X_i Z_i)$ and $\mathbf{Rf}_z(X_i Y_i)$ can be expressed analytically in the form:

$$\mathbf{Rf}_y(X_i Z_i) = \begin{bmatrix} 1 & 0 & 0 & 0 \\ 0 & -1 & 0 & 0 \\ 0 & 0 & 1 & 0 \\ 0 & 0 & 0 & 1 \end{bmatrix} \tag{D.51}$$

$$\mathbf{Rf}_z(X_i Y_i) = \begin{bmatrix} 1 & 0 & 0 & 0 \\ 0 & 1 & 0 & 0 \\ 0 & 0 & -1 & 0 \\ 0 & 0 & 0 & 1 \end{bmatrix} \tag{D.52}$$

A linear transformation that reverses direction of the coordinate axis is an *opposite transformation*. Transformation of reflection is an example of *orientation-reversing transformation*.

D.3 Useful Equations

The sequence of the successive rotations can vary depending on the intention of the researcher. Several special types of successive rotations are known, including *Eulerian transformation*, *Cardanian transformation*, two kinds of *Euler-Krylov transformations*, and so forth. The sequence of the successive rotations can be chosen from a total of 12 different combinations. Even though the *Cardanian transformation* is different from the *Eulerian transformation* in terms of the combination of the rotations, they both use a similar approach to calculate the orientation angles.

D.3.1 *RPY*-Transformation

A series of rotations can be performed in the order *roll matrix,* (*R*), by *pitch matrix* (*P*), and finally by *yaw matrix,* (*Y*). The linear transformation of this kind is commonly referred to as *RPY–transformation*. The resultant transformation of this kind can be represented by the homogenous coordinate transformation matrix:

$$\mathbf{RPY}\varphi_x, \varphi_y\varphi_z = \begin{bmatrix} \cos\varphi_y\cos\varphi_z + \sin\varphi_x\sin\varphi_y\sin\varphi_z & \cos\varphi_y\sin\varphi_z - \sin\varphi_x\sin\varphi_y\cos\varphi_z & \cos\varphi_x\sin\varphi_y & 0 \\ -\cos\varphi_x\sin\varphi_z & \cos\varphi_x\cos\varphi_z & \sin\varphi_x & 0 \\ \sin\varphi_x\cos\varphi_y\sin\varphi_z - \sin\varphi_y\cos\varphi_z & -\sin\varphi_x\cos\varphi_y\cos\varphi_z - \cos\varphi_y\sin\varphi_z & \cos\varphi_x\cos\varphi_y & 0 \\ 0 & 0 & 0 & 1 \end{bmatrix}$$

(D.53)

The *RPY–transformation* can be used for solving problems in the field of gear science.

D.3.2 Operator of Rotation About an Axis in Space

A spatial rotation operator for the rotational transformation of a point about a unit axis $\mathbf{a}_0(\cos\alpha, \ \cos\beta, \ \cos\gamma)$ passing through the origin of the coordinate system can be described as follows, with $\mathbf{a}_0 = \mathbf{A}_0/\,|\,\mathbf{A}_0|$ designating the unit vector along the axis of rotation $\mathbf{A}_0$.

Suppose the angle of rotation of the point about $\mathbf{a}_0$ is θ, the *rotation operator* is expressed by:

$$\mathbf{Rt}(\theta, \mathbf{a}_0)$$

$$= \begin{bmatrix} (1-\cos\theta)\cos^2\alpha + \cos\theta & (1-\cos\theta)\cos\alpha\cos\beta - \sin\theta\cos\gamma \\ (1-\cos\theta)\cos\alpha\cos\beta + \sin\theta\cos\gamma & (1-\cos\theta)\cos^2\beta + \cos\theta \\ (1-\cos\theta)\cos\alpha\cos\gamma - \sin\theta\cos\beta & (1-\cos\theta)\cos\beta\cos\gamma + \sin\theta\cos\alpha \\ 0 & 0 \end{bmatrix}$$

$$\begin{matrix} (1-\cos\theta)\cos\alpha\cos\gamma+\sin\theta\cos\beta & 0 \\ (1-\cos\theta)\cos\beta\cos\gamma-\sin\theta\cos\alpha & 0 \\ (1-\cos\theta)\cos^2\gamma+\cos\theta & 0 \\ 0 & 1 \end{matrix}\Bigg] \tag{D.54}$$

Solution to a problem in the field of gearing can be significantly simplified by implementation of the rotational operator $\mathbf{Rt}(\theta, \mathbf{a}_0)$ [see Eq. (D.54)].

D.3.3 Combined Linear Transformation

Suppose point, p, on a rigid body rotates with an angular displacement, θ, about an axis along a unit vector, $\mathbf{a}_0$, passing through the origin of the coordinate system at first, and then followed by a translation at a distance, B, in the direction of a unit vector, $\mathbf{b}$. The linear transformation of this kind can be analytically described by the homogenous matrix:

$$\mathbf{Rt}(\theta_{a0}, B_b)$$

$$= \left[\begin{matrix} (1-\cos\theta)\cos^2\alpha+\cos\theta & (1-\cos\theta)\cos\alpha\cos\beta-\sin\theta\cos\gamma \\ (1-\cos\theta)\cos\alpha\cos\beta+\sin\theta\cos\gamma & (1-\cos\theta)\cos^2\beta+\cos\theta \\ (1-\cos\theta)\cos\alpha\cos\gamma-\sin\theta\cos\beta & (1-\cos\theta)\cos\beta\cos\gamma+\sin\theta\cos\alpha \\ 0 & 0 \end{matrix}\right.$$

$$\left.\begin{matrix} (1-\cos\theta)\cos\alpha\cos\gamma+\sin\theta\cos\beta & B\cos\alpha \\ (1-\cos\theta)\cos\beta\cos\gamma-\sin\theta\cos\alpha & B\cos\beta \\ (1-\cos\theta)\cos^2\gamma+\cos\theta & B\cos\gamma \\ 0 & 1 \end{matrix}\right] \tag{D.55}$$

More operators of particular linear transformations can be found in the literature.

D.4 Chains of Consequent Linear Transformations and Closed Loop of Consequent Coordinate Systems Transformations

Consequent coordinate system transformations form chains (circuits) of linear transformations. The elementary chain of coordinate system transformation is composed of two consequent transformations. Chains of linear transformations play an important role in the theory of gearing.

Two different kinds of chains of consequent coordinate system transformations are distinguished.

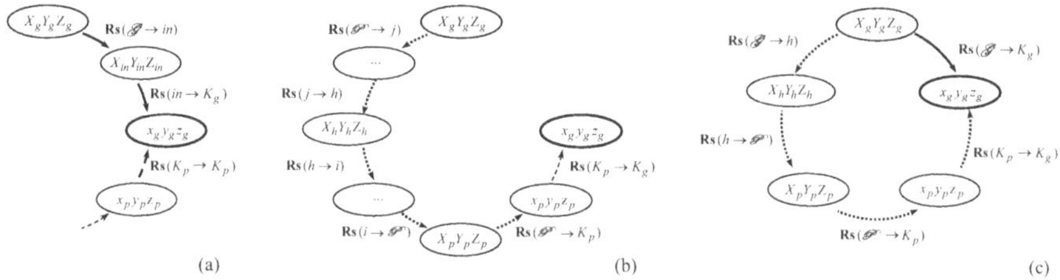

Fig. D.13 An example of direct chain (**a**), of reverse chain (**b**), and a closed loop (**c**) of consequent coordinate system transformations

First, transition from the coordinate system $X_gY_gZ_g$ associated with the gear tooth flank, $\mathcal{G}$, to the local *Cartesian* coordinate system $x_gy_gz_g$ having the origin at contact point, K, of between the gear tooth flank, $\mathcal{G}$, and of the pinion tooth flank, $\mathcal{P}$. This linear transformation is also made up of numerous operators of intermediate coordinate system transformations ($X_{in}Y_{in}Z_{in}$). It forms a chain of direct consequent coordinate system transformations illustrated in Fig. D.13a.

The local coordinate system, $x_gy_gz_g$, is associated with the gear tooth flank, $\mathcal{G}$. The operator $\mathbf{Rs}\left(\mathcal{G} \rightarrow K_g\right)$ of the resultant coordinate system transformation for a direct chain of the linear transformations can be composed using for this purpose a certain number of the operators of translations [see Eq. (D.4) through Eq. (D.6)], and a corresponding number of the operators of rotations [see Eq. (D.10) through Eq. (D.12)].

Second, transition from the coordinate system, $X_gY_gZ_g$, to the local *Cartesian* coordinate system, $x_py_pz_p$, with the origin at point K of contact of the tooth flanks, $\mathcal{G}$ and $\mathcal{P}$. The local coordinate system, $x_py_pz_p$, is associated to pinion tooth flank, $\mathcal{P}$. This linear transformation is also made up of numerous intermediate coordinate system transformations ($X_jY_jZ_j$), for example, transitions from the coordinate system $X_hY_hZ_h$ associated with gear housing, to numerous intermediate coordinate system $X_iY_iZ_i$. The linear transformation of this kind forms a chain of inverse consequent coordinate system transformations shown in Fig. D.13b. The operator $\mathbf{Rs}\left(\mathcal{G} \rightarrow K_p\right)$ of the resultant coordinate system transformations for the inverse chain of transformations can be composed using for this purpose a certain number of the operators of translations [see Eqs. (D.4), (D.5), and (D.6)], and a corresponding number of the operators of rotations [see Eq. (D.10) through Eq. (D.12)].

Chains of the direct and of the reverse consequent coordinate system transformations together with the operator of transition from the local coordinate system, $x_py_pz_p$, to the local coordinate system, $x_gy_gz_g$, (or from the local coordinate system, $x_gy_gz_g$, to the local coordinate system, $x_py_pz_p$) form a closed loop (a closed circuit) of the consequent coordinate system transformations depicted in Fig. D.13c.

If a closed loop of the consequent coordinate system transformations is complete, implementation of a certain number of the operators of translations [see Eqs. (D.4), (D.5), and (D.6)], and a corresponding number of the operators of rotations [see Eq. (D.10) through Eq. (D.12)], returns a result that is identical to the input data. This means that analytical description of a meshing process specified in the original coordinate system remains the same after implementation of the operator of the resultant coordinate system transformations. This condition is the necessary and sufficient condition for existence of a closed loop of consequent coordinate system transformations.

Implementation of the chains, as well as of the closed loops of consequent coordinate system transformations makes possible consideration of the meshing process of the teeth flanks, $\mathcal{G}$ and $\mathcal{P}$, in any and all of reference systems that form up the loop. Therefore, for consideration of a particular problem of part surface generation the most convenient reference system can be chosen.

In order to complete the construction of a closed loop of consequent coordinate system transformations, an operator of transformation from the local coordinate system, $x_p y_p z_p$, to the local coordinate system, $x_g y_g z_g$, must be composed. Usually, the local reference systems, $x_g y_g z_g$ and $x_p y_p z_p$, are the kind of semi-orthogonal coordinate systems. This means that the axis, z_P, is always orthogonal to the coordinate plane, $x_g y_g$, while the axes, x_g and y_g, can be either orthogonal or not orthogonal to each other. The same is valid with respect the local coordinate system, $x_p y_p z_p$.

Two possible ways for performing the required transformation of the local reference systems, $x_g y_g z_g$ and $x_p y_p z_p$, are considered below.

Following the first way, the operator $\mathbf{Rt}_\omega(p \to g)$ of the linear transformation of semi-orthogonal coordinate systems (see Fig. D.14) needs to be composed. The operator $\mathbf{Rt}_\omega(p \to g)$ can be represented in the form of the homogenous matrix:

$$\mathbf{Rt}_\omega(p \to g) = \left[\frac{\sin(\omega_p + \alpha)}{\sin \omega_p} - \frac{\sin(\omega_g - \omega_p - \alpha)}{\sin \omega_p} 00 - \frac{\sin \alpha}{\sin \omega_p} \frac{\sin(\omega_g - \alpha)}{\sin \omega_p} 0000 - 100001 \right] \quad \text{(D.56)}$$

Here is designated:

ω_g – the angle that is formed by the coordinate lines, U_g and V_g, on the gear tooth flank, $\mathcal{G}$

ω_p – the angle that is formed by the coordinate lines, U_p and V_p, on the pinion tooth flank, $\mathcal{P}$

α – the angle that is formed by the axes, x_g and x_p, of the local coordinate systems $x_g y_g z_g$ and $x_p y_p z_p$

The auxiliary angle β in Fig. D.14 is equal to $\beta = \omega_T + \alpha$.

The inverse coordinate system transformation, that is, the transformation from the local coordinate system, $x_g y_g z_g$, to the local coordinate system, $x_p y_p z_p$, is analytically described by the operator $\mathbf{Rt}_\omega(g \to p)$ of the inverse coordinate system transformation. The operator $\mathbf{Rt}_\omega(g \to p)$ can be represented in the form of the homogenous matrix:

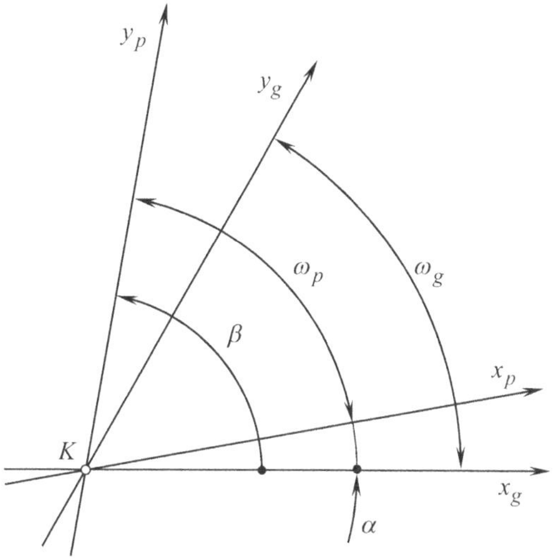

Fig. D.14 Local coordinate systems, $x_g y_g z_g$ and $x_p y_p z_p$, that originate at contact point, K

$$\mathbf{Rt}_{\omega}(g \rightarrow p)$$

$$= \left[\frac{\sin\left(\omega_g - \alpha\right)}{\sin \omega_g} \; \frac{\sin\left(\omega_g - \omega_p - \alpha\right)}{\sin \omega_g} 00 \frac{\sin \alpha}{\sin \omega_g} \frac{\sin\left(\omega_p + \alpha\right)}{\sin \omega_g} 0000 - 100001 \right] \tag{D.57}$$

Following the second way of transformation of the local coordinate systems, the auxiliary orthogonal local coordinate system must be constructed.

Let's consider an approach, according to which a closed loop (a closed circuit) of the consequent coordinate system transformations can be composed.

In order to construct an orthogonal normalized basis of the coordinate system $x_g y_g z_g$, an intermediate coordinate system $x_1 y_1 z_1$ is used. Axis x_1 of the coordinate system $x_1 y_1 z_1$ is pointed out along the unit vector $\mathbf{u}_g$ that is tangent to the U_g-coordinate curve (see Fig. D.15). Axis y_1 is pointed along vector $\mathbf{v}_g$ that is tangent to the V_g-coordinate line on the gear tooth flank, $\mathcal{G}$. The axis, z_1, aligns with unit normal vector, $\mathbf{n}_g$, and is pointed outward from the gear tooth body.

For an orthogonally parameterized gear tooth flank, $\mathcal{G}$, (for which $F_g = 0$, and therefore $\omega_g = \pi/2$), analytical description of coordinate system transformations is significantly simpler. Further simplification of the coordinate system transformation is possible when the coordinate U_g-, and V_g-lines are congruent to the lines of curvature on the part surface $\mathcal{G}$. Under such a scenario, the local coordinate system is represented by the corresponding *Darboux frame.*

In order to construct *Darboux frame*, principal directions on the gear tooth flank, $\mathcal{G}$, must be calculated. Determination of the unit tangent vectors, $\mathbf{t}_{1.\,g}$ and $\mathbf{t}_{2.\,g}$, of the principal directions on the gear tooth flank, $\mathcal{G}$, is considered in Appendix B.

In the common tangent plane, orientation of the unit vector, $\mathbf{t}_{1.\,g}$, of the first principal direction on the gear tooth flank, $\mathcal{G}$, can be uniquely specified by the

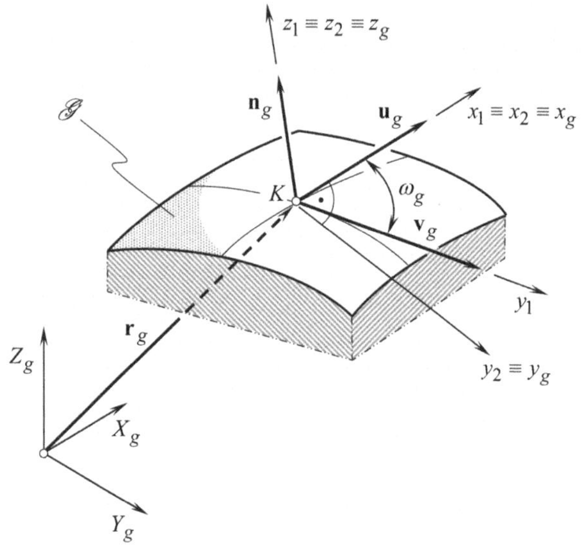

Fig. D.15 Local coordinate system, $x_g y_g z_g$, associated with the gear tooth flank, $\mathcal{G}$

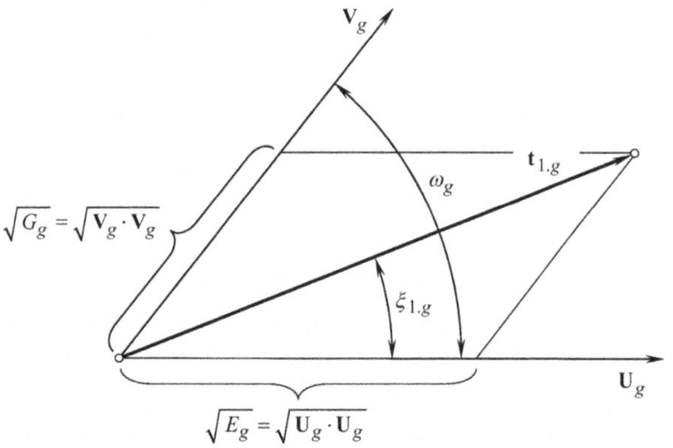

Fig. D.16 Differential relationships between the tangent vectors, $\mathbf{U}_g$ and $\mathbf{V}_g$, the fundamental magnitudes of the first order, the included angle, $\xi_{1.\ g}$, and the direction of the unit tangent vector, $\mathbf{t}_{1.\ g}$, at point of the gear tooth flank, $\mathcal{G}$

included angle, $\xi_{1.\ g}$, that the unit vector, $\mathbf{t}_{1.\ g}$, forms with the U_g—coordinate curve. This angle depends on both on the gear tooth flank, $\mathcal{G}$, geometry and on the gear tooth flank, $\mathcal{G}$, parameterization. Depicted in Fig. D.16, is the relationship between the tangent vectors $\mathbf{U}_g$ and $\mathbf{V}_g$, and the included angle $\xi_{1.\ g}$. From the law of sine's,

$$\frac{\sqrt{G_g}}{\sin \xi_{1.g}} = \frac{\sqrt{E_g}}{\sin\left[\pi - \xi_{1.g} - \left(\pi - \omega_g\right)\right]} = \frac{\sqrt{F_g}}{\sin\left(\omega_g - \xi_{1.g}\right)} \tag{D.58}$$

Here, $\omega_g = \cos^{-1}\left(F_g/\sqrt{E_g G_g}\right)$.

Solving the expression above for the included angle, $\xi_{1.\ g}$, results in:

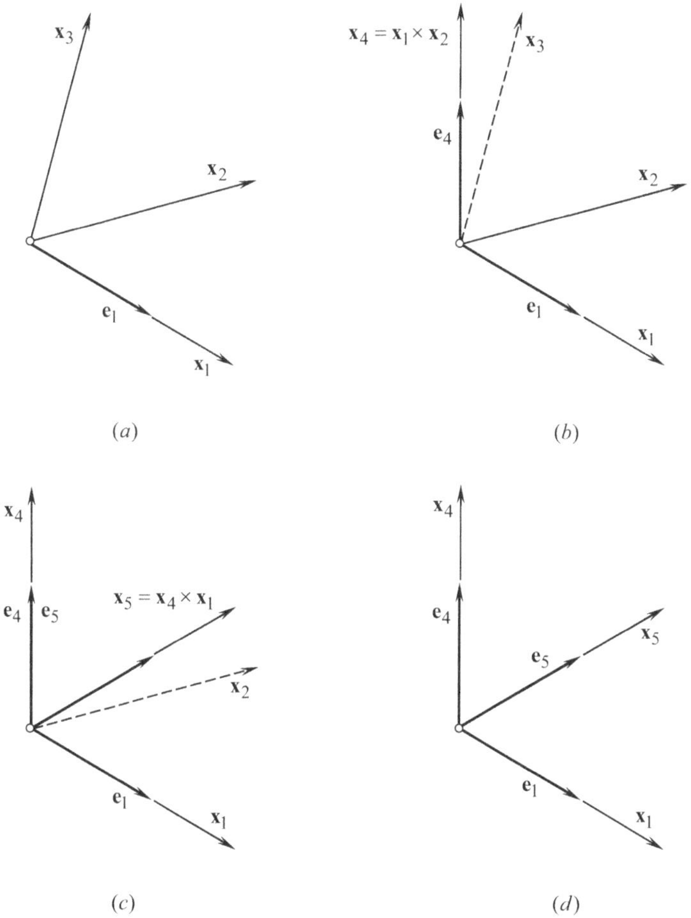

Fig. D.17 A normalized and orthogonally parameterized basis, $\mathbf{e}_1\mathbf{e}_4\mathbf{e}_5$, that is constructed based on arbitrary basis, $\mathbf{x}_1\mathbf{x}_2\mathbf{x}_3$

$$\xi_{1.g} = \tan^{-1}\frac{\sqrt{E_g G_g - F_g^2}}{E_g + F_g} \tag{D.59}$$

Another possible way of constructing of orthogonal local basis of the local *Cartesian coordinate system*, $x_P y_P z_P$, is based on the following consideration.

Consider an arbitrary non-orthogonal and not normalized basis, $\mathbf{x}_1\mathbf{x}_2\mathbf{x}_3$ (see Fig. D.17a). Let's construct an orthogonal and normalized basis based on the initially given basis, $\mathbf{x}_1\mathbf{x}_2\mathbf{x}_3$.

Cross product of any two of three vectors, $\mathbf{x}_1$, $\mathbf{x}_2$, $\mathbf{x}_3$, for example, cross product of the vectors $\mathbf{x}_1 \times \mathbf{x}_2$, determines a new vector, $\mathbf{x}_4$ (see Fig. D.17b). Evidently, the vector, $\mathbf{x}_4$, is orthogonal to the coordinate plane, $\mathbf{x}_1\mathbf{x}_2$. Then, use the calculated vector, $\mathbf{x}_4$, and one of two original vectors, $\mathbf{x}_1$ or $\mathbf{x}_2$, for instance, use the vector, $\mathbf{x}_2$. This yields calculation of a new vector, $\mathbf{x}_5 = \mathbf{x}_4 \times \mathbf{x}_2$ (see Fig. D.17c). The calculated basis, $\mathbf{x}_1\mathbf{x}_4\mathbf{x}_5$, is orthogonal. In order to convert it into a normalized basis, each of the vectors, cos $\angle\,(Y^{(K)}, X_{д}) =$ [cos $\angle\,(X^{(K)}, Y_{д})$ cos $\angle\,(Z^{(K)}, Z_{д})$ − cos $\angle\,(X^{(K)}$, $Z_{д})$ cos $\angle\,(Z^{(K)}, Y_{д})$];, cos $\angle\,(Y^{(K)}, Y_{д}) =$ [cos $\angle\,(X^{(K)}, X_{д})$ cos $\angle\,(Z^{(K)}, Z_{д})$ − cos $\angle\,(X^{(K)}, Z_{д})$ cos $\angle\,(Z^{(K)}, X_{д})$];, and cos $\angle\,(Y^{(K)}, Z_{д}) =$ [cos $\angle\,(X^{(K)}$, $X_{д})$ cos $\angle\,(X^{(K)}, Y_{д})$ − cos $\angle\,(X^{(K)}, Y_{д})$ cos $\angle\,(Z^{(K)}, X_{д})$]. must be divided by its magnitude:

$$\mathbf{e}_1 = \frac{\mathbf{x}_1}{|\mathbf{x}_1|} \tag{D.60}$$

$$\mathbf{e}_4 = \frac{\mathbf{x}_4}{|\mathbf{x}_4|} \tag{D.61}$$

$$\mathbf{e}_5 = \frac{\mathbf{x}_5}{|\mathbf{x}_5|} \tag{D.62}$$

The resultant basis $\mathbf{e}_1\mathbf{e}_4\mathbf{e}_5$ (see Fig. D.17d) is always orthogonal, as well as it is always normalized.

In order to complete the analytical description of a closed loop of consequent coordinate system transformations, it is necessary to compose the operator **Rs** $(K_p \rightarrow K_g)$ of transformation from the local reference system, $x_p y_p z_p$, to the local reference system, $x_g y_g z_g$ (see Fig. D.13c).

In the case under consideration, the axes, z_g and z_p, align with the common unit normal vector, $\mathbf{n}_g$. The axis, z_g, is pointed out from the bodily side to the void side of the gear tooth flank, $\mathcal{G}$. The axis, z_g, is pointed oppositely. Due to that, the following equality is observed:

$$\mathbf{Rs}\left(K_p \rightarrow K_g\right) = \mathbf{Rt}\left(\varphi_z,\ z_p\right) \tag{D.63}$$

The inverse coordinate system transformation is analytically described by the operator:

$$\mathbf{Rs}\left(K_g \rightarrow K_p\right) = \mathbf{Rs}^{-1}\left(K_p \rightarrow K_g\right) = \mathbf{Rt}\left(-\varphi_z,\ z_p\right) \tag{D.64}$$

Implementation of the discussed results allows for:

(a) representation of the gear tooth flank, $\mathcal{G}$, and the pinion tooth flank, $\mathcal{P}$, as well as their relative motion, in a common coordinate system, and
(b) consideration meshing of the gear tooth flank, $\mathcal{G}$, and the pinion tooth flank, $\mathcal{P}$, in any desirable coordinate systems that is a component of the chain and/or the closed loop of consequent coordinate system transformations.

Transition from one coordinate system to another coordinate system can be performed in both of two feasible directions, say in direct as well as in inverse directions.

D.5 Impact of Coordinate Systems Transformations onto Fundamental Forms of Surface

Every coordinate system transformation entails a corresponding alteration to equation of the gear tooth flank, $\mathcal{G}$, and/or pinion tooth flank, $\mathcal{P}$. Because of this, it is often necessary to recalculate coefficients of the first $\Phi_{1.\ g}$ and of the second $\Phi_{2.\ g}$ fundamental forms of the gear tooth flank, $\mathcal{G}$ (as many times as the coordinate system transformation is performed). This routing and time-consuming operation can be eliminated if the operators of coordinate system transformations are used directly to the fundamental forms $\Phi_{1.\ g}$ and $\Phi_{2.\ g}$.

After being calculated in an initial reference system, fundamental magnitudes E_g, F_g, and G_g of the first, $\Phi_{1.\ g}$, and fundamental magnitudes L_g, M_g, and N_g of the second, $\Phi_{2.\ g}$, fundamental forms can be determined in any new coordinate system using for this purpose the operators of translation, of rotation and of resultant coordinate system transformation. Transformation of such kind of the fundamental forms, $\Phi_{1.\ g}$ and $\Phi_{2.\ g}$, becomes possible due to implementation of a formula, that is derived immediately below.

Let's consider a gear tooth flank, $\mathcal{G}$, that is given by equation $\mathbf{r}_g = \mathbf{r}_g(U_g,\ V_g)$, where $\left(U_g, V_g\right) \in \mathcal{G}$.

For the analysis below, it is convenient to use the equation of the first fundamental form, $\Phi_{1.\ g}$, of the gear tooth flank, $\mathcal{G}$, represented in matrix form:

$$\left[\Phi_{1.g}\right] = \left[dU_g \quad dV_g \quad 0 \quad 0\right] \cdot \begin{bmatrix} E_g & F_g & 0 & 0 \\ F_g & G_g & 0 & 0 \\ 0 & 0 & 1 & 0 \\ 0 & 0 & 0 & 1 \end{bmatrix} \cdot \begin{bmatrix} dU_g \\ dV_g \\ 0 \\ 0 \end{bmatrix} \tag{D.65}$$

Similarly, equation of the second fundamental form $\Phi_{2.\ g}$ of the gear tooth flank, $\mathcal{G}$, can be given by:

$$\left[\Phi_{2.g}\right] = \left[dU_g \quad dV_g \quad 0 \quad 0\right] \cdot \begin{bmatrix} L_g & M_g & 0 & 0 \\ M_g & N_g & 0 & 0 \\ 0 & 0 & 1 & 0 \\ 0 & 0 & 0 & 1 \end{bmatrix} \cdot \begin{bmatrix} dU_g \\ dV_g \\ 0 \\ 0 \end{bmatrix} \tag{D.66}$$

The coordinate system transformation that is performed by the operator of linear transformation $\mathbf{Rs}\,(1 \rightarrow 2)$ transfers the equation $\mathbf{r}_g = \mathbf{r}_g(U_g,\ V_g)$ of the gear tooth flank, $\mathcal{G}$, initially given in $X_1Y_1Z_1$, to the equation $\mathbf{r}^*_g = \mathbf{r}^*_g\left(U^*_g,\ V^*_g\right)$ of that same gear tooth flank, $\mathcal{G}$, in a new coordinate system $X_2Y_2Z_2$. It is clear that $\mathbf{r}_g \neq \mathbf{r}^*_g$.

In the new coordinate system, the gear tooth flank, $\mathcal{G}$, is analytically described by the following expression:

$$\mathbf{r}_g^*\left(U_g^*,\ V_g^*\right) = \mathbf{Rs}\,(1 \to 2) \cdot \mathbf{r}_g\left(U_g,\ V_g\right) \tag{D.67}$$

The column matrices of variables in Eqs. (D.65) and (D.66) cast the operator of the resultant coordinate system transformation $\mathbf{Rs}\,(1 \to 2)$ to the form:

$$\left[dU_g^* \quad dV_g^* \quad 0 \quad 0\right]^T = \mathbf{Rs}\,(1 \to 2) \cdot \left[dU_g \quad dV_g \quad 0 \quad 0\right]^T. \tag{D.68}$$

Substitution of Eq. (D.68) into Eqs. (D.65) and (D.66) makes possible the expressions for the fundamental forms, $\Phi_{1.g}^*$ and $\Phi_{2.g}^*$, in the new coordinate system:

$$\begin{aligned}\left[\Phi_{1.g}^*\right] = &\left[\mathbf{Rs}\,(1 \to 2) \cdot \left[dU_g \quad dV_g \quad 0 \quad 0\right]^T\right]^T \cdot \left[\Phi_{1.g}\right] \cdot \mathbf{Rs}\,(1 \to 2)\\ &\cdot \left[dU_g \quad dV_g \quad 0 \quad 0\right]^T\end{aligned} \tag{D.69}$$

$$\begin{aligned}\left[\Phi_{2.g}^*\right] = &\left[\mathbf{Rs}\,(1 \to 2) \cdot \left[dU_g \quad dV_g \quad 0 \quad 0\right]^T\right]^T \cdot \left[\Phi_{2.g}\right] \cdot \mathbf{Rs}\,(1 \to 2)\\ &\cdot \left[dU_g \quad dV_g \quad 0 \quad 0\right]^T\end{aligned} \tag{D.70}$$

The following equation is valid for multiplication:

$$\begin{aligned}\left[\mathbf{Rs}\,(1 \to 2) \cdot \left[dU_g \quad dV_g \quad 0 \quad 0\right]^T\right]^T = \mathbf{Rs}^T(1 \to 2)\\ \cdot \left[dU_g \quad dV_g \quad 0 \quad 0\right]\end{aligned} \tag{D.71}$$

Therefore,

$$\begin{aligned}\left[\Phi_{1.g}^*\right] = \left[dU_g \quad dV_g \quad 0 \quad 0\right]^T \cdot \left\{\mathbf{Rs}^T(1 \to 2) \cdot \left[\Phi_{1.g}\right] \cdot \mathbf{Rs}\,(1 \to 2)\right\}\\ \cdot \left[dU_g \quad dV_g \quad 0 \quad 0\right]\end{aligned} \tag{D.72}$$

$$\begin{aligned}\left[\Phi_{2.g}^*\right] = \left[dU_g \quad dV_g \quad 0 \quad 0\right]^T \cdot \left\{\mathbf{Rs}^T(1 \to 2) \cdot \left[\Phi_{2.g}\right] \cdot \mathbf{Rs}\,(1 \to 2)\right\}\\ \cdot \left[dU_g \quad dV_g \quad 0 \quad 0\right]\end{aligned} \tag{D.73}$$

It can be easily shown that the matrices $\left[\Phi_{1.g}^*\right]$ and $\left[\Phi_{2.g}^*\right]$ in Eqs. (D.72) and (D.73) represent quadratic forms with respect to dU_g and dV_g.

Transformation of the first, $\Phi_{1.\,g}$, and the second, $\Phi_{2.\,g}$, fundamental forms of the gear tooth flank, $\mathcal{G}$, from the initial coordinate system $X_1Y_1Z_1$ to the new coordinate system, $X_2Y_2Z_2$, is executed by the operator of transformation $\mathbf{Rs}\,(1 \to 2)$. This results in that in the new coordinate system the corresponding fundamental forms are expressed by the expressions:

$$\left[\Phi^*_{1.g}\right] = \mathbf{Rs}^T(1 \rightarrow 2) \cdot \left[\Phi_{1.g}\right] \cdot \mathbf{Rs}(1 \rightarrow 2) \tag{D.74}$$

$$\left[\Phi^*_{2.g}\right] = \mathbf{Rs}^T(1 \rightarrow 2) \cdot \left[\Phi_{2.g}\right] \cdot \mathbf{Rs}(1 \rightarrow 2) \tag{D.75}$$

Equations (D.74) and (D.75) reveal that after the coordinate system transformation is completed, the first $\Phi^*_{1.g}$ and the second $\Phi^*_{2.g}$ fundamental forms of the gear tooth flank, $\mathcal{G}$, in the coordinate system $X_2Y_2Z_2$ are expressed in terms of the first, $\Phi_{1.g}$, and the second, $\Phi_{2.g}$, fundamental forms initially represented in the coordinate system $X_1Y_1Z_1$. In order to do that, the corresponding fundamental form (either the form, $\Phi_{1.g}$ or the form, $\Phi_{2.g}$) must be pre-multiplied by $\mathbf{Rs}(1 \rightarrow 2)$ and after that it been post-multiplied by $\mathbf{Rs}^T(1 \rightarrow 2)$.

Implementation of Eqs. (D.74) and (D.75) significantly simplifies formulae transformations.

Equations similar to those above Eqs. (D.74) and (D.75) are valid with respect to pinion tooth flank, $\mathcal{P}$, as well as to any and all quadratic forms.

In case of use of the third, $\Phi_{3.g}$, and of the fourth, $\Phi_{4.g}$, fundamental forms, their coefficients can be expressed in terms of the fundamental magnitudes of the first and of the second order.

Appendix E: Contact Geometry of Gear and Mating Pinion Tooth Flanks

In the theory of gearing, the kinematics of gearing is considered the prime element of the gear pair. Other important elements of gearing, namely:

(a) the shape and the geometry of the gear tooth flank, $\mathcal{G}$, and
(b) the shape and the geometry of the mating pinion tooth flank, $\mathcal{P}$ (as well as numerous others)

are considering the secondary elements of gearing. This does not mean that the importance of the secondary elements is lower than that of the primary element. No, this is incorrect. This means just that the most favorable parameters of the secondary elements can be expressed in terms of the parameters of the prime elements. Ultimately, the entire gear pair can be synthesized on the premise just of the prime elements, that is, on the premise of the desirable kinematics of the gear pair. In other words, having just the desirable kinematics of the gear pair to be designed, it is possible to synthesize the rest of the design parameters of the gear pair. Only the kinematics of gearing is used for the purposes of synthesizing of the best possible gear pair for transmitting the input rotation and torque.

In order to solve the problem of synthesizing the most favorable gear pair, an appropriate analytical description of contact geometry of the gear tooth flank, $\mathcal{G}$, and the mating pinion tooth flank, $\mathcal{P}$, is required. The problem of analytical description of contact geometry between two smooth regular surfaces in the first order of tangency is a sophisticated one.

Investigation into contact geometry of curves and surfaces can be traced back to the eighteenth century. In considerable detail the study of the contact of curves and surfaces has been undertaken by *J.L. Lagrange*[16] in his "*Theorié des Fonctions Analytiques*" (1797), and *A.L. Cauchy*[17] in his "*Leçons sur les Applications du Calcul Infinitésimal á la Geometrie*" (1826). Later on, in the twentieth century an investigation in the realm of contact geometry of curves and surfaces has been undertaken by *J. Favard*[18] in his "*Course de Gèomètrie Diffèrentialle Locale*" (1957). A few more names of the researchers in the field are to be mentioned.

The results of the research obtained in the field of contact geometry of two smooth regular surfaces are widely used in the theory of gearing. The problem of synthesizing the most favorable gear pair can be solved on the premise of the analysis of topology of the contacting surfaces in differential vicinity of the point of their contact.

Various methods for analytical description of contact geometry between two smooth regular surfaces are developed by now. Latest achievements in the field are discussed in plurality of papers and monograph available in the public domain.

An in-detail analysis of known methods of analytical description of the geometry of contact between two smooth regular surfaces uncovered poor capability of known methods for solving problems in the field of designing efficient gear pairs. Therefore, an accurate method for analytical description of contact geometry between two smooth regular surfaces, $\mathcal{G}$ and $\mathcal{P}$, in the first order of tangency, which fits the needs of the theory of gearing, is necessary. Such a method is worked out in this section of the book.

It is convenient to begin the discussion starting from analytical description of local relative orientation of the gear tooth flank, $\mathcal{G}$, and the mating pinion tooth flank, $\mathcal{P}$. The proposed analytical description is relevant to differential vicinity of the point of contact, K, of the tooth flanks, $\mathcal{G}$ and $\mathcal{P}$.

E.1 Local Relative Orientation at Point of Contact of Gear and Mating Pinion Tooth Flanks

When the gears rotate, a gear tooth flank, $\mathcal{G}$, and the mating pinion tooth flank, $\mathcal{P}$, are in permanent tangency with one another. The contacting surfaces, $\mathcal{G}$ and $\mathcal{P}$, can be locally approximated by the corresponding quadrics as schematically illustrated in Fig. E.1. Certain restrictions on the relative configuration (location and orientation) of the tooth flanks, $\mathcal{G}$ and $\mathcal{P}$, and on their instant relative motion are imposed by the requirement to be permanently in tangency to each other.

---

[16] *Joseph-Louis* ***Lagrange*** (January 25, 1736 – April 10, 1813) – an Italian born [born *Giuseppe Lodovico (Luigi)* ***Lagrangia***] famous French mathematician and mechanician.

[17] *Augustin-Louis* ***Cauchy*** (August 21, 1789 – May 23, 1857) – a famous French mathematician.

[18] *Jean* ***Favard*** (August 28, 1902 – January 21, 1965) – a French mathematician.

In the theory of gearing, a quantitative measure of relative orientation of the gear tooth flank, $\mathcal{G}$, and of the mating pinion tooth flank, $\mathcal{P}$, is established.

Relative orientation of the gear tooth flank, $\mathcal{G}$, and of the mating pinion tooth flank, $\mathcal{P}$, is specified by the angle, μ, of local[19] relative orientation of the surfaces. By definition, angle μ is equal to the angle that the unit tangent vector, $\mathbf{t}_{1.\ g}$, of the first principal direction of the gear tooth flank, $\mathcal{G}$, forms with the unit tangent vector, $\mathbf{t}_{1.\ p}$, of the first principal direction of the mating pinion tooth flank, $\mathcal{P}$. That same angle, μ, can also be defined as the angle that make the unit tangent vectors, $\mathbf{t}_{2.\ g}$ and $\mathbf{t}_{2.\ p}$, of the second principal directions of the surfaces, $\mathcal{G}$ and $\mathcal{P}$, at contact point, K. This immediately yields equations for the calculation of the angle, μ:

$$\sin\mu = |\mathbf{t}_{1.g}\times\mathbf{t}_{1.p}| = |\mathbf{t}_{2.g}\times\mathbf{t}_{2.p}|, \tag{E.1}$$

$$\cos\mu = \mathbf{t}_{1.g}\cdot\mathbf{t}_{1.p} = \mathbf{t}_{2.g}\cdot\mathbf{t}_{2.p}, \tag{E.2}$$

$$\tan\mu = \frac{|\mathbf{t}_{1.g}\times\mathbf{t}_{1.p}|}{\mathbf{t}_{1.g}\cdot\mathbf{t}_{1.p}} \equiv \frac{|\mathbf{t}_{2.g}\times\mathbf{t}_{2.p}|}{\mathbf{t}_{2.g}\cdot\mathbf{t}_{2.p}} \tag{E.3}$$

Here is designated:

$\mathbf{t}_{1.\ g}$, $\mathbf{t}_{2.\ g}$ – are the unit vectors of principal directions on the gear tooth flank, $\mathcal{G}$, measured at contact point, K

$\mathbf{t}_{1.\ p}$, $\mathbf{t}_{2.\ p}$ – are the unit vectors of principal directions on the mating pinion tooth flank, $\mathcal{P}$, at that same contact point, K, of the tooth flanks, $\mathcal{G}$ and $\mathcal{P}$

Directions of the unit tangent vectors, $\mathbf{t}_{1.\ g}$ and $\mathbf{t}_{2.\ g}$, of the principal directions on the gear tooth flank, $\mathcal{G}$ (as well as directions of the unit tangent vectors $\mathbf{t}_{1.\ p}$ and $\mathbf{t}_{2.\ p}$ of the principal directions on the pinion tooth flank, $\mathcal{P}$) can be specified in terms of the ratio dU_g/dV_g (or in terms of the ratio dU_p/dV_p in case of the pinion tooth flank, $\mathcal{P}$). The corresponding values of the ratio, $dU_{g(p)}/dV_{g(p)}$, are calculated as roots of the quadratic equation:

$$\begin{vmatrix} E_{g(p)}\dfrac{dU_{g(p)}}{dV_{g(p)}} + F_{g(p)} & F_{g(p)}\dfrac{dU_{g(p)}}{dV_{g(p)}} + G_{g(p)} \\ L_{g(p)}\dfrac{dU_{g(p)}}{dV_{g(p)}} + M_{g(p)} & M_{g(p)}\dfrac{dU_{g(p)}}{dV_{g(p)}} + N_{g(p)} \end{vmatrix} = 0 \tag{E.4}$$

In case of point of contact of the surfaces, $\mathcal{G}$ and $\mathcal{P}$, the actual value of the angle, μ, is calculated at the contact point, K. When the tooth flanks, $\mathcal{G}$ and $\mathcal{P}$, are in line contact, then actual value of the angle, μ, can be calculated at every point within the

[19] The surfaces orientation is *local* by nature, as it relates only to the differential vicinity of point, K, of contact of the tooth flanks, $\mathcal{G}$ and $\mathcal{P}$.

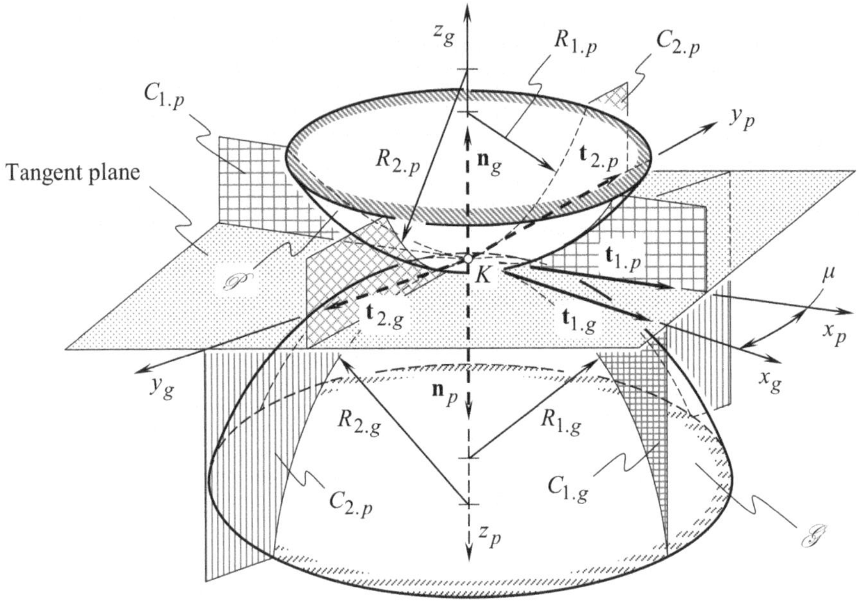

Fig. E.1 Local configuration of two quadrics that are tangent to a gear tooth flank $\mathcal{G}$ and to a mating pinion tooth flank $\mathcal{P}$ at contact point K. (After Prof. *S.P. Radzevich*, ~2008)

line of contact[20]. The line of contact of the tooth flanks, $\mathcal{G}$ and $\mathcal{P}$, is commonly referred to *characteristic line*, $\mathcal{E}$, or just *characteristic*, $\mathcal{E}$, for simplicity.

The determination of the angle, μ, of local relative orientation of the tooth flanks, $\mathcal{G}$ and $\mathcal{P}$, of a gear and a mating pinion is illustrated in Fig. E.1.

In order to calculate the actual value of the angle, μ, of local relative orientation of the tooth flanks, $\mathcal{G}$ and $\mathcal{P}$, the unit vectors of the principal directions, $\mathbf{t}_{1.\,g}$ and $\mathbf{t}_{1.\,p}$, are employed.

Consider tooth flanks, $\mathcal{G}$ and $\mathcal{P}$, in point of contact, which are represented in a common reference system. The surfaces make contact at point, K. For further analysis, an equation of the common tangent plane to the tooth flanks, $\mathcal{G}$ and $\mathcal{P}$, at the contact point, K, is necessary (see Figs. E.1 and E.2).

$$\left(\mathbf{r}_{tp} - \mathbf{r}_K\right) \cdot \mathbf{u}_g \cdot \mathbf{v}_g = 0 \tag{E.5}$$

Here:

$\mathbf{r}_{tp}$ – is the position vector of point of the common tangent plane
$\mathbf{r}_K$ – is the position vector of the contact point, K

[20] It is worthy pointing out here that in a case of line contact, relative orientation of two surfaces, $\mathcal{G}$ and $\mathcal{P}$, is predetermined in *global* sense. However, actual value of the angle, μ, of the surface *local* relative orientation at different points of the characteristic, $\mathcal{E}$, is different.

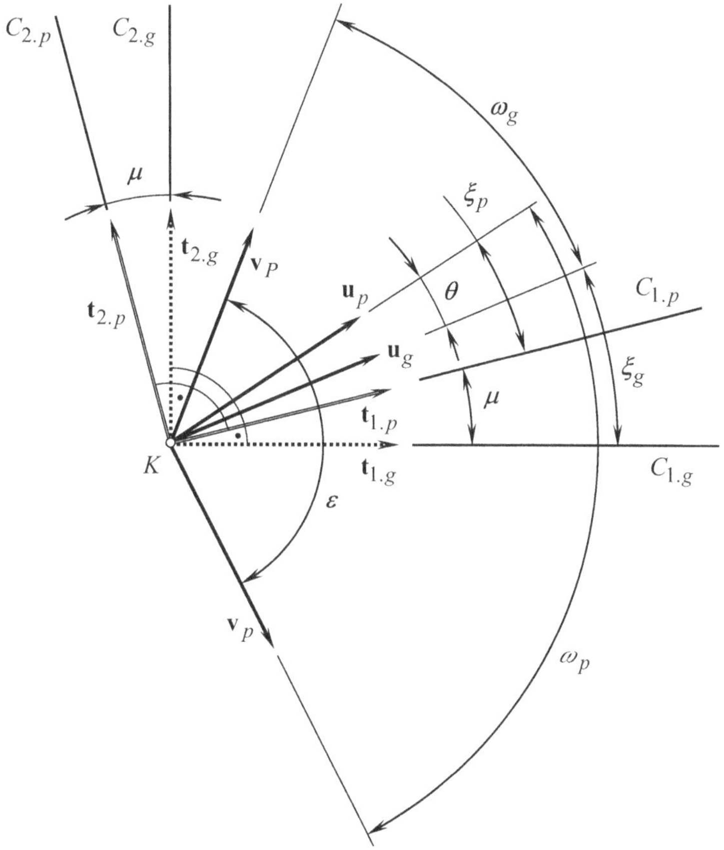

Fig. E.2 Local *relative* orientation at contact point of tooth flanks of a gear, $\mathcal{G}$, and a mating pinion, $\mathcal{P}$, considered in a common tangent plane

$\mathbf{u}_g$ and $\mathbf{v}_g$ – are unit vectors that are tangent to U_g-, and V_g-coordinate lines on the gear tooth flank, $\mathcal{G}$ at the contact point, K

The angle ω_g is the angle that is formed by the unit vectors, $\mathbf{u}_g$ and $\mathbf{v}_g$. The actual value of the angle, ω_g, can be calculated from one of the following equations:

$$\sin \omega_g = \frac{\sqrt{E_g G_g - F_g^2}}{\sqrt{E_g G_g}} \tag{E.6}$$

$$\cos \omega_g = \frac{F_g}{\sqrt{E_g G_g}} \tag{E.7}$$

$$\tan \omega_g = \frac{\sqrt{E_g G_g - F_g^2}}{F_g} \tag{E.8}$$

Equations similar to Eq. (E.6) through Eq. (E.8) are also valid for the calculation of the angle, ω_p, at point on the pinion tooth flank, $\mathcal{P}$.

Tangent directions, $\mathbf{u}_g$ and $\mathbf{v}_g$, to the U_g- and V_g-coordinate lines at point on the gear tooth flank, $\mathcal{G}$, as well as tangent directions, $\mathbf{u}_p$ and $\mathbf{v}_p$ to the U_p- and V_p- coordinate at point lines on the pinion tooth flank, $\mathcal{P}$, are specified in terms of the angles, θ and ε. For the calculation of actual values of the angles, θ and ε, the following equations are: used:

$$\cos\theta = \mathbf{u}_g \cdot \mathbf{u}_p \tag{E.9}$$

$$\cos\varepsilon = \mathbf{v}_g \cdot \mathbf{v}_p \tag{E.10}$$

The angle, ξ_g, is the angle that the first principal direction, $\mathbf{t}_{1.\,g}$, on the gear tooth surface, $\mathcal{G}$, forms with the unit tangent vector, $\mathbf{u}_g$ (see Fig. E.2). The equation for the calculation of an actual value of the angle, ξ_g, is derived by Prof. *S.P. Radzevich*:

$$\sin\xi_g = \frac{\eta_g}{\sqrt{\eta_g{}^2 - 2\eta_g\cos\omega_g + 1}}\sin\omega_g \tag{E.11}$$

where η_g designates the ratio $\eta_g = \frac{\partial U_g}{\partial V_g}$.

In the event $F_g = 0$, the following equality, $\tan\xi_g = \eta_g$, is observed. Here, the ratio, η_g, is equal to the root of the quadratic equation:

$$\left(F_g L_g - E_g M_g\right)\eta_g^2 + \left(G_g L_g - E_g N_g\right)\eta_g + \left(G_g M_g - F_g N_g\right) = 0 \tag{E.12}$$

which immediately follows from the equation:

$$\begin{vmatrix} E_g\,dU_g + F_g\,dV_g & F_g\,dU_g + G_g\,dV_g \\ L_g\,dU_g + M_g\,dV_g & M_g\,dU_g + N_g\,dV_g \end{vmatrix} = 0 \tag{E.13}$$

The equation for the calculation of actual value of the angle, ξ_g, allows for another representation. Following chain rule, $d\,\mathbf{r}_g$ can be represented in the form:

$$d\,\mathbf{r}_g = \mathbf{U}_g\,dU_g + \mathbf{V}_g\,dV_g \tag{E.14}$$

By definition, $\tan\xi_g = \frac{\sin\xi_g}{\cos\xi_g}$. The functions, $\sin\xi_g$ and $\cos\xi_g$, yield representation as:

$$\sin\xi_g = \frac{|\,\mathbf{U}_g \times d\,\mathbf{r}_g\,|}{|\mathbf{U}_g|\cdot|\,d\,\mathbf{r}_g\,|} \tag{E.15}$$

$$\cos \xi_P g = \frac{\mathbf{U}_g \cdot d\,\mathbf{r}_g}{|\mathbf{U}_g| \cdot |\, d\,\mathbf{r}_g \,|} \tag{E.16}$$

The last expressions yield:

$$\tan \xi_g = \frac{\sin \xi_g}{\cos \xi_g} = \frac{|\,\mathbf{U}_g \times d\,\mathbf{r}_g\,|}{\mathbf{U}_g \cdot d\,\mathbf{r}_g} = \frac{|\,\mathbf{U}_g \times d\,\mathbf{r}_g\,|}{\mathbf{U}_g \cdot (\mathbf{U}_g dU_g + \mathbf{V}_g dV_g)}$$

$$= \frac{|\,\mathbf{U}_g \times d\,\mathbf{r}_g\,| \cdot dV_g}{\mathbf{U}_g \cdot \mathbf{U}_g dU_g + \mathbf{U}_g \cdot \mathbf{V}_g dV_g} \tag{E.17}$$

By definition:

$$\mathbf{U}_g \cdot \mathbf{U}_g = E_g \tag{E.18}$$

$$\mathbf{U}_g \cdot \mathbf{V}_g = F_g \tag{E.19}$$

$$|\mathbf{U}_g \times \mathbf{V}_g| = \sqrt{E_g G_g - F_g^2} \tag{E.20}$$

Equation (E.14) through Eq. (E.20) yield the formula:

$$\tan \xi_g = \frac{\sqrt{E_g G_g - F_g^2}}{\eta_g \cdot E_g + F_g} \tag{E.21}$$

for the calculation of actual value of the angle ξ_g.

Equations similar to those above Eqs. (E.11) and (E.21) are also valid for the calculation of actual value of the angle, ξ_p, that the first principal direction, $\mathbf{t}_{1.\,p}$, at point on the pinion tooth flank, $\mathcal{P}$, forms with the unit tangent vector, $\mathbf{u}_p$.

The performed analysis makes possible the following equations for the calculation of the unit vectors of principal directions, $\mathbf{t}_{1.\,g}$ and $\mathbf{t}_{2.\,g}$:

$$\mathbf{t}_{1.g} = \mathbf{Rt}\left(\xi_g,\ \mathbf{n}_g\right) \cdot \mathbf{u}_g \tag{E.22}$$

$$\mathbf{t}_{2.g} = \mathbf{Rt}\left[\left(\xi_g + \frac{\pi}{2}\right),\ \mathbf{n}_g\right] \cdot \mathbf{u}_g \tag{E.23}$$

for the gear tooth flank, $\mathcal{G}$, and similar equations for the calculation of the unit vectors of principal directions, $\mathbf{t}_{1.\,p}$ and $\mathbf{t}_{2.\,p}$:

$$\mathbf{t}_{1.p} = \mathbf{Rt}\left(\xi_p,\ \mathbf{n}_g\right) \cdot \mathbf{u}_p \tag{E.24}$$

$$\mathbf{t}_{2.p} = \mathbf{Rt}\left[\left(\xi_p + \frac{\pi}{2}\right), \mathbf{n}_g\right] \cdot \mathbf{u}_p \tag{E.25}$$

for the pinion tooth flank, $\mathcal{P}$.

The operator of rotation $\mathbf{Rt}(\varphi_A, A_0)$ through an angle φ_A about an axis A_0 is employed for the calculation of the operators of rotation in Eq. (E.22) through Eq. (E.25).

E.2 The Second Order Analysis: Planar Characteristic Images

For a more accurate analytical description of contact geometry of the gear tooth flank, $\mathcal{G}$, and the pinion tooth flank, $\mathcal{P}$, consideration of the second order parameters is desirable. The second order analysis incorporates elements both of the first order and of the second order analysis. For performing the second order analysis, familiarity with *Dupin indicatrix* is highly desirable[21]. *Dupin indicatrix* is a perfect start-point for consideration of the second order analysis.

E.2.1 Preliminary Remarks: *Dupin indicatrix*

At any point of a smooth regular gear tooth flank, $\mathcal{G}$ (as well as at any point of a smooth regular pinion tooth flank, $\mathcal{P}$) a corresponding *Dupin indicatrix* can be constructed. *Dupin indicatrix, Dup* $(\mathcal{G})$ at a point of a gear tooth flank, $\mathcal{G}$, as well as *Dupin indicatrix, Dup* $(\mathcal{P})$ at a point of the pinion tooth flank, $\mathcal{P}$, are planar characteristic curves of the second order. They are used for graphical interpretation of the distribution of normal radii of curvature of a surface in the differential vicinity of a surface point.

Dupin indicatrices at a point of the tooth flank, $\mathcal{G}$, and at a point of the tooth flank, $\mathcal{P}$, are of critical importance in the theory of gearing. Generation of this planar characteristic curve is illustrated with a diagram shown in Fig. E.3.

A plane W through the unit normal vector, $\mathbf{n}_g$, to the gear tooth flank, $\mathcal{G}$, at a point, m, is rotating about the unit normal vector, $\mathbf{n}_g$. While rotating, the plane occupies consecutive positions, W_1, W_2, W_3, and others. The radii of normal curvature of the line of intersection of the gear tooth flank, $\mathcal{G}$, by normal planes, W_1, W_2, W_3, are equal $R_{g,\,1}$, $R_{g,\,2}$, $R_{g,\,3}$, and so forth. The gear tooth flank, $\mathcal{G}$, is intersected by a plane, Q. The plane, Q, is orthogonal to the unit normal vector, $\mathbf{n}_g$. This plane is at a certain small distance, δ, from the point, m. When the distance, δ, approaches a zero ($\delta \to 0$), and when the scale of the line of intersection of the gear tooth flank, $\mathcal{G}$, by the plane, Q, approaches an infinity, then the line of intersection of the gear surface, $\mathcal{G}$, by the plane, Q approaches to the planar characteristic curve that is commonly referred to as *Dupin indicatrix, Dup* $(\mathcal{G})$.

In differential geometry a surface is construed as a zero-thickness film. Because of this, *Dupin indicatrices* of the following five different types are distinguished in differential geometry of surfaces (see Fig. E.4):

---

[21] *Fransua Pier Charles* **Dupin** (October 6, 1784 – January 18, 1873) – a French mathematician.

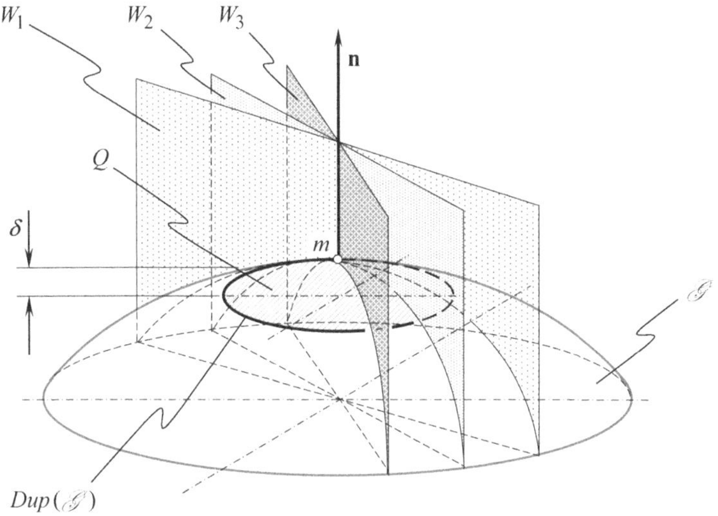

Fig. E.3 *Dupin indicatrix* at point of smooth regular gear tooth flank, $\mathcal{G}$

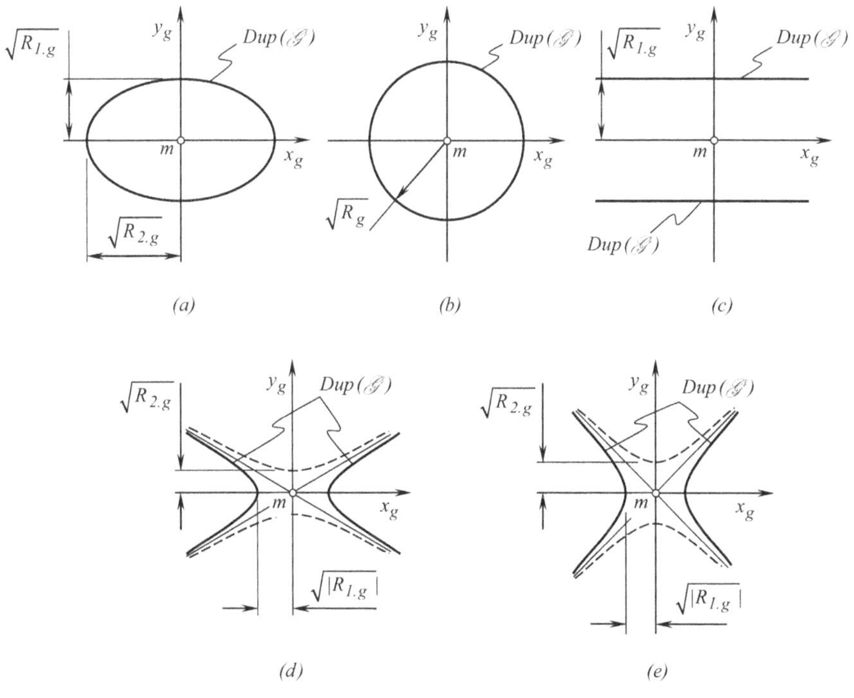

Fig. E.4 Five different kinds of *Dupin indicatrix*, $Dup(\mathcal{G})$, at point, m, of a smooth regular gear tooth flank, $\mathcal{G}$

- elliptic (see Fig. E.4a)
- umbilic (see Fig. E.4b)
- parabolic (see Fig. E.4c)
- hyperbolic (see Fig. E.4d)
- minimal (see Fig. E.4e).

Dupin indicatrix for a plane local surface patch doesn't exist. In the case of plane, all points of the *Dupin indicatrix* are remote to infinity.

For local surface patches having negative full curvature ($\mathcal{G}_g < 0$), phantom branches (that is, the branches that are not intersected by a plane perpendicular to the unite normal vector, $\mathbf{n}_g$, to the gear tooth flank, $\mathcal{G}$, at point, m) of the characteristic curve, $Dup(\mathcal{G})$, are shown in dashed lines (see Fig. E.4d, e).

An easy way to derive an equation of the characteristic curve, $Dup(\mathcal{G})$, is discussed immediately below.

Euler formula:

$$k_{1.g} \cos^2\varphi + k_{2.g} \sin^2\varphi = k_g \tag{E.26}$$

yields representation in the form:

$$\frac{k_{1.g}}{k_g} \cos^2\varphi + \frac{k_{2.g}}{k_g} \sin^2\varphi = 1 \tag{E.27}$$

Transition from polar coordinates to *Cartesian coordinates* is performed using well-known formulas:

$$x_g = \rho \cos\varphi \tag{E.28}$$

$$y_g = \rho \sin\varphi \tag{E.29}$$

These formulas make possible the following expressions for $\cos^2\varphi = x_g^2/\rho^2$ and $\sin^2\varphi = y_g^2/\rho^2$. After substituting the later formulas into the above equation Eq. (E.27), one can come up with the equation:

$$\frac{k_{1.g}}{k_g} \cdot \frac{x_g^2}{\rho^2} + \frac{k_{2.g}}{k_g} \cdot \frac{y_g^2}{\rho^2} = 1. \tag{E.30}$$

It is convenient to designate $\rho = \sqrt{k_g^{-1}}$. Principal curvatures $k_{1.g}$ and $k_{2.g}$ are the roots of the quadratic equation:

$$\begin{vmatrix} L_g - E_g k_g & M_g - F_g k_g \\ M_g - F_g k_g & N_g - G_g k_g \end{vmatrix} = 0 \tag{E.31}$$

Substituting the calculated values of the principal curvatures $k_{1.\ g}$ and $k_{2.\ g}$ into Eq. (E.30), and after performing necessary formulae transformations, an equation[22] for the *Dupin indicatrix, Dup* $(\mathcal{G})$ can be represented in the form:

$$k_{1.g}x_g^2 + k_{2.g}y_g^2 = 1 \tag{E.32}$$

Equation (E.32) describes a particular case of *Dupin indicatrix*, which is represented in *Darboux frame*[23].

General form of equation of *Dupin indicatrix* at point m of a gear tooth flank, $\mathcal{G}$, is often represented as:

$$Dup\,(\mathcal{G}) \quad \Rightarrow \quad \frac{L_g}{E_g}x_g^2 + \frac{2M_g}{\sqrt{E_g G_g}}x_g y_g + \frac{N_g}{G_g}y_g^2 = 1 \tag{E.33}$$

In Eq. (E.33), the characteristic curve, $Dup\,(\mathcal{G})$, is expressed in terms of the fundamental magnitudes, E_g, F_g, and G_g of the first, $\Phi_{1.\ g}$, and in terms of the fundamental magnitudes, L_g, M_g, and N_g of the second order, $\Phi_{2.\ g}$, at a point of the gear tooth flank, $\mathcal{G}$.

E.2.2 Matrix Representation of Equation of *Dupin Indicatrix* at Point of Gear Tooth Flank

Like any other quadratic form, equation of *Dupin indicatrix* of the gear tooth flank, $\mathcal{G}$, can be represented in matrix form:

$$Dup\ (\mathcal{G}) \quad \Rightarrow \quad \begin{bmatrix} x_g & y_g & 0 & 0 \end{bmatrix} \cdot \begin{bmatrix} \frac{L_g}{E_g} & \frac{2M_g}{\sqrt{E_g G_g}} & 0 & 0 \\ \frac{2M_g}{\sqrt{E_g G_g}} & \frac{N_g}{G_g} & 0 & 0 \\ 0 & 0 & \mp 1 & 0 \\ 0 & 0 & 0 & 1 \end{bmatrix} \cdot \begin{bmatrix} x_g \\ y_g \\ 0 \\ 0 \end{bmatrix} = \pm 1 \tag{E.34}$$

In *Darboux frame* this equation reduces to:

---

[22] The same equation of the *Dupin indicatrix* could be derived in another way. *Coxeter* considers a pair of conics obtained by expanding an equation in *Monge*'s form $z = z(x, y)$ in a *McLaurin series*:

$$z = z(0, 0) + z_1 x + z_2 y + \frac{1}{2}\left(z_{11}x_1^2 + 2z_{12}xy + z_{22}y^2\right) + \ldots = \frac{1}{2}\left(b_{11}x^2 + 2b_{12}xy + b_{22}y^2\right)$$

This gives the equation $(b_{11}x^2 + 2b_{12}xy + b_{22}y^2) = \pm 1$ of the *Dupin indicatrix*.

[23] *Jean Gaston* **Darboux** (August 13, 1842 – February 23, 1917), a French mathematician.

$$Dup\ (\mathcal{G}) \quad \Rightarrow \quad \begin{bmatrix} x_g & y_g & 0 & 0 \end{bmatrix} \cdot \begin{bmatrix} L_g & M_g & 0 & 0 \\ M_g & N_g & 0 & 0 \\ 0 & 0 & \mp 1 & 0 \\ 0 & 0 & 0 & 1 \end{bmatrix} \cdot \begin{bmatrix} x_g \\ y_g \\ 0 \\ 0 \end{bmatrix} = \pm 1 \tag{E.35}$$

It is convenient to implement matrix representation of equation of the *Dupin indicatrix* (see above), for instance, when investigating spatial gearings, that is, crossed-axes gearing, when multiple coordinate system transformations are required.

The equation of *Dupin indicatrix* can be represented in the form:

$$r_{Dup}(\varphi) = \sqrt{|R_g(\varphi)|} \cdot \operatorname{sgn} \Phi_{2.g}^{-1} \tag{E.36}$$

The latter equation reveals that the position vector of point of *Dupin indicatrix*, $Dup\,(\mathcal{G})$, in any direction is equal to the square root of the radius of curvature in that same direction[24].

E.3 Degree of Conformity at Point of Contact of Gear and Mating Pinion Tooth Flanks (in the First Order of Tangency)

For an accurate analytical description of contact geometry of the gear and the mating pinion tooth flanks in the first order of tangency, a higher order analysis is necessary to be done.

The discussed below method of a higher order analysis is targeting the development of an analytical description of degree of conformity of the pinion tooth flank, $\mathcal{P}$, to the gear tooth flank, $\mathcal{G}$, at a current point, K of their contact. The higher the degree of conformity of the tooth flanks, $\mathcal{G}$ and $\mathcal{P}$, the closer these surfaces to each other in differential vicinity of the point, K. This qualitative (*intuitive*) definition of degree of conformity of two smooth regular surfaces needs in a corresponding quantitative measure.

---

[24] Similar to *Dupin indicatrix, Dup* $(\mathcal{G})$, a planar characteristic curve of another type can be introduced. This planar characteristic curve is referred to *curvature indicatrix*. Equation of this characteristic curve can be postulated in the form: $r_{Dup.k}(\varphi) = \sqrt{|\,k_g(\varphi)\,|} \cdot \operatorname{sgn} \Phi_{2.g}^{-1}$. Application of *curvature indicatrix* in the form, $r_{Dup.\,k}(\varphi)$, makes possible avoiding uncertainty in cases of plane surface. For plane surface, the characteristic curve, $Dup\,(\mathcal{G})$, does not exist, while, $r_{Dup.\,k}(\varphi)$, exists. It shrinks to the point, m, on the gear tooth flank, $\mathcal{G}$.

E.3.1 Preliminary Remarks

Implementation of the resultant deviation, l_{cnf} (see Fig. E.5), of two smooth regular surfaces in contact for the analytical description of contact geometry of two surfaces in contact is a type of straightforward solution to the problem under consideration. This approach is proven to be computationally ineffective. However, the approach gives an insight for how an effective method for solving the problem under consideration can be developed.

As seen in Fig. E.5, three geometrical parameters are interrelated when a deviation of a surface from the tangent plane is considered in differential vicinity of surface point. They are:

(a) the measure of the deviation, $m_g m_g^*$, of a gear tooth flank, $\mathcal{G}$, from the tangent plane, l_{cnf},
(b) the distance, Km_g^*, of current point, m_g, from the contact point, K, and
(c) radius of normal curvature R_g of the gear tooth flank, $\mathcal{G}$, at the contact point, K.

As a consequence from this relationship among the parameters, $m_g m_g^*$, Km_g^*, and R_g, any one of them can be used for the purposes of quantitative evaluation of degree of conformity of the contacting tooth flanks, $\mathcal{G}$ and $\mathcal{P}$, of the gear and the mating pinion. As it is following from Fig. E.5:

$$m_g m_g^* = R_g - \sqrt{R_g^2 - \left(Km_g^*\right)^2}\ \Bigg|_{m_g \to K} \mapsto \quad l_{cnf} \tag{E.37}$$

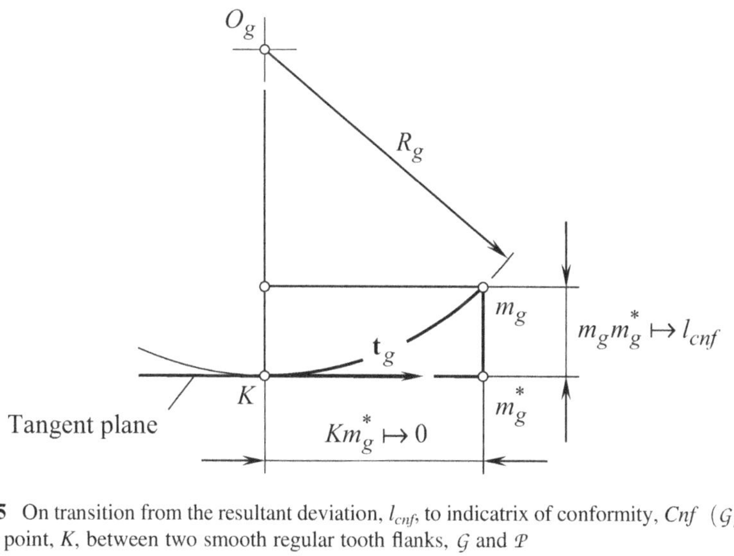

Fig. E.5 On transition from the resultant deviation, l_{cnf}, to indicatrix of conformity, $Cnf\ (\mathcal{G}/\mathcal{P})$, at contact point, K, between two smooth regular tooth flanks, $\mathcal{G}$ and $\mathcal{P}$

Inversely, for the radius of normal curvature, R_g, at a point of the gear tooth flank, $\mathcal{G}$, the following expression is valid:

$$R_g = \frac{\left(m_g m_g^*\right)^2 + \left(K m_g^*\right)^2}{2 \cdot m_g m_g^*} \tag{E.38}$$

Ultimately, one may conclude that any legitimate analytical function of normal radii of curvature, R_g and R_p, at a point of contact of the gear tooth flank, $\mathcal{G}$, and the pinion tooth flank, $\mathcal{P}$, can be used for this particular purpose.

Consider two smooth regular tooth flanks, $\mathcal{G}$ and $\mathcal{P}$, in the first order of tangency that make contact at point, K. Degree of conformity of the tooth flanks, $\mathcal{G}$ and $\mathcal{P}$, can be construed as a function of radii of normal curvature, R_g and R_p, of the contacting surfaces. Radii of normal curvature, R_g and R_p, of the tooth flanks, $\mathcal{G}$ and $\mathcal{P}$, are taken in a common normal plane section through the point, K. For a specified radius of normal curvature, R_g, of the tooth flank, $\mathcal{G}$, the degree of conformity of the tooth flanks depends upon the corresponding value of radius of normal curvature, R_p, of the pinion tooth flank, $\mathcal{P}$.

In most cases of gear meshing, degree of conformity at point of contact of the tooth flanks, $\mathcal{G}$ and $\mathcal{P}$, is not of a constant value, and it is changing as coordinates of the contact point change. Degree of the surfaces conformity to one another depends on orientation of the normal plane section through the contact point, K, and alters as the normal plane section is turning about the common perpendicular, $\mathbf{n}_g$. This statement immediately follows from the above made conclusion according to which degree of conformity at point of contact of the tooth flanks, $\mathcal{G}$ and $\mathcal{P}$, yields interpretation in terms of radii of normal curvature, R_g and R_p.

The change of degree of conformity of a gear tooth flank, $\mathcal{G}$, and a mating pinion tooth flank, $\mathcal{P}$, due to turning of the normal plane section about the common perpendicular, $\mathbf{n}_g$, is illustrated in Fig. E.6. Here, in Fig. E.6, just two-dimensional examples are shown, for which that same normal plane section of the gear tooth flank, $\mathcal{G}$, makes contact with different plane sections, $\mathcal{P}^i$, of the pinion tooth flank, $\mathcal{P}$.

In the example shown in Fig. E.6a, radius of normal curvature, R_p^1, of the convex plane section, $\mathcal{P}^1$, of the pinion tooth flank, $\mathcal{P}$, is positive ($R_p^1 > 0$). Convex normal plane section of the pinion tooth flank, $\mathcal{P}$, makes contact with convex normal plane section ($R_g > 0$) of a gear tooth flank, $\mathcal{G}$. The degree of conformity of the pinion tooth flank, $\mathcal{P}$, to the gear tooth flank, $\mathcal{G}$, in Fig. E.6a is relatively low as both the contacting surfaces are convex.

Another example is shown in Fig. E.6b. The radius of normal curvature, R_p^2, of convex plane section, $\mathcal{P}^2$, of the pinion tooth flank, $\mathcal{P}$, also is of a positive value, ($R_p^2 > 0$). However, its value exceeds the value, R_p^1, of the radius of normal curvature in the first example ($R_p^2 > R_p^1$). This results in that that degree of conformity of the pinion toot flank, $\mathcal{P}$, to the gear tooth flank, $\mathcal{G}$ (Fig. E.6b), is greater compared to that shown in Fig. E.6a.

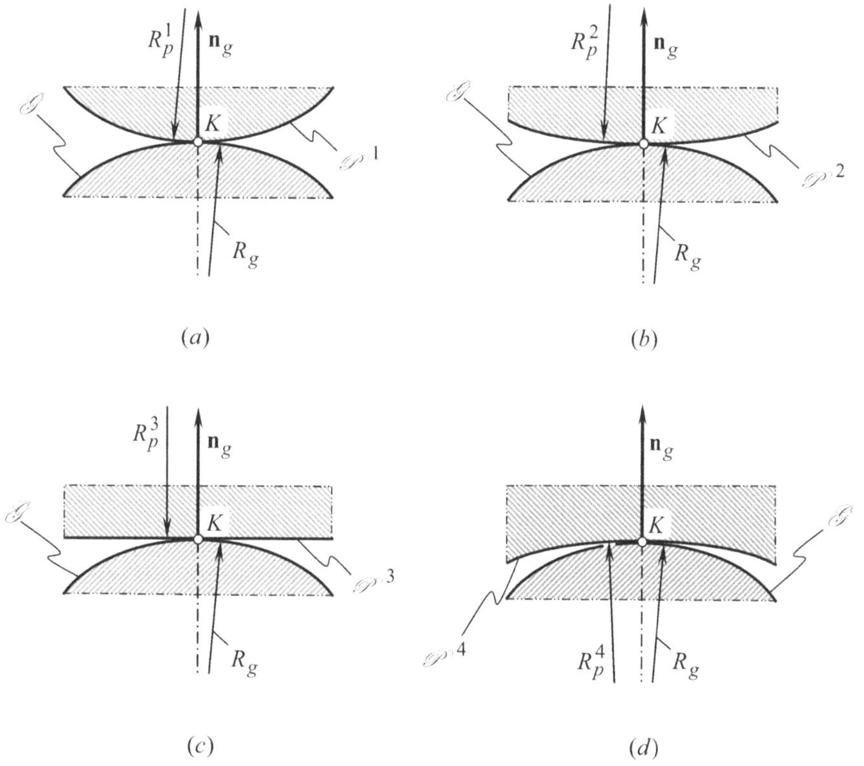

Fig. E.6 Sections of two smooth regular tooth flanks, $\mathcal{G}$ and $\mathcal{P}$, in contact, by a plane through the common perpendicular, $\mathbf{n}_g$

In the next example depicted in Fig. E.6c, the normal plane section $\mathcal{P}^3$ of the pinion tooth flank, $\mathcal{P}$, is represented with locally flatten section. Radius of normal curvature, R_p^3, of the flatten plane section, $\mathcal{P}^3$, approaches an infinity ($R_p^3 \rightarrow \infty$). Thus, the inequality, $R_p^3 > R_p^2 > R_p^1$, is valid. Therefore, degree of conformity of the pinion tooth flank, $\mathcal{P}$, to the gear tooth flank, $\mathcal{G}$ in Fig. E.6c, is also getting greater.

Finally, for a concave normal plane section, $\mathcal{P}^4$, of the pinion tooth flank, $\mathcal{P}$, that is illustrated in Fig. E.6d, radius of normal curvature, R_p^4, is of negative value ($R_p^4 < 0$). In this case, degree of conformity of the pinion tooth flank, $\mathcal{P}$, to the gear tooth flank, $\mathcal{G}$, is the greatest of four examples considered in Fig. E.6.

The examples shown in Fig. E.6, qualitatively illustrate what is already known intuitively regarding the different degree of conformity of two smooth regular surfaces in the first order of tangency. Intuitively one can realize that in the examples shown in Fig. E.6a through Fig. E.6d, degree of conformity at point of contact of two tooth flanks, $\mathcal{G}$ and $\mathcal{P}$, is gradually increased.

A similar observation is made for a given pair of the tooth flanks, $\mathcal{G}$ and $\mathcal{P}$, when different sections of the surfaces by a plane surface through the common

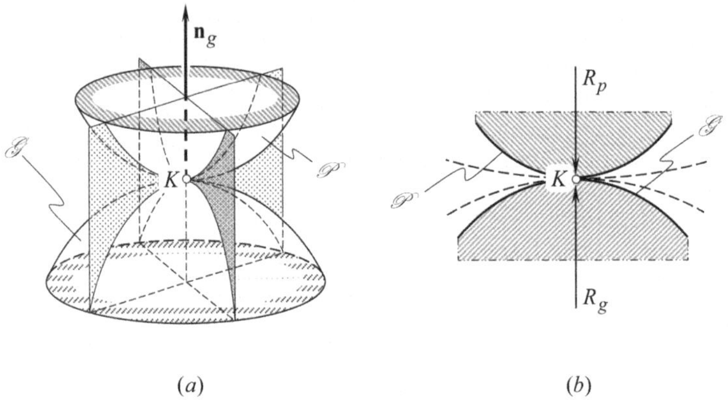

Fig. E.7 On analytical description of contact geometry of two smooth regular tooth flanks, $\mathcal{G}$ and $\mathcal{P}$, of a gear and a mating pinion

perpendicular, $\mathbf{n}_g$, are considered (see Fig. E.7a). When rotating the plane section about the common perpendicular, $\mathbf{n}_g$, it can be observed that degree of conformity of the gear and the pinion tooth flanks, $\mathcal{G}$ and $\mathcal{P}$, is different in different configurations of the cross-sectional plane (see Fig. E.7b).

The above examples provide an intuitive understanding of what the degree of conformity at point of contact of two smooth regular tooth flanks, $\mathcal{G}$ and $\mathcal{P}$, means. The examples cannot be employed directly for the purpose of evaluating in quantities the degree of conformity at point of contact of two smooth regular tooth flanks, $\mathcal{G}$ and $\mathcal{P}$. The next necessary step to be made up is to introduce an appropriate quantitative evaluation of the degree of conformity of two smooth regular surfaces in the first order of tangency. In other words, how can a certain degree of conformity of two smooth regular surfaces being described analytically?

E.3.2 Indicatrix of Conformity at Point of Contact of Tooth Flanks of Gear and Mating Pinion

This section is aiming introduction of a quantitative measure of degree of conformity at point of contact between two smooth regular surfaces. The degree of conformity at point of contact of two tooth flanks, $\mathcal{G}$ and $\mathcal{P}$, indicates how the pinion tooth flank, $\mathcal{P}$, is close to the gear tooth flank, $\mathcal{G}$, in the differential vicinity of point, K, of their contact, say how much the surface, $\mathcal{P}$, is *congruent* to the surface, $\mathcal{G}$, in differential vicinity of the contact point, K. This particular type of congruency between the contacting surfaces, $\mathcal{G}$ and $\mathcal{P}$, can also be construed as the *local congruency* of the contacting surfaces.

Quantitatively, the degree of conformity at point of contact of a smooth regular surface, $\mathcal{P}$, to another surface, $\mathcal{G}$, can be expressed in terms of the difference between the corresponding radii of normal curvature of the contacting surfaces. In order to

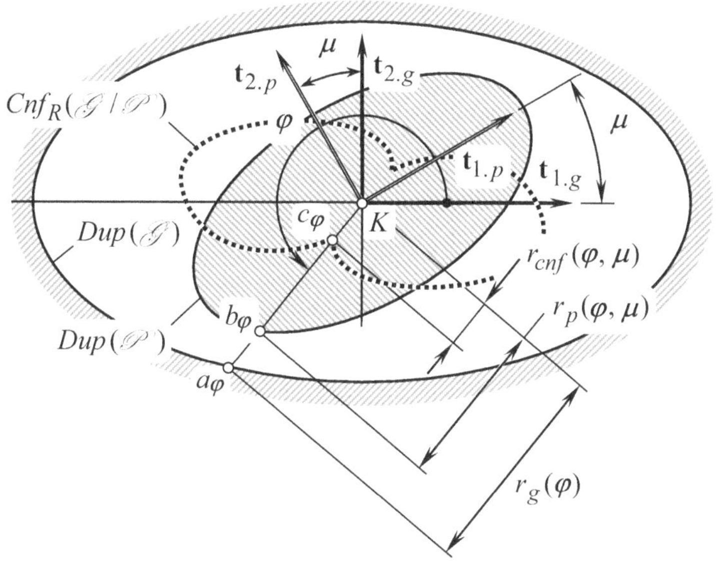

Fig. E.8 On the definition of the indicatrix of conformity, $Cnf_R(\mathcal{P} \mapsto \mathcal{G})$, at point of contact of tooth flanks, $\mathcal{G}$ and $\mathcal{P}$. (After Prof. *S.P. Radzevich*: Radzevich, S.P., *Differential-Geometric Method of Surface Generation*, Dr.Sci. Thesis, Tula, Tula Polytechnic Institute, 1991, 300 pages)

develop a quantitative measure of the degree of conformity of the tooth flanks, $\mathcal{G}$ and $\mathcal{P}$, it is convenient to implement *Dupin indicatrices*, $Dup(\mathcal{G})$ and $Dup(\mathcal{P})$, constructed at point of contact of the gear tooth flank, $\mathcal{G}$, and the pinion tooth flank, $\mathcal{P}$, correspondingly.

It is natural to assume that the smaller difference between the normal curvatures of the surfaces, $\mathcal{G}$ and $\mathcal{P}$, in a common cross-section by a plane through the common normal vector, $\mathbf{n}_g$, results in the greater degree of conformity at point of contact of the tooth flanks, $\mathcal{G}$ and $\mathcal{P}$.

Dupin indicatrix, $Dup(\mathcal{G})$ indicates the distribution of radii of normal curvature at point of the gear tooth flank, $\mathcal{G}$, as it had been shown, for example, for a concave elliptic patch of the surface, $\mathcal{G}$ (see Fig. E.8). For a gear tooth flank, $\mathcal{G}$, equation of this characteristic curve in polar coordinates can be represented in the form:

$$Dup(\mathcal{G}) \quad \Rightarrow \quad r_g(\varphi_g) = \sqrt{|\, R_g(\varphi_g)\,|} \tag{E.39}$$

where:

r_g – is the position vector of point of the *Dupin indicatrix*, $Dup(\mathcal{G})$ at point of the gear tooth flank, $\mathcal{G}$

φ_g – is the polar angle of the indicatrix, $Dup(\mathcal{G})$

The similar is true with respect to the *Dupin indicatrix*, $Dup(\mathcal{G})$ at point of the pinion tooth flank, $\mathcal{P}$, as it had been shown, for instance, for a convex elliptical patch of the pinion tooth flank, $\mathcal{P}$ (see Fig. E.8). Equation of this characteristic curve in polar coordinates can be represented in the form:

$$Dup\,(\mathcal{P}) \quad \Rightarrow \quad r_p(\varphi_p) = \sqrt{|\,R_p(\varphi_p)\,|} \tag{E.40}$$

where:

r_p – is the position vector of point of the *Dupin indicatrix*, $Dup\,(\mathcal{G})$ at point of the pinion tooth flank, $\mathcal{P}$
φ_p – is the polar angle of the indicatrix, $Dup\,(\mathcal{P})$

In the coordinate plane, $x_g y_g$, of the local reference system, $x_g y_g z_g$, the equalities, $\varphi_g = \varphi$ and $\varphi_p = \varphi + \mu$, are valid. Therefore, in the coordinate plane, $x_g y_g$, Eq. (E.39) and Eq. (E.40) cast into:

$$Dup\,(\mathcal{G}) \quad \Rightarrow \quad r_g(\varphi) = \sqrt{|\,R_g(\varphi)\,|} \tag{E.41}$$

$$Dup\,(\mathcal{P}) \quad \Rightarrow \quad r_p(\varphi,\ \mu) = \sqrt{|\,R_p(\varphi,\ \mu)\,|} \tag{E.42}$$

When degree of conformity at point of contact of the to the gear tooth flank, $\mathcal{G}$, is greater, then the difference between the functions $r_g(\varphi)$ and $r_p(\varphi,\ \mu)$ becomes smaller and vice versa. The last makes valid the following conclusion:

Conclusion E.1 *The distance between the corresponding*[25] *points of the Dupin indicatrices,* $Dup\,(\mathcal{G})$ *and* $Dup\,(\mathcal{P})$, *constructed at point of contact of a gear tooth flank,* $\mathcal{G}$, *and a mating pinion tooth flank,* $\mathcal{P}$, *can be employed for the purpose of indication of the degree of conformity at point of contact of the gear tooth flank,* $\mathcal{G}$, *and of the pinion tooth flank,* $\mathcal{P}$, *at the contact point,* K.

The equation of the *indicatrix of conformity*, $Cnf_R(\mathcal{G}/\mathcal{P})$ at point of contact of a gear tooth flank, $\mathcal{G}$, and a mating pinion tooth flank, $\mathcal{P}$, is defined of the following structure:

$$\begin{aligned} Cnf_R(\mathcal{G}/\mathcal{P}) \quad \Rightarrow \quad r_{cnf}(\varphi,\ \mu) &= \sqrt{|\,R_g(\varphi)\,|}\,\operatorname{sgn} R_g(\varphi) \\ &+ \sqrt{|\,R_p(\varphi,\ \mu)\,|}\,\operatorname{sgn} R_p(\varphi,\ \mu) = \\ &= r_g(\varphi)\operatorname{sgn} R_g(\varphi) + r_p(\varphi,\ \mu)\operatorname{sgn} R_p(\varphi,\ \mu) \end{aligned} \tag{E.43}$$

[25] Corresponding points of the *Dupin indicatrices*, $Dup\,(\mathcal{G})$ and $Dup\,(\mathcal{P})$, share the same straight line through the contact point, K, of the tooth flanks, $\mathcal{G}$ and $\mathcal{P}$, and are located at the same side of the point, K.

Because of the location of point, a_φ, of the *Dupin indicatrix, Dup* $(\mathcal{G})$ at point of the gear tooth flank, $\mathcal{G}$, is specified by the position vector, $r_g(\varphi)$, and the location of point, b_φ, of the *Dupin indicatrix, Dup* $(\mathcal{P})$ at point of the pinion tooth flank, $\mathcal{P}$, is specified by the position vector, $r_p(\varphi,\ \mu)$, then the location of a point, c_φ (see Fig. E.8) of the *indicatrix of conformity, Cnf* $_R(\mathcal{G}/\mathcal{P})$ at point of contact, K, of the tooth flanks, $\mathcal{G}$ and $\mathcal{P}$, is specified by the position vector $r_{cnf}(\varphi,\ \mu)$. Therefore, the equality $r_{cnf}(\varphi,\ \mu) = Kc_\varphi$ is observed, and the length of the straight-line segment, Kc_φ, is equal to the distance, $a_\varphi b_\varphi$.

Here, in Eq. (E.43) is designated:

$r_g = \sqrt{\mid R_g \mid}$ – is the position vector of point of *Dupin indicatrix* of the gear tooth flank, $\mathcal{G}$, at point K of contact with pinion tooth flank, $\mathcal{P}$

$r_p = \sqrt{\mid R_p \mid}$ – is the position vector of corresponding point of the *Dupin indicatrix* of the pinion tooth flank, $\mathcal{P}$

Here, in Eq. (E.43), the multipliers $\operatorname{sgn} R_g(\varphi)$ and $\operatorname{sgn} R_p(\varphi,\ \mu)$ are assigned to each of the functions, $r_g(\varphi) = \sqrt{\mid R_g(\varphi) \mid}$ and $r_p(\varphi,\ \mu) = \sqrt{\mid R_p(\varphi,\ \mu) \mid}$, accordingly, just for the purpose to remain the corresponding sign of the functions, that is, to remain that same sign that the radii of normal curvature, $R_g(\varphi)$ and $R_p(\varphi,\ \mu)$, have.

Ultimately, one can conclude that position vector, r_{cnf}, of point of the *indicatrix of conformity, Cnf* $_R(\mathcal{G}/\mathcal{P})$ can be expressed in terms of position vectors, r_g and r_p, of the *Dupin indicatrices*, *Dup* $(\mathcal{G})$ and *Dup* $(\mathcal{P})$.

For the calculation of a current value of the radius of normal curvature, $R_g(\varphi)$, at point of the gear tooth flank, $\mathcal{G}$, the equality:

$$R_g(\varphi) = \frac{\Phi_{1.g}}{\Phi_{2.g}} \tag{E.44}$$

is used.

Similarly, for the calculation of the current value of the radius of normal curvature, $R_p(\varphi,\ \mu)$, at point of pinion tooth flank, $\mathcal{P}$, the equality:

$$R_p(\varphi,\ \mu) = \frac{\Phi_{1.p}}{\Phi_{2.p}} \tag{E.45}$$

can be employed.

Use of the angle, μ, of local relative orientation of the tooth flanks, $\mathcal{G}$ and $\mathcal{P}$, indicates that the radii of normal curvature, $R_g(\varphi)$ and $R_p(\varphi,\ \mu)$, are taken in a common normal plane section through the contact point, K.

Further, it is well-known that the inequalities, $\Phi_{1.\,g} \geq 0$ and $\Phi_{1.\,p} \geq 0$, are always valid. Therefore, Eq. (E.43) can be rewritten in the following form:

$$r_{cnf} = r_g(\varphi) \operatorname{sgn} \Phi_{2.g}^{-1} + r_p(\varphi,\ \mu) \operatorname{sgn} \Phi_{2.p}^{-1} \tag{E.46}$$

For the derivation of an equation of the *indicatrix of conformity, $Cnf_R(\mathcal{G}/\mathcal{P})$*, it is convenient to use the *Euler equation* for normal radius of curvature, $R_g(\varphi)$, at a point of the gear tooth flank, $\mathcal{G}$:

$$R_g(\varphi) = \frac{R_{1.g} \cdot R_{2.g}}{R_{1.g} \cdot \sin^2\varphi + R_{2.g} \cdot \cos^2\varphi} \tag{E.47}$$

Here, the radii of principal curvature, $R_{1.\,g}$ and $R_{2.\,g}$, are the roots of the quadratic equation:

$$\begin{vmatrix} L_g \cdot R_g - E_g & M_g \cdot R_g - F_g \\ M_g \cdot R_g - F_g & N_g \cdot R_g - G_g \end{vmatrix} = 0 \tag{E.48}$$

Recall, that the inequality, $R_{1.\,g} < R_{2.\,g}$, is always observed.

Equations (E.47) and (E.48) allow for expression of the radius of normal curvature, $R_g(\varphi)$, at point of the gear tooth flank, $\mathcal{G}$, in terms of the fundamental magnitudes of the first order, E_g, F_g, and G_g, and of the fundamental magnitudes of the second order, L_g, M_g, and N_g.

A similar consideration is applicable for the pinion tooth flank, $\mathcal{P}$. Omitting routing analysis, one can conclude that the radius of normal curvature, $R_p(\varphi,\ \mu)$, of the at point of the pinion tooth flank, $\mathcal{P}$, can be expressed in terms of the fundamental magnitudes of the first order, E_p, F_p, and G_p, and of the fundamental magnitudes of the second order, L_p, M_p, and N_p.

Finally, on the premise of the above performed analysis, the following equation for the *indicatrix of conformity, $Cnf_R(\mathcal{G}/\mathcal{P})$* at point of contact of the tooth flanks, $\mathcal{G}$ and $\mathcal{P}$, can be derived:

$$r_{cnf}(\varphi,\ \mu) = \sqrt{\left|\frac{E_g G_g}{L_g G_g \cos^2\varphi - M_g\sqrt{E_g G_g}\sin 2\varphi + N_g E_g \sin^2\varphi}\right|}\ \mathrm{sgn}\ \Phi_{2.g}^{-1} +$$
$$+\sqrt{\left|\frac{E_p G_p}{L_p G_p \cos^2(\varphi+\mu) - M_p\sqrt{E_p G_p}\sin 2(\varphi+\mu) + N_p E_p \sin^2(\varphi+\mu)}\right|}\ \mathrm{sgn}\ \Phi_{2.p}^{-1} \tag{E.49}$$

Equation (E.49) of the characteristic curve[26] *$Cnf_R(P/T)$* is known from the late 1970s.

[26]Equation of this characteristic curve is known from:

(a) Pat. No.1249787, USSR, *A Method of Sculptured Part Surface Machining on a Multi-Axis NC Machine, S.P. Radzevich*, B23C 3/16, Filed: December 27, 1984, and (in hidden form) from:
(b) Pat. No.1185749, USSR, *A Method of Sculptured Part Surface Machining on a Multi-Axis NC Machine, S.P. Radzevich*, B23C 3/16, Filed: October 24, 1983.

Analysis of Eq. (E.49) reveals that *indicatrix of conformity*, $Cnf_R(\mathcal{G}/\mathcal{P})$ at point of contact of a gear tooth flank, $\mathcal{G}$, and a mating pinion tooth flank, $\mathcal{P}$, is represented by a planar centro-symmetrical curve of the fourth order. In particular cases, this characteristic curve possesses also a property of mirror symmetry. Mirror symmetry of indicatrix of conformity is observed, for example, when the angle, μ, of local relative orientation of the tooth flanks, $\mathcal{G}$ and $\mathcal{P}$, is equal $\mu = \pm\pi \cdot n/2$, where n designates an integer number.

It is important to note here, that even for the most general case of gearing, the position vector of point, $r_{cnf}(\varphi, \mu)$, of *indicatrix of conformity*, $Cnf_R(\mathcal{G}/\mathcal{P})$ is not dependent on the fundamental magnitudes, F_g and F_p. Independence of *indicatrix of conformity*, $Cnf_R(\mathcal{G}/\mathcal{P})$ of the fundamental magnitudes, F_g and F_p, is because of the following.

Coordinate angle, ω_g, at point of the gear tooth flank, $\mathcal{G}$, is calculated from the formula:

$$\omega_g = \arccos \frac{F_g}{\sqrt{E_g G_g}} \tag{E.50}$$

It is natural that the position vector, $r_{cnf}(\varphi, \mu)$, of point of indicatrix of conformity, $Cnf_R(\mathcal{G}/\mathcal{P})$, is not a function of the coordinate angle, ω_g. Although the position vector, $r_{cnf}(\varphi, \mu)$, depends on the fundamental magnitudes, E_g, G_g and E_p, G_p, the above analysis makes it evident why the position vector, $r_{cnf}(\varphi, \mu)$, does not depend upon the fundamental magnitudes F_g and F_p.

Two illustrative examples of *indicatrix of conformity*, $Cnf_R(\mathcal{G}/\mathcal{P})$ at point of contact of a gear tooth flank, $\mathcal{G}$, and a mating pinion tooth flank, $\mathcal{P}$, are shown in Fig. E.9. The first example (see Fig. E.9a) relates to a case of contact of a saddle-like local patch of the tooth surface, $\mathcal{G}$, and of a convex elliptic-like local patch of tooth surface, $\mathcal{P}$. The second one (see Fig. E.9b) is for a case of contact of a convex parabolic-like local patch of the tooth surface, $\mathcal{G}$, and of a convex elliptic-like local patch of tooth, $\mathcal{P}$. For both the cases, the corresponding curvature indicatrices $Crv(\mathcal{G})$ and $Crv(\mathcal{P})$ at the point of contact of the tooth flanks, $\mathcal{G}$ and $\mathcal{P}$, are depicted in Fig. E.9 as well. The imaginary (phantom) branches of the *Dupin indicatrix*, $Dup(\mathcal{G})$ (not labeled in Fig. E.9a) for the saddle-like local patch of the gear tooth flank, $\mathcal{G}$, are shown in dashed line (see Fig. E.9a).

The gear tooth flank, $\mathcal{G}$, and the pinion tooth flank, $\mathcal{P}$, can make contact geometrically however physical conditions of their contact could be violated. Violation of the physical condition of contact results in that the bodies those bounded by the interacting surfaces, $\mathcal{G}$ and $\mathcal{P}$, interfere into one another. Implementation of the *indicatrix of conformity*, $Cnf_R(\mathcal{G}/\mathcal{P})$ immediately uncovers the surfaces interference if there is any. Three illustrative examples of violation of physical condition of contact are illustrated in Fig. E.10. When correspondence between the radii of normal curvature of the contacting tooth flanks, $\mathcal{G}$ and $\mathcal{P}$, is inappropriate, then the *indicatrix of conformity*, $Cnf_R(\mathcal{G}/\mathcal{P})$, either intersects itself (see Fig. E.10a) or all of its diameters become negative (see Fig. E.10b, c).

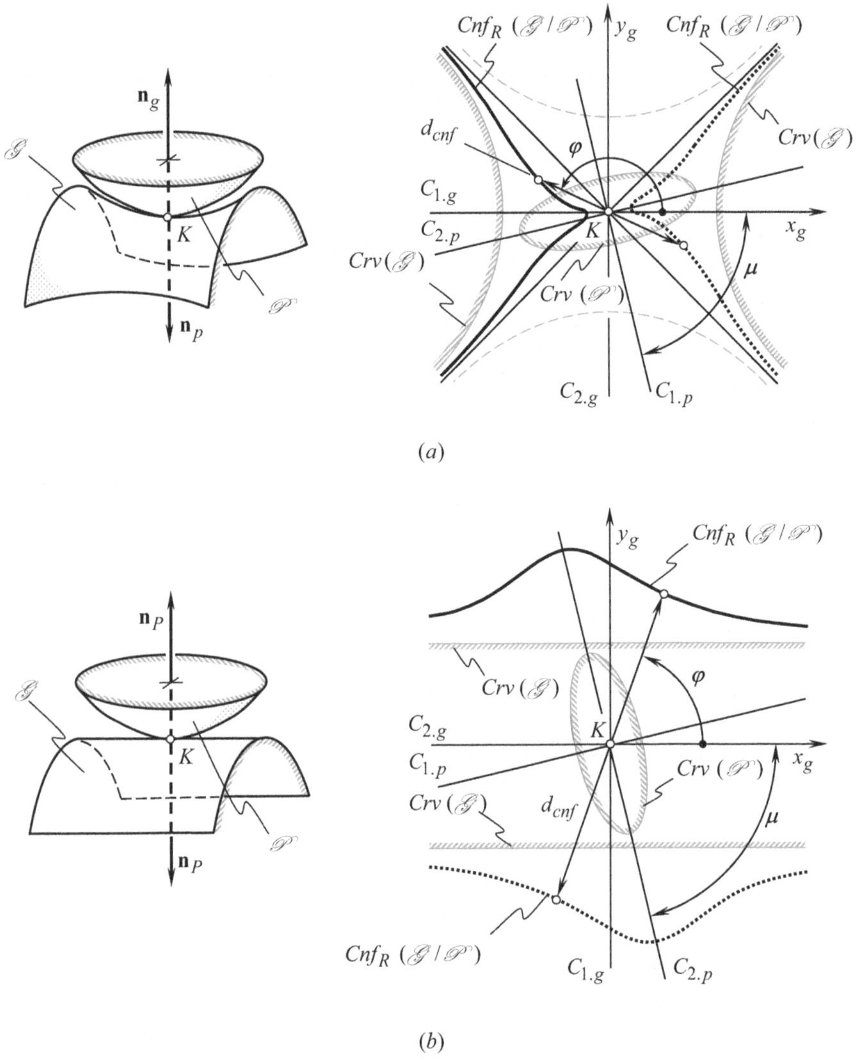

Fig. E.9 Examples of indicatrix of conformity, $Cnf_R(\mathcal{G}/\mathcal{P})$, at point of contact, K, of a smooth regular gear tooth flank, $\mathcal{G}$, and a mating pinion tooth flank, $\mathcal{P}$

Another interpretation of satisfaction and violation of physical condition of contact of two smooth regular tooth flanks, $\mathcal{G}$ and $\mathcal{P}$, is illustrated in Fig. E.11. Condition of physical contact is fulfilled when all diameters of the *indicatrix of conformity,* $Cnf_R(\mathcal{G}/\mathcal{P})$ are of positive values. In this case, the gear tooth flank, $\mathcal{G}$, and the mating pinion tooth flank, $\mathcal{P}$, may contact one another like two rigid bodies do. An example of the *indicatrix of conformity,* $Cnf_R(\mathcal{G}/\mathcal{P})$ for such a case is depicted in Fig. E.11a. In cases when this planar characteristic curve has negative diameters as it is schematically shown in Fig. E.11b physical contact between the tooth flanks, $\mathcal{G}$ and $\mathcal{P}$, is infeasible.

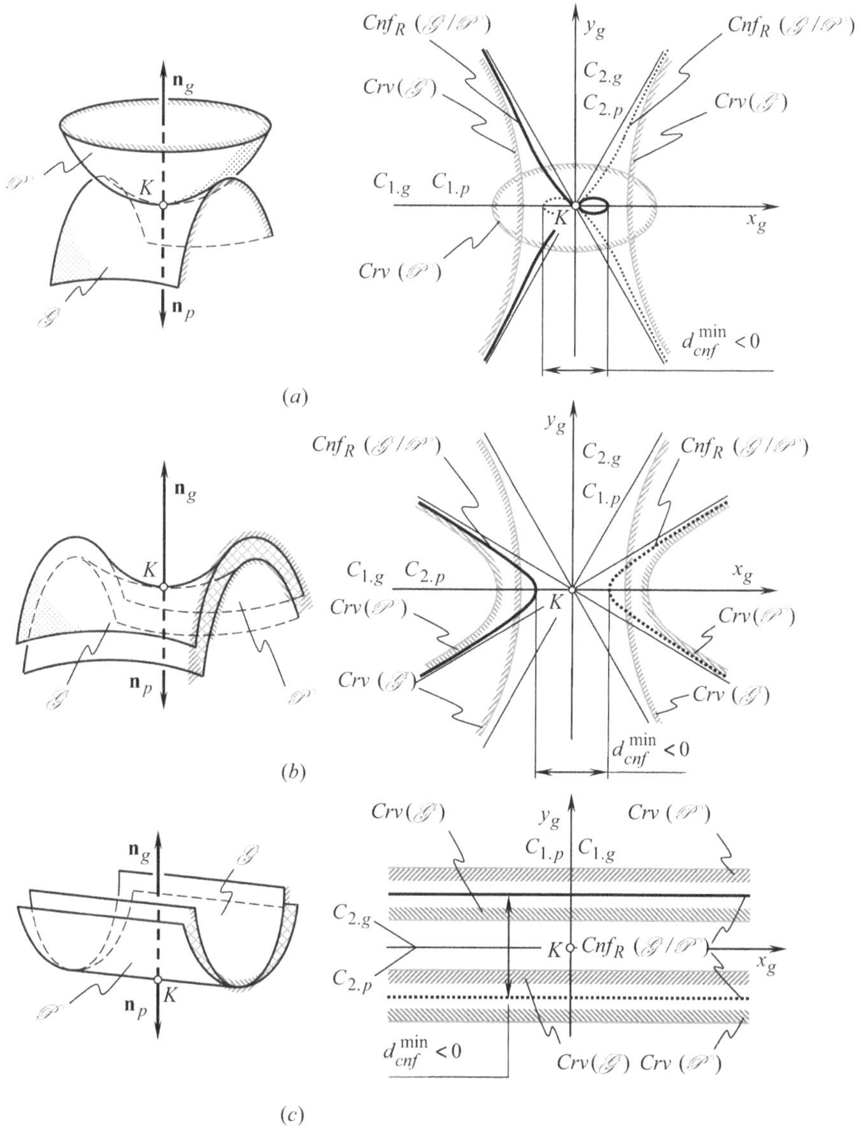

Fig. E.10 Examples of violation of physical condition of contact of a smooth regular gear tooth flank, $\mathcal{G}$ and a mating pinion tooth flank, $\mathcal{P}$

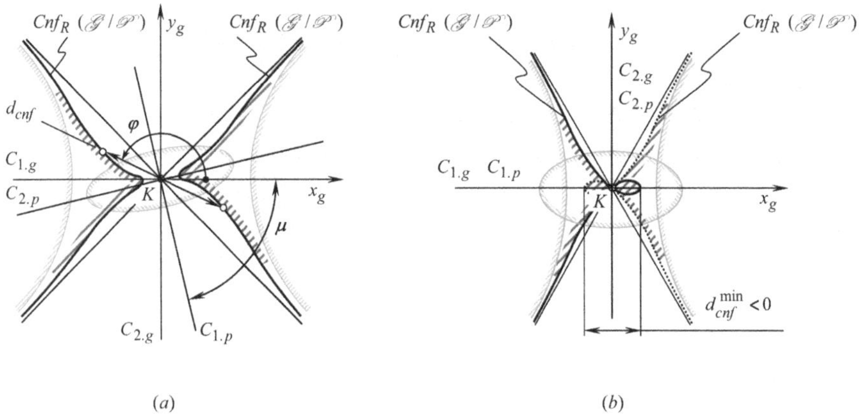

Fig. E.11 Another interpretation of satisfaction (**a**) and of violation (**b**) of condition of physical contact of a smooth regular gear tooth flank, $\mathcal{G}$, and a mating pinion tooth flank, $\mathcal{P}$

The actual value of current diameter[27] d_{cnf} of the *indicatrix of conformity*, $Cnf_R(\mathcal{G}/\mathcal{P})$, indicates degree of conformity to each other of the gear tooth flank, $\mathcal{G}$, and the mating pinion tooth flank, $\mathcal{P}$, in a corresponding cross-section of the surfaces by normal plane through the common perpendicular. The orientation of the normal plane section with respect to the tooth flanks, $\mathcal{G}$ and $\mathcal{P}$, is specified by the corresponding central angle, ϕ.

For the orthogonally parameterized gear tooth flank, $\mathcal{G}$, and the mating pinion tooth flank, $\mathcal{P}$, equation of the *Dupin indicatrices,Dup* $(\mathcal{G})$ and *Dup* $(\mathcal{P})$, is simplified to:

$$L_g x_g^2 + 2M_g x_g y_g + N_g y_g^2 = \pm 1 \tag{E.51}$$

$$L_p x_p^2 + 2M_p x_p y_p + N_p y_p^2 = \pm 1 \tag{E.52}$$

After been represented in a common reference system, use of Eqs. (E.51) and (E.52) makes possible a simplified equation of the *indicatrix of conformity*, $Cnf_R(\mathcal{G}/\mathcal{P})$ at a point of contact of the tooth flanks, $\mathcal{G}$ and $\mathcal{P}$:

$$\begin{aligned} r_{cnf}(\varphi,\ \mu) = \left(L_g \cos^2\varphi - M_g \sin 2\varphi + N_g \sin^2\varphi\right)^{-0.5} \operatorname{sgn}\ \Phi_{2.g}^{-1} + \\ + \left[L_p \cos^2(\varphi+\mu) - M_p \sin 2(\varphi+\mu) + N_p \sin^2(\varphi+\mu)\right]^{-0.5} \operatorname{sgn}\ \Phi_{p.T}^{-1} \end{aligned} \tag{E.53}$$

Equation (E.53) is valid for the orthogonally parameterized tooth flanks, $\mathcal{G}$ and $\mathcal{P}$.

---

[27] Diameter of a symmetrical planar curve that possesses a property of central symmetry can be defined as a distance between two points of the curve, measured along the corresponding straight line through the center of symmetry of the curve.

When designing gears, a reasonable degree of conformity, cnf_{res}, at point of contact of a gear, $\mathcal{G}$, and a mating pinion, $\mathcal{P}$, tooth flanks is a trade-off between the desirable maximal degree of conformity, $cnf_{\max}$ (max bearing capacity), and a minimal permissible degree of conformity, $cnf_{\min}$ (min sensitivity to the axes misalignment), that is, the inequality $cnf_{\min} < cnf_{\text{res}} < cnf_{\max}$ is observed.

E.3.3 Vector and Matrix Representation of Equation of Indicatrix of Conformity

Equation (E.49) of indicatrix of conformity, $Cnf_R(\mathcal{G}/\mathcal{P})$, at point of contact of tooth flanks of a gear, and of a mating pinion, $\mathcal{G}$ and $\mathcal{P}$, yields representation in both forms, in vector form, as well, as in matrix form.

Consider a unit tangent vector, $\mathbf{t}_{cnf}(\varphi)$, through contact point, K, in a common section of the contacting surfaces, $\mathcal{G}$ and $\mathcal{P}$, by a normal plane. The unit tangent vector, $\mathbf{t}_{cnf}(\varphi)$, is entirely situated in a common tangent plane through contact point, K. In a common normal section, radii of curvature of the tooth flanks, $\mathcal{G}$ and $\mathcal{P}$, are designated as R_g and R_p, correspondingly. For convenience, the following designation is used here: $r_g(\varphi) = \sqrt{R_g(\varphi)}$ and $r_p(\varphi + \mu) = \sqrt{R_p(\varphi + \mu)}$.

The position vector, $\mathbf{r}_{cnf}$, of point of indicatrix of conformity at point of contact of tooth flanks, $\mathcal{G}$ and $\mathcal{P}$, of a gear and a mating pinion can be expressed in terms of the parameters r_g and r_g as follows:

$$\mathbf{r}_{cnf}(\varphi + \mu) = \left[r_g(\varphi) - r_p(\varphi + \mu)\right] \cdot \mathbf{t}_{cnf}(\varphi) \tag{E.54}$$

In a conventional equation for $Cnf_R(\mathcal{G} \to \mathcal{P})$, the distance, $r_{cnf}(\varphi)$, can be replaced with, $\mathbf{r}_{cnf}$, after the corresponding expression for $Cnf_R(\mathcal{G} \to \mathcal{P})$ is multiplied by $\mathbf{t}_{cnf}(\varphi)$.

Equation (E.54) casts into:

$$\mathbf{r}_{cnf}(\varphi + \mu) = \mathbf{i} \cdot r_{cnf}(\varphi,\ \mu) \cos\varphi + \mathbf{j} \cdot r_{cnf}(\varphi,\ \mu) \sin\varphi \tag{E.55}$$

The unit vector, $\mathbf{t}_{cnf}(\varphi)$, can be expressed in terms of fundamental magnitudes of the first and the second order calculated at point of contact, K, of tooth flanks, $\mathcal{G}$ and $\mathcal{P}$, of a gear and a mating pinion.

Matrix representation of indicatrix of conformity, $Cnf_R(\mathcal{G} \to \mathcal{P})$, follows immediately:

$$\mathbf{r}_{cnf}(\varphi,\ \mu) = \begin{bmatrix} r_{cnf}(\varphi,\ \mu) \cos\varphi \\ r_{cnf}(\varphi,\ \mu) \sin\varphi \\ 0 \\ 1 \end{bmatrix} \tag{E.56}$$

Equation of indicatrix of conformity, $Cnf_R(\mathcal{G} \rightarrow \mathcal{P})$, in matrix form [see Eq. (E.56)] is preferred when this characteristic curve ($Cnf_R(\mathcal{G} \rightarrow \mathcal{P})$) is involved in a multiple coordinate system transformation.

E.3.4 Directions of Extremum Degree of Conformity at Point of Contact of Tooth Flanks of a Gear and a Mating Pinion

The directions, those along which degree of conformity at point of contact of a gear tooth flank, $\mathcal{G}$, and a mating pinion tooth flank, $\mathcal{P}$, is extremum, that is, degree of conformity reaches either maximal of its value or minimal of its value, are of prime importance for engineering applications. This issue is especially important for synthesizing a favorable gear pair.

The directions of extremal degree of conformity of the interacting smooth regular tooth flanks, $\mathcal{G}$ and $\mathcal{P}$, that is, the directions pointed along the extremal diameters, $d_{cnf}^{\min}$ and $d_{cnf}^{\max}$, of *indicatrix of conformity*, $Cnf_R(\mathcal{G}/\mathcal{P})$, can be found from the equation of the *indicatrix of conformity*, $Cnf_R(\mathcal{G}/\mathcal{P})$. For the reader's convenience, equation of this characteristic curve is represented in the form:

$$r_{cnf}(\varphi,\ \mu) = \sqrt{|\, r_{1.g}\cos^2\varphi + r_{2.g}\sin^2\varphi\,|\operatorname{sgn}\Phi_{2.g}^{-1}} + \sqrt{|\, r_{1.p}\cos^2(\varphi+\mu) + r_{2.p}\sin^2(\varphi+\mu)\,|\operatorname{sgn}\Phi_{2.p}^{-1}} \tag{E.57}$$

Two directions within the common tangent plane are specified by the angles, $\varphi_{\min}$ and $\varphi_{\max}$. These directions feature an extremum degree of conformity of the pinion tooth flank, $\mathcal{P}$, to the gear tooth flank, $\mathcal{G}$. Actually, the angles are the roots of equation:

$$\frac{\partial}{\partial\varphi} r_{cnf}(\varphi,\ \mu) = 0. \tag{E.58}$$

It can be easily proved that in general case of contact of two smooth regular tooth flanks, $\mathcal{G}$ and $\mathcal{P}$, the difference between the angles, $\varphi_{\min}$ and $\varphi_{\max}$, is not equal to 0.5π. This means that the equality

$$\varphi_{\min} - \varphi_{\max} = \pm 0.5\pi n \tag{E.59}$$

is not always observed, and in most cases the relationship:

$$\varphi_{\min} - \varphi_{\max} \neq \pm 0.5\pi n \tag{E.60}$$

is valid (here n is an integer number). The condition [see Eq. (E.59)]: $\varphi_{\min} = \varphi_{\max} \pm 0.5\pi n$ is fulfilled only in cases when the angle, μ, of local relative orientation of the contacting surfaces, $\mathcal{G}$ and $\mathcal{P}$, is equal to $\mu = \pm 0.5\pi n$, and thus the principal directions, $\mathbf{t}_{1.\ g}$ and $\mathbf{t}_{2.\ g}$, of the gear tooth flank, $\mathcal{G}$, and the principal directions, $\mathbf{t}_{1.\ p}$ and $\mathbf{t}_{2.\ p}$, of the mating pinion tooth flank, $\mathcal{P}$, either aligned to one another or they are pointed oppositely.

This enables one making the following statement:

Statement E.1 *In general case of contact of two smooth regular tooth flanks, the directions along which degree of conformity of the tooth flanks, $\mathcal{G}$ and $\mathcal{P}$, is extremal are not orthogonal to one another.*

This statement is important for engineering applications.

Solution to equation, $\partial\, r_{rel}(\varphi)/\partial\varphi = 0$, returns two extremal angles, $\varphi_{\min}$ and $\varphi_{\max} = \varphi_{\min} + 90^{\circ}$ [here $r_{rel}(\varphi)$ denotes position vector of a point of *Dupin indicatrix* at point of the surface of relative curvature]. Equation (E.58) allows for two solutions, $\varphi^{*}_{\min}$ and $\varphi^{*}_{\max}$. Therefore, the extremal difference:

$$\Delta\varphi_{\min} = \varphi_{\min} - \varphi^{*}_{\min} \tag{E.61}$$

as well as the extremal difference:

$$\Delta\varphi_{\max} = \varphi_{\max} - \varphi^{*}_{\max} \tag{E.62}$$

can be easily calculated.

Generally speaking, neither the extremal difference, $\Delta\varphi_{\min}$ nor the extremal difference, $\Delta\varphi_{\max}$, is equal to zero. They are equal to zero only in particular cases, say when the angle, μ, of local relative orientation of the surfaces, $\mathcal{G}$ and $\mathcal{P}$, fulfills the relationship $\mu = \pm 0.5\pi\, n$.

Let us consider an example that illustrates the Statement E.1.

Example. As an illustrative example, let us describe analytically contact geometry of two convex parabolic patches of the contacting tooth flanks, $\mathcal{G}$ and $\mathcal{P}$ (see Fig. E.12). The example pertains to finishing a helical involute gear by a disk-type shaving cutter. In the example under consideration, the design parameters of the gear and of the shaving cutter, along with the specified the gear and the shaving cutter configuration yield the following numerical data for the calculation. At the point, K, of the tooth flanks contact, principal curvatures of the gear tooth flank, $\mathcal{G}$, are equal $k_{1.\ g} = 4mm^{-1}$ and $k_{2.\ g} = 0$. Principal curvatures of the mating pinion tooth flank, $\mathcal{P}$, are equal $k_{1.\ p} = 1mm^{-1}$ and $k_{2.\ p} = 0$. The angle, μ, of local relative orientation of the tooth flanks, $\mathcal{G}$ and $\mathcal{P}$, is equal $\mu = 45^{\circ}$.

Two approaches can be implemented for the analytical description of the contact geometry of the tooth flanks, $\mathcal{G}$ and $\mathcal{P}$. The first one is based on implementation of *Dupin indicatrix* of the surface of relative curvature. The second one is based on application of the *indicatrix of conformity, $Cnf_R(\mathcal{G}/\mathcal{P})$* constructed at a contact point, K, of the interacting tooth flaks, $\mathcal{G}$ and $\mathcal{P}$.

The first approach. In the case under consideration, normal curvature $k_{\mathcal{R}}$ of the surface of relative curvature, $\mathcal{R}$, can be analytically expressed as:

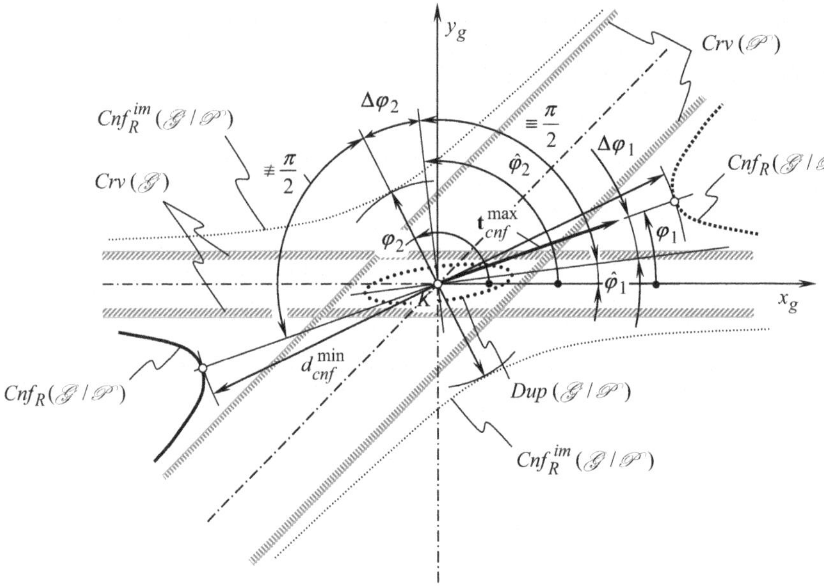

Fig. E.12 Example: Determination of the optimum instantaneous kinematics in a gear shaving operation

$$k_{\mathcal{R}} = k_{1.g}\cos^2\varphi - k_{1.p}\cos^2(\varphi+\mu) \tag{E.63}$$

Therefore, the following equality:

$$\frac{\partial k_{\mathcal{R}}}{\partial \varphi} = -2k_{1.g}\sin\varphi\cos\varphi + 2k_{1.p}\sin(\varphi+\mu)\cos(\varphi+\mu) = 0 \tag{E.64}$$

is valid for the directions of the extremum degree of conformity of the tooth flaks, $\mathcal{G}$ and $\mathcal{P}$, at every point of their contact.

For the directions of the extremal degree of conformity at the point of contact of the gear tooth flank, $\mathcal{G}$, and the mating pinion tooth flank, $\mathcal{P}$, Eq. (E.64) yields calculation the extremal values $\varphi_{\min} = 7^\circ$ and $\varphi_{\max} = \varphi_{\min} + 90^\circ = 97^\circ$of the angles $\varphi_{\min}$ and $\varphi_{\max}$.

The direction that is specified by the angle $\varphi_{\min} = 7^\circ$, indicates the direction of the minimal diameter of the *Dupin indicatrix* of the surface of relative curvature. That same direction corresponds to the maximal degree of conformity at the point of contact of the tooth flanks, $\mathcal{G}$ and $\mathcal{P}$. Another direction, that is specified by the angle $\varphi_{\max} = 97^\circ$, indicates the direction of the minimum degree of conformity of the contacting tooth flanks, $\mathcal{G}$ and $\mathcal{P}$, at that same contact point.

The second approach. For the case under consideration, use of Eq. (E.50) of the *indicatrix of conformity*, $Cnf_R(\mathcal{G}/\mathcal{P})$ at point of contact of the gear tooth flank, $\mathcal{G}$, and the mating pinion tooth flank, $\mathcal{P}$, makes possible calculation of the extremal angles $\varphi^*_{\min} = 19^\circ$ and $\varphi^*_{\max} = 118^\circ$.

Imaginary branches of the *indicatrix of conformity*, $Cnf_R(\mathcal{G}/\mathcal{P})$ at the point of contact of the tooth flanks, $\mathcal{G}$ and $\mathcal{P}$, in Fig. E.12 are depicted in dashed line.

It is important to stress the readers' attention here onto two issues.

First, the extremal angles, $\varphi_{\min}$ and $\varphi_{\max}$, that are calculated using the first approach, are not equal to the corresponding extremal angles, $\varphi^*_{\min}$ and $\varphi^*_{\max}$, that are calculated using the second approach. The relationships, $\varphi_{\min} \neq \varphi^*_{\min}$ and $\varphi_{\max} \neq \varphi^*_{\max}$, are generally observed.

Second, the difference, $\Delta\varphi^*$, between the extremal angles, $\varphi^*_{\min}$ and $\varphi^*_{\max}$, is not equal to half of π. Therefore, the relationship, $\varphi^*_{\max} - \varphi^*_{\min} \not\equiv 90°$, between the extremal angles, $\varphi^*_{\min}$ and $\varphi^*_{\max}$, is observed. In general case of contact of two sculptured surfaces the directions of the extremal degree of conformity of the gear tooth flank, $\mathcal{G}$, and the mating pinion tooth flank, $\mathcal{P}$, are not orthogonal to one another.

The discussed example reveals that in general cases of contact of two smooth regular tooth flanks, *indicatrix of conformity*, $Cnf_R(\mathcal{G}/\mathcal{P})$ can be implemented for the purpose of accurate analytical description of the contact geometry of the surfaces. The *Dupin indicatrix* of the surface of relative curvature can be implemented for this purpose only in particular cases of the surfaces, $\mathcal{G}$ and $\mathcal{P}$, relative orientation. Application of the *Dupin indicatrix* of the surface of relative curvature enables only approximate analytical description of geometry of contact of the surfaces. The *Dupin indicatrix* of the surface of relative curvature could be equivalent to indicatrix of conformity only in degenerate cases of contact of the surfaces. Advantages of indicatrix of conformity over the *Dupin indicatrix* of the surface of relative curvature are due to that this characteristic curve, $Cnf_R(\mathcal{G}/\mathcal{P})$, is a curve of the fourth order.

E.3.5 Important Properties of Indicatrix of Conformity $Cnf_R(\mathcal{G}/\mathcal{P})$ at Point of Contact of Tooth Flanks of a Gear and a Mating Pinion

The performed analysis of Eq. (E.50) of *indicatrix of conformity*, $Cnf_R(\mathcal{G}/\mathcal{P})$ at point of contact of a gear tooth flank and a mating pinion tooth flank reveals that this characteristic curve possesses the following important properties.

1. *Indicatrix of conformity*, $Cnf_R(\mathcal{G}/\mathcal{P})$ at point of contact of the tooth flanks, $\mathcal{G}$ and $\mathcal{P}$, is a planar characteristic curve of the fourth order. It possesses the property of central symmetry, and, in particular cases, it also possesses the property of mirror symmetry.
2. *Indicatrix of conformity*, $Cnf_R(\mathcal{G}/\mathcal{P})$ is closely related to the surfaces, $\mathcal{G}$ and $\mathcal{P}$, second fundamental forms, $\Phi_{2.\,g}$ and $\Phi_{2.\,p}$. This characteristic curve is invariant with respect to the kind of parameterization of the tooth flanks, $\mathcal{G}$ and $\mathcal{P}$, but its equation does. A change in the surfaces, $\mathcal{G}$ and $\mathcal{P}$, parameterization leads to that that the equation of *indicatrix of conformity*, $Cnf_R(\mathcal{G}/\mathcal{P})$ also changes, while the shape and parameters of this characteristic curve are remained unchanged.

3. The characteristic curve, $Cnf_R(\mathcal{G}/\mathcal{P})$, is independent of actual value of the coordinate angle, ω_g, that is formed by the coordinate lines, U_g and V_g, on the gear tooth flank, $\mathcal{G}$. It is also independent of actual value of the coordinate angle, ω_p, that is formed by the coordinate lines, U_p and V_p, on the mating pinion tooth flank, $\mathcal{P}$. However, parameters of *indicatrix of conformity*, $Cnf_R(\mathcal{G}/\mathcal{P})$ are depending upon the angle, μ, of local relative orientation of the tooth flanks, $\mathcal{G}$ and $\mathcal{P}$. Therefore, for a given pair of the tooth flanks, $\mathcal{G}$ and $\mathcal{P}$, the actual value of the degree of conformity of the surface varies correspondingly to variation of the angle, μ, while the pinion tooth flank, $\mathcal{P}$, is spinning around the unit vector of the common perpendicular.

More properties of the *indicatrix of conformity*, $Cnf_R(\mathcal{G}/\mathcal{P})$ at point of contact of a gear tooth flank, $\mathcal{G}$, and a mating pinion tooth flank, $\mathcal{P}$, can be outlined.

E.3.6 Converse Indicatrix of Conformity at Point of Contact of Tooth Flanks of a Gear and a Mating Pinion

For the *Dupin indicatrix*, $Dup(\mathcal{G}/\mathcal{P})$ at point of the surface of relative curvature, $\mathcal{R}$, there exists a corresponding inverse *Dupin indicatrix*, $Dup_k(\mathcal{G}/\mathcal{P})$. Similarly, for the indicatrix of conformity, $Cnf_R(\mathcal{G}/\mathcal{P})$, at a point of contact of the tooth flanks, $\mathcal{G}$ and $\mathcal{P}$, there exists a corresponding *converse indicatrix of conformity*, $Cnf_k(\mathcal{G}/\mathcal{P})$. This characteristic curve can be expressed directly in terms of the surfaces, $\mathcal{G}$ and $\mathcal{P}$, normal curvatures, k_g and k_p:

$$Cnf_k(\mathcal{G}/\mathcal{P}) \quad \Rightarrow \quad r^{cnv}_{cnf}(\varphi,\ \mu) = \sqrt{|\,k_g(\varphi)\,|} \cdot \operatorname{sgn} \Phi^{-1}_{2.g} - \sqrt{|\,k_p(\varphi,\ \mu)\,|} \cdot \operatorname{sgn} \Phi^{-1}_{2.p} \tag{E.65}$$

For derivation of an equation of the *converse indicatrix of conformity*, $Cnf_k(\mathcal{G}/\mathcal{P})$, the *Euler formula* for a surface normal curvature is used in the following representation:

$$k_g(\varphi) = k_{1.g} \cos^2\varphi + k_{2.g} \sin^2\varphi \tag{E.66}$$

$$k_p(\varphi,\ \mu) = k_{1.p} \cos^2(\varphi + \mu) + k_{2.p} \sin^2(\varphi + \mu) \tag{E.67}$$

Here in Eqs. (E.66) and (E.67), the principal curvatures of the gear tooth flank, $\mathcal{G}$, are designated as $k_{1.\,g}$ and $k_{2.\,g}$, while $k_{1.\,p}$ and $k_{2.\,p}$ designate the principal curvatures of the mating pinion tooth flank, $\mathcal{P}$.

After substituting Eq. (E.66) and Eq. (E.67) into Eq. (E.65), one can come up with the equation:

$$r_{cnf}^{cnv}(\varphi,\ \mu)=\sqrt{|\,k_{1.g}\cos^2\varphi+k_{2.g}\sin^2\varphi\,|\operatorname{sgn}\Phi_{2.g}^{-1}}$$
$$-\sqrt{|\,k_{1.p}\cos^2(\varphi+\mu)+k_{2.p}\sin^2(\varphi+\mu)\,|\operatorname{sgn}\Phi_{2.p}^{-1}} \qquad \text{(E.68)}$$

for the *converse indicatrix of conformity*, $Cnf_k(\mathcal{G}/\mathcal{P})$ at point of contact of the tooth flanks $\mathcal{G}$ and $\mathcal{P}$ in the first order of tangency.

Here, in Eq. (E.68), principal curvatures $k_{1.\,g}$, $k_{2.\,g}$ and $k_{1.\,p}$, $k_{2.\,p}$ can be expressed in terms of the corresponding fundamental magnitudes E_g, F_g, and G_g of the first and L_g, M_g, and N_g of the second order of the gear tooth flank $\mathcal{G}$, and in terms of the corresponding fundamental magnitudes E_p, F_p, and G_p of the first and L_p, M_p, and N_p of the second order of the mating pinion tooth flank $\mathcal{P}$. Following this way, Eq. (E.68) of the *converse indicatrix of conformity*, $Cnf_k(\mathcal{G}/\mathcal{P})$ can be cast to the form that is similar to Eq. (E.50) of the ordinary *indicatrix of conformity*, $Cnf_R(\mathcal{G}/\mathcal{P})$ at a point of contact of the tooth flanks, $\mathcal{G}$ and $\mathcal{P}$.

It can be shown that similar to the *indicatrix of conformity*, $Cnf_R(\mathcal{G}/\mathcal{P})$, the characteristic curve $Cnf_k(\mathcal{G}/\mathcal{P})$ also possess the property of central symmetry. In particular cases of the surfaces contact, it also possesses the property of mirror symmetry. The directions of the extremal degree conformity of the gear tooth flank, $\mathcal{G}$, and the mating pinion tooth flank, $\mathcal{P}$, are orthogonal to one another only in degenerate cases of the surfaces contact.

Equation (E.68) of the *converse indicatrix of conformity*, $Cnf_k(\mathcal{G}/\mathcal{P})$ is convenient for implementation when:

(a) either the gear tooth flank $\mathcal{G}$ or
(b) mating pinion tooth flank $\mathcal{P}$ or
(c) both of them

feature point(s) or line(s) of inflection. In the point(s) or (line(s)) of inflection, radii of normal curvature, $R_{g(p)}$ of the surface, $\mathcal{G}$ and $\mathcal{P}$, are equal to infinity. Points/lines of inflection cause indefiniteness when calculating the position vector, $r_{cnf}(\varphi,\ \mu)$, of point of the characteristic curve, $Cnf_R(\mathcal{G}/\mathcal{P})$. Equation (E.68) of the *converse indicatrix of conformity*, $Cnf_k(\mathcal{G}/\mathcal{P})$ is free of the disadvantages of such kind and therefore it is recommended for practical applications.

Appendix F: The Closest Distance of Approach of Tooth Flanks of a Gear and a Mating Pinion

Generally, the problem of the calculation of the closest distance of approach between two smooth regular surfaces is sophisticated and challenging. Per the author's knowledge, no general solution to the problem of calculation of the closest distance of approach between two smooth regular surfaces is available in the public domain. For the purpose of calculation of the deviation, δ_g, of the actual gear tooth flank,

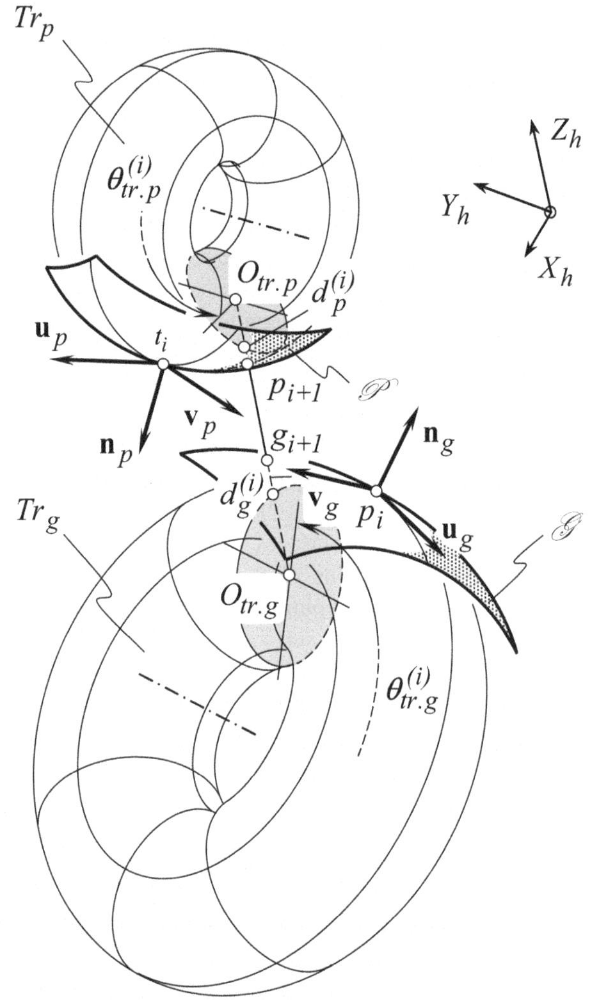

Fig. F.1 Computation of the closest distance of approach between gear tooth flank, $\mathcal{G}$, and mating pinion tooth flank, $\mathcal{P}$

$\mathcal{G}_{ac}$, with respect to the desired (nominal) gear tooth flank, $\mathcal{G}_{nom}$, the problem under consideration can be reduced to the problem of computation of the closest distance of approach between two torus surfaces, Tr_g and Tr_p.

Consider a gear tooth flank, $\mathcal{G}$, and a mating pinion tooth flank, $\mathcal{P}$, that initially are given in a common coordinate system, $X_hY_hZ_h$, associated with the gear housing, as illustrated in Fig. F.1. The tooth flanks, $\mathcal{G}$ and $\mathcal{P}$, are locally approximated by portions of torus surfaces, Tr_g and Tr_p, respectively. Again, not all points of the torus surfaces, Tr_g and Tr_p, can be used for the local approximation of the gear and the pinion tooth flanks, $\mathcal{G}$ and $\mathcal{P}$. Only points that are located either within the greatest meridian or within the shortest meridian of the torus surface are employed for this purpose.

The points, K_g and K_p^*, are chosen as the first guess points on the torus surfaces, Tr_g and Tr_p. For the analysis below, it is convenient to relabel the points, K_g and K_p^*, to g_i and p_i, correspondingly.

For a given configuration of the torus surfaces, Tr_g and Tr_p, the closest distance of approach between these surfaces can be used as a first approximation to the closest distance of approach between the original gear and the pinion tooth flanks, $\mathcal{G}$ and $\mathcal{P}$.

The closest distance of approach between the torus surfaces, Tr_g and Tr_p, is measured along the common perpendicular to these surfaces. The following equations can be composed on the premise of this property of the closest distance of approach.

Unit normal vector, $\mathbf{n}_{Tr.\ g}$, to the torus surface, Tr_g, is located within a plane through the axis of rotation of the surface, Tr_g. In the coordinate system $X_{tr.\ g}Y_{tr.\ g}Z_{tr.\ g}$ that is associated with the torus surface, Tr_P, the equation of a plane through the axis of rotation of the torus surface, Tr_g, can be expressed in the form:

$$\left[\mathbf{r}_{\tau g} - \mathbf{r}_{tr.g}^{(0)}\right] \times \mathbf{k}_{tr.g} \times \mathbf{R}_{tr.g} = 0 \tag{F.1}$$

Here is designated:

$\mathbf{r}_{\tau g}$ – is the position vector of point of the plane through the axis of rotation of the torus, Tr_g

$\mathbf{r}_{tr.g}^{(0)}$ – is the position vector of point within the plane, $\mathbf{r}_{\tau g}$ (it is assumed below that this point coincides with the origin of the coordinate system, $X_{tr.\ g}Y_{tr.\ g}Z_{tr.\ g}$)

$\mathbf{k}_{tr.\ g}$ – is the unit vector of the $Z_{tr.\ g}$– axis

Equation (F.1) is expressed in terms of the radius, $\mathbf{R}_{tr.\ g}$. This indicates that the set of all planes through the fixed $Z_{tr.\ g}$–axis forms a pencil of planes. The equation of the pencil of planes, $\mathbf{r}_{\tau g}$, in the common coordinate system, $X_hY_hZ_h$, can be represented in the form:

$$\mathbf{r}_{\tau g}\left(Z_{tr.g},\ V_{\tau r.g},\ \theta_{tr.g}\right) = \mathbf{Rs}\left(Tr_g \mapsto h\right) \cdot \begin{bmatrix} V_{tr.g} \cdot \cos\theta_{tr.g} \\ V_{tr.g} \cdot \sin\theta_{tr.g} \\ Z_{tr.g} \\ 1 \end{bmatrix} \tag{F.2}$$

The unit normal vector, $\mathbf{n}_{Tr.\ p}$, to the torus surface, Tr_p, is located within a plane through the axis of rotation of the surface, Tr_p. In the coordinate system, $X_{tr.\ p}Y_{tr.\ p}Z_{tr.\ p}$, that is associated with the surface, Tr_p, equation of a plane through the axis of rotation of the torus surface, Tr_p, can be represented in the form:

$$\left[\mathbf{r}_{\tau p} - \mathbf{r}_{tr.p}^{(0)}\right] \times \mathbf{k}_{tr.p} \times \mathbf{R}_{tr.p} = 0 \tag{F.3}$$

Here is designated:

$\mathbf{r}_{\tau p}$ – is the position vector of point of the plane through the torus, Tr_p, axis of rotation

$\mathbf{r}_{tr.p}^{(0)}$ – is the position vector of point within the plane, $\mathbf{r}_{\tau p}$ (it is assumed below that this point coincides with the origin of the coordinate system, $X_{tr.\ p}Y_{tr.\ p}Z_{tr.\ p}$)

$\mathbf{k}_{tr.\ p}$ – is the unit vector of the $Z_{tr.\ p}$– axis

Equation (F.3) is expressed in terms of the radius, $\mathbf{R}_{tr.\ p}$. This indicates that the set of all planes through the fixed $Z_{tr.\ p}$–axis forms a pencil of planes. The equation of this pencil of planes, $\mathbf{r}_{\tau p}$, in the common coordinate system, $X_hY_hZ_h$, can be represented in the form:

$$\mathbf{r}_{\tau p}\left(Z_{tr.p},\ V_{tr.p},\ \theta_{tr.p}\right) = \mathbf{Rs}\left(Tr_p \mapsto h\right) \cdot \begin{bmatrix} V_{tr.p} \cdot \cos\theta_{tr.p} \\ V_{tr.p} \cdot \sin\theta_{tr.p} \\ Z_{tr.p} \\ 1 \end{bmatrix} \tag{F.4}$$

A straight line through the points, $d_g^{(i)}$ and $d_p^{(i)}$, along which the shortest distance of approach, $d_{gp}^{\min}$, of the torus surfaces, Tr_g and Tr_p, is measured, is the line of intersection of the planes, $\mathbf{r}_{\tau g}$ and $\mathbf{r}_{\tau p}$. Therefore, this line, $d_{gp}^{\min}$, must be aligned with both unit normal vectors, $\mathbf{n}_{tr.\ g}$ and $\mathbf{n}_{tr.\ p}$.

In the coordinate system, $X_hY_hZ_h$, the equation for the unit normal vector, $\mathbf{n}_{tr.\ g}$, to the torus surface, Tr_g, yields representation in matrix form:

$$\mathbf{n}_{tr.g} = \mathbf{Rs}\left(Tr_g \mapsto h\right) \cdot \begin{bmatrix} \left(C_{tr.g} + \cos\varphi_{tr.g}\right) \cdot \cos\varphi_{tr.g} \cdot \cos\theta_{tr.g} \\ \left(C_{tr.g} + \cos\varphi_{tr.g}\right) \cdot \cos\varphi_{tr.g} \cdot \sin\theta_{tr.g} \\ \left(C_{tr.g} + \cos\varphi_{tr.g}\right) \cdot \sin\varphi_{tr.g} \\ 1 \end{bmatrix} \tag{F.5}$$

were $C_{tr.\ g}$ designates the parameter $C_{tr.g} = 1 - \frac{R_{2.g}}{R_{1.g}}$.

Similarly, in the coordinate system, $X_hY_hZ_h$, the equation for the unit normal vector, $\mathbf{n}_{tr.\ p}$, to the torus surface, Tr_p, yields matrix representation in the form:

$$\mathbf{n}_{tr.p} = \mathbf{Rs}\left(Tr_p \mapsto h\right) \cdot \begin{bmatrix} \left(C_{tr.p} + \cos\varphi_{tr.p}\right) \cdot \cos\varphi_{tr.p} \cdot \cos\theta_{tr.p} \\ \left(C_{tr.p} + \cos\varphi_{tr.p}\right) \cdot \cos\varphi_{tr.p} \cdot \sin\theta_{tr.p} \\ \left(C_{tr.p} + \cos\varphi_{tr.p}\right) \cdot \sin\varphi_{tr.p} \\ 1 \end{bmatrix} \tag{F.6}$$

were $C_{tr.\ p}$ designates the parameter $C_{tr.p} = 1 - \frac{R_{2.p}}{R_{1.p}}$.

Evidently, the points $O_{tr.\ g}$, $O_{tr.\ p}$, $d_g^{(i)}$, and $d_p^{(i)}$ (see Fig. F.1), are located within the straight line through the centers, $O_{tr.\ g}$ and $O_{tr.\ p}$. The position vector, $\mathbf{r}_{cd}$, of this straight line can be calculated from the equation:

$$\left(\mathbf{r}_{cd}-\mathbf{r}_{cg}\right)\times\left(\mathbf{r}_{cp}-\mathbf{r}_{cg}\right)=0 \tag{F.7}$$

Here is designated:

$\mathbf{r}_{cg}$ – is the position vector of a point on the circle of a radius, $R_{tr.\ g}$
$\mathbf{r}_{cp}$ – is the position vector of a point on the circle of a radius, $R_{tr.\ p}$

It is necessary that the straight line, $\mathbf{r}_{cd}$, be along the unit normal vectors, $\mathbf{n}_{tg}$ and $\mathbf{n}_{tp}$, to the torus surfaces, Tr_g and Tr_p.

Considered together, Eqs. (F.2), (F.4), and (F.7) make possible calculation of the closest distance of approach between the torus surfaces, Tr_g and Tr_p. Then, the straight line, $d_{gp}^{\min}$, intersects the gear tooth flank, $\mathcal{G}$, and the pinion tooth flank, $\mathcal{P}$, at the points, g_{i+1} and p_{i+1}, correspondingly. The points, g_{i+1} and p_{i+1}, serve as the second guess to the closest distance of approach between the surfaces, $\mathcal{G}$ and $\mathcal{P}$.

The cycle of the recursive calculations is repeated as many times, as necessary for making the deviation of the calculation of the closest distance of approach between the surfaces, $\mathcal{G}$ and $\mathcal{P}$, smaller than the maximal permissible value.

There is an alternative approach for the calculation of the closest distance of approach between two torus surfaces. The direction of the unit normal vector to an offset surface to, Tr_g, is identical to the direction of the unit normal vector, $\mathbf{n}_{tr.\ g}$, to the torus surface, Tr_g. This statement is also valid for the unit normal vector $\mathbf{n}_{tr.\ T}$ to the torus surface, Tr_p. This property of the unit normal vectors, $\mathbf{n}_{tr.\ g}$ and $\mathbf{n}_{tr.\ p}$, can be used for the modification of the method of calculation of the closest distance of approach between two torus surfaces.

The equation of the circle of radius, $R_{tr.\ g}$, yields matrix representation:

$$\mathbf{r}_{cg}\left(\theta_{tr.g}\right)=\mathbf{Rs}\left(\mathcal{G}\mapsto h\right)\cdot\begin{bmatrix} R_{tr.g}\cdot\cos\theta_{tr.g} \\ R_{tr.g}\cdot\sin\theta_{tr.g} \\ 0 \\ 1 \end{bmatrix} \tag{F.8}$$

The equation of the circle of radius, $R_{tr.\ p}$, can be analytically described in the similar way:

$$\mathbf{r}_{cp}\left(\theta_{tr.p}\right)=\mathbf{Rs}\left(\mathcal{P}\mapsto h\right)\cdot\begin{bmatrix} R_{tr.p}\cdot\cos\theta_{tr.p} \\ R_{tr.p}\cdot\sin\theta_{tr.p} \\ 0 \\ 1 \end{bmatrix} \tag{F.9}$$

The distance, d_{gp}, between two arbitrary points on the circles, $\mathbf{r}_{cg}(\theta_{tr.\ g})$ and $\mathbf{r}_{cp}(\theta_{tr.\ p})$, equals:

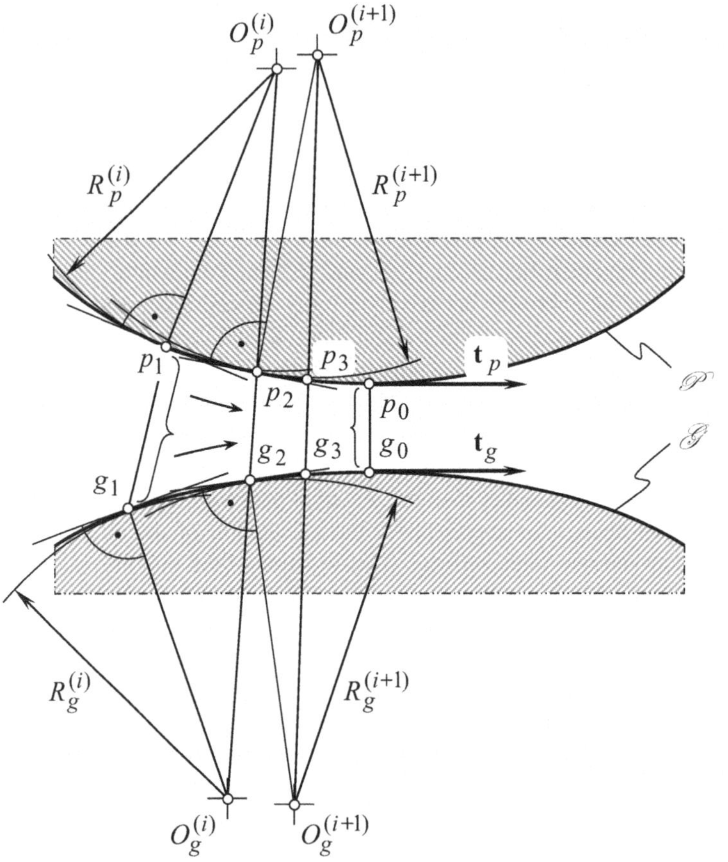

Fig. F.2 Convergence of the methods of calculation of the closest distance of approach of a gear tooth flank, $\mathcal{G}$, and a mating pinion tooth flank, $\mathcal{P}$

$$d_{gp}\left(\theta_{tr.g},\ \theta_{tr.p}\right) = \left|\ \mathbf{r}_{cg}\left(\theta_{tr.g}\right) - \mathbf{r}_{cp}\left(\theta_{tr.p}\right)\ \right| \tag{F.10}$$

The distance, d_{gp}, is minimal for a specific (optimal) combination of the parameters, $\theta_{tr.\ g}$ and $\theta_{tr.\ p}$. The favorable values of the parameters, $\theta_{tr.\ g}$ and $\theta_{tr.\ p}$, can be calculated on solution of the set of two equations:

$$\frac{\partial}{\partial \theta_{tr.g}} \mathbf{r}_{cg}\left(\theta_{tr.g}\right) = 0 \tag{F.11}$$

$$\frac{\partial}{\partial \theta_{tr.p}} \mathbf{r}_{cp}\left(\theta_{tr.p}\right) = 0 \tag{F.12}$$

On solution of Eqs. (F.11) and (F.12), the optimal values, $\theta_{tr.g}^{(opt)}$ and $\theta_{tr.p}^{(opt)}$, can be determined. These angels specify the direction of the closest distance of approach of the torus surfaces, Tr_g and Tr_p.

Following this method, the three-dimensional problem of calculation of the closest distance of approach of two torus surfaces is reduced to the problem of calculation of the closest distance of approach between two circles. Under a certain scenario, the last approach could possess an advantage over the previous approach.

The convergence of the disclosed algorithms for the computation of the closest distance of approach between two smooth regular surfaces is illustrated in Fig. F.2. The computation procedure is convergent regardless of the actual location of the first guess points on the surfaces, $\mathcal{G}$ and $\mathcal{P}$.

It is instructive to draw attention here to the similarities between the disclosed iterative method for the computation of the closest distance of approach between two smooth regular surfaces, and between the *Newton-Raphson*'s *method*, the iterative method of chords, and so forth. Many similarities can be found out on this comparison.

Appendix G: Engineering Formulae for Specification of Gear Tooth Flank

The engineering representation of gear tooth flank can be converted into scientific its representation and vice versa. For the conversion, it is convenient to use the so-called *Fundamental Gear Equations* listed below (Table G.1). More useful equations can be found out from many advanced sources.

Formulae used for the conversion from English (pitch) system to metric system are summarized below (Table G.2)

The formula $P = \frac{25.4}{m}$ is used to express diametral pitch P in terms of module m. The expression:

$$m = \frac{25.4}{P} \tag{G.1}$$

is used for the inverse conversion. For the correspondence between millimeters and inches, the ratio:

$$\text{Millimeters } (mm) = \frac{\text{Inches}}{.03937} = 25.4 \text{ Inches} \tag{G.2}$$

$$\text{Inches} = .03937 \ \ mm = \frac{mm}{25.4} \tag{G.3}$$

is valid.

Table G.1 Fundamental gear equations

Name of the parameter	Formulae
Transverse diametral pitch, P_t	$P_t = P_n \cos\psi$
Pitch diameter, D	$D = \frac{N}{P_t}$
Standard addendum, a	$a = \frac{1}{P_n}$
Outside diameter, D_o	$D_o = D + 2 \cdot a$
Transverse pressure angle, ϕ_t	$\tan\phi_t = \frac{\tan\phi_n}{\cos\psi}$
Base diameter, d_b	$d_b = D \cdot \cos\phi_t$
Lead, L	$L = \pi \cdot D \cdot \cot\psi = \frac{\pi \cdot D}{\tan\psi}$ $L = \pi \cdot d_b \cdot \cos\psi_b$
Normal circular pitch, p_n	$p_n = \frac{\pi}{P_n}, \quad p_n = \frac{\pi \cdot D \cdot \cos\psi}{N}$
Standard normal circular thickness, t_n	$t_n = \frac{p_n}{2}, \quad t_n = t \cdot \cos\psi$
Axial pitch, p_x	$p_x = \frac{\pi}{P_n \cdot \sin\psi} = \frac{p_n}{\sin\psi} = \frac{L}{N}$
Transverse circular pitch, p_t	$p_t = \frac{\pi}{P_t} = \frac{p_n}{\cos\psi}$
Helix angle, when the given center distance is standard, ψ	$\cos\psi = \frac{N_p + N_g}{2 \cdot P_n \cdot C}$
Operating pitch diameter (pinion), D_{rp}; with nonstandard center-distance, C	$D_{rp} = \frac{2 \cdot C \cdot N_p}{N_p + N_g}$
Operating pressure angle, ϕ_{rt}; with nonstandard center-distance C,	$\sin\phi_{rt} = \frac{d_{bp} + d_{bg}}{2 \cdot C}$
Normal diametral pitch, P_n	$P_n = P_t \cdot \sec\psi$
Helix angle, ψ	$\cos\psi = \frac{N}{D \cdot P_n}, \quad \sin\psi = \frac{\pi \cdot N}{P_n \cdot L}$
Helix angle, ψ_y at any diameter, d_y	$\tan\psi_y = \frac{d_y}{d_1} \cdot \tan\psi_1$
Transverse circular pitch, p_{t2} at any diameter, d_y	$p_{t2} = \frac{\pi \cdot d_y}{N}$
Involute function of pressure angle	$\text{inv}\phi = \tan\phi - \breve{\phi}$
Normal pressure angle, ϕ_n	$\tan\phi_n = \tan\phi_t \cdot \cos\psi$ $\sin\phi_n = \sin\phi \cdot \cos\psi_b$ $\cos\phi_n = \sin\psi_b \cdot \csc\psi$
Transverse pressure angle ϕ_{ty}, at any diameter, d_y	$\cos\phi_{ty} = \frac{d_b}{d_y}$
Base helix angle, ψ_b	$\cos\psi_b = \frac{\cos\psi \cdot \cos\phi_n}{\cos\phi_t} = \frac{\sin\phi_n}{\sin\phi_t}$ $\sin\psi_b = \sin\psi \cdot \cos\phi_n$ $\tan\psi_b = \tan\psi \cdot \cos\phi_t$
Base pitch, p_b (here ρ designates radius of profile curvature)	$p_b = \frac{\pi \cdot d_b}{N} = \frac{\pi \cdot D \cdot \cos\phi_t}{N} = \rho \cdot \cos\phi_t$

A brief analysis of the types of gears used in industry, together with the analysis of local topology of gear tooth flanks to be machined, is necessary for the design of gear cutting tools, especially gear cutting tools for machining/finishing of precision gears.

Table G.2 Formulae for the conversion from pitch system to metric system

Name of the parameter	English (inches)	Metric (millimeters)
Pitch diameter, D	$D = \frac{N}{P}$	$D = m \cdot N$
Addendum, a	$a = \frac{1}{P}$	$a = m$
Standard outside diameter, D_o	$D_o = D + 2 \cdot a$	$D_o = D + 2 \cdot m$
Base diameter, d_b	$d_b = D \cdot \cos \phi$	
Circular pitch, p	$p = \frac{\pi}{P}$	$p = \pi \cdot m$
Standard circular tooth thickness, t	$t = \frac{p}{2}$	
Average backlash per pair, B	$B = \frac{.040}{P}$	$B = .040 \cdot m$

Appendix H: On the Inadequacy of the Terms *Wildhaber-Novikov Gearing* and *W-N Gearing*

Wildhaber Helical Gearing [1] and *Novikov Gearing* [2] are briefly discussed below aiming to illustrate inadequacy of the terms *Wildhaber-Novikov Gearing* and *W-N Gearing*. As it follows from the discussion, helical gearing proposed by Dr. *E. Wildhaber* must be referred to as *Wildhaber Helical Gearing*, or so. Helical gearing proposed by Dr. *M. Novikov* must be referred to as *Novikov Gearing*, or so. The extensively used terms *Wildhaber-Novikov Gearing* and *W-N Gearing* are meaningless, and commonly are used by poorly knowledgeable people.

H.1 Wildhaber's Helical Gearing

Helical gearing with circular arc tooth profile [1] targets an improved power capacity of the gear pair. The invention relates to the tooth shape of gears, which run on parallel axes, and may be applied to helical gears, such as single helical gears and double helical gears or herringbone gears.

The purpose of the invention is threefold:

1. To provide helical gearing with an improved tooth contact, so as to lessen surface stresses and wear;
2. To provide helical gearing, which is capable of rapid and accurate production, and which may be ground without difficulty, if so desired.
3. To provide accurate gearing of circular tooth profile.

The invention is illustratively exemplified in the accompanying drawings, in which, Fig. H.1a is a side elevational view of the proposed gear showing parts thereof in section; Fig. H.1d is a normal sectional view if Fig. H.1a, taken on the lines 2-2 of the latter figure; Fig. H.1g is a side elevational view of a pair of gears constructed in accordance with the invention; Fig. H.1h is a sectional view taken through a pair of gears; Fig. H.1b, c are sectional views of milling cutters used in the manufacture of gear of the proposed design; Fig. H.1e, f are elevational views of corresponding tools of rack shape, to be used in reciprocating machines for cutting helical gears in accordance with the invention; Fig. H.1k, l are side elevational views of the improved gear showing a pair of grinding wheels in different operating positions, the wheels being set to grind opposite tooth surfaces; Fig. H.1m is a view of a gear taken in normal section and showing the grinding wheels in operation position; Fig. H.1n is a view of a mate pinion sowing the grinding wheels in operating position; Fig. H.1o is a view of modified form of gear made in accordance with the invention; Fig. H.1j is a sectional view taken through an internal gear and its pinion; Fig. H.1 is a normal section through helical teeth of composite outline, constructed from the invention; Fig. H.1p is a view of reciprocating tool of rack shape in operating position; and Fig. H.1s is a view of a modified type of reciprocating tools, in position to start a cut on a herringbone gear.

Referring to the drawings, and particularly to Fig. H.1a, d, 1 denotes a helical gear having teeth 2 in contact with the teeth 3 of a mating pinion 4. In order to clearly illustrate the degree of contact between the teeth of the gear and pinion the tooth 4 is shown in section in Fig. H.1a.

It is customary to analyze helical gearing with reference to a normal section, that is, line 2-2 of Fig. H.1a, line 2-2 being normal to the helix of the pitch circle. Figure H.1d illustrates the said normal section 2-2 for both pinion 4 and gear 1.

It has been assumed as an example that the tooth profiles 6 of gear 1 are circular arcs of radii 7 and centers 8, in the shown normal section. Centers 8 are situated close to the pitch circle 9 of the gear[28]. The corresponding teeth of pinion 4 are so shaped as to allow rolling of the pitch circles 9 and 10 on each other, as well known to those skilled in the art.

When the gear tooth 2 is in the position shown in Fig. H.1a, d, and its center is at 8, then it contacts with tooth 3 at point 11, which may be determined by a perpendicular to tooth 2 through point 12, point 12 being the contact point between the two pitch circles 9 and 10. The said perpendicular is in the present case the connecting line between point 12 and centers 8 of the tooth profile.

Another position $2'$ of the gear tooth, and $3'$ of the corresponding pinion tooth are shown in dotted lines in Fig. H.1d. The tooth profiles contact here at a point $11'$, which can be determined like point 11. It will be noted that the contact point has traveled from 11 to $11'$ during a small angular motion of the gears[29]. The contact point has passed practically over the whole active profile during a turning angle 13 of the gear, which angle corresponds to a fraction only of the normal pitch14, $14'$. The said normal pitch equals the circle pitch of the shown normal section.

---

[28] Centers 8 need to be situated along the line of action, *LA*, that is a must. Otherwise, the condition of contact ($\mathbf{n} \cdot \mathbf{V}_{\Sigma} = 0$) is violated.

[29] Once the contact point has traveled from 11 to 11' during a small angular motion of the gears, this immediately reveals that the transverse contact ratio, m_p, in the *Wildhaber's Helical Gearing* is greater than 0, that is, $m_p > 0$, which is not permissible.

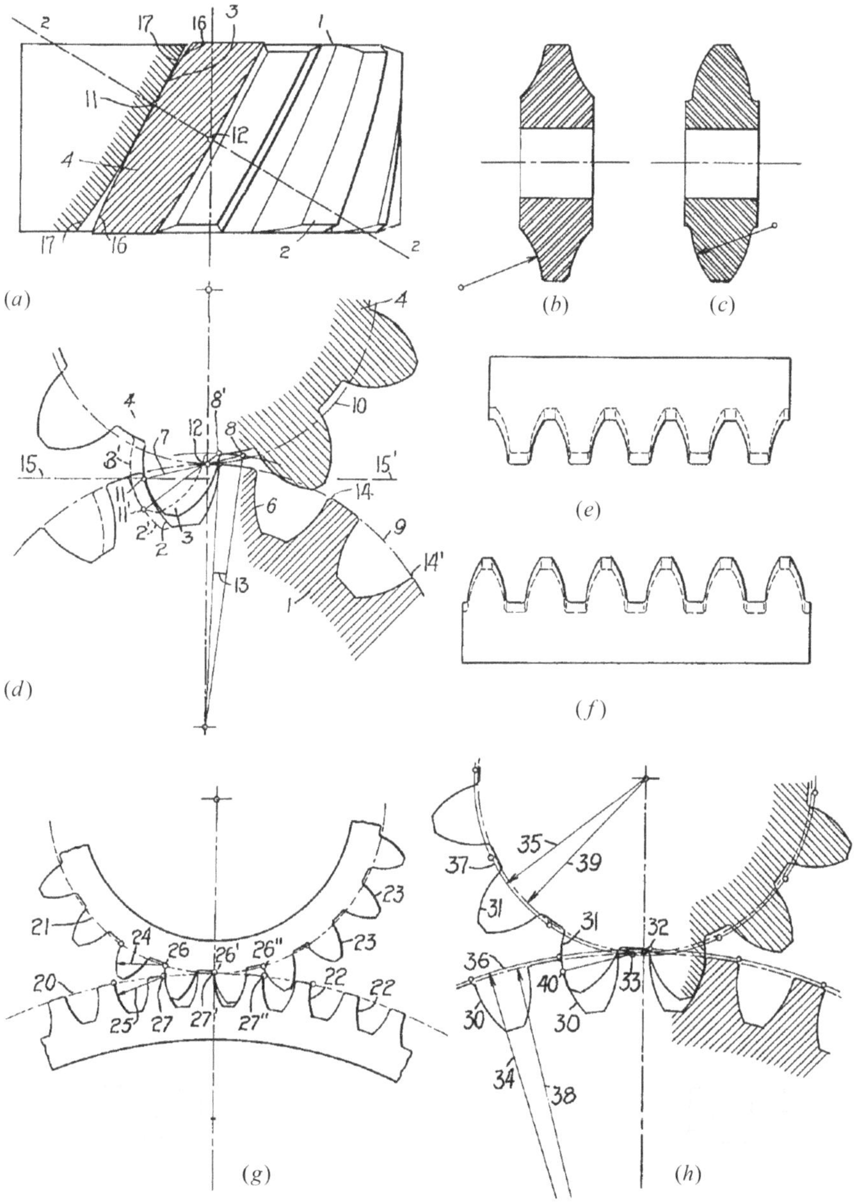

Fig. H.1 *Helical Gearing*. (After E. Wildhaber., US Patent 1,601,750, Patented: October 5, 1926, Filed: November 2, 1923)

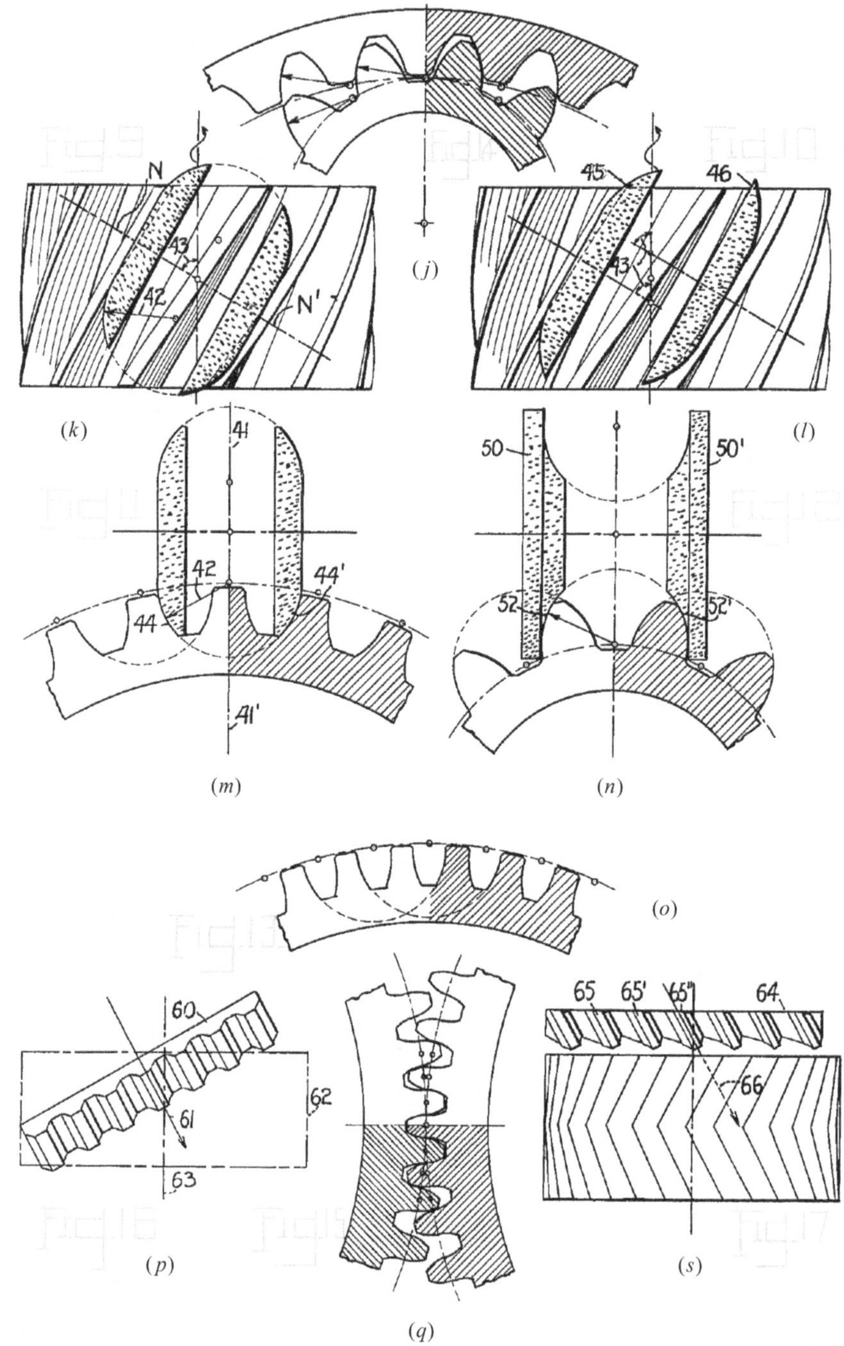

Fig. H.1 (continued)

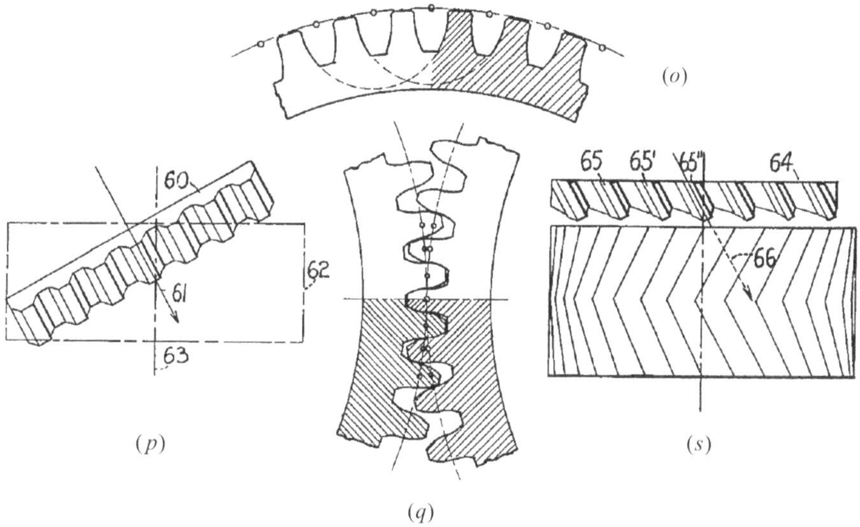

Fig. H.1 (continued)

In gearing now in use, however, the tooth outline and the tooth proportions are so selected that the contact of corresponding normal profiles lasts for an angle, which, as a rule, corresponds to more than one full pitch.

In gearing according to the invention, the contact point between two normal profiles passes over the whole active profile during a turning angle, which corresponds to less than one half the normal pitch, and usually to much less than that.

Gearing designed according to the invention, allows the teeth to come into better contact with each other, inasmuch as the tooth surfaces remain much closer to each in a direction perpendicular to the contact line between two mating gears. This is illustrated by a section taken in direction of lines 15, 15′ of Fig. H.1d. In Fig. H.1a the lateral profile 16 of tooth 3 and profile 17 of tooth 2 of said section are shown to contact at point 11, and to remain close to each other on their whole length. The same holds true for other sections, taken parallel to section 15, 15′.

Close contact between teeth is well known to reduce wear and to improve the efficiency of the gears.

Although a circular arc is shown as the normal tooth profile of gear 1, in Fig. H.1d, it will be understood that this is not the only shape to affect the stated purpose, of increasing the speed, at which the contact point travels over the tooth profile of a normal section. As a rule, however, the shape can be approximated by a circle, whose center is close to the pitch center.

The gearing according to the present invention is strictly a gearing for helical teeth. It would not be advisable on straight teeth, on account of the explained short duration of contact between tooth profiles. This would cause intermittent action, whereas on helical gears similar parts of the teeth are always in contact, on account of the twisted nature of the tooth surfaces.

Figure H.1g may be considered as a view taken in the direction of the axes of a pair of gears. The tooth profiles are the circles in a section, which is perpendicular to the axes. The gear is provided with helical teeth, with working faces below the pitch circle 20, while the pinion teeth have working faces above the pitch circle 21 only. The working profiles 22 of the gear are concave and circular, and their centers are substantially situated on the pitch circle 20. The convex working profiles 23 of the pinion are also of circular shape. Their radii 24 are substantially the same as the radii 25 of the mate profiles. The centers 26, 26′, 26″ are similarly situated on pitch circle 21. Profile centers 27, 27′, 27″ of pitch circle 20, and profile centers 26, 26′, 26″ of pitch circle 21 correspond to each other. They coincide during the mesh, which takes place on the whole tooth profile at once.

Figure H.1g may also be considered as a section perpendicular to the helical teeth, and shows then the normal tooth profiles.

Figure H.1h shows a refinement of the preferred embodiments of the invention. It is a normal section through the helical teeth, but can also be considered as a section perpendicular to the axes. Corresponding profiles 30 and 31 are circular, as in Fig. H.1g, but in this case the radius of the concave circular profile 30 is made a trifle larger than the radius of the convex circular profile 31. Consequently, the profile centers 32 and 33 do not exactly coincide during the mesh. The radii 34 and 35 of the circles 36 and 37, constituted by the profile centers 32 and 33 respectively, are not accurately identical with the pitch radii 38 and 39 of the two gears. The sum of the radii 34 and 35 is a trifle larger than the sum of the pitch radii. The radii 34 and 35 are so selected, that the main tooth pressure runs about in a direction 33, 40.

The slight difference of the radii of profiles 30 and 31 facilitates the tooth contact, and allows for small errors in making and assembling.

Figure H.1b, c show a pair of milling cutters for milling gear teeth. The cutter may be applied in the usual manner, their axis being inclined in correspondence with the tooth inclination, that is, with the helix angle of teeth. It will be found that the cutters to be inclined for an angle, which is a trifle smaller than the helix angle in the pitch circle, for producing most accurate results.

In Fig. H.1e, f a pair of rack shaped cutters is shown[30]. These cutters are for use in a reciprocating machine. The teeth of these tools are relieved inwardly, in the usual manner, as evident by the dotted lines.

The convex grinding wheels shown in Fig. H.1k are illustrated in their operating positions, in a view which is taken perpendicular to the axis of the gear blank as well as to the axis of the grinding wheels, that is, in a view along the gear radii 41, 41′ of Fig. H.1m. The wheels which are to produce concave circular teeth profiles in a normal section are of convex circular profile, its radius 42 being the same as the radius of the concave circular profile. The grinding wheels are inclined for an angle 43, which equals the helix angle of the teeth, in the pitch circle. The wheels grind along their profiles indicated in dotted lines 44 and 44′, which are located in a normal

[30]Conformal gears cannot be cut by rack cutters, and, more generally, they cannot be machined in any gear generating process.

section. As shown in Fig. H.1k, the two grinding wheels are coaxially arranged with respect to each other.

The device shown in Fig. H.1l corresponds to that shown in Fig. H.1k, with the exception that the grinding wheels 45 and 46 are not coaxially arranged. Although the arrangement shown in Fig. H.1k imposes certain restrictions on the tooth design, it is frequently preferred. The arrangement of Fig. H.1l is advantageous, when grinding wheels are not free to run out, for instance when they must clear against a shoulder, or in the case of herringbone teeth.

Referring particularly to Fig. H.1m, a normal section is illustrated and taken along lines N, N′ of Fig. H.1k. In this view, the axis of the coaxially arranged grinding wheels is situated in the said normal section. The wheels grind along the profiles 44 and 44′ of the shown normal section, while the blank performs a translator motion in the direction of its axis, and, in timed relation thereto, a turning motion about its axis. In other words, the blank is screwed past the grinding wheels.

Figure H.1n discloses a normal section through the teeth of the mating gear or pinion. Grinding wheels 50 and 50′ are provided with concave circular profiles 52 and 52′ with which they grind the convex gear teeth.

It will be understood, that milling cutters might be used instead of the grinding wheels shown in Fig. H.1k–n; and also that grinding wheels of a shape shown in Fig. H.1b, c might be used, if so desired.

The teeth ground according to Fig. H.1k, m, n are preferably so designed that the centers of opposite tooth arcs 44 and 44′, 52 and 52′, respectively, in Fig. H.1n coincide. In Fig. H.1m, n, the tooth arcs of every third tooth side have a common center.

The tooth arcs of every fifth tooth side have a common center in the normal section shown in Fig. H.1o.

In Fig. H.1m the common center of opposite tooth arcs of alternate teeth is situated on the centerline of the intermediate tooth. The corresponding pinion shows convex circular profiles, of which opposite tooth sides of adjacent teeth have common centers in the middle of the intermediate tooth space.

The normal section shown in Fig. H.1j shows an internal gear and its mate pinion, constructed in accord ace with the concave tooth profiles. In external gears similarly preference is given to providing the larger gear with concave tooth profiles.

The normal section through a pair of helical gears shown in Fig. H.1q, discloses opposite tooth profiles, the addendum being convex and the dedendum concave.

A rack-shaped planing tool is illustrated in operating position in Fig. H.1p. Tools of this kind have been shown in another view in the Fig. H.1e, f. The reciprocatory tool 60 moves in the direction 61, at an inclination, which equals the helix angle of the teeth. Gear 62, with its axis 63, is shown in dotted and dash lines. In order to cut the proper tooth shape, gear blank 62 after every cut is slightly fed in a rolling generating motion with respect to a rack which is embodied by tool 60.

Another reciprocatory tool 64 is shown in Fig. H.1s; the tool in this case being provided with stepped teeth 65, 65′, 65″, which allow it to clear shoulders and herringbone teeth. The tool moves in the direction 66 of the helical teeth, which cuts.

Other ways of producing gearing according to the invention, that is, hobbing, planing with a pinion cutter, rolling and casting, may be contemplated, but it is not deemed necessary at this time to give a detailed explanation of the mechanism used in connection therewith.

Briefly stated, the invention consists in providing helical gearing of such profile, that the tooth contact passes rapidly over the normal profile of the teeth. This has been found to result in close contact between helical mate teeth. In a direction at right angles to the contact line, the mate teeth recede from each other only slightly, and thus provide a tooth contact, which is not very far from surface contact.

It is claimed[31], the *Wildhaber's gearing* features a nonzero transverse contact ratio ($m_p > 0$), and a nonzero face contact ratio $m_F > 0$. The total contact ratio, m_t, is greater than one ($m_t \equiv m_p + m_F > 1$).

The *Wildhaber's gearing* (see Fig. H.1) does not meet the condition of contact. The condition of conjugacy of the tooth flanks, and the condition of equality of the base pitches of the gear and the pinion to the operating base pitch are not fulfilled in this design of gearing. This immediately yields a conclusion that *Wildhaber's gearing* is not workable at all.

The invention by *Wildhaber* is explained to the best possible extent in [3, 4].

H.2 Novikov Gearing

Novikov Gearing [2] is an example of conformal helical gearing. For a long while *Novikov*'s patent (S.U. Patent No. 109,113 of 1956) was not available to most of the gear experts.

Known designs of gearing, those featuring point system of meshing, feature low contact strength and are not widely used in practice.

The contact strength of known designs of gearing with a line system of meshing, including the widely used involute gearing, is limited.

The proposed design of gearing[32] [2] features higher contact strength due to favorable curvatures of the interacting tooth flanks. Under equivalent contact stress, similar dimensions, and comparable remaining design parameters, greater circular forces are permissible by the proposed gearing. Lower sensitivity to manufacturing errors and to deflections under the load is the other advantage of the proposed gearing.

The proposed design of gearing can be designed either with parallel, intersecting, or crossing axes of rotations of the gears. External gearing as well as internal gearing

[31] Wildhaber, E., *Helical Gearing*, US Patent 1,601,750, Patented: October 5, 1926, Filed: November 2, 1923.

[32] Pat. No. 109,113, (USSR). *Gear Pairs and Cam Mechanisms Having Point System of Meshing.* / M.L. Novikov, National Classification 47h, 6; Filed: April 19, 1956, published in Bull. of Inventions No.10, 1957.

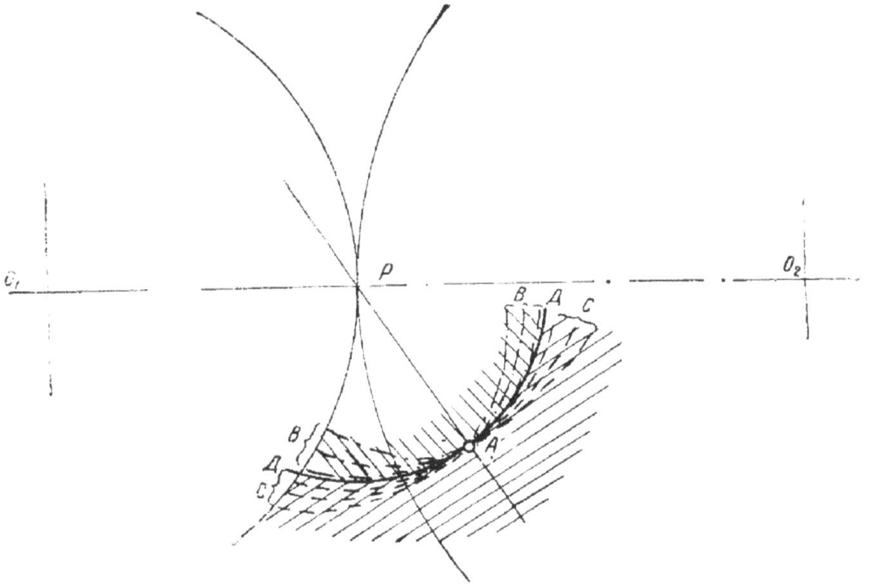

Fig. H.2 *Helical gearing* (After Dr. *M.L. Novikov*, Pat. No. 109,113, (USSR). *Gear Pairs and Cam Mechanisms Having Point System of Meshing*. National Classification 47h, 6; Filed: April 19, 1956, published in Bull. of Inventions No.10, 1957)

of the proposed system of meshing is possible. The tooth ratio of the proposed gearing can be either of constant value or it can be variable, and time dependent. The proposed concept of gearing can be utilized in the design of cam mechanisms.

In Fig. H.2, possible tooth profiles in the cross-section of tooth flanks by a plane that is perpendicular to the instant axis of relative rotation through the current point of contact are illustrated.

Here, the point of intersection of the planar cross-section by the axis of instant relative rotation is denoted by *P*.

O_1 and O_2 are the points of intersection of the planar cross-section by the axes of the gear and the pinion.

A is the point of meshing (in its current location).

PA denotes the line of action.

ДАД is the circle centered at the point *P* which corresponds to the limit case of the tooth profiles (in the case the profiles are aligned to each other).

Several curves, *BAB*, represent examples of the tooth profiles of one of the mating gears. The curves *BAB* are arbitrary smooth curves, which are located inside of the circular arc *ДАД* (that is, the arcs are located within the bodily side of the limit tooth flank of one of the gears). The curves *BAB* are located close to the circular arc *ДАД* and they feature high degree of conformity to the circular arc.

Several curves, *CAC*, represent examples of the tooth profiles of the second of the mating gears. The curves *CAC* are arbitrary smooth curves, which are located outside of the circular arc *ДАД* (that is, the arcs are located within the bodily side of the limit tooth flank of another of two gears). The curves *CAC* are also located close to the circular arc *ДAД* and they feature high degree of conformity to the circular arc.

The entity of the invention is disclosed below in detail.

The location and orientation either of a straight path of contact or a smooth curved path of contact is specified in space in which the location and orientation of the axes of rotations of the gear and the pinion are given. The path of contact is located reasonably close to the axis of instant relative rotation of the gears. Either constant or time-dependent (smoothly varying in time) speed of motion of the contact point along the path of contact is assigned. A coordinate system is associated with the gear, and a corresponding coordinate system is associated with the pinion. In the coordinate systems the moving contact point traces the so-called *contact lines*[33]. One of the *paths of contact* is associated with the gear and another one is associated with the pinion. Certain smooth regular surfaces through the *pseudo-paths of contact* can be used as the tooth flanks of the gear and the pinion. The following requirements should be fulfilled in order to the surfaces can be used as the tooth flanks:

- At every location of the contact point the tooth flanks should have a common perpendicular, and thus the requirements of the main theorem of meshing should be fulfilled.
- The curvatures of the tooth profiles should correspond to each other.
- No tooth flanks interference occurs within the working portions of the surfaces.

The tooth flanks of the proposed geometry fulfill the above-listed requirements and allow for high contact strength of the gear teeth.

Consider a plane through the current contact point, which is perpendicular to the instant axis of relative rotation. Then, two circular arcs are constructed. The circular arcs are centered at points within the straight line through the pitch point and the contact point. The arcs centers are located close to the pitch point. The constructed circular arcs can be considered an example of the tooth profiles of the gear and the pinion. The tooth flanks are generated as loci of the tooth profiles constructed for all possible locations of the contact point. The working portion of one of two tooth flanks is convex, while the working portion of another tooth flank is concave (in the direction towards the axis of instant relative rotation). In a particular case, the radii of tooth profiles could be of the same magnitude and equal to the distance from the contact point to the axis of instant relative rotation. The centers of both profiles in this particular case are located at the axis of instant relative rotation. Under such a scenario, point meshing reduces to a special line meshing. This would require an extremely high accuracy of the center distance and independence of it from operation conditions, which is impractical. Point meshing is preferred when designing tooth profiles. A small difference between the radii of curvature of the tooth profiles is desirable. It should be kept in mind that during run-in period of time, point meshing

[33] The *contact line* is a term used by Dr. *M.L. Novikov*. Actually, the *contact line* is the *pseudo-path of contact*.

of the gear teeth will transform to the above-mentioned *locally line* meshing of the tooth profiles. However, the theoretical point of contact of the tooth flanks will be retained.

Tooth profiles can differ from the circular arcs. However, the tooth profiles of other geometries (those always passing through the contact point) should be located (for one gear) within the interior of the above-mentioned circular arc profile that centers at the point within the axis of instant relative rotation as shown in the Fig. H.2. For another gear, the tooth profile should be located outside the circular arc.

The law of motion of the contact point (that is, speed of the point and its trajectory) should be chosen so as to minimize the friction and wear losses. The friction and wear loss are proportional to the relative sliding velocity in the gear mesh. Therefore, it is desirable to reduce the sliding velocity as much as possible. For this purpose, the path of contact should be located not far from the axis of instant relative rotation. On the other hand, a too-close location of the path of contact to the axis of instant relative rotation is also not desirable as that reduces the contact strength of the gear tooth flanks. In addition, it is recommended to ensure favorable angles between the common perpendicular (along which the tooth flanks of one of the gears acts against the tooth flank of another gear) and between the axes of rotations of the gears.

Opposite sides of tooth profiles are designed in a similar manner to that just discussed. Tooth thicknesses and angular pitch are assigned to ensure the required bending tooth strength.

The face width of the gear or the length of the gear teeth should correlate to their pitch to ensure the required value of the face contact ratio. Gear pairs can feature either one point of contact (when working portions of the tooth flank contact each other just in one point, excluding the phases of the teeth reengagement), or they can feature multiple contact points when the tooth flanks contact each other at several points simultaneously.

For parallel-axes gear pairs, it is preferred to use a straight line as the path of contact, which is parallel to the axes of rotations of the gear and the pinion. The speed of the contact point along the straight path of contact can be of a constant value. In this particular case, the radii of curvature of the tooth profiles in all cross-sections by planes are equal to each other. Tooth flanks in this case are regular screw surfaces. Gears that feature tooth flanks of such a geometry are easy to manufacture, and they can be cut on machine tools available on the market.

An example of parallel-axes gearing with limit geometry of the tooth profiles is illustrated in the Fig. H.2. Point of contact of the tooth flanks in this particular case is transformed to almost line contact of the tooth flanks. The curved contact line is located across the tooth profile. When axial thrust in the gear pair is strongly undesirable, herring-bone gears can be used instead.

Novikov gearing (see Fig. H.2) meets the condition of contact, as well as it meets the condition of conjugacy of the tooth surfaces, and the condition of equality of the base pitches of the gear and the pinion to the operating base pitch. *Novikov gearing* is a kind of perfect gearing[34]. *Novikov gearing* is the only feasible type of gearing with point of contact of the tooth flanks that is capable of transmitting a rotation smoothly. As it is evident now, *Novikov gearing* is a reduced type of involute gearing.

[34]It needs to be noticed here that Dr. *E. Wildhaber* did not recognize the difference between *Novikov gearing* and the helical gearing proposed by him [1]. This conclusion immediately follows from the *Wildhaber*'s statement: "*I may say also that it gives me satisfaction to see the original concept vindicated through the Russian reinvention and effort and through subsequent efforts and articles*". This *Wildhaber*'s statement can be found on page 949 in the paper by T. Allan [Allan, T., "Some Aspects of the Design and Performance of Wildhaber-Novikov Gearing", *Proc. Inst. Mech. Engrs, Part I*, v. 179, n. 30, 1964/1965, pp. 931-954.] (see the *Communications* section). Other evidences in this concern are also known.

References

1. Pat. No. 1,601,750, (USA). *Helical Gearing*. /E. Wildhaber, Filed: November 2, 1923, published October 5, 1926.
2. Pat. No. 109,113, (USSR). *Gear Pairs and Cam Mechanisms Having Point System of Meshing*. / M.L. Novikov, National Classification 47h, 6; Filed: April 19, 1956, published in Bull. of Inventions No.10, 1957.
3. Radzevich, S. P. (2015). High-Conformal Gearing: A new look at the concept of *Novikov gearing*. In *Proceedings of international conference on gears 2015*, October 5–7, 2015, Technische Universität München (TUM), Garching (near Munich), Germany, 2015, pp. 457–470.
4. Radzevich, S. P. (2020). *High-conformal gearing: kinematics and geometry* (2nd edn, p. 506). Elsevier [1st edition: Radzevich, S.P., *High-Conformal Gearing: Kinematics and Geometry*, 1nd edition, CRC Press, Boca Raton, Florida, 2015, 332 pages].

S. P. Radzevich (ed.), *Gear Accuracy*, Mechanical Engineering Series,
https://doi.org/10.1007/978-3-031-74798-4

Index

S. P. Radzevich (ed.), *Gear Accuracy*, Mechanical Engineering Series,
https://doi.org/10.1007/978-3-031-74798-4

The manufacturer's authorised representative in the EU is Springer Nature Customer Service Centre GmbH, Europaplatz 3, 69115 Heidelberg, Germany. If you have any concerns regarding our products, please contact ProductSafety@springernature.com

Printed and bound by CPI Group (UK) Ltd, Croydon, CR0 4YY
30/09/2026
02225600-0012